INTEGRATED
MATHEMATICS
C O U R S E   I

BUMBY

KLUTCH

COLLINS

EGBERS

GLENCOE
McGraw-Hill

New York, New York   Columbus, Ohio   Mission Hills, California   Peoria, Illinois

Send all inquiries to:
Glencoe/McGraw-Hill
936 Eastwind Drive
Westerville, Ohio 43081

ISBN: 0-02-824566-0 (Student Edition)
ISBN: 0-02-824567-9 (Teacher's Annotated Edition)

Printed in the United States of America.

2 3 4 5 6 7 8 9 10 RRD-C/LH-P 03 02 01 00 99 98 97 96 95

Authors

Douglas R. Bumby is chairman of the mathematics department at Scarsdale High School, Scarsdale, New York. Dr. Bumby taught mathematics at Hunter College High School and was Clinical Associate in Mathematics Education at Teachers College, Columbia University, where he taught graduate courses and advised doctoral candidates. Dr. Bumby's research interests are in the areas of mathematics curriculum development and readability of mathematics textbooks. He was instrumental in developing a model integrated mathematics program at Hunter College High School in New York City.

Richard J. Klutch taught mathematics at Freeport High School in New York and now teaches mathematics at the Hunter College Campus Schools of the city University of New York. He has participated in the development of major mathematics curriculum projects and played a major role in developing such programs at Hunter College High School and elsewhere. He has served as consultant to the Bureau of Mathematics Education of the State of New York. He is one of the authors of the new syllabus for the three-year sequence for high school mathematics for the State of New York.

Donald W. Collins is an assistant professor of mathematics education at Texas Tech University in Lubbock, Texas. Dr. Collins taught mathematics at Elmhurst Public Schools in Elmhurst, Illinois and Abilene Public School in Abilene, Texas. He has many years of experience in the field of mathematical education, including over 20 years of work in the mathematics textbook publishing industry.

Elden B. Egbers was the Supervisor of Mathematics for the state of Washington. Mr. Egbers taught mathematics and was the department chairman at Queen Anne High School in Seattle, Washington. Most recently, he directed the writing of the Washington State Curriculum Guidelines. Mr. Egbers was active in giving workshops on integrated mathematics to mathematics teachers and advisors.

This edition of Integrated Mathematics is
dedicated to the memory of Elden B. Egbers.

Reviewers

Carole Bickel
Mathematics Department Chair
Gahanna Lincoln High School
Gahanna, Ohio

Sue Boice
Mathematics Teacher
Wilson Magnet School
Rochester, New York

Bill Bonney
Mathematics Teacher
Ballard High School
Seattle, Washington

Bill Collins
Supervisor: Mathematics and
    Computer Training
Syracuse City School District
Syracuse, New York

Thomas Ford
Mathematics Teacher
Alhambra High School
Martinez, California

Barbara Lapetina
Mathematics Department Head
Brentwood High School
Brentwood, New York

Jane S. Prochazka
Mathematics Teacher
Calabasas High School
Calabasas, California

Donald G. Sexton
Assistant Principal
    Supervision — Mathematics
Adlai E. Stevenson High School
Bronx, New York

William K. Somerville
Mathematics Teacher
Massapequa High School
Massapequa, New York

Jeanette Tomasullo
Assistant Principal
    Supervision — Mathematics
Eastern District High School
Brooklyn, New York

Table of Contents

v

High Interest Features

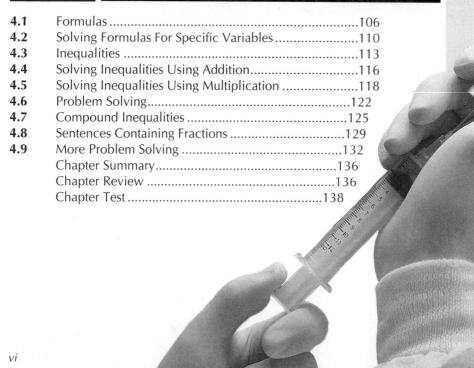

Chapter 5 | Aspects of Geometry

Chapter 6 | Geometric Relationships

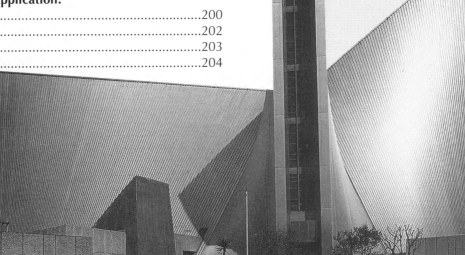

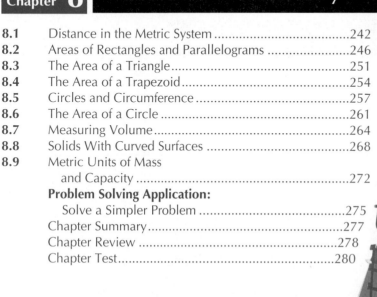

High Interest Features

Journal Entries
310, 323

Mixed Review
288, 302, 312, 345

Mathematical Excursions
294, 306, 335, 336, 341

Portfolio Suggestions
315, 346

Performance Assessment
315, 346

Manipulative Lab Activities and Extended Projects

Did you know you can show how to add integers without knowing the algebraic rules? You can represent many mathematical concepts by using manipulatives. A manipulative can be a colored counter, an algebra tile, a folded piece of paper, or even a graphing calculator. Pages A1–A16 contain 12 mathematics labs covering various concepts in this book.

Manipulative Lab Activities

What does President Woodrow Wilson have in common with solving equations? Turn to page B2 and you will find a project to work on with others in your class that explores all of the different aspects of balance, including the answer to this question. There are four Extended Projects on pages B1–B16 that use the mathematics you will learn over the year to investigate specific topics from history, science, geometry, and modern lifestyles.

Extended Projects

Introduction to
Logic

Application in Politics

Candidates for political office try to use **logical reasoning** to persuade voters to vote for them. Sometimes their reasoning is correct. Other times, there are flaws in their thinking. Informed voters use rules of logic and make intelligent decisions.

During her campaign for city council, Marie Hernandez promises that if she is elected, she will vote for park improvements. In order for the improvements to be approved, just one of the new council members must vote for the improvements. What can you conclude will happen if Ms. Hernandez is elected? What can you conclude will happen if she is not elected?

Group Project: *Literature*

Choose a chapter other than the first one in *Alice in Wonderland* by Lewis Carroll. Read the chapter and find two examples of how Mr. Carroll uses logic in the story. Report to the class about the examples you found.

1.1 Statements and Sentences

Most discussions, or arguments, which use logic begin with a statement. Consider the following statements.

1. Lima is the capital of Delaware.
2. Lima is not the capital of Delaware.
3. The average snowflake weighs 20 pounds.
4. $30 \div 2 = 15$
5. The Twelfth Amendment to the Constitution of the United States was ratified in 1842.

Statements may be true, such as 2 and 4, or they may be false such as 1, 3, and 5.

Definition of Statement and Truth Value

A **statement** is any sentence which is true or false, but *not* both. The truth or falsity of a statement is called its truth value.

A statement also may be called an <u>assertion</u> or <u>closed sentence</u>.

The truth value of statements 2 and 4 is true, and the truth value of statements 1, 3, and 5 is false.

Questions are *not* statements. Consider the following.

Was President James Buchanan ever married?

The question has an answer which is true or false, but the question itself *cannot* be described as true or false.

Similarly, truth values *cannot* be assigned to a command such as "Clean your room" or an exclamation such as "Right on!" Therefore, commands and exclamations are *not* statements.

Consider the following.

He was the fifteenth President of the United States.
$$x + 3 = 12$$

You *cannot* determine the truth value of such sentences unless you know what the placeholders *he* and *x* represent. For example, if $x = 6$, the sentence $x + 3 = 12$ is false; if $x = 9$, the sentence is true.

The placeholders in mathematical sentences are called **variables.** An **open sentence** contains one or more placeholders or variables. A variable may be replaced by any member of the **domain** or **replacement set.** The set of all replacements from the domain which make an open sentence true is called the **solution set** of the open sentence.

A set of numbers is indicated by braces. {1, 2, 3} is the set of numbers 1, 2, 3.

Example

1 **Find the solution set for the open sentence 8 − 5 = x if the domain is {0, 1, 2, 3}.**

The only replacement which makes the sentence 8 − 5 = x true is 3. Thus, the solution set is {3}.

In this book we will use special symbols for some familiar sets of numbers.

$$\mathcal{N} = \{1, 2, 3, 4, \ldots\} \text{ or the set of natural numbers}$$
$$\mathcal{W} = \{0, 1, 2, 3, \ldots\} \text{ or the set of whole numbers}$$

Examples

2 **Find the solution set for each open sentence if the domain is $\mathcal{W}$.**

    a. 8 − 3 = x   Replacing x by 5 produces the true statement 8 − 3 = 5. Therefore, the solution set is {5}.

    b. 2y > 4

        For y = 0, is 2 × 0 > 4?   *No*
        For y = 1, is 2 × 1 > 4?   *No*
        For y = 2, is 2 × 2 > 4?   *No*
        For y = 3, is 2 × 3 > 4?   *Yes*
        For y = 4, is 2 × 4 > 4?   *Yes*

        The sentence 2y > 4 is true when y is replaced by any number greater than 2. Thus, the solution set is {3, 4, 5, . . .}.

3 **Find the solution set for 8 − x = 3 if the domain is the set of even natural numbers, {2, 4, 6, 8, . . .}.**

For x = 2, is 8 − 2 = 3? *No*   For x = 4, is 8 − 4 = 3? *No*
If the pattern is continued you can see that there are *no* replacements from the domain which make the sentence true. Therefore, the solution set is the empty set.

The **empty set,** or **null set,** is symbolized by { } or Ø.

▰▰▰ Exercises ▰▰▰

Exploratory   State whether each of the following sentences is a statement. If it is a statement, determine its truth value. If it is *not* a statement, state why.

1. Shut the door!
2. Mars has two moons.
3. Is today Monday?
4. $x + 4 = 10$
5. $2x = 6$ is an equation.
6. Enough of this!
7. Is 3^2 equal to 6?
8. $0.05 = 5\%$
9. Monte Carlo is the capital of Nevada.
10. He was President of the United States during the War of 1812.
11. Zero is a natural number.
12. Are all squares also rectangles?
13. There is a whole number x such that $x + 4 = 12$.
14. She is a member of the Society of Actuaries.
15. Nigeria is a country in East Africa.
16. *Mano* is the Spanish word for *man*.

Written   Find the solution set for each of the following open sentences. The domain is {California, Georgia, Idaho, Massachusetts, Nebraska, New York, Pennsylvania}.

1. Its capital is Harrisburg.
2. It is a landlocked state.
3. It is a New England state.
4. It is located east of the Mississippi River.
5. It is a country in South America.
6. It is located in the United States.

Find the solution set for each of the following open sentences. The domain is $\mathcal{W}$.

7. $x + 2 = 14$
8. $x + 15 = 25$
9. $x - 4 = 11$
10. $h - 11 = 14$
11. $8 - y = 2$
12. $12 - y = 7$
13. $x + x = 10$
14. $\dfrac{n}{2} = 8$
15. $\dfrac{y}{4} = \dfrac{1}{2}$
16. $x < 10$
17. $x + x + x = 3x$
18. $2x < 8$
19. $2n = 6$
20. $3x > 1$
21. $n + n = 8$
22. $8 - y < 6$
23. $y - 5 > 1$
24. $m + 0 = m$
25. $8 + x = 15$
26. $y > 9$
27. $x + 3 = 1$
28. $3x = 12$
29. $2x + 1 = 5$
30. $x^2 > 5$

Challenge   The set of integers, $\mathcal{Z}$, is $\{\ldots, {}^-3, {}^-2, {}^-1, 0, 1, 2, 3, \ldots\}$. Find the solution set for each of the following open sentences if the domain is $\mathcal{Z}$.

1. $x + 3 = 1$
2. $y + 5 = 6$
3. $8 + x = {}^-2$
4. ${}^-3 + y = 2$
5. ${}^-7 + x = {}^-6$
6. $2y + 1 = {}^-7$
7. $3x - 2 = 10$
8. $4 + 2x = 8$
9. $x^2 = 4$

1.2   Negations

Statements are the building blocks in a logical system. A convenient way of referring to a specific statement is to represent the statement with a letter such as p or q.

Let p represent the following statement.

Arizona was admitted to the Union in 1892.

Suppose we want to say

"It is false that Arizona was admitted to the Union in 1892"

or, more simply,

"Arizona was not admitted to the Union in 1892."

We can write *not p*. The statement represented by *not p* is called the **negation** of p.

Definition of Negation

If a statement is represented by p, then *not p* is the negation of that statement. Likewise, p is the negation of *not p*. The symbol $\sim$ is used for *not*. Thus, $\sim p$ means *not p*.

Example

1 **Let p represent "it is snowing." Let q represent "$9 - 7 = 2$." Write the statements represented by each of the following.**

a. $\sim p$ "It is false that it is snowing" or "It is not snowing."

b. $\sim q$ "$9 - 7 \neq 2$"

c. $\sim(\sim p)$ This is the negation of $\sim p$. Thus, $\sim(\sim p)$ is "It is false that it is not snowing" or "It is snowing."

Is $\sim(\sim p)$ always the same as p?

Let *p* represent the statement, "Arizona was admitted to the Union in 1892." If *p* is a *true* statement, then ~*p* represents a *false* statement. However, if *p* represents a *false* statement, then ~*p* will represent a *true* statement. This example illustrates the principle that *the negation of a true statement is false and the negation of a false statement is true.*

Truth Table for Negation

p	~*p*
T	F
F	T

Sometimes it is convenient to organize the truth values of statements in a table like the one shown at the left. This particular table shows how the truth values of a statement and its negation are related. A table such as this is called a **truth table.**

Reading down the first column, we place the possible truth values that can be assigned to any statement represented by *p*, either T (true) or F (false). Reading down the second column, we place the corresponding truth values for the statement represented by ~*p*. Thus, by reading across the first row, we see that when *p* represents a true statement, ~*p* represents a false statement and vice versa.

Exercises

Exploratory   Write the negation of each of the following statements.

1. $7 = 4 + 2$

2. $5 \div 2 = 7$

3. $9 \neq 3 + 2$

4. $0.2 \times 0.2 = 0.4$

5. The product of 8 and 11 is an odd number.

6. Water freezes at 0°C.

7. Maine is the smallest state in area in the United States.

8. The speed of light is about 186,000 miles per second.

9. It is false that Dolley Madison was the wife of President Madison.

10. It is not true that Labor Day is celebrated in March.

State the truth value of each statement and the truth value of its negation in each of the following exercises.

11. Exercise 1	**12.** Exercise 2	**13.** Exercise 3	**14.** Exercise 4
15. Exercise 5	**16.** Exercise 6	**17.** Exercise 7	**18.** Exercise 8
19. Exercise 9	**20.** Exercise 10		

Written   Let p and q represent the following statements.

                 p: "8 is a prime number."

                 q: "Lincoln was a U.S. President."

Write each of the following statements in symbolic form.

1. 8 is a prime number.

2. It is false that 8 is a prime number.

3. Lincoln was not a U.S. President.

4. Lincoln was a U.S. President.

Using statements p and q above, write the statement symbolized by each of the following.

5. p **6.** $\sim p$ **7.** $\sim q$ **8.** $\sim(\sim p)$ **9.** $\sim(\sim q)$

Copy and complete each of the following truth tables.

10.

p	$\sim p$

11.

q	$\sim q$	$\sim(\sim q)$

12.

p	$\sim p$	$\sim(\sim p)$	$\sim(\sim(\sim p))$

If p represents any statement, state the relationship of the truth values between each of the following.

13. p and $\sim p$ **14.** $\sim p$ and $\sim(\sim p)$ **15.** $\sim p$ and $\sim(\sim(\sim p))$

16. Suppose a statement is represented by $\sim(\sim(\sim(. . .\sim p). . .))$ with the symbol $\sim$ appearing 47 times. What is the simplest statement that has the same truth value?

17. Suppose a statement is represented by $\sim(\sim(\sim(. . .\sim p). . .))$ with the symbol $\sim$ appearing 916 times. What is the simplest statement that has the same truth value?

18. What general rule can you state about the truth value of a statement in which $\sim$ appears n times?

Find the solution set for each of the following open sentences. The domain is $\mathcal{W}$.

19. $2x = 4$ **20.** $x + 5 = 7$ **21.** $x + x = 2x$

Write the negation of each open sentence in each of the following exercises. Then find the solution set for each negation.

22. Exercise 19   **23.** Exercise 20   **24.** Exercise 21

Mixed Review

Find the solution set for each inequality if the domain is $\mathcal{W}$.

1. $x > 3$ **2.** $y < 5$ **3.** $z < 1$

4. $m > 8$ **5.** $n > 21$ **6.** $t < 18$

7. $d < 0$ **8.** $a > 0$ **9.** $3b > 6$

10. $2s < 8$ **11.** $3x < 18$ **12.** $z + 1 > 2$

13. $r - 2 > 3$ **14.** $a - 3 < 2$ **15.** $3y + 1 < 4$

1.3 Conjunctions

Suppose we have the following statements.

"George Washington was the first President."
"Five is an even number."

These two statements can be connected by the word *and*. The result is a compound statement called a **conjunction.**

"George Washington was the first President and five is an even number."

Definition of Conjunction

A compound statement formed by joining two statements with the word *and* is called a conjunction. Each of the statements is called a **conjunct.**

In logic, the symbol $\wedge$ is used for the word *and*. Therefore, the conjunction of the statements p and q can be written as $p \wedge q$ and read *"p and q."*

Example

1 Let *r* represent "10 + 2 = 12." Let *s* represent "Delaware is the capital of Montana." Write the conjunction represented by each of the following.

a. $r \wedge s$ "10 + 2 = 12 and Delaware is the capital of Montana."

b. $r \wedge \sim s$ "10 + 2 = 12 and Delaware is not the capital of Montana."

c. $\sim r \wedge s$ "10 + 2 ≠ 12 and Delaware is the capital of Montana."

d. $\sim r \wedge \sim s$ "10 + 2 ≠ 12 and Delaware is not the capital of Montana."

Since all statements have truth values, a conjunction must have a truth value. Consider the first conjunction. If it is the case that Washington was the first President and it is also the case that five is an even number, then the conjunction is true. However, if the first statement is false, or if the second statement is false, or if both statements are false, then the conjunction is false.

Furthermore, this reasoning should hold for any statements represented by p, q, and $p \wedge q$. A conjunction, $p \wedge q$, will be true *only* when both p and q represent true statements.

Example

2 Let r represent "10 + 2 = 12," s represent "Delaware is the capital of Montana," and t represent "United States senators serve six-year terms." Determine the truth value of each of the following.

a. $s \wedge \sim t$ This is false because s is false and $\sim t$ is false.

b. $t \wedge (r \wedge \sim s)$ First determine the truth value of the statement in parentheses. In this case $r \wedge \sim s$ is true. Since t also is true, the conjunction $t \wedge (r \wedge \sim s)$ is true.

Truth Table for Conjunction

p	q	$p \wedge q$
T	T	T
T	F	F
F	T	F
F	F	F

A truth table can be used to summarize the truth value of a conjunction. There are four rows in the truth table because there are four possible combinations of truth values for the statements represented by p and q. Row 1 shows the case where both p and q represent true statements. In this case, the conjunction $p \wedge q$ is true. In the other rows, at least one statement is false. Thus, the conjunction is false.

Example

3 Construct a truth table for the conjunction $p \wedge \sim q$.

First, make columns with the headings p, q, $\sim q$, and $p \wedge \sim q$.
Next, list the possible combinations of truth values for p and q.
Then, write the truth values for $\sim q$.
Finally, write the truth values for $p \wedge \sim q$.

p	q	$\sim q$	$p \wedge \sim q$
T	T	F	F
T	F	T	T
F	T	F	F
F	F	T	F

The table shows that $p \wedge \sim q$ is true only when p is true and q is false.

Conjunctions of open sentences also can be formed.

Example

4 **Find the solution set for $(x > 3) \land (x < 7)$ if the domain is $\mathcal{W}$.**

Find all replacements for x which make both $x > 3$ and $x < 7$ true. Study the graph at the right.

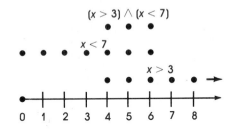

From the graph you can see that the solution set is $\{4, 5, 6\}$.

Exercises

Exploratory State the truth value of each of the following.

1. $3 < 5$ and $2 > 0$

2. $5 > 0$ and $7 = 0$

3. $25\% = 0.25$ and New York City is the capital of Vermont.

4. Cincinnati is located in Ohio and 10% of 320 is 32.

5. Every year has 12 months and July has 30 days.

6. Dallas is the capital of Texas and $5^2 > 2^5$.

Written Let p, q, and r represent the following statements.

    p: "Longfellow was a poet." (true)
    q: "Shakespeare wrote *Hamlet*." (true)
    r: "Mozart was a painter." (false)

Write each of the following statements in symbolic form using the letters above. Then state the truth value of the statement.

1. Longfellow was a poet and Shakespeare wrote *Hamlet*.

2. Mozart was a painter and Longfellow was a poet.

3. Shakespeare wrote *Hamlet* and Mozart was not a painter.

4. Longfellow was not a poet and Mozart was a painter.

5. Mozart was a painter and it is not the case that Longfellow was a poet.

6. Mozart was not a painter and it is false that Shakespeare wrote *Hamlet* and Longfellow was a poet.

Using statements p, q, and r from the previous exercises, write the conjunction represented by each of the following and then determine its truth value.

7. $p \wedge q$　　　　　**8.** $p \wedge r$　　　　　**9.** $p \wedge \sim r$　　　　**10.** $\sim p \wedge q$
11. $\sim p \wedge \sim q$　　　**12.** $q \wedge \sim r$　　　　**13.** $\sim p \wedge \sim r$　　**14.** $\sim q \wedge p$
15. $(p \wedge q) \wedge r$　**16.** $p \wedge (q \wedge \sim r)$　**17.** $\sim (q \wedge r)$　　**18.** $\sim (p \wedge q)$

Copy and complete each of the following truth tables.

19.

p	q	$\sim p$	$\sim p \wedge q$

20.

p	q	$\sim q$	$p \wedge \sim q$

21.

p	q	$\sim q$	$p \wedge \sim q$	$\sim (p \wedge \sim q)$

22.

p	q	$\sim p$	$\sim q$	$\sim p \wedge \sim q$

Find the solution set for each of the following compound sentences if the domain for each variable is $\mathcal{W}$.

23. $x > 3$ and $x < 5$　　　**24.** $t < 11$ and $t > 6$　　　**25.** $y < 4$ and $y < 2$
26. $y < 10$ and $y < 4$　　**27.** $x < 3$ and $x > 8$　　　**28.** $x < 7$ and $x > 2$
29. $t > 3$ and $t > 5$　　　**30.** $3y > 6$ and $y > 5$　　**31.** $2n > 4$ and $3n < 18$

Let s, t, and u represent the following statements.
　s:　"If the domain is $\mathcal{N}$, the solution set for $y > 6$ and $y - 5 < 1$ is $\{6\}$."
　t:　"If the domain is $\mathcal{N}$, the solution set for $x + 2 < 5$ and $2x > 2$ is $\{2\}$."
　u:　"If the domain is $\mathcal{N}$, the solution set for $3z < 6$ and $2z - 7 < 3$ is $\{0\}$."
Determine whether each of the statements represented by s, t, and u is true or false. Then determine the truth value of each of the following.

32. $(s \wedge t) \wedge u$　　　**33.** $s \wedge (t \wedge u)$　　　**34.** $(s \wedge t) \wedge \sim u$
35. $s \wedge (t \wedge \sim u)$　**36.** $(s \wedge t) \wedge \sim u$　　**37.** $\sim s \wedge (t \wedge u)$
38. $\sim (s \wedge t) \wedge u$　**39.** $(\sim s \wedge t) \wedge \sim u$　**40.** $(\sim s \wedge t) \wedge u$

Let A represent "x is a multiple of 9" and B represent "x is a multiple of 12." Determine the truth value of $A \wedge B$ for each value of x.

41. $x = 27$　　　　　**42.** $x = 36$　　　　　**43.** $x = 48$
44. $x = 51$　　　　　**45.** $x = 30$　　　　　**46.** $x = 54$

Challenge　Find the solution set for each of the following compound sentences if the domain for x is the set of integers, $\mathcal{Z}$.

1. $x > {}^-3$ and $x < 4$　　　　　　**2.** $x + 2 < 8$ and $x + 3 < {}^-5$
3. $2x < 10$ and $3x > {}^-12$　　　　**4.** $2x < 10$ and $x = 4$
5. $(x + 4 > 1) \wedge (x < 5)$　　　　**6.** $(2x < {}^-6) \wedge (2x < {}^-100)$

1.4 Disjunctions

Suppose we have the following statements.

> "The sum of two even numbers is even."
> "Harvard is the oldest university in the United States."

These two statements can be connected by the word *or*. The result is a compound statement called a **disjunction.**

> "The sum of two even numbers is even or Harvard is the oldest university in the United States."

Definition of Disjunction

> A compound statement formed by joining two statements with the word *or* is called a disjunction. Each of the statements is called a **disjunct.**

In logic, the symbol $\vee$ is used for the word *or*. Therefore, the disjunction of the statements p and q can be written as $p \vee q$ and read "*p or q*."

Example

1 Let *r* represent **"Shakespeare is the author of *Huckleberry Finn*"** and *s* represent **"A kilometer is approximately $\frac{5}{8}$ of a mile."** Write the disjunction represented by each of the following.

a. $r \vee s$ "Shakespeare is the author of *Huckleberry Finn* or a kilometer is approximately $\frac{5}{8}$ of a mile."

b. $r \vee \sim s$ "Shakespeare is the author of *Huckleberry Finn* or a kilometer is not approximately $\frac{5}{8}$ of a mile."

c. $\sim r \vee \sim s$ "Shakespeare is not the author of *Huckleberry Finn* or a kilometer is not approximately $\frac{5}{8}$ of a mile."

d. $\sim(r \vee s)$ "It is not the case that Shakespeare is the author of *Huckleberry Finn* or a kilometer is approximately $\frac{5}{8}$ of a mile."

A disjunction is false if both of its disjuncts are false. However, if one of its disjuncts is true, then the disjunction is true.

*A **journal** is a record of personal thoughts, impressions, ideas, and observations. In this book, you will find suggestions like the one below for items to write about in your mathematics journal.*

Journal

Complete this sentence: "The thing I like best about math class is...." Tell why.

What happens if both disjuncts are true? If we say that a Republican or a Democrat will be elected mayor, we mean just *one* of them, not both. On the other hand, if we say that it may rain Friday or Saturday, we mean that it could rain both days. We see that the word *or* is used in two different ways in everyday conversation. In logic, the word *or* is used in the second way. So, if *p* and *q* both represent true statements, the disjunction $p \lor q$ is true.

Truth Table for Disjunction

p	*q*	*p* $\lor$ *q*
T	T	T
T	F	T
F	T	T
F	F	F

Thus, *a disjunction is true if at least one of its disjuncts is true.* It is false only if both of its disjuncts are false. The truth table at the left summarizes the truth value of a disjunction.

Examples

2 Let *p* represent "Chicago is in Iowa," *q* represent "Portuguese is the national language of Brazil," and *r* represent "$6^2 = 36$." Determine the truth value of each of the following.

a. $r \lor p$ ⠀⠀⠀⠀⠀⠀⠀This is true since *r* is true.

b. $\sim q \lor p$ ⠀⠀⠀⠀⠀⠀This is false because $\sim q$ is false and *p* is false.

c. $p \lor (q \lor \sim r)$ ⠀⠀The disjunction $q \lor \sim r$ is true. Thus, even though *p* is false, the disjunction $p \lor (q \lor \sim r)$ is true.

3 Construct a truth table for the disjunction $p \lor (\sim q \land p)$.

First, make columns with the headings for *p*, *q*, $\sim q$, $\sim q \land p$, and $p \lor (\sim q \land p)$. Next, list the possible combinations of truth values for *p* and *q*.

p	*q*	$\sim q$	$\sim q \land p$	$p \lor (\sim q \land p)$
T	T	F	F	T
T	F	T	T	T
F	T	F	F	F
F	F	T	F	F

Then, complete the remaining columns.

Example

4 Find the solution set for each of the following disjunctions if the domain is $\mathcal{W}$.

a. $x \geq 5$

Recall that $\geq$ means *is greater than or equal to.* Thus, the solution set is $\{5, 6, 7, \ldots\}$.

b. $(x > 3) \lor (x < 7)$

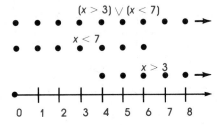

Compare this with Example 4 in Section 1.3.

The solution set is $\{0, 1, 2, 3, \ldots\}$ or $\mathcal{W}$.

Exercises

Exploratory State the truth value of each of the following.

1. $5 < 8$ or $9 < 0$
2. $9 \geq 8$
3. $0.10 > 0.01$ or $5 < 7$
4. $2 > 1$ or $0.5 = 0.50$
5. $75\% \leq 0.70$
6. $14 \leq 15.7$
7. Water boils at 32°F or 15% of 100 is 15.
8. $2^3 = 8$ or Louisville is in Kentucky.

Written Let *p, q,* and *r* represent the following statements.
 p: "The Thames flows through London." (true)
 q: "The Seine flows through Paris." (true)
 r: "The Nile flows through Rome." (false)
Write each of the following statements in symbolic form and state its truth value.

1. The Thames flows through London or the Seine flows through Paris.

2. The Nile flows through Rome or the Thames flows through London.

3. The Seine flows through Paris or the Nile does not flow through Rome.

4. The Thames does not flow through London or the Nile flows through Rome.

5. The Nile flows through Rome or it is not true that the Thames flows through London.

6. The Nile does not flow through Rome and it is false that the Seine flows through Paris or the Thames flows through London.

Using statements *p*, *q*, and *r* above, write the disjunction represented by each of the following and then state its truth value.

7. $p \vee q$ 8. $p \vee r$ 9. $q \vee r$ 10. $\sim p \vee q$
11. $\sim p \vee \sim q$ 12. $q \vee \sim r$ 13. $\sim p \vee \sim r$ 14. $\sim q \vee p$
15. $(p \vee q) \vee r$ 16. $p \vee (q \vee r)$ 17. $\sim(p \vee q)$ 18. $\sim(p \vee \sim r)$

Copy and complete each of the following truth tables.

19.

s	*t*	$\sim s$	$\sim s \vee t$

20.

s	*t*	$\sim s$	$\sim t$	$\sim s \vee \sim t$

Construct truth tables for each of the following.

21. $s \vee \sim t$ 22. $\sim(s \vee \sim t)$ 23. $\sim s \wedge (s \vee t)$
24. $\sim s \vee (s \wedge t)$ 25. $(s \wedge t) \vee s$ 26. $(s \vee t) \vee (s \wedge t)$

Find the solution set for each compound sentence if the domain is $\mathcal{W}$.

27. $x \leq 6$ 28. $y \geq 3$ 29. $n \geq 7$
30. $x \leq 9$ 31. $(x > 2) \vee (x < 5)$ 32. $(x < 6) \vee (x < 2)$
33. $(x \geq 2) \vee (x < 5)$ 34. $(3x < 18) \vee (x = 4)$ 35. $x \neq 3$

Let *u*, *x*, and *w* represent the following statements.

  u: "The product of two odd numbers is an even number."
  x: "The product of two even numbers is an even number."
  w: "The product of an even number and an odd number is an even number."

Determine whether each of the statements represented by *u*, *x*, and *w* is true or false. Then determine the truth value of each of the following.

36. $u \vee (x \wedge w)$ 37. $(u \vee x) \wedge w$ 38. $(u \wedge x) \vee (u \wedge w)$
39. $(u \wedge x) \vee w$ 40. $(u \vee x) \wedge \sim w$ 41. $(u \wedge \sim x) \vee (\sim u \wedge w)$

Let *s* represent "*x* is a multiple of 9." and *t* represent "*x* is a multiple of 12." Determine the truth value of $[(s \vee t) \wedge \sim s] \vee \sim t$ for each value of *x*.

42. $x = 27$ 43. $x = 36$ 44. $x = 48$ 45. $x = 51$ 46. $x = 30$ 47. $x = 54$

Challenge   Find the solution set for each compound sentence if the domain is $\mathcal{Z}$.

1. $(x < {}^-3) \wedge (x < {}^-7)$ 2. $(x < 3) \wedge (x > 0)$ 3. $(x^2 = 9) \wedge (x > 0)$
4. $(x^2 = 9) \wedge (x < 0)$ 5. $(3x < 18) \vee (x > 4)$ 6. $[(x + 5) > 3] \vee (x = 2)$

7. In a class of 100 students, 39 take Spanish, 47 take French, and 5 take both languages. How many students are not taking either language?

8. In a class of 250 students, 106 take Chemistry, 14 take both Chemistry and Biology, and 23 take neither class. How many students take Biology?

1.5 The Conditional

We can form a new statement by connecting two statements with the words *if . . ., then*. The new statement is called a **conditional.**

Definition of Conditional

> A compound statement formed by joining two statements with the words *if . . ., then* is called a conditional.

Sometimes a conditional is called an implication.

In logic, the symbol → is used to represent a conditional. Therefore, the conditional "if *p*, then *q*" can be written as $p \rightarrow q$ and read *"p implies q"* or *"if p then q."* Statement *p* is called the **antecedent** of the conditional and statement *q* is called the **consequent** of the conditional.

If you want a healthy cat, start with a healthy kitten.

The statement at the left is a conditional with *then* omitted.

Conditional statements are written in many different ways, but all conditional statements can be written in *if-then* form. For example, the following statements could *all* be written "If you are a genius, then you are left-handed."

All geniuses are left-handed.
You are left-handed if you are a genius.
Being a genius implies you are left-handed.

Example

1 Let *r* represent "Ron lives in San Francisco" and *s* represent "Ron lives in California." Write the conditional represented by each of the following.

a. $r \rightarrow s$ "If Ron lives in San Francisco, then Ron lives in California."

b. $s \rightarrow \sim r$ "If Ron lives in California, then Ron does not live in San Francisco."

▀▄ Example

2 Write the following statement in symbolic form using the letters after the given statement.

"If it is not raining tomorrow, then I will either jog or go hiking." *(R, J, H)*

This statement is represented by $\sim R \rightarrow (J \vee H)$.

The antecedent, p, is "you pass the final exam." The consequent, q, is "you pass the course."

Both p and q are true.

In this case, p is true and q is false.

p is false.

Truth Table for the Conditional

p	q	$p \rightarrow q$
T	T	T
T	F	F
F	T	T
F	F	T

When is a conditional true? Suppose a teacher, Mr. Kelly, makes the following statement.

"If you pass the final exam, then you pass the course."

Suppose that you *do* pass the final exam and you *do* pass the course. Certainly, Mr. Kelly's statement was true.

Suppose that after passing the final exam, you do *not* pass the course. This time Mr. Kelly's statement was false.

What happens if you do *not* pass the final exam? Mr. Kelly's statement does not say what will happen if you do *not* pass the final exam. Maybe you could pass the course anyway by doing an extra project. Or you could fail the course. In this case, we cannot say that Mr. Kelly's statement is false.

We will consider a conditional, $p \rightarrow q$, to be *true* whenever p is false regardless of the truth value of q. The truth values of a conditional are summarized in the truth table at the left.

▀▄ Example

3 Determine the truth value of each of the following statements.

a. "If Denver is not the capital of Vermont, then $3^2 = 10$." The antecedent is true, but the consequent is false. Thus, the conditional is false.

b. "If $3^2 = 11$, then Denver is the capital of Vermont." Since the antecedent is false, the conditional is true.

Example

4 **Construct a truth table for the conditional $\sim p \rightarrow q$.**

First, make columns with the headings
p, q, $\sim p$, and $\sim p \rightarrow q$.
Next, list the possible combinations of
truth values for p and q.
Then, complete the remaining
columns.

p	q	$\sim p$	$\sim p \rightarrow q$
T	T	F	T
T	F	F	T
F	T	T	T
F	F	T	F

Exercises

Exploratory **State the antecedent and consequent of each of the following conditionals.**

1. If it rains today, then I will stay home.

2. If $x + 3 = 7$, then $x = 4$.

3. If Joe earns an A in math, he will be on the honor roll.

4. The team will win if Bill makes the free throw.

5. Garnet will date Paul if Mark is out of town.

6. The team has a good chance of winning if either Ann or Pam pitches.

Determine the truth value of each of the following statements.

7. If $3 > 2$, then $1 = 2$.

8. If $3 \neq 4$, then $5 < 7$.

9. $2^3 = 8$ implies $5 < 0$.

10. $0.01 > 0.1$ if $9 \leq 9$.

11. If today is Wednesday, then tomorrow is Friday.

12. Two is a prime number if air is mostly nitrogen.

Written **Let p, q, and r represent the following statements.**

> p: "June has 30 days."
> q: "1983 is a leap year."
> r: "Labor Day is in September."

Write the statement represented by each of the following and state its truth value.

1. $p \rightarrow q$

2. $q \rightarrow p$

3. $p \rightarrow r$

4. $p \rightarrow \sim q$

5. $p \rightarrow \sim r$

6. $\sim p \rightarrow \sim q$

7. $\sim p \rightarrow q$

8. $(p \wedge r) \rightarrow q$

9. $(p \wedge q) \rightarrow r$

10. $\sim p \rightarrow \sim r$

11. $(p \wedge q) \rightarrow \sim r$

12. $(p \vee q) \rightarrow r$

13. $p \rightarrow (q \wedge \sim r)$

14. $r \rightarrow (p \wedge \sim q)$

15. $q \rightarrow (\sim p \vee r)$

Write each of the following statements in symbolic form using the letters after each statement.

16. Mozart or Haydn composed that symphony. *(M, H)*

17. If Mozart composed that symphony, then Haydn did not. *(M, H)*

18. If bears hibernate in winter, then they do not eat Crunchy-Wunchies for breakfast. *(H, C)*

19. If Jim does not eat Crunchy-Wunchies for breakfast, he is a fool. *(J, F)*

20. If I do not eat Crunchy-Wunchies for breakfast and if I do not hibernate in winter, then I will not go to Harvard. *(C, W, H)*

21. If I eat Crunchy-Wunchies for breakfast, or if I graduate from Harvard, then I am bound to be successful. *(C, H, S)*

22. If Sally did not attend Yale or if bears do not hibernate in winter, then either Shakespeare ate Crunchy-Wunchies for breakfast, or he was a fool. *(S B, C, F)*

23. If Shakespeare did not write Hamlet, then if Francis Bacon did, he was very clever. *(S, F, C)*

Construct a truth table for each of the following.

24. $r \rightarrow s$ **25.** $p \rightarrow \sim q$ **26.** $\sim p \rightarrow \sim q$ **27.** $p \rightarrow (p \vee q)$

The first two sentences in each problem have the indicated truth value. Determine, if possible, the truth value of the third sentence.

28. **(i)** If Kent dates Maria, then Nancy is happy.   (true)
    (ii) Kent dates Maria.   (true)
    (iii) Nancy is happy.

29. **(i)** If Bruce does not pass French, his mother is not happy.   (true)
    (ii) Bruce's mother is not happy.   (true)
    (iii) Bruce does not pass French.

30. **(i)** Kent is healthy if he eats Crunchy-Wunchies.   (true)
    (ii) Kent is not healthy.   (true)
    (iii) Kent eats Crunchy-Wunchies.

31. **(i)** If Crunchy-Wunchies contain iron, then sugar prices increase. (true)
    (ii) Sugar prices do not increase.   (false)
    (iii) Crunchy-Wunchies do not contain iron.

If p represents a false statement, q represents a true statement, and r represents a false statement, determine the truth value of each of the following statements.

32. $\sim p \rightarrow [q \vee (q \wedge r)]$

33. $\sim p \rightarrow [q \vee (\sim q \vee r)]$

34. $\sim p \rightarrow [\sim q \wedge (q \vee r)]$

35. $(p \vee q) \rightarrow [p \vee (p \vee r)]$

36. $[p \vee (q \wedge \sim r)] \rightarrow r$

37. $(p \rightarrow r) \rightarrow (p \rightarrow q)$

1.6 Tautologies

Consider the following statement.

"If McKinley was elected to the Presidency in 1896, then McKinley was elected to the Presidency in 1896."

Certainly, you would agree that this is a true statement as is the following.

"If McKinley was elected to the Presidency in 1980, then McKinley was elected to the Presidency in 1980."

These sentences about McKinley are constructed in such a way that they would be true no matter what facts they contain. In general, for any statement p, it must always be the case that the conditional $p \rightarrow p$ is true. This conditional is an example of a statement called a **tautology.**

Definition of Tautology

A tautology is a compound statement which is true regardless of the truth values of the statements of which it is composed.

Example

1 Write each of the following statements in symbolic form and determine if each is a tautology.

a. "The rose is yellow or it is not yellow."
If p represents "the rose is yellow," then $p \lor \sim p$ represents statement **a.** Since statement **a** is always true, it is a tautology.

b. "Each statement is true or it is false."
Likewise, statement **b** can be represented by $p \lor \sim p$ and is a tautology.

One way to determine if a statement is a tautology is to use a truth table. Consider the following statement which in symbolic form is written $p \rightarrow (p \vee q)$.

"If it is raining, then it is raining or it is snowing."

Is this a true statement? Is any statement which fits the pattern $p \rightarrow (p \vee q)$ true regardless of the truth values of p and q? In other words, is $p \rightarrow (p \vee q)$ a tautology?

p	q	$p \vee q$	$p \rightarrow (p \vee q)$
T	T	T	
T	F	T	
F	T	T	
F	F	F	

To investigate these questions, we can use a truth table as shown at the left. To complete the final column of the table, we recognize that the statement $p \rightarrow (p \vee q)$ is a conditional. The *only* way this conditional would be false would be if p, the antecedent, were true and $(p \vee q)$, the consequent, were false. However, this is *never* the case and we complete the table as shown at the left.

p	q	$p \vee q$	$p \rightarrow (p \vee q)$
T	T	T	T
T	F	T	T
F	T	T	T
F	F	F	T

The T's in the final column of the table show us that any statement which can be symbolized by $p \rightarrow (p \vee q)$ must *always* be true. We conclude that the statement $p \rightarrow (p \vee q)$ is a tautology.

Example

2 **Determine if $p \rightarrow \sim (\sim p \vee q)$ is a tautology by constructing a truth table.**

p	q	$\sim p$	$\sim p \vee q$	$\sim(\sim p \vee q)$	$p \rightarrow \sim (\sim p \vee q)$
T	T	F	T	F	F
T	F	F	F	T	T
F	T	T	T	F	T
F	F	T	T	F	T

To find the truth value of $\sim(\sim p \vee q)$, first find the truth value of $\sim p \vee q$.

The final column shows that $p \rightarrow \sim(\sim p \vee q)$ can be false.

Thus, it is *not* a tautology.

Exercises

Exploratory If p represents a true statement, q represents a true statement and r represents a false statement, determine the truth value of each of the following statements.

1. $p \wedge q$
2. $p \wedge \sim q$
3. $\sim(p \vee q)$
4. $p \rightarrow \sim r$
5. $r \rightarrow (p \vee q)$
6. $\sim p \rightarrow (r \wedge p)$
7. $(r \rightarrow q) \rightarrow p$
8. $(p \wedge r) \rightarrow \sim q$
9. $(\sim r \rightarrow p) \wedge q$
10. $(\sim p \wedge q) \vee (p \vee r)$
11. $[p \vee (r \wedge q)] \rightarrow r$
12. $\sim r \rightarrow [p \vee (q \wedge r)]$
13. $(r \vee \sim r) \rightarrow (p \wedge \sim q)$
14. $(p \vee \sim r) \rightarrow (p \rightarrow r)$
15. $[r \wedge (p \rightarrow \sim r)] \rightarrow p$
16. $r \wedge [(p \rightarrow \sim r) \rightarrow p]$

Written Construct a truth table for each of the following statements and then determine whether each statement is a tautology.

1. $(p \wedge q) \rightarrow (p \vee q)$
2. $(p \wedge q) \rightarrow (p \vee \sim q)$
3. $(p \wedge \sim p) \rightarrow q$
4. $(p \vee \sim p) \vee q$
5. $p \vee (q \vee \sim q)$
6. $(p \vee q) \rightarrow \sim q$
7. $\sim(p \vee q) \rightarrow \sim p$
8. $\sim p \rightarrow (\sim p \wedge r)$
9. $(p \vee q) \rightarrow (q \vee p)$
10. $\sim(p \wedge q) \rightarrow (\sim p \wedge \sim q)$
11. $q \rightarrow (p \vee \sim p)$
12. $(p \wedge \sim q) \rightarrow \sim q$
13. $\sim(p \vee q) \rightarrow (p \rightarrow q)$
14. $[p \wedge (p \rightarrow q)] \rightarrow q$
15. $(\sim p \vee q) \rightarrow (p \rightarrow q)$
16. $(q \vee \sim q) \rightarrow (p \vee q)$
17. $(p \vee q) \rightarrow (p \rightarrow q)$
18. $[p \vee (\sim p \wedge q)] \vee (\sim p \wedge \sim q)$

State what conclusion, if any, can be made from the given statements.

19. In a certain mathematical system it is known that $4 + 3 = 5$ or $4 + 3 = 6$, but it can be shown that in this system $4 + 3 \neq 6$.

20. In the Jones murder case, the police have determined that the butler did not murder Dr. Jones. However, it has also been shown that the butler or the nurse is the murderer.

Answer each of the following.

21. What are the truth values for every tautology?

22. Does each individual statement in a tautology need to be true?

23. Is it possible to have a tautology in which each of the individual statements is false? If so, give an example.

24. A compound statement which is always false, regardless of the truth value of its individual statements, is called a **contradiction.** Give an example of a contradiction.

Mathematical Excursions

April Barlow is a paralegal aide for the Tucson City prosecuting attorney's office. She is gathering information regarding a shoplifting case. She must find evidence to prove that Val Deeds did shoplift at Levy's Department Store.

From the police report and witnesses, she has the following statements.

A. Val Deeds was in the store at the time of the shoplifting.
B. Val Deeds is 16 years old.
C. Val Deeds has no receipt for the merchandise.
D. Val Deeds attends high school.
E. Val Deeds has blue eyes.
F. Val Deeds was in the parking lot with the merchandise.

Suppose the definition of shoplifting is *the removal of merchandise from a store by an individual(s) without proof of payment*. The information in statements **A, C,** and **F** can be used as evidence of shoplifting.

Exercises For each of the following, tell which statements can be used as evidence.

1. A *speeding violation* occurs when a motorist drives at a speed over the posted limit.

   A. Bell Wilkins was driving 30 mph in a 35 mph zone.

   B. Jon Bunch was found driving 45 mph in a 20 mph zone.

2. An *isosceles triangle* is a triangle with at least two sides congruent.

   A. Triangle *ABC* has three congruent sides.

   B. Triangle *RST* has three congruent angles.

3. A *good swimmer* is a person who can do the 100 meter backstroke in less than 01:02.00.

   A. Kornelia Ender can do the 100 meter freestyle in 00:55.65.

   B. Ulriche Richter can do the 100 meter backstroke in 01:01.51.

4. *Supplementary angles* are two angles whose degree measures total 180.

   A. The degree measure of $\angle C$ is 50, and the degree measure of $\angle D$ is 40.

   B. The degree measure of $\angle R$ is 125, and the degree measure of $\angle S$ is 55.

1.7 Converse, Inverse, and Contrapositive

The conditional $p \rightarrow q$ has three related statements, each of which is also a conditional.

The **converse** of a conditional is formed by switching the order of p and q of the original conditional. Thus, the converse of $p \rightarrow q$ is $q \rightarrow p$.

Original: If $\underbrace{\text{today is Wednesday}}_{p}$ then $\underbrace{\text{tomorrow is Thursday.}}_{q}$

Converse: If $\underbrace{\text{tomorrow is Thursday}}_{q}$ then $\underbrace{\text{today is Wednesday.}}_{p}$

In this example, both the original and converse are true. Is this always the case? Consider the following example.

If the original statement is true, the converse may or may not be true. Why?

Original: If it's raining, then I stay home.
Converse: If I stay home, then it's raining.

Consider another example.

Assume all horses have four legs.

Original:
If Charlie is a horse, then Charlie has four legs.

Converse:
If Charlie has four legs, then Charlie is a horse.

Certainly, the original is a true statement, but the converse is definitely false. Thus, the converse of a true statement may be true or it may be false.

The **inverse** of a conditional is formed by negating both the antecedent and the consequent. Thus, the inverse of $p \rightarrow q$ is $\sim p \rightarrow \sim q$.

Original: If $\underbrace{\text{today is Wednesday}}_{p}$ then $\underbrace{\text{tomorrow is Thursday.}}_{q}$

Inverse: If $\underbrace{\text{today is not Wednesday}}_{\sim p}$ then $\underbrace{\text{tomorrow is not Thursday.}}_{\sim q}$

In the example above, the original and the inverse are true. But consider the following example.

Is the inverse true?

Original:
If Charlie is a horse, then Charlie has four legs.

Inverse:
If Charlie is not a horse, then Charlie does not have four legs.

Thus, the inverse of a true statement may be true or it may be false.

The **contrapositive** of a conditional is formed by negating both p and q *and* switching their order. Thus, the contrapositive of $p \rightarrow q$ is $\sim q \rightarrow \sim p$.

True Original: If <u>today is Wednesday</u> then <u>tomorrow is Thursday</u>.

p q

True Contrapositive: If <u>tomorrow is not Thursday</u> then <u>today is not Wednesday</u>.

$\sim q$ $\sim p$

Consider the following.

Do you agree that in this case, <u>both</u> the original and its contrapositive are true?

Original:  Contrapositive:

If Charlie is a horse, then Charlie has four legs. If Charlie does not have four legs, then Charlie is not a horse.

What happens if the original conditional is false?

Original: If $2 + 2 = 4$, then $3 + 3 = 7$. *False*
Contrapositive: If $3 + 3 \neq 7$, then $2 + 2 \neq 4$. *False*

We can use truth tables to see if a conditional and its contrapositive always have the same truth value. The truth table for a conditional $p \rightarrow q$ is shown below on the left. To its right is the truth table for the contrapositive $\sim q \rightarrow \sim p$.

Make sure you understand how each entry in this table was obtained.

p	q	$p \rightarrow q$
T	T	T
T	F	F
F	T	T
F	F	T

p	q	$\sim q$	$\sim p$	$\sim q \rightarrow \sim p$
T	T	F	F	T
T	F	T	F	F
F	T	F	T	T
F	F	T	T	T

Compare the last columns of each table.

Notice that the truth value of the contrapositive is always the same as the truth value of the original conditional.

The following table summarizes the related statements of a conditional and their truth value.

Definitions and Truth Values of Converse, Inverse, and Contrapositive

Conditional	Relation to Original	Truth Value Compared to Original
$p \rightarrow q$	original	
$q \rightarrow p$	converse	Converse of true conditional may be true or false.
$\sim p \rightarrow \sim q$	inverse	Inverse of true conditional may be true or false.
$\sim q \rightarrow \sim p$	contrapositive	Always the same truth value as the original.

Examples

1 Write the converse, inverse, and contrapositive of the conditional $r \rightarrow {\sim}s$.

Converse: ${\sim}s \rightarrow r$
Inverse: ${\sim}r \rightarrow {\sim}({\sim}s)$ or ${\sim}r \rightarrow s$
Contrapositive: ${\sim}({\sim}s) \rightarrow {\sim}r$ or $s \rightarrow {\sim}r$

2 Write the converse, inverse, and contrapositive of the conditional, "If $3 + 2 = 6$, then Mars is a planet." Then determine the truth value of each statement.

Let p represent "$3 + 2 = 6$" and q represent "Mars is a planet."

Converse: $q \rightarrow p$ or "If Mars is a planet, then $3 + 2 = 6$." Since q is true and p is false, the converse is false.

Inverse: ${\sim}p \rightarrow {\sim}q$ or "If $3 + 2 \neq 6$, then Mars is not a planet." The inverse is false.   *Why?*

Contrapositive: ${\sim}q \rightarrow {\sim}p$ or "If Mars is not a planet, then $3 + 2 \neq 6$." Since the antecedent is false, the contrapositive is true.

Exercises

Exploratory   State the converse of each of the following conditionals.

1. $p \rightarrow q$ **2.** ${\sim}r \rightarrow s$ **3.** ${\sim}m \rightarrow {\sim}t$ **4.** $(p \vee q) \rightarrow r$

State the inverse of each of the following conditionals.

5. $p \rightarrow q$ **6.** $r \rightarrow {\sim}s$ **7.** ${\sim}m \rightarrow {\sim}p$ **8.** ${\sim}m \rightarrow t$

State the contrapositive of each of the following conditionals.

9. $p \rightarrow q$ **10.** ${\sim}r \rightarrow s$ **11.** ${\sim}m \rightarrow {\sim}t$ **12.** $s \rightarrow {\sim}t$

Written   Write the converse of each of the following conditionals.

1. If $2^2 = 4$, then $3^2 = 9$.

2. If $2 = 2$, then 7 is even.

3. If $2 + 2 = 4$, then Mark Twain wrote *Tom Sawyer*.

4. If $2 + 2 \neq 4$, then Jane Austen wrote *Romeo and Juliet*.

5. If you pass GO, you collect $200.

6. If you do not vote for Snurr for Senator, the economy will collapse.

7. Louisa Alcott wrote *Little Women* if 2 is not a prime number.

8. Hog prices will skyrocket if you vote for Snurr for Senator.

Write the inverse of each conditional in exercises 1-8.

9. Exercise 1   10. Exercise 2   11. Exercise 3   12. Exercise 4
13. Exercise 5   14. Exercise 6   15. Exercise 7   16. Exercise 8

Write the contrapositive of each conditional in exercises 1-8.

17. Exercise 1   18. Exercise 2   19. Exercise 3   20. Exercise 4
21. Exercise 5   22. Exercise 6   23. Exercise 7   24. Exercise 8

Copy and complete the following truth table.

25.

p	q	$\sim p$	$\sim q$	$p \rightarrow q$	$q \rightarrow p$	$\sim p \rightarrow \sim q$	$\sim q \rightarrow \sim p$

Determine the truth value of each conditional, its converse, its inverse, and its contrapositive.

26. If water freezes at 0°C, then 2 + 3 = 5.

27. If Detroit is in Michigan, then the sun is cold.

28. If the Empire State Building has 60 floors, then 2 ÷ 2 = 1.

29. If a monkey is an animal, then a monkey is a mammal.

30. If 1984 is a leap year, then February will have 29 days.

31. Robert Frost was a novelist if 3 is an even number.

Suppose a teacher says, "If you do not complete the project, you will fail the course." Assume the statement is true and answer each of the following.

32. You fail the course. Did you complete the project?

33. You complete the project. Will you pass the course?

34. You do not complete the project. Will you pass the course?

35. You pass the course. Did you complete the project?

Suppose Fred and Jill agree that they will not go to the movies if it does not rain. Answer each of the following.

36. Fred and Jill did not go to the movies. Did it rain?

37. It did not rain. Did Fred and Jill go to the movies?

38. Fred and Jill went to the movies. Did it rain?

39. It rained. Did Fred and Jill go to the movies?

For each of the following, assume that statement (i) is true. Then determine if you can logically conclude that statement (ii) must also be true.

40. (i) If sugar prices are up, then cookies are more expensive.
    (ii) If cookies are more expensive, then sugar prices are up.

41. (i) If Rodriquez is elected, the substation will be built.
    (ii) If the substation is not built, then Rodriquez was not elected.

42. (i) If Sally uses White-on toothpaste, her teeth will not decay.
    (ii) If Sally does not use White-on toothpaste, her teeth will decay.

1.8   Biconditionals and Equivalent Statements

We know that the converse of a true conditional may or may not be true. Sometimes it is helpful to know when a given conditional, $p \rightarrow q$, and its converse, $q \rightarrow p$, are both true. In other words, we need to know when the conjunction $(p \rightarrow q) \wedge (q \rightarrow p)$ is true.

What is the truth value of this statement?

"If a figure is a square then it is a rectangle and if a figure is a rectangle then it is a square."

p	q	$p \rightarrow q$	$q \rightarrow p$	$(p \rightarrow q) \wedge (q \rightarrow p)$
T	T	T	T	T
T	F	F	T	F
F	T	T	F	F
F	F	T	T	T

To see when a statement of the form $(p \rightarrow q) \wedge (q \rightarrow p)$ is true we can construct a truth table.

Notice that $(p \rightarrow q) \wedge (q \rightarrow p)$ is true in those cases, and only those cases, when p and q are *both* true or *both* false. This type of statement is called a **biconditional.**

Definition of Biconditional

A statement formed by the conjunction of the conditionals $p \rightarrow q$ and $q \rightarrow p$ is a biconditional.

Truth Table for the Biconditional

p	q	$p \leftrightarrow q$
T	T	T
T	F	F
F	T	F
F	F	T

In logic, the symbol $\leftrightarrow$ is used to write the biconditional. The biconditional $(p \rightarrow q) \wedge (q \rightarrow p)$ can be written as $p \leftrightarrow q$. The truth table for the biconditional is shown at the left.

Example

1 Let p represent "$2 + 2 = 4$" and q represent "March has 31 days." Determine the truth value of the biconditional represented by each of the following.

a. $p \leftrightarrow q$ This is true since p and q have the same truth value.

b. $p \leftrightarrow \sim q$ This is false since p and $\sim q$ have different truth values.

c. $\sim p \leftrightarrow \sim q$ This is true since $\sim p$ and $\sim q$ have the same truth value.

The biconditional $p \leftrightarrow q$ can be read *"p if and only if q."* Thus, $p \leftrightarrow q$ in example 1 is read "2 + 2 = 4 if and only if March has 31 days."

Example

2 **Let *r* represent "a year has 4 seasons" and *s* represent "the sun is hot." Write the biconditional represented by each of the following.**

a. $p \leftrightarrow \sim q$ "A year has 4 seasons if and only if the sun is not hot."

b. $\sim p \leftrightarrow \sim q$ "A year does not have 4 seasons if and only if the sun is not hot."

Whenever two statements are *always* either both true or both false, the two statements are **equivalent.** Thus, biconditionals which are true are called **equivalences.** Another example of statements which are equivalent is a conditional and its contrapositive. This can be symbolized as follows.

$$(p \rightarrow q) \leftrightarrow (\sim q \rightarrow \sim p).$$

Exercises

Exploratory State the truth value of each of the following.

1. Detroit is the capital of Kansas if and only if 2 + 2 > 7.

2. A triangle has 3 sides if and only if a pentagon has 6 sides.

3. Water boils at 100°C if and only if 31 is a prime number.

4. Chicago is the capital of Belgium if and only if it is false that 2 < 1.

5. Jack Nicklaus is a golfer if and only if January is a summer month.

6. A robin is a mammal if and only if Venus is a planet.

7. $0.2 \times 0.2 = 0.04$ if and only if 9 is an odd number or a prime number.

8. 18 is even if and only if 27 is divisible by 3 or 5.

Written   Let *p, q,* and *w* represent the following statements.

  p: New Mexico is in Central America.
  q: 3 + 4 < 8.
  w: Pink is a primary color.

Write the biconditional represented by each of the following and then determine its truth value.

1. $p \leftrightarrow q$	**2.** $p \leftrightarrow w$	**3.** $q \leftrightarrow w$
4. $q \leftrightarrow \sim w$	**5.** $\sim p \leftrightarrow \sim w$	**6.** $(p \wedge q) \leftrightarrow w$
7. $(p \vee w) \leftrightarrow q$	**8.** $p \leftrightarrow (q \wedge \sim w)$	**9.** $(p \vee q) \leftrightarrow (p \vee w)$

Write each of the following statements in symbolic form using the letters after each statement.

10. Sally dates Tom if and only if Ted is out of town.   *(S, F)*

11. Renee eats Crunchy-Wunchies for breakfast if and only if she does not have pancakes and is not in a hurry.   *(R, P, H)*

12. Earle does not take a foreign language and does not take art if and only if he is taking two sciences.   *(E, A, S)*

13. If and only if there is a taxi strike does Mr. Oglethorpe ride the bus.   *(T, B)*

14. Kevin practices his piano lesson or does not walk the dog if and only if it is raining.   *(P, W, R)*

15. Fred has French toast for breakfast if and only if he wakes up before dawn and he wakes up before dawn if and only if he sets his alarm clock. *(F, W, A)*

Construct a truth table for each of the following. Then determine if the statement is a tautology.

16. $\sim p \leftrightarrow q$	**17.** $(p \vee q) \leftrightarrow q$
18. $(\sim p \vee q) \leftrightarrow p$	**19.** $\sim (p \vee q) \leftrightarrow (\sim p \wedge \sim q)$
20. $\sim (p \wedge q) \leftrightarrow (\sim p \vee \sim q)$	**21.** $[(\sim p \vee q) \wedge p] \leftrightarrow (p \wedge q)$

Determine if each of the following pairs of statements is equivalent by constructing truth tables.

22. the converse of $p \rightarrow q$ and the inverse of $p \rightarrow q$

23. $p \rightarrow q$ and $\sim p \vee q$

Use the results of exercise 23 to write each of the following conditionals as a disjunction. (i.e., *p* → *q* is equivalent to ∼*p* ∨ *q*)

24. If Jay dates Mary, then Ellen is out of town.

25. If John lives in Cleveland, then he lives in Ohio.

26. If Marcos is a biologist, then he specializes in zoology or botany.

27. $\sim p \rightarrow [q \vee (\sim r \wedge s)]$

Problem Solving Application: Using Logical Reasoning

Many real-world problems are solved by translating a verbal problem into mathematical language and then solving the problem. To do this, a four-step problem-solving plan can be used.

Explore   Read the problem carefully and explore the situation described. Identify what information is given and what you are asked to find.

Plan   Plan how to solve the problem. Your plan may involve solving equations or inequalities, drawing a picture, using deductive reasoning, or making a guess.

Solve   Carry out your plan for solving the problem.

Examine   Check your results. Be sure your solution satisfies the conditions of the given problem.   *Ask yourself, "Is the answer reasonable?"*

Some verbal problems can be solved by using logical reasoning. Study the following problem.

> **Suppose Detective Clouseau knows that the following statements are true.**
> • **If Mr. Bell was in Brooklyn, he did not commit the crime.**
> • **Mr. Bell was in Brooklyn.**
> **From this information, can Detective Clouseau conclude that Mr. Bell did not commit the crime?**

Use the four-step problem-solving plan to solve the problem.

Explore   Identify the information given in the problem.
   Let p represent "Mr. Bell was in Brooklyn."
   Let q represent "He did not commit the crime."
   Then $p \rightarrow q$ represents "If Mr. Bell was in Brooklyn, then he did not commit the crime."
   Now we need to know if $p \rightarrow q$ and p lead to the conclusion, q.

Plan   To solve the problem, use the truth table for the conditional $p \rightarrow q$.

Solve   Only the first row of this truth table has both $p \rightarrow q$ true and p true. In this row, q is also true. Thus, we can conclude Mr. Bell did not commit the crime.

p	q	$p \rightarrow q$
T	T	**T**
T	F	F
F	T	T
F	F	T

Examine   Since we know Mr. Bell was in Brooklyn, and we know he could not commit the crime if he was in Brooklyn, the conclusion that he did not commit the crime does follow logically.

Exercises

Exploratory The statements in each problem have the indicated truth value. Use truth tables to determine what conclusions can be made from these statements.

1. $p \wedge q$ is true.

2. $p \vee q$ is false.

3. $r \rightarrow s$ is false.

4. $m \wedge n$ is false.
  m is true.

5. $d \vee e$ is true.
  d is false.

6. $m \rightarrow s$ is true.
  s is false.

7. $a \leftrightarrow c$ is true.
  c is false.

8. $p \leftrightarrow q$ is false.
  p is true.

9. $n \rightarrow t$ is false.
  $p \leftrightarrow t$ is false.

Written The statements in each problem have the indicated truth value. Use truth tables to determine what conclusions, if any, can be made from these statements.

1. $p \wedge q$ is false.
  q is false.

2. $A \wedge B$ is false.
  A is false.

3. $q \vee t$ is true.
  q is true.

4. $c \vee d$ is true.
  c is false.

5. $n \rightarrow p$ is true.
  n is false.

6. $r \rightarrow t$ is true.
  t is true.

7. $b \leftrightarrow k$ is true.
  k is true.

8. $r \rightarrow s$ is true.
  $r \wedge q$ is true.

9. $k \vee m$ is false.
  $\sim m \rightarrow p$ is true.

10. The inverse of $p \rightarrow q$ is true.
  p is false.

11. The converse of $n \rightarrow t$ is false.
  The contrapositive of $s \rightarrow n$ is true.

12. Sarah jogs or plays racquetball every day. On the days that Sarah jogs, she does not do aerobics. Sarah always swims after playing racquetball. On the days that Sarah does aerobics, does she do any other activities?

13. Alan, Bill, and Cathy each had different lunches. One had soup, one had a sandwich, and one had a salad. Alan did not have a sandwich. Bill did not have soup or sandwich. Determine what item each person had for lunch.

14. Jay Rd. is perpendicular to 1st St. If streets run North-South, avenues run East-West, and roads run in either direction, is Jay Rd. parallel to or perpendicular to 2nd Ave?

15. Hill Ave. is parallel to Adams St. Davis Blvd. is perpendicular to River Rd. If River Rd. is parallel to Adams St., is Davis Blvd. parallel to or perpendicular to Hill Ave?

16. While hiking with his dog, cat, and bird, Jeremiah came to a rope bridge. He can only carry one animal at a time across the bridge. He cannot leave the dog and cat or the cat and bird together or they will fight. How can he get all the animals across the bridge without having any fights among the animals?

Portfolio Suggestion

A portfolio contains representative samples of your work, collected over a period of time. Begin your portfolio by selecting an item that shows something new you learned in this chapter.

Performance Assessment

If Ed has to stay home, Shar will not go to the dance. If Shar goes to the dance, then either Ed or Eli will have to stay home. If Suzi stays home, then Eli will not have to stay home. However, Shar will go to the dance. Who will go to the dance and who will stay home? Include an explanation of your solution.

Chapter Summary

1. A **statement** is any sentence which is true or false, but *not* both. The truth or falsity of a statement is its **truth value.** (2)
2. The placeholder in a mathematical sentence is called a **variable.** An **open sentence** contains one or more placeholders or variables. A variable may be replaced by any member of a set called the **domain** or **replacement set.** The set of all replacements from the domain which make an open sentence true is called the **solution set** of the open sentence. (2)
3. If a statement is represented by *p,* then *not p* is the **negation** of that statement. In symbols, *not p* is written $\sim p$. (5)
4. **Truth tables** can be used to find the truth values of statements. (6)
5. A compound statement formed by joining two statements with the word *and* is a **conjunction.** Each of the statements is a **conjunct.** In symbols, the conjunction of *p* and *q* is written $p \wedge q$. (8)
6. A compound statement formed by joining two statements with the word *or* is a **disjunction.** Each of the statements is a **disjunct.** In symbols, the disjunction of *p* and *q* is written $p \vee q$. (12)
7. A compound statement formed by joining two statements with the words *if . . . , then* is a **conditional.** In symbols, the conditional, *if p then q,* is written $p \rightarrow q$. Statement p is the **antecedent** of the conditional and statement *q* is the **consequent** of the conditional. (16)
8. A **tautology** is a compound statement which is true regardless of the truth values of the statements of which it is composed. (20)

9.

Conditional	Relation to Original	Truth Value Compared to Original
$p \rightarrow q$	original	
$q \rightarrow p$	**converse**	Converse of true conditional may be true or false.
$\sim p \rightarrow \sim q$	**inverse**	Inverse of true conditional may be true or false.
$\sim q \rightarrow \sim p$	**contrapositive**	Always the same truth value as the original.

(24, 25)

10. A statement formed by the conjunction of the conditionals $p \rightarrow q$ and $q \rightarrow p$ is a **biconditional.** In symbols, the biconditional can be written as $(p \rightarrow q) \wedge (q \rightarrow p)$ or $p \longleftrightarrow q$. (28)

11. Whenever two statements are always either both true or both false, the two statements are **equivalent.** (28)

 # Chapter Review

1.1 State whether each of the following is *true, false,* or an *open sentence.*

1. $8 + 2 > 18$ **2.** $3x + 1 = 7$

3. $3 \times 2 + 1 = 7$ **4.** It is hot.

5. Bach was a sculptor.   **6.** It was signed in Paris in 1853.

Find the solution set for each of the following. The domain is $\mathcal{W}$.

7. $x + 3 = 14$ **8.** $3x < 12$ **9.** $x^2 = 8$ **10.** $2y + 1 > 10$

1.2 Write the negation of each statement.

11. Lincoln was not a U.S. President.   **12.** There are 24 hours in a day.

13. $8 - 6 \neq 5$ **14.** $2 + 5 > 7$

State the truth value of each statement and the truth value of its negation in exercises 11-14.

15. Exercise 11   **16.** Exercise 12   **17.** Exercise 13   **18.** Exercise 14

1.3 and 1.4 Let *s, t,* and *w* represent the following statements.

s: "Water boils at 100°C."   (true)

t: "France is not in Europe."   (false)

w: "Mozart wrote *Hamlet*."   (false)

Write the compound sentence represented by each of the following and then determine its truth value.

19. $s \wedge t$ **20.** $t \wedge \sim w$ **21.** $\sim s \wedge t$ **22.** $\sim s \wedge \sim w$

23. $t \vee w$ **24.** $\sim s \vee w$ **25.** $\sim t \vee \sim w$ **26.** $\sim (t \vee w)$

Copy and complete each of the following truth tables.

27.

p	q	p $\wedge$ q

28.

p	q	~q	p $\vee$ ~q	~ (p $\vee$ ~q)

Construct a truth table for each of the following.

29. ~s $\vee$ t   **30.** q $\wedge$ ~r   **31.** r $\vee$ (~r $\wedge$ s)

Find the solution set for each of the following. The domain is $\mathcal{W}$.

32. x ≤ 3   **33.** y ≥ 6   **34.** (x < 6) $\wedge$ (x > 2)

1.5   Let *s, t,* and *w* represent the statements used for exercises 19-26. Write the conditional represented by each of the following and then determine its truth value.

35. s → t   **36.** ~t → s   **37.** (s $\vee$ t) → w   **38.** (s $\wedge$ ~t) → w

Construct a truth table for each of the following.

39. ~p → q   **40.** ~r → ~s   **41.** (~r $\vee$ s) → s   **42.** (~r $\wedge$ ~s) → r

Write each of the following statements in symbolic form using the letters after each statement.

43. If Ryan is elected, then either wheat prices will rise or there are too many hogs in Kansas. *(R,W,H)*

44. If there are too many hogs in Kansas, and if Mary attended Yale, then Sam does not hibernate in winter. *(H,Y,S)*

45. Sam will study sculpture if plastics are expensive and chalk becomes scarce. *(S,P,C)*

46. If ink is scarce, then it is false that if students do not do their homework, they are lazy. *(I,H,L)*

1.6   Construct a truth table for each of the following statements and then determine whether each statement is a tautology.

47. (p $\wedge$ q) → p

48. p → (p $\vee$ ~q)

49. p → (q $\vee$ ~q)

50. ~(p $\wedge$ q) → (~p $\wedge$ ~q)

1.7   Let *s, t,* and *w* represent the statements used for exercises 19-26. Write the converse, inverse, and contrapositive of each of the following in symbolic form. Then determine the truth value of each.

51. s → t   **52.** t → ~w   **53.** ~s → ~t   **54.** ~t → ~s

Write the converse, inverse, and contrapositive of each statement.

55. If gas is expensive, then I will drive less.

56. You are a champion if you eat Crunchy-Wunchies.

1.8   Let *s, t,* and *w* represent the statements used for exercises 19-26. Write the biconditional represented by each of the following and then determine its truth value.

57. s ↔ t   **58.** ~t ↔ w   **59.** ~w ↔ (s $\wedge$ t)

Chapter Test

Let *r* represent "Mel attends Yale" and *s* represent "Flora does not like spinach." Write the statement represented by each of the following.

1. $\sim r$ **2.** $r \wedge s$ **3.** $\sim r \vee s$ **4.** $r \rightarrow s$ **5.** $r \leftrightarrow \sim s$

If *p* is true, *q* is false, and *r* is true, determine the truth value of each of the following.

6. $\sim p$ **7.** $p \wedge r$ **8.** $q \wedge r$ **9.** $p \rightarrow r$

10. $p \vee q$ **11.** $p \vee r$ **12.** $\sim p \vee q$ **13.** $q \rightarrow \sim p$

14. $\sim q \rightarrow \sim r$ **15.** $\sim (p \rightarrow q)$ **16.** $p \leftrightarrow r$ **17.** $\sim p \leftrightarrow q$

18. $(p \vee q) \rightarrow \sim r$ **19.** $(p \vee q) \leftrightarrow \sim r$ **20.** $(\sim p \wedge q) \leftrightarrow (\sim q \vee r)$

Find the solution set for each of the following open sentences. The domain is $\mathcal{W}$.

21. $x + 4 = 10$ **22.** $y - 1 > 4$ **23.** $3n \leq 15$

24. $4x - 1 = 11$ **25.** $b + b + b = 3b$ **26.** $5y = 4$

27. $y > 6$ **28.** $x \leq 3$ **29.** $(x < 3) \vee (x \leq 6)$

Copy and complete each of the following truth tables.

30.

p	*q*	$p \vee q$

31.

p	*q*	$\sim p$	$\sim q$	$\sim p \vee q$	$(\sim p \vee q) \wedge \sim q$

Write the converse, inverse, and contrapositive of each of the following.

32. If 2 is a prime number, then 2 is an odd number.

33. I will go swimming if it is not raining.

Let *p* represent "*x* is divisible by 3." and *q* represent "*x* is divisible by 4." For the given value of *x*, determine the truth value of each of the following.

34. $x = 16; p \wedge q$ **35.** $x = 54; p \vee q$

36. $x = 35; p \vee q$ **37.** $x = 45; p \wedge q$

For each of the following, select the correct answer.

38. Which of the following is equivalent to $m \leftrightarrow t$?

a. $(m \rightarrow t) \wedge (t \rightarrow m)$ **b.** $(m \rightarrow t) \vee (t \rightarrow m)$ **c.** $(m \rightarrow t) \wedge (\sim m \rightarrow t)$

39. Suppose the sentence "If I study, then I do well" is true. Which of the following also must be true?

a. If I do not study, then I do not do well.

b. If I do well, then I study.

c. If I do not do well, then I do not study.

40. Construct a truth table for the statement $(\sim p \rightarrow q) \leftrightarrow (p \vee q)$. Then determine if the statement is a tautology.

Operations and Numbers

Application in Ecology

Ecologists use **algebraic techniques** to determine how much beach will be washed away in a certain number of years. The study of these algebraic techniques begins with a knowledge of numbers and operations and their properties.

The Saleh family has owned a piece of lake-front property for many years. The width of their property has already lost 7 inches due to erosion. Ecologists tell them that they will continue to lose an average of 1.5 inches per year if they do not install a sea wall or take other protective measures. The expression $7 + 1.5x$ represents the number of inches the property will lose after x years. Evaluate this expression if x equals 15 years.

Group Project:
History

Make a large time line and mark significant events in the history of mathematics on the time line. Include the development of the abacus and the introduction of modern computers. Each group member should select the event that seems most significant and write a report telling why.

2.1   Variables

Consider the following sentences. Which are true? Which are false? Which are neither true nor false?

It is a state.   He is 14 years old.
8 − ☐ = 1 $x = 17 − 10$

To answer the questions above, you must know what name or names can replace *It, He,* ☐, and *x*. For example, if 7 replaces ☐, the sentence 8 − ☐ = 1 is true.

The words *it* and *he* and symbols ☐ and *x* are **variables.** As in logic, a statement containing a variable is an **open sentence.** The set of replacements for a variable is the **domain** of the variable. The set of replacements from the domain that makes an open sentence true is called the **solution set.** The replacements are called **solutions.**

Example

1 **Find the solution set for the sentence *x* < 5 when the domain is $\mathcal{W}$.**

Recall that $\mathcal{W}$ = {0, 1, 2, 3, . . . }.
Any number less than 5 in the domain will make the sentence $x < 5$ true. Thus, the solution set is {0, 1, 2, 3, 4}.

Expressions such as $x + 1$ and $bc − 2$ are examples of **algebraic expressions.** An algebraic expression contains variables and operations such as addition and division.

You can **evaluate** an algebraic expression when the value of each variable is known.

Example

2 **Evaluate *n* + 6 if *n* = 5.**

$n + 6 = 5 + 6$ *Replace n by 5.*
$\qquad\; = 11$

Example

3 Evaluate $\dfrac{a + b}{2}$ if $a = 3$ and $b = 7$.

$$\dfrac{a + b}{2} = \dfrac{3 + 7}{2} \qquad \textit{Replace a by 3 and b by 7.}$$

$$= \dfrac{10}{2} \text{ or } 5$$

4(3) means 4 × 3.
4 • 3 means 4 × 3.

6ab means 6 • a • b.

Suppose 4 is to be multiplied by 3. You may write this 4 × 3. Sometimes parentheses or a dot are used instead of × to indicate multiplication. When variables are used to represent factors, the multiplication sign is usually omitted.

Count the factors on the left side of each equation below.

$$7 = 7^1$$
$$7 \cdot 7 = 7^2$$
$$7 \cdot 7 \cdot 7 = 7^3$$
$$7 \cdot 7 \cdot 7 \cdot 7 = 7^4$$

The number of factors on the left side is recorded at the upper right on the other side of each equation.

The numerals at the upper right in expressions like 7^1, 2^2, n^3, and $6x^4$ are called **exponents.** The exponent is the number of times the **base** is used as a factor. **Powers** are numbers written with exponents.

7^1 means 7.

2^2 means $2 \cdot 2$.

n^3 means $n \cdot n \cdot n$.

$6x^4$ means $6 \cdot x \cdot x \cdot x \cdot x$

7^1 is read *seven to the first power.*

2^2 is read *two to the second power* or *two squared.*

n^3 is read *n to the third power* or *n cubed.*

$6x^4$ is read *six times x to the fourth power.*

Example

4 Evaluate ab if $a = 8$ and $b = 9$.

$$ab = 8 \cdot 9 \qquad \textit{Replace a by 8 and b by 9.}$$
$$= 72$$

Example

5 Evaluate $5n^3$ if $n = 2$.

$5n^3 = 5(2^3)$ *Replace n by 2.*

    $= 5 \cdot 8$ *2^3 is $2 \cdot 2 \cdot 2$ or 8.*

    $= 40$

Exercises

Exploratory   **Write each expression using exponents.**

1. $3 \cdot 3 \cdot 3$
2. $7 \cdot 7 \cdot 7 \cdot 7$
3. $5 \cdot 5$
4. $x \cdot x$
5. $y \cdot y \cdot y \cdot y \cdot y$
6. $m \cdot m \cdot m$
7. $4 \cdot n \cdot n$
8. $6 \cdot x \cdot x \cdot x$
9. $18 \cdot z \cdot z \cdot z \cdot z$
10. $3 \cdot 3 \cdot x \cdot x$
11. $5 \cdot 5 \cdot b \cdot b \cdot b$
12. $4 \cdot 4 \cdot 4 \cdot c \cdot c \cdot c \cdot c$

Evaluate.

13. 8^2
14. 4^2
15. 5^4
16. 10^4
17. 4^5
18. 6^3
19. $2 + 2^2$
20. $3^2 + 3$

Written   **Evaluate if $a = 6$, $k = 8$, $m = 5$, $p = 7$, and $s = 11$.**

1. $9 + k$
2. $s - 7$
3. $p - 3$
4. $m + 5$
5. $7 - m$
6. $s + 19$
7. $a + k$
8. $s - p$
9. $k + p$
10. $a + k + m$
11. $s + a - m$
12. $s + k - m$
13. $a + m + p$
14. $p + s + m$
15. $s + s + s + s$
16. $a + m + m$
17. $k + m$
18. $m + k$
19. $s - s$
20. $k + k + k$
21. $s + 0$
22. $0 + a$
23. $m + p + s$
24. $s - m$
25. $\dfrac{p + s}{3}$
26. $\dfrac{m + p}{4}$
27. $\dfrac{k}{s - p}$
28. $\dfrac{a + 4}{m}$
29. $16a$
30. ak
31. $2(5k)$
32. $p \cdot m$
33. $m \cdot 5$
34. mk
35. s^2
36. 55^2
37. $4m^2$
38. pm^2
39. p^2m
40. s^2m^2
41. m^2a^2
42. m^2a^3
43. $2(4s)$
44. $6ma$
45. $9k^5$
46. m^5a^6
47. $5m^3k^4$
48. $11p^2s^3$

Find the solution set for each of the following open sentences. The domain of the variable is $\mathcal{W}$.

49. $x + 6 = 10$
50. $2 + y = 8$
51. $t < 5$
52. $y - 4 = 2$
53. $12 - y = 4$
54. $y > 6$
55. $3y = 6$
56. $x^2 = 4$

2.2   Order of Operations

How do you evaluate $4 + 5 \cdot 2$? You might evaluate the expression in one of two ways.

Add 4 and 5. Multiply by 2.	*Multiply 5 by 2. Add 4.*
$4 + 5 \cdot 2 = 9 \cdot 2$	$4 + 5 \cdot 2 = 4 + 10$
$= 18$	$= 14$

Which answer is correct?

Each numerical expression should have a *unique* value. To find this value, you must know the order for doing operations.

Order of Operations

> 1. Evaluate all powers.
> 2. Then do all multiplications and/or divisions from left to right.
> 3. Then do all additions and/or subtractions from left to right.

Using these rules we see that $4 + 5 \cdot 2 = 14$.

Example

1 Evaluate $a + 3ab^2$ if $a = 5$ and $b = 2$.

$a + 3ab^2$	$= 5 + 3 \cdot 5 \cdot 2^2$	*Replace a by 5 and b by 2.*
	$= 5 + 3 \cdot 5 \cdot 4$	*Evaluate 2^2.*
	$= 5 + 15 \cdot 4$	*Multiply 3 by 5.*
	$= 5 + 60$	*Multiply 15 by 4.*
	$= 65$	*Add 5 and 60.*

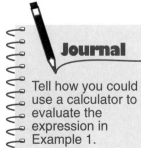

Journal

Tell how you could use a calculator to evaluate the expression in Example 1.

You can indicate a change in the order of the operations by using parentheses.

$3 \cdot (4 + 8) = 3 \cdot 12$	*First add 4 and 8.*
$= 36$	*Then multiply 3 by 12.*

Brackets may be used in place of a second set of parentheses. Parentheses and brackets are called **grouping symbols.** When evaluating expressions, start with the operation in the *innermost* set of grouping symbols. Follow the order of operations in doing this.

Example

2 Evaluate $4 + 2[8 + (3 - 1)^2]$.

$$\begin{aligned}
4 + 2[8 + (3 - 1)^2] &= 4 + 2[8 + 2^2] &&\text{\textit{First subtract 1 from 3}}\\
&= 4 + 2[8 + 4] &&\text{\textit{Then evaluate }} 2^2.\\
&= 4 + 2[12] &&\text{\textit{Add 8 and 4.}}\\
&= 4 + 24 &&\text{\textit{Multiply 2 by 12.}}\\
&= 28 &&\text{\textit{Add 4 and 24.}}
\end{aligned}$$

Exercises

Exploratory State how to evaluate each expression. Do *not* evaluate.

1. $7 + 6 - 4 + 3$
2. $6 \cdot 2 - 4 \cdot 2$
3. $6 \cdot 4 + 2 \cdot 2$
4. $4^2 - 3^2$
5. $5 + 3 \cdot 6$
6. $42 - 54 \div 6$
7. $32 \div 4 \cdot 2$
8. $24 \div 6 - 3^2 \div 3$
9. $56 \div 4 \div 2$
10. $3 \cdot (4 + 6)$
11. $7(9 - 2)$
12. $(2 \cdot 4) - 1$
13. $(4 \cdot 4) - 2$
14. $(8 - 5)(4 + 2)$
15. $(4 + 2)(8 + 3)$

Written Evaluate.

1. $8 + 5 + 6 \div 3 - 2$
2. $7 + 4 - 3 + 1$
3. $11 \cdot 11 - 10^2$
4. $6 \div 2 + 5 \cdot 4$
5. $6 \div 2 + 8 \div 2$
6. $6 \cdot 3 \div 9 - 1$
7. $5^2 - 4 \cdot 4 + 5^2$
8. $12 \cdot 6 \div 3 \cdot 2 \div 8$
9. $2^6 \div 2 \div 2^3 \div 2^2$
10. $24 - 3 + 9 \div 3 \div 3$
11. $(21 + 18) \div 3$
12. $3(62 - 11)$
13. $(25 \cdot 4) + (15 \cdot 4)$
14. $(12 \cdot 4) \div (3 \cdot 2)$
15. $144 \div 3(8 + 4)$
16. $[2(12 - 3) \div 3] \div 2$
17. $[(2 \cdot 12 - 3) \div 3] \div 7)$
18. $[2 \cdot 12 - (3 \div 3)] \div 23$
19. $\dfrac{84 + 12}{13 + 11}$
20. $\dfrac{4^3 + 8}{2^4 - 7}$
21. $\dfrac{(4 \cdot 9) - 2^2}{(11 - 3) - 4}$

Evaluate if $a = 2$, $b = 4$, and $c = 3$.

22. $16a$
23. $b \cdot c$
24. $3bc$
25. $4c^2$
26. $c^2 a^2$
27. $4c^2 b$
28. $9b^2$
29. $a^2 b^3 c^4$
30. $2a^2 - a$
31. $b^2 - 4b + 2$
32. $3a^2 2b^2 - c$
33. $b^2 + c^2 - a^2$
34. $ab + ac - b$
35. $ab + c - a^3$
36. $c^3 - a + b^2$
37. $a^2 + 2ab + b^2 - c$
38. $a + (b + c)$
39. $a + (b - c)$
40. $7a - (2b + c)$
41. $\dfrac{4(b - a)}{c - 1}$
42. $\dfrac{3(4b + a)}{9}$
43. $(5b - 2c) - c^2$
44. $[(4a^2 + 3b^2) - 4c] \div 4$
45. $4a^2 + 3b^2 - (5c \div 5)$

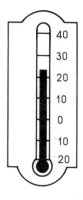

2.3   The Integers

In winter the temperature may go *below* zero. In various parts of the world, land may be *above* sea level or *below* sea level. These situations require a new type of number. In mathematics a **number line** can be used to show these numbers.

Points to the right and left of 0 may be named using numbers and signs, + and −. The sign of a **negative number** is −. The sign of a **positive number** is +.

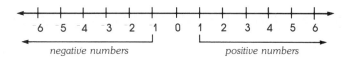

A raised positive sign may or may not be used with a positive number.

⁻4 is read *negative four*.   ⁺5 or 5 is read *positive five*.

The set of numbers used to name the points on the number line above is called the **integers.** In this book, the symbol $\mathbb{Z}$ will be used to name the set of integers. Thus,

$$\mathbb{Z} = \{ \ldots, {}^-3, {}^-2, {}^-1, 0, 1, 2, 3, \ldots \}.$$

Zero is neither positive nor negative.

To **graph** a set of numbers means to locate the points named by those numbers on the number line. The number that corresponds to a point on the number line is called the **coordinate** of the point.

▚Examples▚

1 **Graph {⁻5, ⁻2, 0, 3} on a number line.**

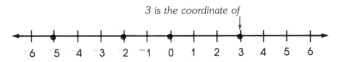

3 is the coordinate of

2 **Name the set of numbers graphed and then name the coordinates of *A* and *B*.**

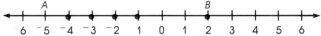

The set of numbers graphed is {⁻4, ⁻3, ⁻2, ⁻1, 2}.
The coordinate of *A* is ⁻5 and the coordinate of *B* is 2.

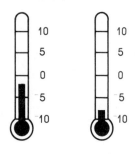

Which temperature is greater, ⁻8°C or ⁻2°C? The thermometer shows that ⁻2°C is greater than ⁻8°C. Notice that ⁻2 is to the right of ⁻8 on the number line.

Any number on the number line is *greater* than every number *to its left*. Any number on the number line is *less* than every number *to its right*.

Exercises

Exploratory   State the coordinate of each point.

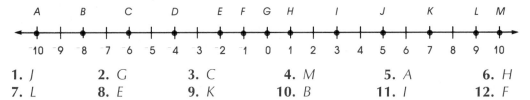

1. J	2. G	3. C	4. M	5. A	6. H
7. L	8. E	9. K	10. B	11. I	12. F

Name the set of numbers graphed.

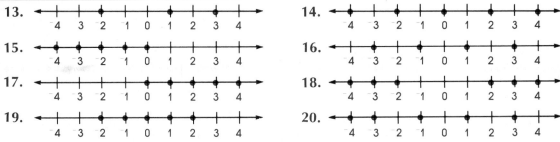

Written   Write an integer to describe each situation.

1. 10 seconds before blast off
2. 15 seconds after blast off
3. $500 profit
4. $600 loss
5. up 8 floors
6. down 6 floors
7. 5 feet forward
8. 3 feet backward
9. 8° above zero
10. 8° below zero
11. 280 m below sea level
12. 540 m above sea level

Graph each set of numbers on a number line.

13. {⁻4, ⁻1, 1}   **14.** {⁻2, 1, 3}   **15.** {⁻5, ⁻3, ⁻2, 0, 1}

16. {0, 2, 4}   **17.** {⁻1, 1, 3}   **18.** {⁻3, 0, 2}

19. {⁻3, ⁻1, 1, 3}   **20.** {8,7}   **21.** {0, 1, 2, 3, . . .}

22. {⁻3, ⁻2, ⁻1, . . .}   **23.** {5, 6, 7, . . .}   **24.** {. . . , ⁻3, ⁻2, ⁻1}

Determine whether each of the following is *true* or *false*.

25. A whole number is an integer.   **26.** Zero is a positive integer.

27. An integer is a whole number.   **28.** One-half is not an integer.

29. If x is a positive integer, then $x \geq 0$.   **30.** If x is a negative integer, then $x < 0$.

31. $5 > 4$ **32.** $^-3 > 4$ **33.** $^-5 < ^-8$ **34.** $0 < ^-6$

35. $(6 < 0) \wedge (4 > ^-4)$ **36.** $(17 > 14) \vee (14 > ^-17)$

37. $(^-1 < ^-3) \vee (^-12 < ^-10)$ **38.** $(^-5 < 0) \wedge (^-3 > 3)$

If each of the following exists, name it.

39. the greatest negative integer   **40.** the least positive integer

41. the least negative integer   **42.** the greatest positive integer

Challenge **Find the solution set for each compound sentence if the domain is** $\mathbb{Z}$.

1. $(x < 7) \wedge (x > ^-2)$ **2.** $(y < ^-12) \vee (y < 3)$

3. $(n > ^-6) \vee (2n > ^-16)$ **4.** $(4a > 8) \wedge (a > ^-13)$

List the next three numbers in each pattern.

5. 33, 25, 17, . . .   **6.** ⁻10, ⁻7, ⁻4, . . . ;   **7.** ⁻8, ⁻4, ⁻2, . . .

8. 1, 3, 9, 27, . . .   **9.** 1, 4, 9, 16, . . .   **10.** 1, 1, 2, 3, 5, 8, . . .

════ Mixed Review ════

State the truth value of each statement.

1. An integer is a whole number or ⁻5 > ⁻3.

2. 0 is a positive integer if and only if ⁻1 < 1.

3. If 0.5 is an integer, then $0 < ^-6$.

4. $^-2 > ^-7$ and ⁻12 is a positive integer.

5. If $^-3 > ^-1$, then 1 is a negative integer or ⁻1 is a negative integer.

Evaluate if $a = 5$, $b = 2$, $c = 3$, and $d = 6$.

6. $c + bd^2 - 2ca \div d$ **7.** $c + (bd)^2 - 2ca \div d$ **8.** $(c + bd)^2 - 2ca \div d$

9. $(c + b)d^2 - 2ca \div d$ **10.** $c + (bd^2 - 2ca) \div d$ **11.** $c + b(d^2 - 2ca) \div d$

12. $c + (bd^2 - 2c)a \div d$ **13.** $c + (bd^2 - 2)ca \div d$

14. $c + b(d^2 - 2ca \div d)$ **15.** $(c + b)(d^2 - 2ca) \div d$

2.4 Addition on a Number Line

The number line can be used to show the addition of integers. For example, to show $^-4 + 7$ follow these steps.

Step 1 Draw an arrow from 0 to $^-4$.

Go to the left for negative numbers.

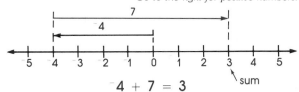

Step 2 Starting at $^-4$ draw an arrow 7 units to the right.

Go to the right for positive numbers.

Step 3 The sum is shown at the head of the second arrow.

$$^-4 + 7 = 3$$

Examples

1 Find $4 + 2$.

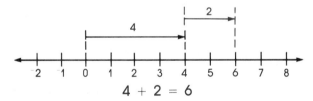

$$4 + 2 = 6$$

2 Find $3 + ^-5$.

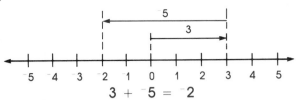

$$3 + ^-5 = ^-2$$

3 Find $^-3 + ^-6$.

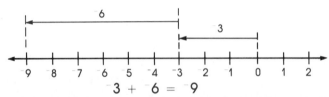

$$^-3 + ^-6 = ^-9$$

Exercises

Exploratory   State the corresponding addition sentence for each diagram.

1.
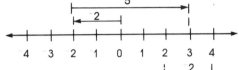

2.
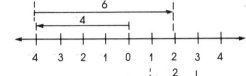

3.
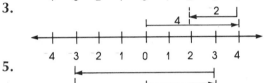

4.

5.

6.
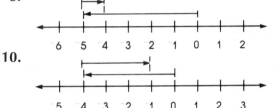

7.
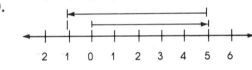

8.

9.

10.

Written   Use a number line to find each sum.

1. 8 + 5
2. 7 + 9
3. ⁻10 + ⁻12
4. ⁻8 + ⁻9
5. ⁻6 + 0
6. 0 + ⁻9
7. ⁻8 + 3
8. 6 + ⁻8
9. 4 + ⁻4
10. ⁻8 + 8
11. 9 + ⁻5
12. ⁻3 + 9
13. 2 + ⁻9
14. 5 + ⁻7
15. 3 + 8
16. 5 + 11

Find each sum.

17. ⁻2 + 11
18. ⁻12 + 11
19. 9 + ⁻4
20. ⁻13 + 7
21. 2 + ⁻2
22. ⁻15 + 15
23. 0 + 7
24. ⁻8 + 0
25. ⁻6 + ⁻11
26. ⁻11 + ⁻13
27. ⁻15 + ⁻10
28. ⁻19 + ⁻17

Write an addition sentence for each problem.

29. In football, a gain of 4 yards is followed by a loss of 7 yards.

30. The temperature goes up 7°C and then drops 11°C.

31. A withdrawal of $8 is followed by a deposit of $14.

32. A person loses $483 and then wins $176.

33. A stock dropped $11 one day and increased by $7 the next day.

34. A person loses 15 pounds, gains 6 pounds, and then loses 10 pounds.

35. Jill's stock went up 6 points last month and down 4 points this month.

36. A submarine dove from the surface to a depth of 200 m. After leveling off, the submarine rose 50 m.

2.5   Adding Integers

3 units   3 units

On the number line you can see that ⁻3 and 3 are each 3 units from 0. We say that the **absolute value** of ⁻3 is 3. The absolute value of 3 is also 3. The symbol for absolute value is a vertical bar on each side of the numeral.

$$|^-3| = 3 \qquad |3| = 3$$

Definition of Absolute Value

> The absolute value of a number is the number of units it is from 0 on the number line.

Calculator Hint

The key for the negative sign on a calculator is not the same as the key for subtraction. Look for a [+/−] or [(−)] key to enter a negative number.

Absolute value can be used in finding the sum of integers. Verify each addition sentence below using a number line.

$$^+8 + {}^+5 = {}^+13$$

Notice that the integers being added have positive signs. Is the sum also positive?

$$^-4 + {}^-6 = {}^-10$$

Notice that the integers being added have negative signs. Is the sum also negative?

These and other similar examples suggest the following rule.

Adding Integers with the Same Sign

> To add integers with the same sign, add their absolute values. The sum has the same sign as the integers.

Lab Activity

You can learn how to use counters to add integers in Lab 1 on pages A2-A3.

$$^-6 + 4 = {}^-2$$

Notice that the difference between 6 and 4 is 2. Which integer being added has the greater absolute value? The sum has the same sign as this integer.

$$5 + {}^-8 = {}^-3$$

Notice that the difference between 5 and 8 is 3. Which integer being added has the greater absolute value? The sum has the same sign as this integer.

These and other similar examples suggest the following rule.

Adding Integers with Different Signs

> To add integers with different signs, find the difference of their absolute values. The sum has the same sign as the integer with the greater absolute value.

Examples

1 Find $^-18 + {}^-41$.

$^-18 + {}^-41 = {}^-(|{}^-18| + |{}^-41|)$ *Both integers are negative.*
$\qquad\qquad\;\; = {}^-(18 + 41)$ or $^-59$ *The sum is negative.*

2 Find $45 + {}^-66$.

$45 + {}^-66 = {}^-(|{}^-66| - |45|)$ $^-66$ *has the greater absolute value.*
$\qquad\qquad = {}^-(66 - 45)$ or $^-21$ *The sum is negative.*

Exercises

Exploratory   State the absolute value of each integer.

1. 4   **2.** $^-5$ **3.** $^-12$ **4.** $^-6$ **5.** 8
6. 0   **7.** $^-14$ **8.** 3   **9.** 9   **10.** $^-13$

State the sign of each sum.

11. $7 + 5$ **12.** $^-6 + {}^-5$ **13.** $^-4 + 3$ **14.** $8 + {}^-6$
15. $^-3 + 7$ **16.** $11 + {}^-20$ **17.** $16 + {}^-5$ **18.** $^-9 + 17$

Written   Find each of the following.

1. $|8| + |3|$ **2.** $|9| - |2|$ **3.** $|{}^-8| + |2|$ **4.** $|{}^-12| + |{}^-6|$
5. $|{}^-13| - |{}^-5|$ **6.** $|{}^-6| - |6|$ **7.** $|10 - 10|$ **8.** $|4 + 9|$

Find each sum.

9. $8 + 7$ **10.** $12 + 16$ **11.** $^-4 + {}^-7$ **12.** $^-8 + {}^-9$
13. $^-1 + 12$ **14.** $5 + {}^-9$ **15.** $^-5 + 7$ **16.** $^-10 + 7$
17. $^-18 + 13$ **18.** $88 + {}^-72$ **19.** $^-33 + 48$ **20.** $101 + {}^-101$
21. $345 + 267$ **22.** $^-347 + {}^-460$ **23.** $^-258 + 379$ **24.** $1429 + {}^-1906$

Evaluate each expression if $x = {}^-2$ and $y = 5$.

25. $x + 4$ **26.** $y + {}^-4$ **27.** $^-6 + x$ **28.** $y + 2$
29. $x + {}^-1$ **30.** $y + {}^-3$ **31.** $x + y$ **32.** $|x| + |y|$

Solve each problem.

33. Mike deposited $250 in his savings account last year. He withdrew $178 this year. What is the gain or loss for this period?

34. The R & D Company lost $64,000 last year. They made $58,000 this year. What is their gain or loss for this period?

2.6   Subtracting Integers

opposites

The opposite of 0 is 0.

Notice that ⁻3 and 3 are the same distance but in *opposite* directions from 0 on the number line. We say that the *opposite* of 3 is ⁻3 and the *opposite* of ⁻3 is 3. A centered negative sign is used to write the opposite of a number. But, -3 names the same number as ⁻3. Thus, ⁻3 $= -3$. To simplify notation, we will use the centered negative sign for both negative numbers and opposites from now on.

Example

1 **If $x = -7$, find $-x$.**
Since x is -7, the opposite of x is 7.
That is, if $x = -7$, then $-x = -(-7)$ or 7.

What is the sum of opposites?

Lab Activity

You can learn how to use counters to subtract integers in Lab 1 on pages A2-A3.

In general, if y is any integer, then $-(-y) = y$.

You can use opposites to subtract one integer from another. Observe the following pattern.

Subtraction	**Addition**
$8 - 5 = 3$	$8 + (-5) = 3$
$10 - 2 = 8$	$10 + (-2) = 8$
$25 - 10 = 15$	$25 + (-10) = 15$

These and other similar examples suggest the following rule.

Subtracting Integers

To subtract an integer, add its opposite. If a and b are integers, then $a - b = a + (-b)$.

Example

2 **Find $8 - 11$.**
$8 - 11 = 8 + (-11)$ *To subtract 11, add -11.*
$ = -3$

Example

3 Find $-[-14 - (-3)]$.

$$-[-14 - (-3)] = -(-14 + 3)$$
$$= -(-11)$$
$$= 11$$

Exercises

Exploratory   State the opposite of each of the following.

1. 7   **2.** -7 **3.** 8   **4.** -16 **5.** -13

6. b **7.** $-b$ **8.** $-8x$ **9.** 0   **10.** $5y$

State the simplest name for each of the following.

11. $-(-4)$ **12.** $-(6)$ **13.** $-(-15)$ **14.** $8 - 2$ **15.** $-(-(10))$

Written   Write each of the following as an addition expression.

1. $10 - 2$ **2.** $9 - 6$ **3.** $12 - 7$ **4.** $8 - 10$

5. $-2 - 11$ **6.** $4 - (-15)$ **7.** $-8 - (-6)$ **8.** $-4 - (-9)$

Find each difference.

9. $9 - 5$ **10.** $50 - 30$ **11.** $36 - 20$ **12.** $11 - 13$

13. $17 - 20$ **14.** $5 - 13$ **15.** $-[10 - 23]$ **16.** $-4 - 8$

17. $69 - (-95)$ **18.** $-6 - (-17)$ **19.** $-17 - (-23)$ **20.** $-[78 - (-19)]$

Evaluate each expression if $x = -4$ and $y = 6$.

21. $y - 6$ **22.** $2y - 1$ **23.** $x - 5$ **24.** $-(x + y)$

25. $3y - x$ **26.** $-y - (-12)$ **27.** $x - y$ **28.** $-(x - y)$

State whether each statement is *always, sometimes,* or *never* true.

29. If x is an integer then $-x$ is a negative number.

30. If x is a negative integer then $-x$ is a positive integer.

31. The opposite of an integer is a negative integer.

32. If x is an integer then $-x \neq x$.

Challenge   Solve each problem.

1. State a rule for finding
$-(-(-\ldots-(a)\ldots))$.

2. Relate the rule in the previous problem to the rule for $\sim(\sim(\sim\ldots\sim p\ldots))$ in logic.

2.7 Multiplying and Dividing Integers

How can you find 3×-4? Recall that multiplication is just repeated addition. The product 3×-4 is the same as the sum $-4 + (-4) + (-4)$.

$$-4 + (-4) + (-4) = -12 \quad \blacksquare\!\!\!\Rightarrow \quad 3 \times -4 = -12$$

The order of factors does *not* change the product. Therefore, if $3 \times -4 = -12$, then $-4 \times 3 = -12$. Notice that in each multiplication sentence, the factors have different signs. What is the sign of each product? These and other similar examples suggest the following rule.

Multiplying Two Integers with Different Signs

> The product of two integers with different signs is negative.

What is the sign of the product when both factors have the same sign? From arithmetic you know that the product of two positive integers is positive.

Both factors are positive. $3 \times 4 = 12$ The product is also positive.

Now study the following pattern.

Look at the first factor in each multiplication sentence. As you go down the list, each of these factors is 1 less than the one before it.

$$3 \times -4 = -12$$
$$2 \times -4 = -8$$
$$1 \times -4 = -4$$
$$0 \times -4 = 0$$
$$-1 \times -4 = \square$$
$$-2 \times -4 = \square$$

Look at the product in each multiplication sentence. As you go down the list, each product is 4 more than the one before it.

What numerals should replace each $\square$ for the pattern to continue?

$$-1 \times -4 = 4 \qquad -2 \times -4 = 8$$

These and other similar examples suggest the following rule.

Multiplying Two Integers with the Same Sign

> The product of two integers with the same sign is positive.

Examples

1 Find $(-8)(6)$.

$(-8)(6) = -(48)$ *The signs are different.*

$\quad\quad\quad = -48$ *The product is negative.*

2 Find $(-8)(2) + (-5)(-2)$.

$(-8)(2) + (-5)(-2) = -16 + (10)$

$\quad\quad\quad\quad\quad\quad\quad = -6$

3 Evaluate $4xy$ if $x = -3$ and $y = -2$.

$4xy = 4(-3)(-2)$ *Replace x by -3 and y by -2.*

$\quad = (-12)(-2)$ *Multiply from left to right.*

$\quad = 24$

Multiplication and division are *inverse operations*. You can use this fact to find rules for dividing integers.

$$2 \times 4 = 8 \quad\Longrightarrow\quad 8 \div 4 = 2$$

same signs   positive

$$3 \times -9 = -27 \quad\Longrightarrow\quad -27 \div -9 = 3$$

In each division problem, the signs of the first two integers are the same. What is the sign of each quotient? These and other similar examples suggest the following rule.

Dividing Two Integers with the Same Sign

> The quotient of two integers with the same sign is positive.

$$-8 \times 5 = -40 \quad\Longrightarrow\quad -40 \div 5 = -8$$

different signs   negative

$$-3 \times -6 = 18 \quad\Longrightarrow\quad 18 \div -6 = -3$$

In each division problem, the signs of the first two integers are different. What is the sign of each quotient? These and other similar examples suggest the following rule.

Dividing Two Integers with Different Signs

The quotient of two integers with different signs is negative.

Examples

4 Find $-24 \div -6$.

$-24 \div -6 = +(24 \div 6)$ *The signs are the same.*
$\qquad\qquad = 4$ *The quotient is positive.*

5 Simplify $\dfrac{54}{(-9)}$.

$\dfrac{54}{(-9)} = 54 \div (-9)$ *Fractions indicate division.*

$\qquad = -(54 \div 9) \text{ or } -6$

Suppose you are asked to find $8 \div 0$.

$$8 \div 0 = x \quad \blacksquare\!\!\!\rightarrow \quad x \cdot 0 = 8$$

But any number times 0 equals 0, *not* 8. These and other similar examples suggest that division by 0 is *undefined*.

Exercises

Exploratory   State the sign of each product.

1. $(-8)(7)$ **2.** $6(-5)$ **3.** $(-9)(-6)$ **4.** $4(-9)$
5. $(-3)(12)$ **6.** $(-6)(-13)$ **7.** $3(-10)$ **8.** $(-8)(-9)$

State the sign of each quotient.

9. $8 \div 4$ **10.** $24 \div (-6)$ **11.** $(-14) \div 7$ **12.** $(-18) \div (-6)$
13. $(-32) \div 4$ **14.** $(-42) \div (-7)$ **15.** $16 \div 2$ **16.** $(-144) \div 12$

State the related multiplication sentence for each division sentence.

17. $9 \div 3 = 3$ **18.** $-12 \div 4 = -3$ **19.** $12 \div (-4) = -3$ **20.** $-48 \div (-6) = 8$

Written   Find each product.

1. $-8 \cdot 9$ 2. $-9 \cdot 11$ 3. $(-8)6$ 4. $7(-5)$
5. $-16 \cdot 6$ 6. $(-7)(-13)$ 7. $(-5)(-17)$ 8. $8(-19)$
9. $(11)(-14)$ 10. $10(-5)$ 11. $-11(-6)$ 12. $2 \cdot 3 \cdot (-5)$
13. $3(7)(-5)(2)$ 14. $(-5) \cdot 3 \cdot (-4) \cdot 6$ 15. $6 \cdot 5 \cdot (-4) \cdot 6$ 16. $3 \cdot 4 \cdot (-5) \cdot 2 \cdot (-1)$

Find each quotient.

17. $8 \div 2$ 18. $16 \div 8$ 19. $24 \div (-4)$ 20. $56 \div (-7)$
21. $-32 \div 4$ 22. $-42 \div (-6)$ 23. $26 \div (-2)$ 24. $90 \div (-15)$
25. $(-120) \div (-8)$ 26. $-52 \div 4$ 27. $-39 \div (-3)$ 28. $-55 \div 11$
29. $\dfrac{14}{7}$ 30. $\dfrac{-40}{5}$ 31. $\dfrac{-36}{-9}$ 32. $\dfrac{42}{-14}$

Evaluate each of the following if $x = -4$ and $y = -6$.

33. $3x$ 34. $4y$ 35. xy 36. x^2
37. $5x - 1$ 38. $3x - 2y$ 39. $y \div 2$ 40. $x \div 4$
41. $y \div (-3)$ 42. $\dfrac{-2y}{3}$ 43. $\dfrac{-48}{y}$ 44. $\dfrac{3x + 2}{5}$

Evaluate each of the following.

45. $(-4)(8 - 10)$ 46. $-4(2 + 6)$ 47. $-6 + 4(-2)$
48. $8 - 54 \div (-6)$ 49. $(-12) \div 3 + 6 - 9$ 50. $(-12)(6) \div (3)(-2)$
51. $-3[2 - (4 - 5)]$ 52. $(-6 - 12) \div 3^2$ 53. $24 - 3 + 9 \div 3$
54. $2[-3 + (-6) \div (-3)]$ 55. $(-3)^2 - [-60 \div (-12)]$

Assume x is negative and y is positive. State whether each sentence is *always, sometimes,* or *never* true.

56. xy is positive.   57. $x + y$ is positive.
58. $-x$ is positive.   59. $y - x$ is positive.
60. $x - y$ is negative.   61. $(x + y)(x - y)$ is negative.
62. $x \div y$ is negative.   63. $-x - y$ is negative.

Assume x and y represent nonzero integers. State whether each sentence is *always, some-times,* or *never* true.

64. $(-1)(x) = -x$ 65. $x(-y) = (-x)y$ 66. $-(xy) = (-x)y$
67. $-(xy) = x(-y)$ 68. $|x| = -x$ 69. $|x| = x$
70. $|x| < x$ 71. $|x| > x$ 72. $\dfrac{-x}{y} = \dfrac{x}{-y}$
73. $-\dfrac{x}{y} = \dfrac{-x}{y}$ 74. $-\dfrac{x}{y} = \dfrac{x}{-y}$ 75. $\dfrac{-x}{-y} = \dfrac{x}{y}$

Challenge   Find the solution set for each of the following if the domain is $\mathbb{Z}$.

1. $2y = -6$ 2. $3x = -12$ 3. $-4m = -20$ 4. $-5y = 30$
5. $-6t = -24$ 6. $10y = -60$ 7. $y \div 7 = 6$ 8. $r \div (-5) = 9$

2.8 Fractions and Decimals

The quotient of two integers may be an integer. Sometimes it is *not* an integer. Fractions can be used to express these quotients.

$$3 \div 4 = \frac{3}{4} \qquad -5 \div 8 = \frac{-5}{8} \qquad 9 \div -4 = \frac{9}{-4}$$

**Definition of
Rational Number**

A rational number is any number that can be written in the form $\frac{a}{b}$ where a is any integer and b is any integer except 0.

The set of rational numbers is represented by Q.

You have used the number line to compare integers. You can also use the number line to compare fractions.

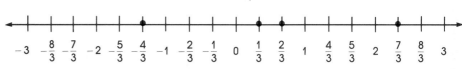

$\frac{7}{3}$ is graphed to the *right* of $\frac{-4}{3}$ so $\frac{7}{3} > \frac{-4}{3}$.

$\frac{1}{3}$ is graphed to the *left* of $\frac{7}{3}$ so $\frac{1}{3} < \frac{7}{3}$.

Infinitely many rational numbers exist between −3 and 3.

Calculator Hint

Some calculators have a fraction key like $\boxed{a^{b}/_{c}}$ that enables you to perform operations on fractions in fraction form.

When two fractions have the same denominator, it is easy to use the number line to compare them. When the denominators are *not* the same, it is difficult to use the number line.

Suppose you want to compare $\frac{3}{8}$ to $\frac{1}{3}$. One method for doing this is to find a common denominator. Any positive integer with 3 and 8 as factors will be a common denominator.

Common denominators for 3 and 8: 24, 48, 72, 96, . . .

The **least common denominator (LCD)** should be used. The least common denominator for 3 and 8 is 24.

$$\frac{3}{8} = \frac{3 \times 3}{8 \times 3} \text{ or } \frac{9}{24} \qquad \text{and} \qquad \frac{1}{3} = \frac{1 \times 8}{3 \times 8} \text{ or } \frac{8}{24}$$

$$\frac{8}{24} < \frac{9}{24} \text{ so } \frac{1}{3} < \frac{3}{8}$$

Examples

1 Compare $-\dfrac{7}{12}$ and $-\dfrac{5}{9}$.

To find the LCD for 12 and 9, use their prime factorizations.

$12 = 2 \cdot 2 \cdot 3$ *Use each factor the greatest*
$9 = 3 \cdot 3$ *number of times it appears.*

The LCD for 12 and 9 is $2 \cdot 2 \cdot 3 \cdot 3 = 36$.

$$-\frac{7}{12} = -\frac{21}{36} \text{ and } -\frac{5}{9} = -\frac{20}{36} \qquad -\frac{20}{36} > -\frac{21}{36} \text{ so } -\frac{5}{9} > -\frac{7}{12}$$

2 Find $-\dfrac{3}{4} + \dfrac{1}{3}$.

$$-\frac{3}{4} + \frac{1}{3} = -\frac{9}{12} + \frac{4}{12} \qquad \text{\textit{The LCD of 4 and 3 is 12.}}$$

$$= \frac{-9 + 4}{12} \qquad \text{\textit{Add numerators.}}$$

$$= \frac{-5}{12} \text{ or } -\frac{5}{12} \qquad \frac{-a}{b} = -\frac{a}{b} = \frac{a}{-b}$$

3 Find $\left(-\dfrac{2}{3}\right)\left(2\dfrac{4}{5}\right)$.

First rename $2\dfrac{4}{5}$ as a fraction.

$$2\frac{4}{5} = \frac{(2 \times 5) + 4}{5} \qquad \text{\textit{Multiply the whole number by the denominator and add the numerator.}}$$

$$= \frac{10 + 4}{5} \text{ or } \frac{14}{5}$$

Then multiply the fractions.

$$\left(-\frac{2}{3}\right)\left(2\frac{4}{5}\right) = \frac{-2}{3} \cdot \frac{14}{5}$$

$$= \frac{-28}{15} \text{ or } -1\frac{13}{15} \qquad \text{\textit{Multiply the numerators and multiply the denominators.}}$$

Decimals also name rational numbers. For example, 4.32 names a rational number since $4.32 = \dfrac{432}{100}$.

The following examples review operating with decimals.

Examples

4 Find $(-4.6) - (3.2)$.
$$-4.6 - (3.2) = -4.6 + (-3.2)$$
$$= -(|-4.6| + |-3.2|) \text{ or } -7.8$$

5 Find $(8.76)(-0.2)$.

$$\begin{array}{r} 8.76 \\ \times\ 0.2 \\ \hline 1.752 \end{array}$$
Thus, $(8.76)(-0.2) = -1.752$

6 Find $(-1.56) \div 0.2$.

$$0.2\overline{)1.5.6}$$ *Multiply the divisor and dividend by 10. Why?*

$$\begin{array}{r} 7.8 \\ 0.2\overline{)1.5.6} \\ \underline{1\,4} \\ 1\,6 \\ \underline{1\,6} \end{array}$$
Thus, $(-1.56) \div 0.2 = -7.8$

Exercises

Exploratory   Name a fraction for each of the following.

1. $3 \div 5$
2. $7 \div (-8)$
3. $(-9) \div (-10)$
4. $-3 \div 4$

State how to express each fraction in two other ways.

Sample: $-\dfrac{4}{5} = \dfrac{-4}{5} = \dfrac{4}{-5}$ $\dfrac{1}{-7} = -\dfrac{1}{7} = \dfrac{-1}{7}$

5. $-\dfrac{2}{3}$
6. $\dfrac{3}{-8}$
7. $\dfrac{-2}{11}$
8. $-\dfrac{7}{9}$

Replace each box with $<$ or $>$ to make a true sentence.

9. $\dfrac{3}{7} \,\square\, \dfrac{5}{7}$
10. $-\dfrac{2}{5} \,\square\, -\dfrac{1}{5}$
11. $\dfrac{5}{8} \,\square\, \dfrac{3}{8}$
12. $\dfrac{4}{11} \,\square\, \dfrac{-4}{11}$

Written   Simplify.

Sample: $-\dfrac{8}{12} = \dfrac{-8 \div 4}{12 \div 4} = \dfrac{-2}{3}$ or $-\dfrac{2}{3}$

1. $\dfrac{6}{8}$ **2.** $\dfrac{12}{15}$ **3.** $\dfrac{6}{12}$ **4.** $\dfrac{9}{27}$ **5.** $\dfrac{12}{18}$

6. $-\dfrac{21}{28}$ **7.** $-\dfrac{18}{10}$ **8.** $-\dfrac{6}{36}$ **9.** $-\dfrac{15}{25}$ **10.** $-\dfrac{27}{45}$

Find each of the following and simplify.

11. $\dfrac{1}{2} + \dfrac{2}{3}$ **12.** $-\dfrac{2}{3} + \dfrac{1}{2}$ **13.** $-\dfrac{4}{9} + \dfrac{2}{3}$ **14.** $-\dfrac{1}{2} + \left(-\dfrac{5}{7}\right)$

15. $\dfrac{1}{3} - \dfrac{5}{8}$ **16.** $-\dfrac{1}{5} - \dfrac{3}{10}$ **17.** $\dfrac{5}{6} - \left(-\dfrac{1}{3}\right)$ **18.** $-\dfrac{3}{7} - \left(-\dfrac{1}{14}\right)$

19. $\dfrac{5}{8} \cdot \dfrac{2}{3}$ **20.** $-\dfrac{1}{2} \cdot \dfrac{3}{4}$ **21.** $\left(\dfrac{5}{6}\right)\left(-\dfrac{1}{4}\right)$ **22.** $(-18)\left(-\dfrac{1}{3}\right)$

23. $2\dfrac{1}{2} + \dfrac{1}{2}$ **24.** $\dfrac{1}{3} - 1\dfrac{1}{2}$ **25.** $\left(-3\dfrac{1}{2}\right)\left(\dfrac{5}{6}\right)$ **26.** $\left(-3\dfrac{1}{2}\right)\left(-2\dfrac{1}{2}\right)$

27. $-4.2 + (-2.3)$ **28.** $-8.2 + (-1.4)$ **29.** $3.6 + (-2.3)$ **30.** $-4.7 - 3.9$

31. $4.3 - 8.4$ **32.** $8.9 - (-11.1)$ **33.** $(-0.5)(9.7)$ **34.** $(-8.16)(-0.75)$

35. $(0.8)(0.5)$ **36.** $0.72 \div (-0.8)$ **37.** $(-18.2) \div (-0.2)$ **38.** $(-4.53) \div 0.03$

Replace each box with $<$ or $>$ to make a true sentence.

39. $\dfrac{2}{3} \square \dfrac{1}{6}$ **40.** $\dfrac{3}{4} \square \dfrac{5}{8}$ **41.** $-\dfrac{7}{8} \square -1$ **42.** $-\dfrac{5}{6} \square -2$

43. $-\dfrac{3}{4} \square -\dfrac{4}{5}$ **44.** $-\dfrac{6}{7} \square \dfrac{-5}{6}$ **45.** $-4\dfrac{1}{3} \square -4$ **46.** $-\dfrac{7}{8} \square \dfrac{-8}{9}$

Arrange each set of numbers from least to greatest.

47. $0, -50, 8$ **48.** $\dfrac{1}{2}, \dfrac{2}{3}, -\dfrac{1}{8}$ **49.** $-\dfrac{1}{2}, -\dfrac{2}{3}, -\dfrac{1}{8}$ **50.** $-3\dfrac{3}{4}, -2\dfrac{1}{2}, -2\dfrac{1}{4}$

Show that each of the following satisfies the definition of a rational number.

51. 15 **52.** -6.2 **53.** 0.2 **54.** -3.14 **55.** $-3.0 \div -0.6$

Determine whether each of the following is *always, sometimes* or *never* true.

56. The sum of two whole numbers is a whole number.

57. The difference of two whole numbers is a whole number.

58. The sum of two rational numbers is a rational number.

59. The difference of two integers is an integer.

60. The product of two integers is an integer.

61. The product of two rational numbers is a rational number.

62. The quotient of two integers is an integer.

63. The quotient of two rational numbers is a rational number.

Evaluate if $a = \dfrac{1}{2}$, $b = -\dfrac{1}{3}$, $c = -0.3$, and $d = 0.4$.

64. $4a - 3b$ **65.** $-8a + 11b$ **66.** $5a + 7b$ **67.** $3c + 8d$
68. $-7c - 2d$ **69.** $16a - 13c$ **70.** $8b + 5d$ **71.** $3ab$
72. $5d^2$ **73.** $ab + 7b$ **74.** $a^2 + 2b$ **75.** $3a^2 - b^2$
76. $3cd - 19d$ **77.** $c^3 + 5d$ **78.** $16c \div d$ **79.** $27bd \div 8ac$

Challenge Find the solution set for each of the following if the domain is $\mathcal{Q}$.

1. $3x = 8$ **2.** $-5z = 12$ **3.** $4m = -11$
4. $-2n = -19$ **5.** $-8a = 7$ **6.** $-10b = -17$
7. $r \div 6 = \dfrac{2}{7}$ **8.** $t \div (-12) = 2\dfrac{5}{9}$ **9.** $y \div 0.3 = 0.5$
10. $q \div 2.3 = -3.7$ **11.** $-0.6c = -0.42$ **12.** $-0.04s = 6.52$

Mixed Review

Let p, q, and r represent the following statements.
p: $-5 - (-15) = -10$
q: $(-5)(-15) = -75$
r: $-15 \div (-5) = 3$
Determine whether each of the statements represented by p, q, and r is true or false. Then determine the truth value of each of the following.

1. $p \rightarrow r$ **2.** $p \wedge q$ **3.** $\sim r \vee q$
4. $\sim r \leftrightarrow p$ **5.** $\sim(q \leftrightarrow \sim p)$ **6.** $(\sim p \vee \sim r) \wedge q$
7. the converse of $q \rightarrow p$ **8.** the inverse of $\sim q \rightarrow \sim r$
9. the contrapositive of $p \rightarrow (r \leftrightarrow q)$

Evaluate each expression if $x = -8$ and $y = 12$.

10. $-(y - x)$ **11.** $2xy$ **12.** $-xy^2$
13. $3x - y$ **14.** $-2y - 5x$ **15.** $4x^2 + y^2$
16. $\dfrac{3x}{y}$ **17.** $\dfrac{4y^2}{x^2}$ **18.** $\dfrac{y - x}{y + x}$

19. Julia started the year with $726 in her savings account. During the year, she withdrew $170 from the account. What was the balance in her account at the end of the year?

20. One morning in White Plains, New York, the temperature was $-4°C$. During the day, the temperature rose $19°C$. What was the temperature then?

2.9   Properties of Rational Numbers

You are familiar with many properties of rational numbers. The sum of two numbers is the same regardless of the order in which they are added.

$$6 + 5 = 5 + 6 \qquad 1\frac{1}{2} + (-3) = -3 + 1\frac{1}{2} \qquad 29.3 + 37.4 = 37.4 + 29.3$$

Likewise, the product of two numbers is the same regardless of the order in which they are multiplied.

$$8 \cdot 7 = 7 \cdot 8 \qquad\qquad 5(-6) = (-6)5 \qquad\qquad \frac{2}{3} \cdot \frac{3}{5} = \frac{3}{5} \cdot \frac{2}{3}$$

Commutative Property for Addition and Multiplication

> For any rational numbers a and b,
> $a + b = b + a$ and $ab = ba$.

How do you evaluate $3 + 8 + 2$?

$$3 + 8 + 2 = (3 + 8) + 2 \qquad 3 + 8 + 2 = 3 + (8 + 2)$$
$$= 11 + 2 \text{ or } 13 \qquad\qquad\qquad = 3 + 10 \text{ or } 13$$

The sum is 13 in each case. The way you group, or *associate*, three numbers does *not* change their sum. In a similar manner, you can see that multiplication is associative.

Associative Property for Addition and Multiplication

> For any rational numbers a and b,
> $a + (b + c) = (a + b) + c$ and $a(bc) = (ab)c$.

Example

1 Evaluate $8\frac{1}{2} + \left(17\frac{5}{8} + 1\frac{1}{2}\right)$.

Group $8\frac{1}{2}$ and $1\frac{1}{2}$ together since their sum is 10.

$$8\frac{1}{2} + \left(17\frac{5}{8} + 1\frac{1}{2}\right) = 8\frac{1}{2} + \left(1\frac{1}{2} + 17\frac{5}{8}\right) \qquad \textit{Commutative Property for Addition}$$

$$= \left(8\frac{1}{2} + 1\frac{1}{2}\right) + 17\frac{5}{8} \qquad \textit{Associative Property for Addition}$$

$$= 10 + 17\frac{5}{8} \text{ or } 27\frac{5}{8}$$

The numbers 0 and 1 are involved in several properties of rational numbers.

Properties of 0 and 1

Name	Property
Additive Identity Property	For any rational number a, $a + 0 = a$.
Multiplicative Property of 0	For any rational number a, $a \cdot 0 = 0$.
Multiplicative Identity Property	For any rational number a, $a \cdot 1 = a$.

Examples of each property are shown in the table below.

Property	Examples
Additive Identity Property	$6 + 0 = 6 \qquad 0 + (-15) = -15$
Multiplicative Property of 0	$7 \cdot 0 = 0 \qquad (-8)(0) = 0$
Multiplicative Identity Property	$14 \cdot 1 = 14 \qquad 1 \cdot (-3) = -3$

Do whole numbers have this property?

You know that every integer has an opposite. Likewise, each rational number has an opposite. The opposite of a number is called the **additive inverse** of the number. The sum of a number and its additive inverse is 0. For example, $4 + (-4) = 0$.

Additive Inverse Property

For every rational number a, there is a rational number $-a$ such that $a + (-a) = 0$.

Zero has no multiplicative inverse.

If the product of two numbers is 1, each number is called the **multiplicative inverse** of the other. For example, since $7 \cdot \dfrac{1}{7} = 1$, 7 and $\dfrac{1}{7}$ are multiplicative inverses of each other. The multiplicative inverse of a number is also called the **reciprocal** of the number.

Multiplicative Inverse Property

For every rational number $\dfrac{a}{b}$, if a and b are not 0, there is a rational number $\dfrac{b}{a}$ such that $\dfrac{a}{b} \cdot \dfrac{b}{a} = 1$.

Example

2 Find $-\dfrac{2}{3} \div \dfrac{3}{4}$.

$$-\frac{2}{3} \div \frac{3}{4} = -\frac{2}{3} \cdot \frac{4}{3}$$

$$= -\frac{8}{9}$$

To find the quotient of two fractions, multiply by the multiplicative inverse of the divisor.

Notice that $\dfrac{3}{4}$ and $\dfrac{4}{3}$ are multiplicative inverses.

If any number is multiplied by -1, the result is the additive inverse of the number.

$$6(-1) = -6 \qquad (-2.3)(-1) = 2.3$$

Multiplicative Property of -1

For every rational number a,
$$(-1)a = -a.$$

Exercises

Exploratory   State the additive inverse of each rational number.

1. -7 **2.** $\frac{2}{3}$ **3.** -2.36 **4.** 0.01 **5.** $-1\frac{1}{4}$

State the multiplicative inverse of each rational number.

6. 3 **7.** -9 **8.** -1 **9.** $\frac{3}{5}$ **10.** $-\frac{2}{3}$

Write each of the following rational numbers as a fraction.

11. $-(-9)$ **12.** $-(\frac{2}{3})$ **13.** $-(3.4)$ **14.** $-(-1\frac{7}{8})$ **15.** $-(-(-2))$

State the property shown in each of the following.

16. $5 + (-6) = (-6) + 5$
17. $-8 + (6 + 4) = (-8 + 6) + 4$
18. $(-7 \cdot 3)4 = -7(3 \cdot 4)$
19. $6(-\frac{1}{2}) = (-\frac{1}{2})6$
20. $13 \cdot 0 = 0$
21. $-2.4 + 0.5 = 0.5 + (-2.4)$
22. $29 + (-2 + 3) = [29 + (-2)] + 3$
23. $8\frac{1}{2} + 0 = 8\frac{1}{2}$
24. $(-3)(1.5 + 2.3) = (1.5 + 2.3)(-3)$
25. $(-11)(1) = -11$

26. Is $8 - 2 = 2 - 8$? Is $15 - 6 = 6 - 15$? Try some other subtraction problems. Is there a commutative property for subtraction?

27. Is $(8 - 2) - 1 = 8 - (2 - 1)$? Is $(20 - 9) - 3 = 20 - (9 - 3)$? Try some other subtraction problems. Is there an associative property for subtraction?

28. Is $24 \div 3 = 3 \div 24$? Is $81 \div 9 = 9 \div 81$? Try some other division problems. Is there a commutative property for division?

29. Is $(5 \div 5) \div 5 = 5 \div (5 \div 5)$? Is $(8 \div 4) \div 2 = 8 \div (4 \div 2)$? Try some other division problems. Is there an associative property for division?

Written   Replace each box with a number to make the sentence true. Then write the property shown by the sentence.

1. $5 \times \square = 4 \times 5$

2. $-7 \times (2 + 12) = (2 + 12) \times \square$

3. $\left(\dfrac{1}{2} + \dfrac{1}{3}\right) + \dfrac{1}{4} = \dfrac{1}{2} + \left(\square + \dfrac{1}{4}\right)$

4. $(5 \cdot 6) + 7 = \square + (5 \cdot 6)$

5. $-6 \cdot 4 = \square \cdot (-6)$

6. $-8 + (-5 + 3) = (-5 + \square) + (-8)$

7. $x \cdot \square = 3x$

8. $x + (y + z) = (x + y) + \square$

9. $3(bc) = (\square \cdot b)c$

10. $-14 + (ab) = (ab) + \square$

11. $5 + \square = 5$

12. $7 + \square = 0$

13. $-8 \cdot \square = -8$

14. $-8 \cdot \square = 0$

15. $\square + x = x$

16. $-3 \cdot \square = 3$

17. $-11 + \square = 0$

18. $3 \cdot \square = 1$

19. $\square \cdot x = 0$

20. $\square \cdot \dfrac{3}{4} = -\dfrac{3}{4}$

21. $\square \cdot \dfrac{1}{2} = \dfrac{1}{2}$

22. $-\dfrac{2}{3} \cdot \square = 1$

Find each of the following.

23. $8 + (14 + 2)$

24. $(72 - 6) + 24$

25. $9(3 + 11)$

26. $17 + 12 + 3 + 6$

27. $(20 + 15) + (-15)$

28. $[6(5 + 2)] - 72$

29. $40 \div [8(-1)]$

30. $9 \div (3 \div 3)$

31. $14(19 - 19)$

32. $\dfrac{5}{6} \div \dfrac{2}{3}$

33. $\dfrac{3}{8} \div \left(-\dfrac{1}{2}\right)$

34. $-\dfrac{5}{6} \div \dfrac{1}{3}$

35. $30 + (-20) + 8 + (-6)$

36. $25 + (-32) + (-61) + 23$

37. $8 + (-9) + 7 + (-6) + (-4)$

38. $-3 + 12 + (-4) + (-3) + 16$

39. $6 - 5 - 7 + 4$

40. $12 - 15 - 11 + 5 - 10$

Challenge   Answer each of the following.

1. a. Find the value of x for which $x - 3 = 0$.

   b. Find the value of x for which the fraction $\dfrac{1}{x - 3}$ is undefined.

   c. Find the value of x for which the multiplicative inverse of $x - 3$ is undefined.

Write the multiplicative inverse for each of the following. Then find the value of x for which the multiplicative inverse is undefined.

2. $x - 2$ **3.** $8 + x$ **4.** $2x$ **5.** $-4 - y$ **6.** $3x - 9$

2.10   The Distributive Property

Julie Fett worked 5 hours Monday at $4.00 an hour. She worked 4 hours Tuesday at $4.00 an hour. There are two ways to figure her total pay.

1. Multiply her total number of hours by 4.
$$4(5 + 4) = 4 \cdot 9 \text{ or } 36$$

2. Add her Monday earnings to her Tuesday earnings.
$$4 \cdot 5 + 4 \cdot 4 = 20 + 16 \text{ or } 36$$

Both methods give the same answer.
$$4(5 + 4) = 4 \cdot 5 + 4 \cdot 4$$

This example illustrates another property of rational numbers.

Distributive Property

> For any rational numbers a, b, and c,
> $a(b + c) = ab + ac$ and $(b + c) a = ba + ca.$

Examples

1 **Find 5(6 + 12).**

$$\begin{aligned} 5(6 + 12) &= 5 \cdot 6 + 5 \cdot 12 \\ &= 30 + 60 \\ &= 90 \end{aligned}$$ *Use $a(b + c) = ab + ac$.*

2 **Find 7 · 9 + 2 · 9.**

$$\begin{aligned} 7 \cdot 9 + 2 \cdot 9 &= (7 + 2) 9 \\ &= 9 \cdot 9 \\ &= 81 \end{aligned}$$ *Use $ba + ca = (b + c) a$.*

Since subtraction sentences can be written as addition sentences, the distributive property applies to subtraction.

The distributive property can be used to simplify expressions containing *like* terms. In $4x + 5x + 3$, for example, $4x$ and $5x$ are like terms. They contain the same variable to the same power.

Example

3 Simplify $-4y + 2y + 3$.

$$-4y + 2y + 3 = (-4 + 2)y + 3$$
$$= -2y + 3$$

Use the distributive property. $-4y$ and $2y$ are like terms.

Exercises

Exploratory Use the distributive property to write each expression in a different form.

1. $4(a + b)$ **2.** $2(x + y)$ **3.** $3(r + s)$

4. $(a + b)6$ **5.** $(c + d)7$ **6.** $3(5 - x)$

7. $-9(5 - x)$ **8.** $2(y - 4)$ **9.** $8(-2 + d)$

10. $(-6 + t)(-8)$ **11.** $-7(-4 - y)$ **12.** $(-3 - t)(-2)$

Simplify.

13. $3x + 5x$ **14.** $4c + 5c$ **15.** $2y + 7y$

16. $20c + 13c$ **17.** $15q - 14q$ **18.** $-13r + 10r$

19. $-6x + 6x$ **20.** $-5x + 7x$ **21.** $8mn + 10mn$

22. $6xy - 28xy$ **23.** $15x^2 + 7x^2$ **24.** $-23y^2 + 32y^2$

Written Use the distributive property to evaluate each expression.

1. $8(9 + 10)$ **2.** $4(6 + 7)$ **3.** $7(2 - 8)$

4. $-3(70 + 5)$ **5.** $4 \cdot 98 + 4 \cdot 2$ **6.** $(-83)(11) + (-17)(11)$

7. $12\frac{3}{4} \cdot 7 + 7\frac{1}{4} \cdot 7$ **8.** $7\frac{1}{4} \cdot 8 + 2\frac{3}{4} \cdot 8$ **9.** $12\frac{1}{4} \cdot 9 - 2\frac{1}{4} \cdot 9$

Simplify.

10. $11y - 5y$ **11.** $-7y + 3y$ **12.** $-12a + 2a$

13. $0.1a + 0.4a$ **14.** $10a + 30a + a$ **15.** $16b + 17b - b$

16. $\frac{1}{2}y + \frac{3}{2}y$ **17.** $\frac{2}{5}t - \frac{3}{5}t$ **18.** $-\frac{3}{8}n - \frac{7}{8}n$

19. $9y^2 + 13y^2 + 3$ **20.** $11a^2 - 11a^2 + 12a^2$ **21.** $5a + 10a + 7b - 5b$

22. $16a + 4a - 8$ **23.** $10c + 4 + 4c$ **24.** $-9r + 4r + 3 + 5r$

25. $5 + 6y + 5y$ **26.** $2(a + b) + b$ **27.** $3(x + 2y) - x$

28. $5(x + y) + 3(x - y)$ **29.** $6(5a + 3b - 2a)$ **30.** $18m^2 + 14n^2 + m^2 + 42$

Problem Solving Application: Guess and Check

Sometimes there will not be one right answer. You may have to find the best answer for that problem.

One important strategy used to solve problems is called **guess and check.** To use this strategy, guess an answer to the problem. Then check whether your guess is correct. If the first guess is incorrect, continue guessing and checking until you find the right answer. Use the results of previous guesses to help you make better guesses. Keep a record of your guesses so you do not make the same guess twice.

Example

1 **Von purchased several pencils for $2.31. If the price of each pencil was greater than 17¢, how many pencils did Von purchase? What was the price of each pencil?**

Explore   Von purchased at least two pencils.
Each pencil cost more than 17¢.
We must find two numbers, one that is greater than or equal to 2 and another that is greater than 17, whose product is 231.

Plan   To solve the problem, divide 231 by numbers greater than 17 until two factors are found. Since 231 is an odd number, we only need to divide by odd numers greater than 17.   *Why?*

Solve   Try 19. $\dfrac{12 \text{ R3}}{19\overline{)231}}$ *A calculator could be used to do these computations.*

Since 19 is not a factor of 231, try the next odd number, 21.

$$\dfrac{11}{21\overline{)231}} \quad \textit{Correct!}$$

One solution is that Von purchased 11 pencils at 21¢ each.

We could continue dividing 231 by odd numbers greater than 21 to determine if there are any other solutions. A better method would be to use the two factors of 231 already determined to find other pairs of numbers whose product is 231.

$$11 \cdot 21 = 231$$
$$11 \cdot (3 \cdot 7) = 231$$
$$(11 \cdot 3) \cdot 7 = 231 \quad \text{or} \quad (11 \cdot 7) \cdot 3 = 231$$
$$33 \cdot 7 = 231 \quad \text{or} \quad 77 \cdot 3 = 231$$

There are two other solutions. Von could have purchased 7 pencils at 33¢ each or 3 pencils at 77¢ each.

Examine In each solution, Von purchased at least two pencils and each pencil cost more than 17¢. Since 11 • 21¢ = $2.31, 7 • 33¢ = $2.31, and 3 • 77¢ = $2.31, each solution is correct.

Exercises

Exploratory Copy each open sentence. Then insert parentheses so that each open sentence is a true equation.

1. $4 • 5 - 2 + 7 = 19$
2. $25 - 4 • 2 + 3 = 5$
3. $10 - 4 • 2 - 1 = 3$
4. $3 + 5 • 8 - 2 = 48$
5. $3 + 6 • 4 • 2 = 54$
6. $2 + 6 ÷ 1 + 3 • 6 = 12$

Written Solve each problem.

1. A bottle and a cork cost $1.10. If the bottle costs $0.96 more than the cork, find the cost of each.

2. Alex has 58¢ in his pocket. If he has exactly six coins, how many of each type of coin does he have?

3. If postage stamps cost 25¢ and postcards cost 18¢, how many of each can be purchased for exactly $1.97?

4. If it costs 5¢ each time you cut and weld a link, what is the minimum cost to make a chain out of 7 links?

5. Gina and Rick raise cows and ducks. They counted all the heads and got 11. They counted all the feet and got 28. How many of each do they have?

6. Canned fruit comes in cases of 24 or 36. If a shipment of 228 cans was packed in 8 cases, how many of these were cases of 36?

7. On a certain test, correct answers are worth 10 points and incorrect answers are worth −5 points. On a 24 question test, how many correct answers are needed to score 150 points?

8. Alice bought a dozen peaches and pears for $1.98. Each peach costs 3¢ more than each pear. What is the cost of each item? How many of each item did Alice buy?

9. Peter will take Jill or Olivia to the prom. Raul will not take Amy if Peter takes Olivia. If Peter takes Jill, then Olivia will go with Don or Raul. Raul said he would not take Olivia or Amy, but he did take Amy. Who was Olivia's date to the prom?

Supply a digit for each letter so that each addition or multiplication problem is correct. Each letter represents a different digit.

10.
```
  SEND
+ MORE
 MONEY
```

11.
```
ABCDE
×    4
EDCBA
```

Portfolio Suggestion

Select one of the assignments from this chapter that you found especially challenging and place it in your portfolio.

Performance Assessment

Ching, Alex, Ellen, and Ashley are playing in a local golf tournament. For the first two days both Ching's scores were under -2. Alex had scores under par, or zero. Ellen's scores were less than Ching's scores and Ashley's scores were par or greater.
a. Draw a graph showing possible scores for Ching.
b. How do Ashley's and Alex's scores compare to Ching's?
c. Were Ellen's scores less than Ashley's?
Include explanations of your solutions.

Chapter Summary

1. **Algebraic expressions** contain variables and operations. They can be **evaluated** when the value of the variables are known. (38)
2. 7^5 is an example of a **power.** 7 is the **base,** and 5 is the **exponent.** (39)
3. **Order of Operations:** (41)
   1. Evaluate all powers.
   2. Then do all multiplications and/or divisions from left to right.
   3. Then do all additions and/or subtractions from left to right.
4. The set of **integers** $\{\dots, -2, -1, 0, 1, 2, \dots\}$ is denoted by $\mathbb{Z}$. (43)
5. The **absolute value** of a number is the number of units it is from 0 on the number line. (48)
6. To add integers with the same sign, add their absolute values. The sum has the same sign as the integers. To add integers with different signs, find the difference of their absolute values. The sum has the same sign as the integer with the greater absolute value. (48)
7. To subtract an integer, add its opposite. If a and b are integers, then $a - b = a + (-b)$. (50)
8. The product of two integers with different signs is negative. The product of two integers with the same sign is positive. (52)

9. The quotient of two integers with the same sign is positive. The quotient of two integers with different signs is negative. (53, 54)

10. A **rational number** is any number that can be written in the form $\frac{a}{b}$ where a is any integer and b is any integer except 0. (56)

11. Properties of Rational Numbers: (Assume a, b and c are rational numbers) (61-63)

Name	Property
Commutative Property	$a + b = b + a \qquad ab = ba$
Associative Property	$a + (b + c) = (a + b) + c$ $a(bc) = (ab)c$
Additive identity Property	$a + 0 = a$
Multiplicative Property of 0	$a \cdot 0 = 0$
Multiplicative Identity Property	$a \cdot 1 = a$
Additive Inverse Property	There exists a rational number $-a$ such that $a + (-a) = 0$.
Multiplicative Inverse Property	If a and b are not 0, then there is a rational number $\frac{b}{a}$ such that $\frac{b}{a} \cdot \frac{a}{b} = 1$.
Multiplicative Property of -1	$a(-1) = -a$
Distributive Property	$a(b + c) = ab + ac$ $(b + c)a = ba + ca$

∃∀ Chapter Review

2.1 **Evaluate if $a = 2$, $b = 3$, $c = 4$, and $d = 6$.**

1. $a + 11$ **2.** $b + c$ **3.** $d - 3$ **4.** $c - a$

5. $4a^2$ **6.** bd **7.** $\frac{c}{2}$ **8.** $\frac{6d}{a^2}$

2.2 **Evaluate.**

9. $7 + 6 - 3 + 2$ **10.** $6 + 5 \cdot 2 - 10 \div 5$ **11.** $5^2 \div 5 + 36 \div 6$
12. $5(6 + 4) \div 2$ **13.** $(6 + 8)(5 + 2) \div 2$ **14.** $8[5 + 3(6 + 8)]$

Evaluate if $a = 2$, $b = 5$, and $c = 1$.

15. $b^2 - a^2 \cdot c^2$ **16.** $3a^2 - (c + b)$ **17.** $\dfrac{3(b - c)}{a^2}$

2.3 **Graph each set of numbers on a number line.**

18. $\{1, 3, 5\}$ **19.** $\{-1, -3, -7\}$ **20.** $\{\ldots, -2, -1, 0, 1\}$

2.4 **Use a number line to find each sum.**

21. $-3 + (-2)$ **22.** $-4 + 1$ **23.** $5 + (-3)$

2.5 **Find each sum.**

24. $-8 + (-3)$ **25.** $8 + (-5)$ **26.** $7 + (-9)$
27. $-9 + 5$ **28.** $-5 + 11$ **29.** $-16 + (-30)$

2.6 **Find each difference.**

30. $15 - 37$ **31.** $-12 - 5$ **32.** $8 - (-46)$
33. $-17 - (-8)$ **34.** $9 - 23$ **35.** $-16 - (-14)$

2.7 **Find each product.**

36. $8 \cdot (-9)$ **37.** $-11(4)$ **38.** $-8 \cdot 5$

Find each quotient.

39. $36 \div (-3)$ **40.** $-48 \div 6$ **41.** $-78 \div (-6)$

2.8 **Find each of the following and simplify.**

42. $\dfrac{5}{16} - \dfrac{9}{16}$ **43.** $\dfrac{5}{7} + \dfrac{3}{14}$ **44.** $\dfrac{7}{10} - \dfrac{3}{20}$

45. $-\dfrac{3}{12} - \left(-\dfrac{4}{18}\right)$ **46.** $\left(\dfrac{4}{9}\right)\left(-\dfrac{3}{8}\right)$ **47.** $\left(-\dfrac{32}{7}\right)\left(-\dfrac{35}{8}\right)$
48. $2.05 - 0.1$ **49.** $6.7 + (-14.2)$ **50.** $(0.7)(-4.5)$

2.9 **State the property shown in each of the following.**

51. $13 + 17 = 17 + 13$ **52.** $11 + 0 = 11$ **53.** $(4 \cdot 5)6 = 4(5 \cdot 6)$
54. $7 \cdot 0 = 0$ **55.** $1 \cdot 17 = 17$ **56.** $3 + (-3) = 0$

2.10 **Simplify using the distributive property.**

57. $16x + 20x$ **58.** $10a + 16 - 3a$
59. $2(x + 4) + 3x$ **60.** $4(2x + 3) + 5(x - 1)$

Chapter Test

Evaluate if $a = 3$, $b = 4$, and $c = 6$.

1. $4a$ **2.** $c - 1$ **3.** $c^2 - 4$ **4.** $\dfrac{24}{b}$

5. $a - (b - c)$ **6.** $a^2 + 2a + b$ **7.** $(a + b)(b - c)$ **8.** $[a + b(c)]a$

Graph each set of numbers on a number line.

9. $\{0, 2, 6\}$ **10.** $\{-4, 0, 1\}$ **11.** $\{-2, -1, 0, \ldots\}$

Find each sum.

12. $-13 + 16$ **13.** $-18 + (-46)$ **14.** $-29 + 13$ **15.** $3.2 + (-4.5)$

Find each difference.

16. $13 - 47$ **17.** $-13 - 28$ **18.** $18 - (-33)$ **19.** $5.7 - 10.4$

Find each product.

20. $-6 \cdot 7$ **21.** $7(-14)$ **22.** $(-8)(-13)$ **23.** $-3.1 \cdot 1.7$

Find each quotient.

24. $84 \div (-12)$ **25.** $-96 \div 8$ **26.** $-144 \div (-24)$ **27.** $-7.82 \div 3.4$

Find each of the following and simplify.

28. $\dfrac{5}{8} + \left(-\dfrac{5}{12}\right)$ **29.** $-\dfrac{2}{7} - \dfrac{4}{21}$ **30.** $\left(\dfrac{4}{5}\right)\left(-\dfrac{25}{36}\right)$ **31.** $-\dfrac{4}{3} \div \left(-\dfrac{16}{9}\right)$

State the property shown in each of the following.

32. $7 \cdot 8 = 8 \cdot 7$ **33.** $3 + (6 + 2) = (3 + 6) + 2$ **34.** $\dfrac{2}{3} \cdot \dfrac{3}{2} = 1$

35. $14 \cdot (-1) = -14$ **36.** $3(4 - 2) = 3 \cdot 4 - 3 \cdot 2$ **37.** $0 + 9 = 9$

Simplify using the distributive property.

38. $8(21 - 4)$ **39.** $13b + 16b - 10$ **40.** $(x + 5)6 + 4(2x - 6)$

Introduction to
Algebra

Group Project:
Measurement

Use a scale to find out how many paper clips balance one pencil. Write an equation showing how the paper clips relate to the pencil. Use the scale and other common objects to write four more equations.

Application in Meteorology

In order to know what to wear or what to plan, people frequently want predictions about the weather. A weather forecaster uses information about atmospheric conditions and weather patterns, as well as **algebraic equations**, to make these predictions.

Julie Connor is a weather forecaster for a TV station. She knows a thunderstorm is approaching New York City and is traveling about 35 miles per hour. At 11:30 A.M. the storm is 105 miles from the city. Write an equation that Julie could use to predict how long before the storm will hit New York. When should she predict the storm will arrive?

3.1   Using Variables

In algebra, we often translate verbal expressions into algebraic expressions. A variable is used to represent an unspecified number. Any letter may be used as a variable.

Verbal Expression	*Algebraic Expression*
7 more than *a number*	$q + 7$
some number decreased by 12	$x - 12$
3 times *a number*	$3n$
one half of *a number*	$\frac{1}{2}y$ or $\frac{y}{2}$
6 less than *a number*	$d - 6$
product of 5 and *some number*	$5t$
some number divided by 8	$\frac{w}{8}$ or $w \div 8$
sum of twice *a number* and 13	$2b + 13$
twice the sum of *a number* and 13	$2(b + 13)$

Sometimes a table is helpful in writing an algebraic expression.

Journal

Variables are used in formulas. Make a list of all the formulas you remember from other classes you've taken.

Example

1 **The population of Pine City is decreasing by 350 people each year. The population is now 100,000. Write an algebraic expression to represent the population *y* years from now.**

Current Population		Decrease Each Year		Number of Years		Population
100,000	−	350	×	0	=	100,000
100,000	−	350	×	1	=	99,650
100,000	−	350	×	2	=	99,300
100,000	−	350	×	3	=	98,950
⋮		⋮		⋮		⋮
100,000	−	350	×	*y*	=	100,000 − 350*y*

Therefore, the expression 100,000 − 350*y* represents the population of Pine City in *y* years.

The terms of an algebraic expression can be any one of the following.

    1. a numeral such as 15

    2. a variable such as x

    3. a product of a numeral and one or more variables such as $-4x$ or $3ab$

    4. a quotient of a numeral and variable such as $\dfrac{x}{8}$

For example, the algebraic expression $6 + 3y - ab$ contains three terms, 6, $3y$, and ab.

The numerical factor in a term is called its **coefficient.** The coefficient in $-7ab$ is -7. The term x has a coefficient of 1 since $1 \cdot x = x$. The expression $-(a + b)$ is the same as $-1(a + b)$.

$$-1(a + b) = (-1)(a) + (-1)(b) \qquad \textit{Use the distributive property.}$$
$$= -a - b$$

Likewise, $-(a - b) = -a + b$.

The properties of operations can be used to simplify algebraic expressions.

Examples

2 **Simplify $2(-3x)$.**

$$2(-3x) = (2 \cdot -3)x \qquad \textit{Use the associative property for multiplication.}$$
$$= -6x$$

3 **Simplify $4b - 3b$.**

$$4b - 3b = (4 - 3)b \qquad \textit{Use the distributive property.}$$
$$= 1 \cdot b \text{ or } b$$

4 **Simplify $5x - (2x - 4)$.**

$$5x - (2x - 4) = 5x - 2x + 4 \qquad \textit{Recall that } -(a - b) = -a + b.$$
$$= 3x + 4$$

5 **Find the perimeter of the figure at the right.**

The perimeter of a polygon is the sum of the lengths of its sides.

$$2x + (x + 4) + 3x + 4x = 2x + x + 3x + 4x + 4$$
$$= 10x + 4$$

The perimeter is $(10x + 4)$ units.

═══ Exercises ═══

Exploratory   State an algebraic expression for each of the following.

1. 7 more than y

2. 8 less than x

3. 4 increased by x

4. y increased by 7

5. the product of 8 and r

6. q times 10

7. n decreased by 3

8. 17 decreased by y

9. the sum of r and 6

10. two-thirds of x

11. one-half of r

12. 7 less than b

13. the product of x and y

14. 18 divided by m

15. x divided by six

16. two times y

State the coefficient in each of the following.

17. $9x$

18. $-4y$

19. $6ab$

20. x

21. 3

22. $\frac{3}{4}x$

23. $-y$

24. $14x^2$

Written   Simplify.

1. $4(2x)$

2. $5(6y)$

3. $-2(4a)$

4. $-3(-4b)$

5. $\frac{1}{2}(2a)$ -

6. $\frac{2}{3}(-6b)$

7. $(-7ab)4$

8. $(-4y)(-2)$

9. $(-6c)(-\frac{2}{3})$

10. $6a + 2a$

11. $8x + 11x$

12. $x + x$

13. $10a - 2a$

14. $3b - b$

15. $-10b + 3b$

16. $8w - 10w$

17. $5y - 11y$

18. $-6b - 3b$

19. $-3d - 14d$

20. $x + x + 2$

21. $4y + 4 + 2y$

22. $y + 2y + 6$

23. $3 + 6y + 2$

24. $2n + 1 + 3n + 5$

25. $4x + 6 - 7x$

26. $8c - 3 - 2c + 2$

27. $2y + 1 - 6y + 4$

28. $\frac{1}{2}x + \frac{3}{2}x - 5$

29. $x + \frac{3}{2}x - 5 - x$

30. $-(3a + 6)$

31. $-(2x - 4)$

32. $-(-x + 2y)$

33. $-(-3y - 4)$

34. $2(x + 3) + 4x$

35. $2(3y - 5) - 7y + 4$

36. $5(2x + 1) + 3(x + 4)$

37. $2(a - 5) + 3(4 - a)$

38. $-3(4 - 2b) + 4b - 2$

39. $-5(2 - 3x) - 2(4x - 2)$

Find the perimeter of each figure in simplest form.

40.

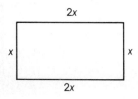

41.

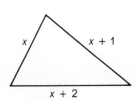

42.

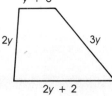

43.

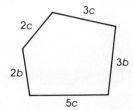

Write an algebraic expression for each of the following.

44. 10 subtracted from twice x

45. 8 more than twice x

46. 2 less than 4 times b

47. 7 times the sum of 5 and y

48. 3 times the sum of r and 4

49. 4 less than the product of 6 and c

50. product of 3 and the sum of x and y

51. twice the sum of n and 5

For each of the following, define a variable. Then write an algebraic expression.

> **Sample:**   Jorge has 12 more stamps than Mike.
> Let x = the number of stamps Mike has.
> Then x + 12 = the number of stamps Jorge has.

52. Michelle has 5 more tapes than Vicki.

53. Tracy is 3 years older than Clint.

54. Steve's wages increased by $4.

55. Sammy is 7 years younger than Suzi.

56. Today is 12° cooler than yesterday.

57. Gary sold 14 more books than Lee.

58. Base A is 15 units longer than base B.

59. Kathy walked 2 miles more than Ann.

60. The length is 2 units more than the width.

61. The width is 5 units less than the length.

Write an algebraic expression for each of the following. A table may be helpful in some cases.

62. number of cents in a dimes and b nickels

63. number of cents in d dimes and d + 5 quarters

64. number of days in x weeks

65. number of centimeters in t meters

66. number of days in b hours

67. number of gallons in s quarts

68. Flora weighs p pounds. If she gains 5 pounds, what is her weight?

69. The width of a rectangle is x feet. If its length is 6 feet more than its width, what is its length?

70. Martin spent $560 for a stereo and a tape deck. If the tape deck costs y dollars, what is the cost of the stereo?

71. The length of a rectangle is 4 units more than twice its width. If the width is d units, what is the length?

72. Sara opens a savings account with an $80 deposit. She then deposits $5 a week. What is the amount in her account after w weeks, excluding interest?

73. An empty book carton weighs 3 kg. A book weighs 1.5 kg. What is the weight of the carton when it is filled with b books?

74. The present population of Bluffton is 47,000. The population increases by 550 each year. What is the population y years from now?

75. A telephone company charges $12 per month plus $0.10 for each local call. What is the monthly bill if c local calls were made?

76. A ranger calculates there are 6,000 deer in a preserve. She also estimates that 75 more deer die than are born each year. How many deer are in the preserve x years from now?

77. The DD Guitar Company produced 600 guitars its first year of business. Each year after that production increased by 100. What was the total number of guitars produced after y years?

3.2 Writing Equations

An **equation** is a statement of equality between two mathematical expressions.

Verbal Sentence	Equation
Seven is equal to four plus three.	$7 = 4 + 3$
A number decreased by 8 is 2.	$y - 8 = 2$
Six times a number is 48.	$6x = 48$
A number divided by 5 is equal to 60.	$\dfrac{n}{5} = 60$

Equations can be used to solve verbal problems.

Examples

1 **Write an equation for this verbal sentence. If a number is multiplied by 5, the result is 78.**

Define a variable.   Let x = the number.

Write an equation.   The number multiplied by 5 is equal to 78.

$$x \qquad \cdot \qquad 5 \quad = \quad 78$$

$$5x = 78$$

2 **Jackie is renting a car for a day. The charge is $17 for the day plus $0.15 for each mile. She wants to spend exactly $40. Write an equation to find how many miles she can drive.**

Define a variable.   Let n = number of miles driven in a day.

Write an equation. $17 for one day plus 0.15 times number of miles is equal to 40

$$17 \quad + \quad 0.15 \quad \cdot \quad n \quad = \quad 40$$

$$17 + 0.15n = 40$$

Exercises

Exploratory State whether each of the following is an algebraic expression, equation, or neither.

1. $5y$
2. $8 + 4 \neq 3$
3. $2y = 9 + y$
4. $3(x + 5)$
5. $4 + 1 < 8 + 2$
6. $x + 1 > 75$
7. $5 + x + y$
8. $ab = 1$
9. $3^2 - 4$
10. $6a = 7 + a$
11. $m + 2n \leq 0$
12. $x^2 \neq 1$

Written Write an equation for each of the following.

1. Five more than x is equal to 15.
2. Three less than y is equal to 20.
3. The product of n and 5 is equal to 35.
4. Three more than twice y is equal to 29.
5. Six less than twice y is equal to -14.
6. The sum of $3y$ and 5 is equal to -13.
7. When 3 is subtracted from the product of y and 4, the result is 5.
8. The width of a rectangle is w units and its length is $2w$ units. The perimeter of the rectangle is 42 units.
9. Meiko weighs x pounds now. If she loses 7 pounds, her weight will be 123 pounds.
10. Buzz has x dimes and $x + 3$ quarters. He has a total of $2.50.

For each problem, define a variable. Then write an equation.

11. Four less than a number is 70.
12. A number increased by 7 is 12.
13. When a number is multiplied by 8, the result is 88.
14. A number decreased by 4 is 12.
15. If 15 is added to a number, the result is 36.
16. Five less than twice a number is 9.
17. A number divided by 6 is 18.
18. One-half of a number is equal to 43.
19. If twice a number is decreased by 10, the result is 26.
20. If eight is added to 3 times a number, the result is 2.
21. The sum of 26 and a number equals 3 times the number.
22. The sum of one-third of a number and 25 is equal to twice the number.
23. The length of a rectangle is twice its width. The perimeter of the rectangle is 60 units.
24. The width of a rectangle is one-half its length. The perimeter of the rectangle is 24 units.
25. Juan has two fewer dimes than quarters. He has a total of $2.25.
26. Jim opens a savings account with a $20 deposit. He then deposits $5 each week. In a certain number of weeks, he has a total of $105 excluding interest.
27. The ABC Car Rental Agency rents cars for $23 a day plus $0.18 a mile. If Rae has exactly $55 to spend she can drive a certain number of miles in a day.
28. The population of Lafayette is 4,800. The population increases by 50 people each year. The population will reach 5,300 in a certain number of years.

3.3 Solving Equations Using Addition

A replacement for the variable which makes an equation true is called a **solution,** or **root,** of the equation. Equations which have the same solution are called **equivalent equations.** For example, $x = 4$ and $2 + 2 = x$ are equivalent.

Study the three scales below. Consider a scale in balance, as shown on the left. If you add weight to only one side, as shown in the center, then the scale is no longer in balance. However, if you add the same weight to both sides, as shown on the right, then the scale will balance.

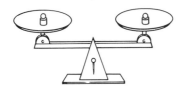  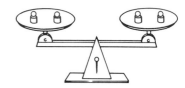

Think of an equation as a scale in balance. If you add the same number to both sides of the equation, then an equivalent equation results.

Addition Property for Equations

If the same number is added to both sides of an equation, the new equation is equivalent to the original.

You can use this property to solve equations involving addition and subtraction.

Example

1 **Solve** $x + 4 = 10$.

$$x + 4 = 10$$
$$x + 4 + (-4) = 10 + (-4) \qquad \textit{Add } -4 \textit{ to both sides.}$$
$$x + 0 = 6$$
$$x = 6$$

Check: $x + 4 = 10$
$$6 + 4 \stackrel{?}{=} 10 \qquad \textit{Replace x by 6 in } x + 4 = 10.$$
$$10 = 10$$

The solution is 6.

Lab Activity

You can learn how to use counters and cups to solve equations involving addition in Lab 2 on pages A4-A5.

Why was the number -4 added to both sides of the equation in example 1 instead of some other number? The additive inverse of 4 is -4. Adding -4 to both sides of $x + 4 = 10$ left just x on the left side of the equation.

Examples

2 **Solve $y - 9 = 14$.**

$$y - 9 = 14$$
$$y + (-9) = 14 \qquad \text{\textit{Subtracting 9 is the same as adding -9.}}$$
$$y + (-9) + 9 = 14 + 9 \qquad \text{\textit{Add 9, the additive inverse of -9, to both sides.}}$$
$$y + 0 = 23$$
$$y = 23$$

Check: $y - 9 = 14$
$$23 - 9 \stackrel{?}{=} 14$$
$$14 = 14$$

The solution is 23.

3 **Solve $8 = x + 2$.**

$$8 = x + 2$$
$$8 + (-2) = x + 2 + (-2) \qquad \text{\textit{Add the additive inverse of 2 to both sides.}}$$
$$6 = x + 0$$
$$6 = x$$

Check: $8 = x + 2$
$$8 \stackrel{?}{=} 6 + 2$$
$$8 = 8$$

The solution is 6.

Recall that adding -2 is the same as subtracting 2. Therefore, the equations in the previous examples also could be solved as follows.

$$y - 9 = 14 \qquad\qquad 8 = x + 2$$
$$y - 9 + 9 = 14 + 9 \qquad\qquad 8 - 2 = x + 2 - 2$$
$$y = 23 \qquad\qquad 6 = x$$

Exercises

Exploratory   State the number you add to both sides of each equation to solve it.

1. $x + 9 = 7$
2. $11 + y = 16$
3. $t + (-7) = 42$
4. $w - 10 = 37$
5. $-12 + m = 13$
6. $-17 + n = 39$
7. $13 = x + 7$
8. $25 = b - 15$
9. $-42 = 18 + c$
10. $37 = x + (-5)$
11. $-46 = -18 + a$
12. $-3 = k - 19$

State whether the given number is a solution of the equation.

13. $y + 5 = 3; -2$
14. $4y = -28; 7$
15. $2y - 1 = 5; 3$
16. $7 = 8 - x; 1$
17. $-9 = \frac{1}{2}y; -18$
18. $3 - 2a = 5; 1$

Written   Solve each equation. Check your answer to see if it is the solution.

1. $x - 1 = 9$
2. $y + (-2) = 5$
3. $b - 6 = 14$
4. $t - 4 = 0$
5. $b - 3 = 9$
6. $q - 2 = 21$
7. $x + 3 = -2$
8. $y - 8 = -4$
9. $m + (-1) = -5$
10. $8 = y - 6$
11. $10 = x - 2$
12. $14 = r - 5$
13. $-3 = x - 7$
14. $-12 = y - 4$
15. $-17 = y - 17$
16. $0 = m - 12$
17. $x - 3 = 8$
18. $y + 7 = 12$
19. $t + 10 = 16$
20. $q + 4 = -7$
21. $x + 1 = -11$
22. $b + 7 = -20$
23. $-8 + y = -5$
24. $6 + m = -2$
25. $-12 + r = -18$
26. $-18 = a + 1$
27. $-11 = 3 + b$
28. $t + 3 = -5$
29. $m + (-7) = 6$
30. $-5 + n = 6$
31. $b + (-6) = -9$
32. $-17 + x = -11$
33. $-33 = k + (-5)$
34. $t + 32 = 15$
35. $y + 876 = -932$
36. $-294 + w = 378$
37. $x + \frac{1}{2} = \frac{1}{2}$
38. $y + \frac{1}{5} = \frac{2}{5}$
39. $x - 1 = -\frac{1}{3}$
40. $5\frac{2}{3} = m + \frac{1}{3}$
41. $y - 11 = 4\frac{1}{3}$
42. $-1\frac{3}{4} = a + \frac{1}{5}$
43. $-1.5 + q = -2.3$
44. $x + 2.3 = 3.5$
45. $t + 1.09 = -2$

Determine whether each pair of equations is equivalent. Write *yes* or *no*.

46. $x + 3 = 7$ and $x = 4$
47. $b = 7$ and $10 - b = 3$
48. $y - 8 = 2$ and $y = 10$
49. $m + 4 = 2$ and $m = -6$
50. $n - 17 = -19$ and $n = 2$
51. $2m + 1 = 7$ and $2m = 8$

Challenge   State whether each of the following biconditionals is *true* or *false*.

1. $[y + 2 = 5] \leftrightarrow [y = 3]$
2. $[x = 2] \leftrightarrow [x - 4 = -2]$
3. $[2y + 1 = 11] \leftrightarrow [2y = 12]$
4. $[a = b] \leftrightarrow [a^2 = b^2]$

Mathematical Excursions

The unit price for a product is found by dividing the selling price by the amount purchased. For example, a box of cereal contains 20 ounces. It sells for $2.99.

$$\text{unit price} = \frac{\text{selling price}}{\text{amount purchased}}$$

$$= \frac{299}{20} \quad \textit{Change \$2.99 to cents.}$$

$$= 14.95$$

The unit price is 14.95¢ per ounce.

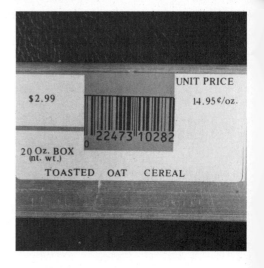

People can use unit prices to compare costs of different sizes of the same product. The size with the lower unit price is the better buy.

Example An 18-ounce jar of grape jelly costs $1.08. A 2-pound jar costs $1.40. Which is the better buy?

18-ounce jar

$$\text{unit price} = \frac{108}{18}$$
$$= 6$$

The unit price is 6¢ per ounce.

2-pound jar

$$\text{unit price} = \frac{140}{32}$$
$$= 4.375$$

Change 2 pounds to ounces. There are 16 ounces in a pound.

The unit price is about 4.4¢ per ounce.

The 2-pound jar is the better buy.

Exercises Use unit prices to determine which items are better buys.

1. a 12-ounce can of frozen orange juice that costs $1.69 or a 16-ounce can that costs $2.29

2. a 15-ounce box of oat cereal that costs $2.19 or a 20-ounce box that costs $2.83

3. an 18-ounce container of oatmeal that costs $1.09 or a 42-ounce container that costs $2.57

4. a 3-pint jar of cranberry juice that costs $2.59 or a 1-gallon jar that costs $6.29

5. an 18-ounce jar of peanut butter that costs $2.57 or a 2.5-pound jar that costs $5.69

6. a 12-ounce container of yogurt that costs $0.69 or a 2-pound container that costs $1.99

3.4 Solving Equations Using Multiplication

Study the three scales below. Consider a scale in balance, as shown on the left. If the weight on each side is doubled (multiplied by 2), as shown in the center, the scale will remain in balance. If the original weight on each side is halved (multiplied by $\frac{1}{2}$), as shown on the right, the scale will remain in balance.

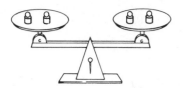 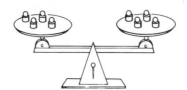 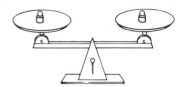

Think of an equation as a scale in balance. If you multiply both sides of an equation by the same number, then an equivalent equation results.

What happens if you multiply both sides of an equation by 0?

$$3n = 27$$
$$3n \cdot 0 = 27 \cdot 0$$

Notice that any number is a solution to the latter equation. Therefore, multiplying both sides of an equation by 0 does *not* produce an equivalent equation. However, a true equation, $0 = 0$, results.

Multiplication Property for Equations

> If each side of an equation is multiplied by the same *nonzero* number, the new equation is equivalent to the original.

You can use this property to solve equations involving multiplication and division.

Example

1 Solve $4x = 36$.

$$4x = 36$$
$$\frac{1}{4} \cdot 4x = \frac{1}{4} \cdot 36 \qquad \textit{Multiply both sides by } \frac{1}{4}.$$
$$1 \cdot x = \frac{36}{4}$$
$$x = 9$$

Lab Activity

You can learn how to use counters and cups to solve equations involving multiplication in Lab 2 on pages A4-A5.

Check: $4x = 36$

$4 \cdot 9 \overset{?}{=} 36$ *Replace x by 9.*

$36 = 36$ The solution is 9.

Why were both sides of the equation in example 1 multiplied by $\frac{1}{4}$ instead of some other number? The multiplicative inverse of 4 is $\frac{1}{4}$. Multiplying both sides of the equation by $\frac{1}{4}$ left just x on the left side of the equation.

Examples

2 Solve $\frac{y}{5} = -3$.

$$\frac{1}{5} \cdot y = -3 \qquad \frac{y}{5} \text{ is the same as } \frac{1}{5} y.$$

$$5 \cdot \frac{1}{5} \cdot y = 5 \cdot -3 \qquad \text{Multiply both sides by 5, the multiplicative inverse of } \frac{1}{5}.$$

$$1 \cdot y = -15$$

$$y = -15$$

Check: $-\dfrac{15}{5} \overset{?}{=} -3$ *Replace y by -15 in $\frac{y}{5} = -3$.*

$-3 = -3$ The solution is -15.

3 Solve $-\dfrac{2x}{3} = \dfrac{1}{2}$.

$$-\frac{2x}{3} = \frac{1}{2}$$

$$\left(-\frac{3}{2}\right)\left(-\frac{2x}{3}\right) = \left(-\frac{3}{2}\right)\left(\frac{1}{2}\right) \qquad \begin{array}{l}\text{Multiply both sides by the} \\ \text{multiplicative inverse of } -\frac{2}{3}.\end{array}$$

$$x = -\frac{3}{4}$$

Check: $\left(-\dfrac{2}{3}\right)\left(-\dfrac{3}{4}\right) \overset{?}{=} \dfrac{1}{2}$

$$\frac{1}{2} = \frac{1}{2} \qquad \text{The solution is } -\frac{3}{4}.$$

Recall that multiplying a number by $-\dfrac{1}{5}$ is the same as dividing that number by -5. Therefore, the equation $-5n = 80$ can be solved as follows.

$$-5n = 80$$

$$\dfrac{-5n}{-5} = \dfrac{80}{-5} \qquad \textit{Divide both sides by } -5.$$

$$n = -16$$

Exercises

Exploratory State the number you multiply both sides of each equation by to solve it.

1. $4n = 8$
2. $6k = 18$
3. $2r = -2$
4. $-y = 7$

5. $\dfrac{m}{3} = 4$
6. $\dfrac{1}{2}k = -5$
7. $\dfrac{p}{-9} = -12$
8. $-11 = \dfrac{m}{10}$

Written Solve each equation. Check your answer to see if it is the solution.

1. $4y = 12$
2. $8m = 80$
3. $3x = 48$
4. $3y = 3$

5. $-30 = 2q$
6. $-86 = 2t$
7. $5r = 0$
8. $11x = -77$

9. $4s = -52$
10. $-84 = 6b$
11. $-8y = 96$
12. $-15u = 45$

13. $126 = -9b$
14. $-12t = 132$
15. $3t = -4$
16. $5r = -4$

17. $4w = \dfrac{1}{2}$
18. $9y = -\dfrac{3}{4}$
19. $16m = 4$
20. $18v = -9$

21. $-9v = -3$
22. $-7x = -1$
23. $-t = 7$
24. $-s = -9$

25. $\dfrac{x}{3} = 8$
26. $\dfrac{d}{5} = 4$
27. $\dfrac{t}{2} = -4$
28. $\dfrac{m}{6} = -7$

29. $\dfrac{1}{8}w = -9$
30. $-10 = \dfrac{1}{4}v$
31. $\dfrac{x}{5} = \dfrac{1}{4}$
32. $\dfrac{3}{4}y = 9$

33. $-6 = \dfrac{3b}{5}$
34. $-\dfrac{4y}{5} = 8$
35. $\dfrac{2}{3}y = \dfrac{4}{9}$
36. $-\dfrac{3}{2}x = -2\dfrac{1}{2}$

37. $1.5x = 0.45$
38. $-4b = 0.8$
39. $-0.3y = -12$
40. $-0.92 = 0.04y$

Challenge Solve each of the following.

1. If $4y = -32$, find the value of $7y$.

2. If $\dfrac{x}{4} = -9$, find the value of $2x$.

3.5   Solving Problems

David Anderson bought a radio on sale and saved $8. The radio regularly sold for $35. What was the sale price of the radio?

Use the four-step problem-solving plan to solve this problem.

Explore   Let x = the sale price of the radio.   *Define a variable.*

Plan   <u>sale price</u> <u>plus</u> <u>discount</u> <u>is equal to</u> <u>regular price</u>   *Write an equation.*

        x   +   8   =   35

Solve
$$x + 8 = 35$$
$$x + 8 - 8 = 35 - 8$$
$$x + 0 = 27$$
$$x = 27$$ The sale price of the radio was $27.

Solve the equation.

Examine   Since $27 is $8 less than $35, the answer is correct.

Example

1 **Ladene weighs 128 pounds. This is 8 times the weight of her saddle. What does her saddle weigh?**

Explore   Let w = the weight of the saddle.   *Define a variable.*

Plan   <u>Ladene's weight</u> <u>is</u> <u>8 times</u> <u>saddle's weight,</u>   *Write an equation.*

       128   = 8 ×   w

Solve
$$128 = 8w$$
$$\frac{128}{8} = \frac{8w}{8}$$
$$16 = w$$ The saddle weighs 16 pounds.

Solve the equation.

Examine   Since 8 times 16 pounds is 128 pounds, the answer is correct.

Sometimes a drawing is helpful in solving a problem.

Example

2 **The perimeter of a square garden is 30 m. Find the length of each side.**

Explore   Let x = the length of each side.   *Define a variable.*

Make a drawing.

Plan $x + x + x + x = 30$ *Write an equation.*

Solve $4x = 30$ *Solve the equation.*

$$\frac{4x}{4} = \frac{30}{4}$$

$x = 7.5$ The length of each side is 7.5 m.

Examine   Since the perimeter of this square is $7.5 + 7.5 + 7.5 + 7.5$ or 30 m, the answer is correct.

Exercises

Exploratory   **For each problem, define a variable. Then state an equation. Do *not* solve the equation.**

1. The sum of 53 and some number is 96. Find the number.

2. A number decreased by 20 equals 47. Find the number.

3. The product of a number and 8 is -112. Find the number.

4. When a number is divided by 6, the result is 90. Find the number.

5. Chris traveled 300 kilometers altogether. He traveled 100 kilometers by plane and the rest by car. How far did he travel by car?

6. Mark's bank charges 8¢ for each check he writes. Last month he wrote 17 checks. How much did the bank charge him?

7. There are 96 students signed up for soccer. How many teams of 8 players each can be formed?

8. Peggy earns $4.50 per hour. How many hours will she need to work to earn $90?

9. One dozen eggs costs $0.96. What is the cost per egg?

10. In 9 years, Marie will be 23 years old. How old is she now?

11. A book contains 329 pages. Art has 161 pages left to read. How many pages has he read?

12. Mr. Oney bought a case of 24 colas. After the party there were 17 colas left. How many colas were consumed?

Written Solve the equation and answer the problem in each of the following Exploratory Exercises.

1. Exercise 1   **2.** Exercise 2   **3.** Exercise 3
4. Exercise 4   **5.** Exercise 5   **6.** Exercise 6
7. Exercise 7   **8.** Exercise 8   **9.** Exercise 9
10. Exercise 10   **11.** Exercise 11   **12.** Exercise 12

For each problem, define a variable. Then write and solve an equation and answer the problem.

13. At Truex Junior High School, there are 325 students taking Spanish. The number of students taking Spanish is 153 more than the number taking French. How many students are taking French?

14. Lynn wants to buy a car stereo that sells for $98. She has already saved $43. How much more must she save to buy the car stereo?

15. Dave decides to save $3.75 a week. How many weeks will it take him to save $120?

16. Francine saves $\frac{1}{4}$ of her allowance. If she saves $0.85, how much is her allowance?

17. Rosita ran around the track in 58.2 seconds. This is 2 seconds slower than the school record. What is the school record?

18. The Mustangs scored 3 runs in the sixth inning. They then led 8 to 7. What was the score before that inning?

19. The length of a rug is 3 times its width. It is 24m long. How wide is the rug?

20. Terry bought 4 tires for his car. The total cost was $196. What was the cost of each tire?

21. A jockey weighs 110 pounds. In the handicap race her mount can carry 134 pounds. How much weight is allowed for the saddle?

22. A motorboat rents for 7 times as much per hour as a canoe. A canoe rents for $3.50 per hour. What is the hourly rent for a motorboat?

23. Eagle's Bluff is about 143 meters above sea level. The distance from the peak of the bluff to the floor of Sulpher Springs Canyon is 217 meters. How far below sea level is Sulpher Springs Canyon?

24. Four cave explorers descended to a depth of 112 meters below the cave entrance. They disscovered a cavern whose ceiling was 27 meters above them. At what depth below the cave entrance was the cavern ceiling?

25. The two candidates for class president received a total of 116 votes. The winner received 3 times as many votes as the runner-up. How many votes did each candidate receive?

26. Robert scored two-thirds as many points as Jamie during the basketball season. Robert and Jamie scored a total of 400 points. How many points did each boy score?

27. The length of a rectangle is twice its width. The perimeter of the rectangle is 246 cm. What are the length and width of the rectangle?

28. The length of a rectangle is half its width. The perimeter of the rectangle is 78 m. What are the length and width of the rectangle?

3.6 Equations with More Than One Operation

Many equations contain more than one operation. Recall the order of operations used to evaluate a mathematical expression. To solve an equation with more than one operation, undo the operations in reverse order. Study the following examples.

Examples

1 Solve $2x - 1 = 7$.

$$2x - 1 = 7$$
$$2x - 1 + 1 = 7 + 1 \qquad \text{Add 1 to both sides.}$$
$$2x = 8$$
$$\frac{2x}{2} = \frac{8}{2} \qquad \text{Divide both sides by 2.}$$
$$x = 4$$

Check:
$$2x - 1 = 7$$
$$2(4) - 1 \overset{?}{=} 7$$
$$8 - 1 \overset{?}{=} 7$$
$$7 = 7 \qquad \text{The solution is 4.}$$

Lab Activity

You can learn how to use counters and cups to solve equations with more than one operation in Lab 3 on page A6.

2 Solve $\dfrac{y}{5} + 8 = 7$.

$$\frac{y}{5} + 8 = 7$$
$$\frac{y}{5} + 8 - 8 = 7 - 8 \qquad \text{Subtract 8 from both sides.}$$
$$\frac{y}{5} = -1$$
$$5 \cdot \frac{y}{5} = 5 \cdot -1 \qquad \text{Multiply both sides by 5.}$$
$$y = -5$$

Check:
$$\frac{y}{5} + 8 = 7$$
$$\frac{-5}{5} + 8 \overset{?}{=} 7$$
$$-1 + 8 \overset{?}{=} 7$$
$$7 = 7 \qquad \text{The solution is } -5.$$

Examples

3 Solve $4 - 2y = -8$.

$$4 - 2y = -8$$
$$-4 + 4 - 2y = -4 - 8 \qquad \text{Add } -4 \text{ to both sides.}$$
$$-2y = -12$$
$$\frac{-2y}{-2} = \frac{-12}{-2} \qquad \text{Divide both sides by } -2.$$
$$y = 6$$

Check: $\quad 4 - 2(6) \overset{?}{=} -8$
$$4 - 12 \overset{?}{=} -8$$
$$-8 = -8 \qquad \text{The solution is 6.}$$

4 Solve $\dfrac{x - 5}{3} = 5$.

$$\frac{x - 5}{3} = 5$$
$$3 \cdot \frac{(x - 5)}{3} = 3 \cdot 5 \qquad \text{Multiply both sides by 3.}$$
$$x - 5 = 15$$
$$x = 20 \qquad \text{Add 5 to both sides.}$$

Check: $\quad \dfrac{20 - 5}{3} \overset{?}{=} 5$

$$\frac{15}{3} \overset{?}{=} 5$$
$$5 = 5 \qquad \text{The solution is 20.}$$

Exercises

Exploratory　State the steps you would use to solve each equation.

1. $4x + 3 = 19$ 　　　**2.** $5 + 2y = 7$ 　　　**3.** $2x - 8 = 12$

4. $6x - 5 = 19$ 　　　**5.** $-3t - 5 = 10$ 　　　**6.** $-4x + 4 = -12$

7. $\dfrac{x}{3} + 5 = 8$ 　　　**8.** $6 + \dfrac{n}{2} = -5$ 　　　**9.** $\dfrac{n}{4} - 5 = 7$

10. $\dfrac{b + 2}{5} = 4$ 　　　**11.** $\dfrac{a - 3}{4} = 5$ 　　　**12.** $\dfrac{2m + 5}{7} = 9$

Written   Solve each equation. Check your answer to see if it is the solution.

1. $2y - 1 = 9$
2. $3m - 2 = 16$
3. $3x + 5 = 38$
4. $2y + 1 = -31$
5. $4 + 8t = 20$
6. $2m + 5 = -29$
7. $19 = 4a + 3$
8. $2b - 8 = 12$
9. $-32 = -2 + 5y$
10. $-11 = 3d - 2$
11. $4t - 8 = 0$
12. $0 = 11y + 33$
13. $8 = -24 + 4r$
14. $-8 + 6m = -50$
15. $13 = -7 + 2y$
16. $\dfrac{x}{4} + 6 = 10$
17. $\dfrac{y}{3} + 4 = 7$
18. $0 = -6 + \dfrac{3}{7}m$
19. $5 + \dfrac{t}{2} = 3$
20. $6 + \dfrac{s}{5} = 0$
21. $2 + \dfrac{x}{5} = -7$
22. $\dfrac{1}{3}y + 6 = -9$
23. $\dfrac{1}{4}t - 2 = 1$
24. $\dfrac{2}{3}x - 1 = 3$
25. $\dfrac{3}{2}x - 12 = 18$
26. $\dfrac{2}{3}y + 11 = 33$
27. $-6 = 4 + \dfrac{5}{2}y$
28. $3y + 4 = -11$
29. $11 - y = 10$
30. $6 - m = -4$
31. $-y + 12 = 4$
32. $8 = -2 - m$
33. $2.8 - y = 6.9$
34. $\dfrac{1}{2} - y = \dfrac{5}{2}$
35. $0 = -a + \dfrac{3}{4}$
36. $-b + \dfrac{1}{4} = \dfrac{1}{2}$
37. $1 - 3x = 7$
38. $10 - 2x = -2$
39. $-3 = -5a + 7$
40. $\dfrac{y - 3}{4} = 5$
41. $\dfrac{x + 2}{5} = 4$
42. $\dfrac{6 + b}{3} = -5$
43. $\dfrac{d + 5}{2} = -4$
44. $8 = \dfrac{x - 5}{2}$
45. $\dfrac{x - 3}{-4} = -5$
46. $\dfrac{y - 6}{-3} = 4$
47. $\dfrac{2y + 7}{9} = -7$
48. $\dfrac{-10 + 5a}{7} = 5$

Find the length of each side in each of the following figures. The perimeter of each figure is given.

49.

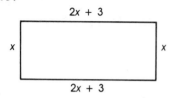

2x + 3

x   x

2x + 3

Perimeter = 126 m

50.

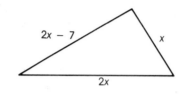

2x − 7   x

2x

Perimeter = 48 ft

51.

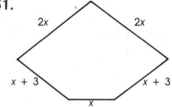

2x   2x

x + 3   x + 3

x

Perimeter = 41 cm

Challenge   Solve each equation. Check your answer to see if it is the solution.

1. $2b - 8b + 29 = 5$
2. $5 - 6d + 2d = -17$
3. $5(x + 3) + 4 = 14$

3.7   More Problem Solving

Rachel has 6 more than twice as many newspaper customers than when she started. She now has 98 customers. How many did she have when she started?

To find how many customers she had when she started, an equation can be written and solved.

Explore   Let c = the original number of customers.   *Define a variable.*

Plan

present number of customers	is	6	more than	twice	original number of customers
98	= 6		+	2 ·	c

Write an equation.

Solve
$$98 = 6 + 2c \qquad \text{Solve the equation.}$$
$$92 = 2c \qquad \text{Subtract 6 from both sides.}$$
$$46 = c \qquad \text{Divide both sides by 2.}$$

Rachel had 46 customers when she started delivering newspapers.

Examine   Since six more than twice 46 is 6 + 92 or 98, the answer is correct.

Example

1   Tom can have 500 calories for lunch. A hamburger without a roll has 320 calories and an average french fry has 15 calories. If he eats the hamburger, how many french fries can he eat and have a total of 500 calories?

Explore   Let n = the number of french fries he can eat.   *Define a variable.*

Plan

Hamburger	plus	15 times number of french fries	is	500.
320	+	15 · n	=	500

Write an equation.

Solve
$$320 + 15n = 500 \qquad \text{Solve the equation.}$$
$$15n = 180 \qquad \text{Add } -320 \text{ to both sides.}$$
$$n = 12 \qquad \text{Divide both sides by 15.}$$

Tom can eat 12 french fries.

Examine   Since 320 calories more than 12 times 15 calories is 320 + 180 or 500 calories, the answer is correct.

Example

2 The length of a rectangle is 3 feet longer than twice its width. Find the length and width if the perimeter is 30 feet.

Explore Let w = the width.   *Define a variable.*
Then $2w + 3$ = the length.

Plan   *The perimeter*   is *30 feet.*   *Write an equation.*
$$w + w + 2w + 3 + 2w + 3 = 30$$

Solve $6w + 6 = 30$ *Solve the equation.*
$$6w = 24$$
$$w = 4$$

The width is 4 feet and the length is 2(4) + 3 or 11 feet.

Examine   Since the perimeter of this rectangle is 4 + 11 + 4 + 11 or 30 ft, the answer is correct.

Exercises

Exploratory   For each problem, define a variable. Then state an equation. Do *not* solve the equation.

1. When twice a number is increased by 3, the result is 29. Find the number.

2. Six less than twice a number is -14. Find the number.

3. The sum of 5 and three times a number is -13. Find the number.

4. If twice a number is decreased by 10, the result is 16. Find the number.

5. Jo wants to buy a 10-speed bike which costs $129. This is $24 more than 3 times what she saved last month. How much did she save last month?

6. One season, Pete Rose scored 3 fewer runs than twice the number of runs he batted in. He scored 103 runs that season. How many runs did he bat in?

7. The length of a certain rectangle is 5 cm more than 3 times its width, w. Write expressions to represent the length and the perimeter of this rectangle.

Written   Solve the equations and answer the problems in Exploratory Exercises 1-6.

1. Exercise 1   **2.** Exercise 2   **3.** Exercise 3
4. Exercise 4   **5.** Exercise 5   **6.** Exercise 6

For each problem, define a variable. Then state and solve an equation and answer the problem.

7. Marilyn bought a used bike for $8 more than half its original price. Marilyn paid $50 for the bike. What was the original price of the bike?

8. Benjamin's bowling handicap is 7 less than half his average. His handicap is 53. What is Benjamin's bowling average?

9. Eric opens a savings account with $50. Each week after that he deposits $8. In how many weeks will he have saved $450?

10. Thelma rents a car for one day. The charge is $18 plus $0.12 per mile. She wants to spend exactly $30. How many miles can she drive?

11. A soccer field is 75 yards shorter than 3 times its width. Its perimeter is 370 yards. Find its length and width.

12. The length of a rectangular garden is 40 m less than 2 times its width. Its perimeter is 220 m. Find its length and width.

13. Emilio earns a flat salary of $140 per week for selling records. He receives an extra 30¢ for each record he sells. How many records must he sell in one week to earn a total of $180.50?

14. An investment of $7,500 in solar heating equipment results in a yearly savings of $450 on heating bills. In addition, the owner receives a $2,100 tax credit from the government the first year. How many years will it take for the savings to equal the investment?

15. One side of a triangle is 3 cm more than twice as long as the shortest side of the triangle. The third side is 11 cm longer than the shortest side. The perimeter of the triangle is 102 cm. Find the length of each side.

16. One side of a triangle is 7 cm more than half as long as the longest side of the triangle. The third side is 6 cm shorter than the longest side. The perimeter of the triangle is 161 cm. Find the length of each side.

17. The winner of an election received 7 more votes than the runner-up. The runner-up received twice as many votes as the third candidate. How many votes did each candidate receive if 292 votes were cast?

18. Last season, Tony had 14 fewer than twice as many hits as Kal. Nick had half as many hits as Tony. Tony, Kal, and Nick combined for a total of 371 hits last season. How many hits did each of them have?

A hamburger contains 80 calories per ounce. An average french fry contains 15 calories. A hamburger roll contains 180 calories. Use this information to solve each problem.

19. Beth has a 4-oz hamburger without a roll. How many french fries can she eat to have a total of 860 calories?

20. Toshio has a 6-oz hamburger without a roll. How many french fries can he eat to have a total of 1,110 calories?

21. Ralph has a 4-oz hamburger with a roll. How many french fries can he eat to have a total of 860 calories?

22. Willa has 20 french fries and a hamburger with a roll. How many ounces must the hamburger be to have a total of 880 calories?

3.8   Consecutive Integers

Consecutive integers are integers in counting order. For example, 6, 7, and 8 are three consecutive integers as are -4, -3, and -2. Therefore, if x represents an integer, then x, $x + 1$, $x + 2$, . . . represent consecutive integers.

Example

1   **The sum of three consecutive integers is 219. Find the integers.**

Define a variable.   Let $x =$ the least integer.
Then $x + 1 =$ the next greater integer
and $x + 2 =$ the greatest integer.

Write an equation.   The sum of the three integers is 219.
$x + (x + 1) + (x + 2) = 219$

Solve the equation.

$$x + x + 1 + x + 2 = 219$$
$$3x + 3 = 219 \qquad \text{Combine like terms.}$$
$$3x + 3 - 3 = 219 - 3 \qquad \text{Subtract 3 from each}$$
$$3x = 216 \qquad \text{side.}$$
$$\frac{3x}{3} = \frac{216}{3} \qquad \text{Divide each side by 3.}$$
$$x = 72$$

Answer the problem.   Since $x + 1 = 73$ and $x + 2 = 74$, the integers are 72, 73, and 74.

Beginning with an even integer and counting by *two's* gives **consecutive even integers.** For example, -4, -2, 0, 2, 4 are consecutive even integers. Beginning with an odd integer and counting by *two's* gives **consecutive odd integers.** For example, -3, -1, 1, 3 are consecutive odd integers.

Example

2   **The sum of the three consecutive even integers is -12. Find the integers.**

Let $x =$ the least even integer.
Then $x + 2 =$ the next greater even integer
and $x + 4 =$ the greatest even integer.

$$x + (x + 2) + (x + 4) = -12$$
$$3x + 6 = -12 \qquad \text{Combine like terms.}$$
$$3x = -18 \qquad \text{Subtract 6 from both sides.}$$
$$x = -6 \qquad \text{Divide both sides by 3.}$$

Therefore, $x + 2 = -4$ and $x + 4 = -2$. The integers are -6, -4, and -2.

Exercises

Exploratory   **For each sentence, define a variable. Then state an equation. Do *not* solve the equation.**

1. The sum of two consecutive integers is 17.
2. The sum of two consecutive integers is -9.
3. The sum of two consecutive even integers is -34.
4. The sum of three consecutive even integers is 48.
5. The sum of three consecutive integers is 39.
6. The sum of four consecutive integers is -46.
7. The sum of two consecutive odd integers is 36.
8. The sum of three consecutive odd integers is -75.

Written   **Solve each equation and find the integers in Exploratory Exercises 1-8.**

1. Exercise 1
2. Exercise 2
3. Exercise 3
4. Exercise 4
5. Exercise 5
6. Exercise 6
7. Exercise 7
8. Exercise 8

For each problem, define a variable. Then write and solve an equation and answer each problem.

9. Find two consecutive integers whose sum is 43.
10. Find two consecutive integers whose sum is -47.
11. Find three consecutive integers whose sum is 72.
12. Find four consecutive integers whose sum is 26.
13. Find two consecutive even integers whose sum is 46.
14. Find three consecutive even integers whose sum is 102.
15. Find three consecutive even integers whose sum is -36.
16. Find three consecutive even integers whose sum is -96.
17. Find two consecutive odd integers whose sum is 40.
18. Find two consecutive odd integers whose sum is 196.
19. Find three consecutive odd integers whose sum is 129.
20. Find three consecutive odd integers whose sum is -81.

3.9 Variables on Both Sides of the Equation

Some equations have variables on both sides. To solve this type of equation, first eliminate the term containing the variable on one of the sides by using inverse operations.

Example

1 Solve $3x - 5 = 7x - 21$.

$$3x - 5 = 7x - 21$$
$$-3x + 3x - 5 = -3x + 7x - 21 \qquad \textit{Add } -3x \textit{ to both sides.}$$
$$-5 = 4x - 21 \qquad \textit{Combine like terms.}$$
$$16 = 4x \qquad \textit{Add 21 to both sides.}$$
$$4 = x \qquad \textit{Divide both sides by 4.}$$

The solution set of an equation with no solution is the empty set, or $\emptyset$.

Sometimes, an equation has *no* solution. Other equations may be true for *every* replacement of the variable. This type of equation is called an **identity.**

Examples

2 Solve $3x + 7 = 3(x + 2)$.

$$3x + 7 = 3(x + 2)$$
$$3x + 7 = 3x + 6 \qquad \textit{Use the distributive property.}$$
$$7 = 6 \qquad \textit{Add } -3x \textit{ to both sides.}$$

Since $7 \neq 6$, the equation $3x + 7 = 3(x + 2)$ has no solution.

3 Solve $2x + 9 = 2(x + 4) + 1$.

$$2x + 9 = 2(x + 4) + 1$$
$$2x + 9 = 2x + 8 + 1 \qquad \textit{Use the distributive property.}$$
$$2x + 9 = 2x + 9 \qquad \textit{Combine like terms.}$$
$$9 = 9 \qquad \textit{Subtract 2x from both sides.}$$

The equation $9 = 9$ is equivalent to the original equation. It is always true. Thus, every real number is a solution and $2x + 9 = 2(x + 4) + 1$ is an identity.

Many problems can be solved by using equations with variables on both sides.

Examples

4 **The greater of two numbers is three times the lesser. If the greater number is decreased by 12, the result is 6 more than the lesser. Find the numbers.**

Define a variable. Let x = the lesser number.
Then $3x$ = the greater number.

Write an equation. The greater decreased by 12 is 6 more than the lesser.

$$3x \quad - \quad 12 = 6 \quad + \quad x$$

Solve the equation.
$$3x - 12 = 6 + x$$
$$2x - 12 = 6 \qquad \text{Subtract } x \text{ from both sides.}$$
$$2x = 18 \qquad \text{Add 12 to both sides.}$$
$$x = 9 \qquad \text{Divide both sides by 2.}$$

Answer the problem. Since $3 \cdot 9$ is 27, the numbers are 9 and 27.

5 **The Doneys budget a certain amount of their weekly income of $350 to pay their rent. Half of the remaining income exceeds the amount budgeted for rent by $25. How much money do they budget for rent each week?**

Define a variable. Let x = amount budgeted for rent.
Then $350 - x$ = remaining income.

Write an equation. Half of the remaining exceeds amount for rent by 25.

$$\frac{1}{2}(350 - x) = x + 25$$

Solve the equation.
$$175 - \frac{1}{2}x = x + 25 \qquad \text{Use the distributive property.}$$
$$175 = \frac{3}{2}x + 25 \qquad \text{Add } \frac{1}{2}x \text{ to both sides.}$$
$$150 = \frac{3}{2}x \qquad \text{Subtract 25 from both sides.}$$
$$100 = x \qquad \text{Multiply both sides by } \frac{2}{3}.$$

Answer the problem. The amount budgeted for rent each week is $100.

▬▬▬Exercises▬▬▬

Exploratory State how the second equation is obtained from the first in each of the following.

1. $5x = 3x - 4$
$2x = -4$

2. $7y = 2y + 10$
$5y = 10$

3. $5x - 9 = 2x$
$-9 = -3x$

4. $9x = 20 - x$
$10x = 20$

5. $m = 8m - 84$
$-7m = -84$

6. $2b - 9 = 3b + 5$
$-14 = b$

Written Solve each equation.

1. $5x = 8 + 3x$

2. $4t = 10 - t$

3. $16y = 13y + 45$

4. $17a = -8 + 9a$

5. $24x = 27x - 6$

6. $x = 2x - 4.6$

7. $0.4x - 36 = 0.5x$

8. $8x + 9 = 7x + 6$

9. $y + 9 = 7y + 9$

10. $t + 28 = 13t - 8$

11. $8 + 6x = 2x$

12. $-2 + 10x = 8x - 1$

13. $1\frac{1}{2}x = \frac{1}{2}x - 5$

14. $2r + 20 = 55 - 3r$

15. $\frac{1}{4} - \frac{2}{3}y = \frac{3}{4} - \frac{1}{3}y$

16. $5x - 12 = 8x + 24$

17. $2y - 8 = 14 - 9y$

18. $6 - 10y = 15 - 37y$

19. $8x - 4 = 5x - x - 24$

20. $12x - 11 + 2x = -4 - 2x$

21. $10b - 3 = 6b - 5 + 2b$

22. $2(3d - 10) = 40$

23. $-3(y + 5) = 3$

24. $3(y - 5) = 2(2y + 1)$

25. $5(3b - 2) + 11 = 106$

26. $7y - (6y + 3) = 12$

27. $11x - (6x + 6) = 19$

28. $8x - 3 = 5(2x + 1)$

29. $2(3x + 2) + 8 = 6$

30. $3(y + 4) = y + 1 + 2y$

31. $25t - 3(8t - 4) = 16t - 4$

32. $6x - 2x + 15 = 4(x + 1)$

33. $0.4(y - 9) = 0.3(y + 4)$

34. $2x - 4(x + 2) = -2x - 8$

35. $-3(2 + 5d) = 1 - 5(3d - 1)$

36. $4(x + 5) - 5(2x - 1) = x - 24$

For each problem, define a variable. Then state and solve an equation and answer the problem.

37. One number is 8 more than another. Their sum is -12. Find the numbers.

38. The greater of two numbers is 6 times the lesser. The greater number is 25 more than the lesser. Find the numbers.

39. The greater of two numbers is 13 more than 2 times the lesser. Nine times the greater number is 5 more than 4 times the lesser. Find the numbers.

40. The greater of two numbers is 7 more than the lesser. Three times the greater number is 5 more than 4 times the lesser number. Find the numbers.

41. If 7 times a number is decreased by 33, the result is 15 less than 10 times the number. Find the number.

42. Find two consecutive odd integers such that 2 times the lesser is 19 less than 3 times the greater.

43. When 2 times a number is increased by 10, the result is 1 more than 3 times the number. Find the number.

44. Find three consecutive even integers such that 3 times the least is 4 more than twice the greatest.

45. The greater of two numbers is twice the lesser. If the greater is increased by 18, the result is 4 less than 4 times the lesser. Find the numbers.

46. Melvin bought a tape recorder on sale for 80% of its original price. The sale price was $28 less than the original price. Find the original price and the sale price.

47. Find two consecutive even integers such that 4 times the lesser is 8 more than 3 times the greater.

48. Find two consecutive even integers such that 4 times the lesser is 28 more than the greater.

49. If a number is doubled and then increased by 20, the result is 5 less than 3 times the number. Find the number.

50. Find three consecutive integers such that the sum of the first two integers is 13 less than the greatest integer.

51. The greater of two numbers is 1 less than 4 times the lesser. Three times the lesser number is 4 less than the greater. Find the numbers.

52. Matt is now 4 times as old as he was 6 years ago. How old is he now?

53. Alisa earns $4.25 per hour. Her total deductions are $22.00 each week. Her take-home pay is $3.75 per hour. How many hours does she work each week?

54. A coin bank contains 4 times more quarters than dimes. The total amount of money in the bank is $2.20. How many dimes and quarters are in the bank?

55. Jenelle bought a 10-speed bicycle on sale for 75% of its original price. The sale price was $41 less than the original price. Find the original price and the sale price.

56. The Drama Club sold 473 tickets for their last performance. Adult tickets were $2.25 and student tickets were $1.50. The total receipts were $949.50. How many adult tickets and student tickets were sold?

Mixed Review

Choose the best answer.

1. Which equation represents the statement "ten less than twice a number is 17?"
   a. $10x - 2 = 17$ **b.** $10 - 2x = 17$ **c.** $2x - 10 = 17$ **d.** $2x + 17 = 10$

2. Which of the following is an example of the Distributive Property?
   a. $2(7 + 6) = (7 + 6)2$ **b.** $2(7 + 6) = 2(7) + 2(6)$
   c. $2(7 + 6) = 2(6 + 7)$ **d.** $2(7 + 6) = (7 + 6) + (7 + 6)$

3. If $p \rightarrow q$ is false, then which statement must be false?
   a. $p \lor q$ **b.** $\sim p \leftrightarrow q$ **c.** $p \land \sim q$ **d.** $\sim p \lor q$

4. Which expression is the simplest form of $4b - 3(2 - 5b)$?
   a. $-11b - 6$ **b.** $19b - 6$ **c.** $-b - 6$ **d.** $9b - 6$

Portfolio Suggestion

Place your favorite word problem from this chapter in your portfolio with a note explaining why it is your favorite. Be sure to include your solution to the problem.

Performance Assessment

Describe the solution of each equation. Explain how you came to each conclusion.

a. $y = y$ **b.** $x + 2 = 4$ **c.** $x + 2 = x + 3$ **d.** $\dfrac{x}{x} = 1$

Chapter Summary

1. Verbal expressions may be translated into algebraic expressions using variables. (74)
2. The numerical factor in an algebraic term is its **coefficient.** (75)
3. An **equation** is a statement of equality between two mathematical expressions. (78)
4. **Addition Property for Equations:** If the same number is added to both sides of an equation, the new equation is equivalent to the original equation. (80)
5. **Multiplication Property for Equations:** If each side of an equation is multiplied by the same nonzero number, the new equation is equivalent to the original equation. (84)
6. The same methods used to solve equations involving a single operation can be used to solve equations involving more than one operation. (90)
7. **Consecutive integers** are integers in counting order, such as 3, 4, and 5. (96)
8. The same methods used to solve equations with the variables on one side can be used to solve equations with variables on both sides. (98)
9. An equation may have no solution. (98)
10. An equation which is true for every replacement of the variable is an **identity.** (98)

Chapter Review

3.1 For each expression, define a variable. Then write an algebraic expression.

1. a number decreased by 5   **2.** 14 years older than Kari
3. 18 divided by some number   **4.** product of 7 and a number

Simplify.

5. $4(5m)$ **6.** $3x - 4x$ **7.** $-(5y - 2)$
8. $-4x - 5 + 11x + 2$ **9.** $2(x - 5) + 3(2x - 6)$

3.2 For each problem, define a variable. Then write an equation. Do *not* solve the equation.

10. Five times a number is 45.   **11.** Three less than twice a number is 7.

3.3 and 3.4 Solve and check each equation.

12. $x + 18 = -15$ **13.** $-15 + y = -2$ **14.** $8y = -128$

15. $-\dfrac{3}{5} + x = \dfrac{1}{20}$ **16.** $\dfrac{x}{5} = 18$ **17.** $-\dfrac{3}{4}m = -54$

3.5 For each problem, define a variable. Then write and solve an equation and answer the problem.

18. The difference of two integers is 16. The greater integer is 7. What is the other integer?   **19.** John weighs 66 kilograms. This is 3 times Carlos' weight. What is Carlos' weight?

3.6 Solve and check each equation.

20. $3x + 1 = 10$ **21.** $7 - 3y = 20$ **22.** $8 = \dfrac{2y + 2}{5}$

3.7 and 3.8 For each problem, define a variable. Then write and solve an equation and answer the problem.

23. Mr. LeMaster bought a used boat for $10 more than half its original price. He paid $150 for the boat. What was its original price?   **24.** The length of a rectangular picture frame is 3 cm less than twice its width. The perimeter is 78 cm. Find its dimensions.

25. Find two consecutive integers whose sum is -53.   **26.** Find three consecutive integers whose sum is 105.

3.9 Solve and check each equation.

27. $3x + 5 = -2x + 10$ **28.** $6 - x = -3x + 10$

29. $\dfrac{1}{2}(x + 4) = x - 2$ **30.** $-3(2 - x) = 2(x - 4)$

Chapter Test

For each expression, define a variable. Then write an algebraic expression.

1. 9° colder than outside   **2.** four years older than Jack

3. three times a number   **4.** a number divided by 3

Simplify.

5. $\frac{1}{2}(4ab)$

6. $-x + 7 + 8x$

7. $2(x + 3) - 3x + 2$

8. $3y + 1 - y + 5$

Solve and check each equation.

9. $y - 5 = 11$

10. $12y = -6$

11. $2y - 9 = -1$

12. $18 = 3 - 5y$

13. $-\frac{2}{5}y = 15$

14. $9 = \frac{3y - 6}{2}$

15. $4x - 3 = 7x + 18$

16. $-2(3x - 1) - (x + 5) = 11$

17. $\frac{1}{5}x - 3\frac{2}{3} = 1\frac{1}{3} - \frac{3}{5}x$

18. $-3(2y + 1) = 2 - 5(12 - y)$

For each problem, define a variable. Then write and solve an equation and answer the problem.

19. Daryl and Jennifer Hall married 25 years ago. She is now 48. How old was Jennifer when she married?

20. The perimeter of a square patio is 76 m. Find the length of each side.

21. Mrs. Diller bought a microwave oven for $60 more than half its original price. She paid $274 for the oven. What was the original price?

22. Find two consecutive integers whose sum is 37.

23. One number is 4 more than another number. Three times the greater number equals 4 times the lesser number. Find the numbers.

24. Find three consecutive even integers such that 3 times the middle integer is 10 more than the greatest.

25. One side of a triangular garden is 4 m longer than the shortest side of the garden. The longest side is 5 m longer than this side. If the perimeter of the garden is 31 m, find the length of each side of the garden.

Using Formulas and Inequalities

Application in Medicine

Nurses are dedicated to helping others. Their work requires a caring attitude, as well as an understanding of how to use **formulas**.

The following formula gives the flow rate of an intravenous fluid (I.V.) given to a patient.

$$D = \frac{V}{t} \cdot f$$

In this formula, D is the flow rate in drops per minute, V is the volume of the fluid ordered by the doctor in milliliters, t is time during which the infusion takes place in minutes, and f is the drip factor (drops per minute). Find the flow rate, if the doctor orders 1000 milliliters over 24 hours at 20 drops per minute.

Individual Project: *Science*

Make a list of five formulas found in your science book. Explain what each formula means. Make a model or diagram to illustrate one of the formulas.

4.1 Formulas

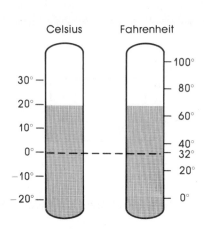

Celsius   Fahrenheit

Some weather reports give the temperature in degrees Fahrenheit as well as degrees Celsius. The following **formula** relates the two temperature scales shown at the left. The Fahrenheit temperature is represented by F and the Celsius temperature by C.

$$F = \frac{9}{5}C + 32$$

Suppose you know the temperature on one scale. Then you can use the formula to find the corresponding temperature on the other scale.

A formula is an equation that states a rule that relates certain quantities. Use the rules of algebra with this formula and other formulas.

Examples

1 **Find the temperature on the Fahrenheit scale that corresponds to 20° Celsius.**

$$F = \frac{9}{5}C + 32$$

$$F = \frac{9}{5} \cdot 20 + 32 \qquad \textit{Replace C by 20.}$$

$$= 36 + 32$$

$$= 68$$

Thus, 68°F corresponds to 20°C.

2 **Find the temperature that corresponds to 50° Fahrenheit.**

$$F = \frac{9}{5}C + 32$$

$$50 = \frac{9}{5}C + 32 \qquad \textit{Replace F by 50.}$$

$$18 = \frac{9}{5}C \qquad \textit{Add } -32 \textit{ to both sides.}$$

$$10 = C \qquad \textit{Multiply both sides by } \frac{5}{9}.$$

Thus, 10°C corresponds to 50°F.

Examples

3 The formula for the perimeter of a rectangle is $P = 2\ell + 2w$. Find ℓ if $P = 68$ and $w = 14$.

$P = 2\ell + 2w$

$68 = 2\ell + 2 \cdot 14$ *Replace P by 68 and*
                           w by 14.

$40 = 2\ell$ *Add −28 to both sides.*

$20 = \ell$ *Divide both sides by 2.*

For this rectangle, ℓ is 20.

ℓ units

w units w units

ℓ units

4 **Suppose the measure of a side of a square is doubled. How does the measure of its perimeter change?**

The formula for the perimeter of a square is $P = 4s$, where s is the measure of a side. Let $x = $ the measure of a side of the original square.
Then, $2x = $ the measure of a side of the new square.

original square   *new square*
      $P = 4s$ $P = 4s$
      $P = 4x$ $P = 4(2x)$ or $8x$

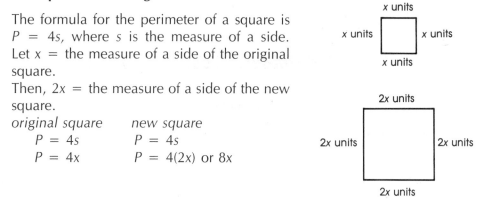

x units

x units   ☐ x units

x units

$2x$ units

$2x$ units $2x$ units

$2x$ units

Compare the two perimeters. Notice that $8x$ is two times $4x$. Thus, if the measure of a side of a square is doubled, the measure of its perimeter also is doubled.

5 **Suppose x is the cost for the first 3 minutes of a long distance call and y is the cost for each additional minute. Write a formula for the cost C of a long distance call lasting 10 minutes.**

Cost of call	is	cost for first 3 minutes	plus	number of additional minutes	times	cost for each additional minute
C	$=$	x	$+$	7	$\times$	y

$C = x + 7y$

Exercises

Exploratory Use the formula $F = \frac{9}{5}C + 32$ to find the temperature that corresponds to each of the following temperatures.

1. $25°C$ 2. $-15°C$ 3. $0°C$ 4. $34.6°C$ 5. $41°F$ 6. $-4°F$

Find how the measure of the perimeter of a square changes when the measure of a side is multiplied by each of the following numbers.

7. 3   8. 4   9. 8   10. 12   11. 0.5   12. 1.5

Find how the measure of the area of a square changes when the measure of a side s is multiplied by each of the following numbers. $(A = s^2)$

13. 2   14. 3   15. 5   16. 10   17. 0.5   18. 1.5

Written Solve each of the following.

1. Suppose $s = a + b + c$. Find s when $a = 25$, $b = 53$, and $c = 32$.

2. Suppose $A = \frac{1}{2}bh$. Find A when $b = 16$ and $h = 4$.

3. Suppose $A = s^2$. Find A when $s = 15$.

4. Suppose $s = \frac{1}{2}gt^2$. Find s when $g = 32$ and $t = 14$.

5. Suppose $q = 2a + b$. Find a when $q = 75$ and $b = 11$.

6. Suppose $V = \ell wh$. Find w when $V = 106$, $\ell = 3$, $h = 2$.

Suppose $S = \frac{n}{2}(f + \ell)$. Solve each of the following.

7. Find n when $f = 2$, $\ell = 18$, and $S = 70$.

8. Find f when $n = 7$, $\ell = \frac{9}{2}$, and $S = 17\frac{1}{2}$.

9. Find ℓ when $n = 21$, $f = -11$, and $S = -168$.

10. Find ℓ when $n = 18$, $f = 4$, and $S = -18$.

Solve each of the following.

11. Find the temperature on the Fahrenheit scale that corresponds to 100° Celsius. Use $F = \frac{9}{5}C + 32$.

12. Find the length of a rectangle with the perimeter of 23 cm and width of 5 cm. Use $P = 2\ell + 2w$.

13. Find the cost C of a 10-minute long distance call if the cost x for the first 3 minutes is $1.60 and each additional minute costs $0.30. Use $C = x + 7y$.

14. Find the cost x for the first 3 minutes of a 10-minute long distance call if each additional minute costs $0.36 and the total cost C of the call is $4.52. Use $C = x + 7y$.

The formula for the area of a rectangle is $A = \ell w$. Find how the area changes in each of the following.

15. The length is doubled and the width remains the same.

16. The length is doubled and the width is multiplied by 3.

17. The length is multiplied by 5 and the width is multiplied by 3.

18. The length is multiplied by 4 and the width is divided by 4.

Write a formula that can be used to find each of the following.

19. What is the amount A collected from the sale of t tickets at \$4.50 each?

20. How many calories C in n chocolate chip cookies if each cookie contains 51 calories?

21. What is the cost C of x 29-cent stamps?

22. What is the cost C of x 29-cent stamps and y 19-cent stamps?

23. How many calories C are in x potato chips and y pretzels if each potato chip has 20 calories and each pretzel has 17 calories?

24. The sum of two numbers is 30. If y is the greater number, find x, the lesser number.

25. Mary is 3 times as old as Howard. Larry is 6 years older than Mary. If Howard is h years old, how old is Larry?

26. Suppose x is the cost for the first minute of a long distance call and y is the cost of each additional minute. What is the cost C of a long distance call lasting 5 minutes?

27. Suppose the charge for an Acme moving van is x dollars plus y dollars for each pound over 1200 lb that is moved. What is the cost C of moving a load of 1700 lb?

28. At Zeb's Copy Center the rate for 1 to 100 copies of a page is x cents per page. The rate is y cents for each copy over 100. What is the cost C of 165 copies?

29. A dealer mixes x lb of cashews costing r cents per pound and y lb of peanuts costing s cents per pound. What is the cost C of the mixture?

30. Jim earns x dollars per hour for a 36-hour week. The overtime rate is y dollars per hour. What is Jim's pay P for a week that he worked 47 hours?

Challenge   Write a formula that can be used to find each of the following.

1. The length of a rectangle exceeds the width by 3 cm. Suppose the length is decreased by 2 cm and the width is doubled. What is the perimeter of the resulting rectangle?

2. Gwen received grades of x, y, z, and w in four French exams. What is her exam average?

3. If p pounds of coffee cost x cents, what is the cost C of m pounds of coffee?

4. If g gallons of oil cost d dollars, what is the cost C of s gallons of oil?

4.2   Solving Formulas For Specific Variables

Suppose you have to convert several temperatures from the Fahrenheit scale to the Celsius. It would be convenient to have a formula that expresses C in terms of F. To find such a formula, solve $F = \frac{9}{5}C + 32$ for C.

$$F = \frac{9}{5}C + 32$$

$$F - 32 = \frac{9}{5}C \qquad \text{\textit{Add }} -32 \text{ \textit{to both sides.}}$$

$$\frac{5}{9}(F - 32) = \frac{5}{9} \cdot \frac{9}{5}C \qquad \text{\textit{Multiply both sides by }} \frac{5}{9}.$$

$$\frac{5}{9}(F - 32) = C \qquad \text{\textit{You can use the formula }} C = \frac{5}{9}(F - 32)$$
$$\text{\textit{to find C when you know F.}}$$

Examples

1 **Find the temperature on the Celsius scale that corresponds to 50°F.**

$$\frac{5}{9}(F - 32) = C \qquad \text{\textit{Use the formula for C in terms of F.}}$$

$$\frac{5}{9}(50 - 32) = C \qquad \text{\textit{Replace F by 50.}}$$

$$\frac{5}{9} \cdot 18 = C$$

$$10 = C$$

The temperature on the Celsius scale that corresponds to $50°F$ is $10°C$.

2 **The formula $d = rt$ relates the distance d traveled by an object moving at a constant rate or speed r in an interval of time t. Solve for r in terms of the other variables.**

$$d = rt$$

$$\frac{d}{t} = \frac{rt}{t} \qquad \text{\textit{Divide both sides by t. Recall that division by 0}}$$
$$\text{\textit{is undefined. Therefore, t may not be 0.}}$$

$$\frac{d}{t} = r \qquad \text{This formula is undefined if } t = 0. \text{ Of course, when } t = 0, \text{ we have the trivial case because either } d \text{ or } r \text{ is also 0.}$$

Examples

3 Solve $a = s(1 - r)$ for r.

$$a = s(1 - r)$$
$$a = s - sr \qquad \text{Use the distributive property.}$$
$$a - s = -sr \qquad \text{Subtract s from both sides.}$$
$$-\frac{a - s}{s} = r \qquad \text{Divide both sides by } -s, \text{ where } s \neq 0.$$

4 Solve $c = 3x - xy$ for x.

$$c = 3x - xy$$
$$c = x(3 - y) \qquad \text{Use the distributive property.}$$
$$\frac{c}{3 - y} = x \qquad \text{Divide both sides by } 3 - y, \text{ where } 3 - y \neq 0.$$

Since $3 - y = 0$ when $y = 3$, we say that $x = \dfrac{c}{3 - y}$ is undefined if $y = 3$.

Exercises

Exploratory   Solve each of the following formulas for the indicated variable.

1. $A = \ell w$ for ℓ

2. $d = rt$ for r

3. $C = \pi d$ for d

4. $V = \ell wh$ for h

5. $P = a + b + c$ for a

6. $\dfrac{a}{p} = q$ for a

Find the value of x for which each of the following expressions is undefined.

7. $\dfrac{2}{x}$

8. $\dfrac{y}{x}$

9. $\dfrac{3}{x - 5}$

10. $\dfrac{y - 7}{x - 7}$

11. $\dfrac{x + 6}{x + 4}$

12. $\dfrac{14}{3x}$

13. $\dfrac{5 + x}{3 + 2x}$

14. $\dfrac{7}{10 - x}$

Written   Find the temperature on the Celsius scale that corresponds to each of the following temperatures.

1. $50°F$ **2.** $59°F$ **3.** $23°F$ **4.** $-13°F$ **5.** $212°F$ **6.** $98.6°F$

In each of the following, find the rate of an object traveling the indicated distance in the indicated time.

7. 60 m, 3 s

8. 400 mi, 8 h

9. 450 km, 5 h

10. 2200 ft, 2 s

11. 100 m, 60 s

12. 10,000 m, 40 min

Solve each of the following. Use $d = rt$ or a formula derived from that formula.

13. What is the distance a train travels in 4 hours at a constant rate of 60 mph?

14. How far can Laura travel in 3 hours if she drives at the constant rate of 80 km/h?

15. What is the speed of an airplane that travels 1960 km in 2 hours?

16. What is the speed of a bullet that travels 1232 feet in 0.88 second?

17. How long does it take Philip to walk 48 km if he walks at a constant rate of 9 km/h?

18. How long does it take Julie to swim 400 meters if she swims at a constant rate of 75 meters per minute?

Solve each of the following formulas for the indicated variable.

19. $i = 12f$ for f

20. $V = \ell wh$ for w

21. $I = prt$ for t

22. $I = prt$ for p

23. $x + 4c = 2b$ for x

24. $3a = 2c - x$ for x

25. $4c - x = a$ for x

26. $bx = 2cd$ for x

27. $w + \ell = 5 - 3\ell$ for w

28. $a + ar = 3$ for r

29. $a + ar = 3$ for a

30. $p - pq = 4$ for q

31. $2x + 3y = 7$ for y

32. $A = \frac{1}{2}bh$ for b

33. $T = \frac{x - m}{g}$ for g

34. $e = i(r + s)$ for r

35. $L = t(a - b)$ for a

36. $P = 2\ell + 2w$ for w

37. $M = \frac{xy - 5}{h}$ for h

38. $A = \frac{h}{2}(b + d)$ for b

39. $H = \frac{2}{3}(k - 5)$ for k

40. $I = \frac{E}{R + r}$ for r

41. $a(\ell - r) = 7$ for r

42. $L = \frac{a - t}{4t}$ for t

Look at the formula you derived in each of the following exercises. State the value of the indicated variable for which this formula is undefined.

43. Exercise 26, b

44. Exercise 28, a

45. Exercise 29, r

46. Exercise 30, p

47. Exercise 33, T

48. Exercise 37, M

49. Exercise 40, I

50. Exercise 41, a

51. Exercise 42, L

Challenge Solve each of the following.

1. If a runner runs a mile in 4 minutes, what is his speed in miles per hour?

2. Millie left Pine City traveling west. An hour later Carlos left Pine City traveling east at the same speed. In 3 hours they were 225 miles apart. What was their speed?

3. Light travels at the rate of about 186,000 mi/s. If the sun is 93,000,000 mi from the earth, about how long does it take the light from the sun to reach the earth in seconds? In minutes?

4. Sound travels through air at about the rate of 1100 ft/s. Assume that Lin saw lightning flash instantaneously. If he heard the thunder 12 s later, how many miles from Lin did lightning strike? (5280 ft = 1 mi)

4.3   Inequalities

You know that 4 is less than 7. This can be expressed by the **inequality** $4 < 7$. The inequality $7 > 4$, read *seven is greater than 4,* also shows this relationship. Some other inequalities are $3n \neq 18$ and $2x + 1 \geq 5$.

≠ means is not equal to.

| **Definition of an Inequality** | A mathematical sentence that contains a symbol such as $<, >, \leq, \geq$, or $\neq$ is called an inequality. |

Consider the open sentence $3x > 12$. Choose several replacements for x.

If $x = 1$, $3 \cdot 1 > 12$ or $3 > 12$.   *The sentence is false.*
If $x = 4$, $3 \cdot 4 > 12$ or $12 > 12$.   *The sentence is false.*
If $x = 4.1$, $3 \cdot 4.1 > 12$ or $12.3 > 12$.   *The sentence is true.*

Unless stated otherwise, we assume the domain is the set of numbers on the number line.

It appears that if x is replaced by any number greater than 4, the inequality is true. We say that all the numbers greater than 4 are solutions of $3x > 12$. The set of solutions is the solution set of $3x > 12$. This solution set can be expressed as {all numbers greater than 4} or $\{x: x > 4\}$.   *$\{x: x > 4\}$ is read the set of all x such that x is greater than 4.*

Examples

1 **Graph the solution set of $3x > 12$.**

The solution set is {all numbers greater than 4}.

The graph of 4 is circled but __not__ darkened. This shows that 4 is __not__ included in the solution set. The heavy arrow shows that all the numbers to the right of 4 are included.

2 **Graph the solution set of $y \leq -2\frac{1}{2}$.**

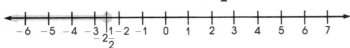

The graph of $-2\frac{1}{2}$ is darkened to show that $-2\frac{1}{2}$ is included in the solution set.

Example

3 When 4 times a number is increased by 2, the result is at least 22. Write an inequality that describes this situation.

The result is at least 22 means the result is *greater than or equal to 22*.

Define a variable.   Let n = the number.

Write an inequality.

4 times	a number	increased by	2	is at least	22.
4 ×	n	+	2	≥	22

$$4n + 2 \geq 22$$

The symbols of inequality may be used to represent different types of verbal sentences.

Verbal Sentence	**Inequality**
x is less than 14.	$x < 14$
x is less than or equal to 14. x is at most 14. x is not greater than 14. The greatest value of x is 14. The maximum value of x is 14.	$x \leq 14$
x is greater than 14.	$x > 14$
x is greater than or equal to 14. x is at least 14. x is not less than 14. The least value of x is 14. The minimum value of x is 14.	$x \geq 14$

Exercises

Exploratory   State whether each of the following sentences is *true* or *false*.

1. $-2 < 2$ **2.** $0 > 3$ **3.** $5 \neq 3$ **4.** $-6 \neq -5$

5. $-6 < -8$ **6.** $-4 > -8$ **7.** $3 \leq 3$ **8.** $-5 \geq -4$

9. $-\frac{2}{4} \leq -\frac{1}{2}$ **10.** $\frac{3}{4} \neq \frac{9}{12}$ **11.** $1\frac{1}{4} \leq 1.2$ **12.** $\frac{1}{3} > \frac{1}{2}$

13. 3 is a solution of $x < 6$.   **14.** 5 is a solution of $2x > 11$.

15. 3 is a solution of $x - 1 < 6$.   **16.** 5 is a solution of $2x + 1 \geq 11$.

Written   Write an inequality for each of the following graphs.

1.

2.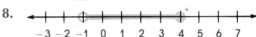

3.
```
   0  1  2  3  4  5  6  7  8  9  10
```

4.
```
  -9 -8 -7 -6 -5 -4 -3 -2 -1  0  1
```

5.
```
  -3 -2 -1  0  1  2  3  4  5  6  7
```

6.
```
  -6 -5 -4 -3 -2 -1  0  1  2  3  4
      -4.25
```

7.
```
  -5 -4 -3 -2 -1  0  1  2  3  4  5
           -1.5
```

8.
```
  -3 -2 -1  0  1  2  3  4  5  6  7
```

Graph the solution set of each of the following inequalities.

9. $y > 2$

10. $x > -3$

11. $m \le -1$

12. $y < 6$

13. $x \ge -5$

14. $t \ge 0$

15. $x \ne 4$

16. $y \ne -2$

17. $x < \dfrac{1}{2}$

18. $y \ge \dfrac{5}{3}$

19. $x \le -2.7$

20. $y \ne -1.5$

For each of the following, define a variable and write an inequality.

21. Two more than a number is greater than 8. Find the numbers.

22. Four more than a number is less than or equal to -6. Find the numbers.

23. Three more than twice a number is greater than 21. Find the numbers.

24. Find two consecutive positive odd integers whose sum is at most 20.

25. Find two consecutive positive even integers whose sum is at least 26.

26. Suppose an empty book crate weighs 15 lb and each book weighs 1.25 lb. What is the greatest number of books that can be packed in the crate if the maximum weight must be 40 lb?

For each inequality, if the domain is $\{-5, -4, -3, -2, -1, 0, 1, 2, 3\}$ find the solution set.

27. $y < 3$

28. $x \ge 0$

29. $x \le -3$

30. $y > 2$

31. $y + 1 > 2$

32. $x - 3 < -4$

33. $3m \ge -9$

34. $4y < -16$

35. $3x - 1 \ge 2$

36. $2 + 5y > -3$

37. $\dfrac{1}{2}x < 2$

38. $\dfrac{2}{3}y > -1$

39. $-y > -2$

40. $-x \le -3$

41. $2y > 6$

Challenge   Graph the solution set of each inequality.

1. $-x > -3$

2. $-y < 5$

3. $-2x < -6$

4. $10 - x \ge -15$

5. $12 < -y + 18$

6. $|y| > 3$

7. $|x| \le 2$

8. $|x| \ge 4$

9. $|y - 3| > 1$

4.4 Solving Inequalities Using Addition

John had $104 in his savings account and Debbie had $95 in her account. They received $25 each for selling newspapers and deposited it in their accounts. Whose account has more money?

John	Debbie

$$\$104 > \$95$$
$$\$104 + \$25 > \$95 + \$25$$
$$\$129 > \$120$$

John had more money before the deposit, and he had more money afterward. Notice that adding the same amount to both sides of the inequality does not change the truth of the inequality.

Subtracting 25 is the same as adding −25.

Suppose John and Debbie each withdrew $25 from their accounts. Subtracting the same amount from both sides of the inequality does *not* change the truth of the inequality.

These examples suggest the following rule which may be used in solving inequalities.

Addition Property for Inequalities

Suppose a, b, and c are any numbers.

If $a < b$, then $a + c < b + c$.

If $a > b$, then $a + c > b + c$.

Example

1 Solve $x - 7 > 1$. Then graph the solution set.

$$x - 7 > 1$$
$$x - 7 + 7 > 1 + 7 \qquad \textit{Add 7 to both sides.}$$
$$x > 8$$

The solution set is {all numbers greater than 8}.
Recall that this may be written $\{x:x > 8\}$.

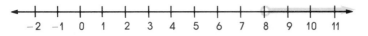

Example

2 **Solve $9n + 7 \leq 8n - 2$. Then graph the solution set.**

$$9n + 7 \leq 8n - 2$$
$$9n + (-8n) + 7 \leq 8n + (-8n) - 2 \qquad \text{Add } -8n \text{ to both sides.}$$
$$n + 7 \leq -2$$
$$n + 7 + (-7) \leq -2 + (-7) \qquad \text{Add } -7 \text{ to both sides.}$$
$$n \leq -9$$

The solution set is $\{n: n \leq -9\}$.

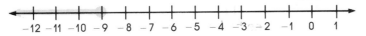

Exercises

Exploratory   State the number you add to both sides to solve each of the following inequalities.

1. $y + 4 < 5$ **2.** $y + 9 < 12$ **3.** $x - 4 > 1$ **4.** $t - 2 > 3$

5. $x - 3 < 0$ **6.** $y + 6 \geq 0$ **7.** $x - 2.3 > 1.4$ **8.** $y - \dfrac{2}{3} \leq \dfrac{1}{6}$

9. $4 + m > -4$ **10.** $-8 + n \leq -5$ **11.** $-2\dfrac{1}{2} + m \geq \dfrac{1}{2}$ **12.** $-8.6 + x < -1.1$

Written   Solve the inequality in each of the following Exploratory Exercises. Then graph the solution set in each exercise.

1. Exercise 1 **2.** Exercise 2 **3.** Exercise 3 **4.** Exercise 4
5. Exercise 5 **6.** Exercise 6 **7.** Exercise 7 **8.** Exercise 8
9. Exercise 9 **10.** Exercise 10 **11.** Exercise 11 **12.** Exercise 12

Solve each of the following inequalities. Then graph the solution set.

13. $2y - y > 4$ **14.** $x - 3 \neq 1$ **15.** $4x - 3x \geq -2$
16. $6x + 2 - 5x \leq -2$ **17.** $-2x + 6 + 3x \geq 6$ **18.** $4 < 2x + 3 - x$
19. $2(x - 5) - x > -6$ **20.** $6x + 4 > 5x$ **21.** $4 + 4x + 2 < -2 + 3x$
22. $7 < 8(x - 1) - 7x$ **23.** $9(2 + x) - 15 \geq 10x$ **24.** $2(3x - 1) - 5x \leq -2$

Challenge   Solve each of the following inequalities for x.

1. $x + c < b$ **2.** $c \geq x - f$
3. $7x - r \leq 6x + 3r$ **4.** $8(x + h) - 7(x - j) \geq 8h + 7j$

4.5 Solving Inequalities Using Multiplication

You know that the inequality $-5 < 3$ is true. What happens if you multiply both sides by 2?

$$-5 < 3$$
$$-5 \cdot 2 \overset{?}{<} 3 \cdot 2$$
$$-10 \overset{?}{<} 6 \qquad \textit{Is this inequality true?}$$

The inequality $-10 < 6$ is true. Try multiplying both sides of any inequality by other positive numbers. The result in each case will be the same—an equality that is true.

What happens if you multiply both sides of $-5 < 3$ by -2?

$$-5(-2) \overset{?}{<} 3(-2)$$
$$10 \overset{?}{<} -6 \qquad \textit{Is this inequality true?}$$

The inequality $10 < -6$ is false, but $10 > -6$ is true. Multiplying both sides of an inequality by the same negative number reverses the direction of the inequality. That is, $>$ replaces $<$ or $<$ replaces $>$.

The examples suggest the following rule which may be used to solve inequalities.

Multiplication Property for Inequalities

Suppose a and b are any numbers.

If $c > 0$ and $a < b$, then $ac < bc$.
If $c > 0$ and $a > b$, then $ac > bc$.
If $c < 0$ and $a < b$, then $ac > bc$.
If $c < 0$ and $a > b$, then $ac < bc$.

Example

1 Solve $5y < -15$. Then graph the solution set.

$$5y < -15$$
$$\frac{1}{5} \cdot 5y < \frac{1}{5}(-15) \qquad \textit{Multiply by } \frac{1}{5}.$$
$$y < -3 \qquad \text{The solution set is } \{y : y < -3\}.$$

More examples follow in which the multiplication property is used to solve inequalities. Sometimes it is necessary to use both multiplication and addition properties.

Examples

2 **Solve $-3x > 15$. Then graph the solution set.**

$$-3x > 15$$
$$-\frac{1}{3}(-3x) < -\frac{1}{3}(15) \qquad \textit{Multiply by } -\frac{1}{3}. \textit{ Reverse the direction of the inequality.}$$
$$x < -5 \qquad \textit{The solution set is } \{x : x < -5\}.$$

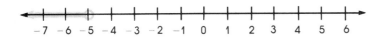

3 **Solve $2x + 1 > -3$. Then graph the solution set.**

$$2x + 1 > -3$$
$$2x + 1 + (-1) > -3 + (-1) \qquad \textit{Add } -1 \textit{ to both sides.}$$
$$2x > -4$$
$$\frac{1}{2}(2x) > \frac{1}{2}(-4) \qquad \textit{Multiply both sides by } \frac{1}{2}.$$
$$x > -2 \qquad \textit{The solution set is } \{x : x > -2\}.$$

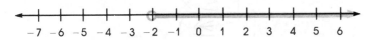

4 **Solve $10x - 13 \le 14x + 3$. Then graph the solution set.**

$$10x - 13 \le 14x + 3$$
$$-13 \le 4x + 3 \qquad \textit{Add } -10x \textit{ to both sides.}$$
$$-16 \le 4x \qquad \textit{Add } -3 \textit{ to both sides.}$$
$$-4 \le x \qquad \textit{Multiply both sides by } \frac{1}{4}.$$

It may be helpful to restate the inequality as $x \ge -4$.
The solution set is $\{x : x \ge -4\}$.

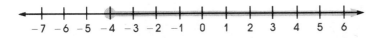

Exercises

Exploratory State the inequality you obtain when you multiply each side of the given inequality by the number written at the right of the inequality.

1. $-7 < -2$; 8

2. $-4 > -100$; -3

3. $6 < 9$; $\dfrac{1}{3}$

4. $-10 < 12$; $-\dfrac{1}{2}$

5. $10x < 30$; -2

6. $-5x < 15$; $-\dfrac{1}{5}$

7. $-\dfrac{1}{3}x > 9$; -3

8. $\dfrac{4}{5}x < 2$; $\dfrac{5}{4}$

9. $-\dfrac{3}{5}y > -\dfrac{1}{10}$; $-\dfrac{5}{3}$

Written Solve each of the following inequalities. Then graph the solution set.

1. $3x > -18$

2. $4s > -8$

3. $\dfrac{x}{2} > 5$

4. $\dfrac{1}{5}y \le -3$

5. $-3x > 21$

6. $-2y < -16$

7. $-5y \ge 100$

8. $24 > -8m$

9. $121 \le -11t$

10. $-\dfrac{3}{4}m \ge 18$

11. $-30 \ge -2t$

12. $-\dfrac{y}{6} \le -\dfrac{1}{2}$

13. $-8 \le t - 3t$

14. $-9 > t - 4t$

15. $2y - 1 \le 5$

16. $2m + 7 > 17$

17. $4x - 2 < 18$

18. $6r + 1 > -11$

19. $7x + 2x - 1 < 29$

20. $8x + 2 - 10x < 20$

21. $24 \le 4 - x - 3x$

22. $x - \dfrac{2}{3}x < -5$

23. $-6 < 2x - 3x + 1$

24. $6 - 3m < -9$

25. $5 - 4m + 8 + 2m > -17$

26. $-4 + y \ge -5 + 3y$

27. $-6y - 4 + y + 10 < -6 - y$

28. $2(r - 3) + 5 \ge 9$

29. $3(2a - 5) + 4(a + 6) \ge -6$

30. $3(3x + 2) > 7x - 2$

31. $2 + 5(2y + 1) < 3(y + 7)$

32. $1 + 2(x + 4) \ge 1 + 3(x + 2)$

33. $5(n + 1) + 2n \le 8n + 10$

34. $4(4z + 5) - 5 > 3(4z - 1)$

35. $9 + 11y > 5y + 21$

36. $2(3m + 4) - 2 \le 3(1 + 3m)$

37. $2(m - 5) - 3(2m - 5) < 5m + 1$

38. $3b - 2(b - 5) < 2(b + 4)$

39. $7 + 3y > 2(y + 3) - 2(-1 - y)$

40. $20\left(\dfrac{1}{5} - \dfrac{w}{4}\right) \ge -2w$

State whether each of the following statements is *always, sometimes,* or *never* true.

41. If $a > b$, then $a + c > b + c$.

42. If $c < d$, then $2c < d + c$.

43. If $c > d$, then $ac > ad$.

44. If $x < y$, then $-x < -y$.

45. $|a + b| = |a| + |b|$

46. $|a + b| > |a| + |b|$

47. If $ab > 0$, then $a > 0$ and $b > 0$.

48. If $ab < 0$, then $a < 0$ and $b > 0$.

49. If $a > b$ and $b > c$, then $a > c$.

50. If $a < b$ and $c < 0$, then $ac > bc$.

51. If $c > d$ and $a < 0$, then $\dfrac{c}{a} < \dfrac{d}{a}$.

52. If $a \ne b$, then $a > b$ or $a < b$.

Mathematical Excursions

Willie and Robyn are jockeys. Suppose you compare their weights. You can make only one of the following statements.

Willie's weight *is less than* Robyn's weight.	Willie's weight *is the same as* Robyn's weight.	Willie's weight *is greater than* Robyn's weight.

Let w = Willie's weight and r = Robyn's weight. Then you can use inequalities and an equation to compare their weights.

$$w < r \qquad w = r \qquad w > r$$

Trichotomy Property

For any two numbers a and b, exactly one of the following statements is true.

$$a < b \qquad a = b \qquad a > b$$

Suppose Willie weighs 108 pounds, Robyn weighs 113 pounds, and Jorge weighs 114 pounds.

Willie's weight is less than Robyn's weight.	Robyn's weight is less than Jorge's weight.

How do Willie's weight and Jorge's weight compare? Willie's weight is less than Jorge's weight. This suggests the following property of inequalities.

Transitive Property of Inequality

For any three numbers a, b, and c, if $a < b$ and $b < c$, then $a < c$.

Exercises

Replace each ☐ with <, >, or = to make each sentence true.

1. 7 ☐ 5

2. 0 ☐ −2

3. 2 ☐ 3

4. −5 ☐ −3

5. 6 ☐ 4 + 2

6. −7 + (−2) ☐ −9

7. 3 ☐ 15 ÷ 3

8. −5 ☐ 0 − 3

9. −6 ☐ −7 + (−4)

10. 10 ☐ 27 ÷ 3

11. 12 ☐ −15 − (−27)

12. −8 ÷ 4 ☐ −2

13. If $z < x$ and $x < y$, then z ☐ y.

14. If $r > s$ and $t < s$, then r ☐ t.

4.6 Problem Solving

Inequalities can be used to solve many verbal problems.

Examples

1 **The sum of two numbers is at least 80. One number is four times the other number. What is the least that each number can be?**

The sum is at least 80 means the sum is greater than or equal to 80.

Explore   Let x = one number.   *Define a variable.*
                Then, $4x$ = the other number.

Plan   The sum of two numbers   is at least 80.   *Write an inequality.*

                    $x + 4x$ $\geq$ 80

Solve $x + 4x \geq 80$ *Solve the inequality.*
                                      $5x \geq 80$
                                      $x \geq 16$

The least that one number can be is 16. The least that the other number can be is 4(16) or 64.

Examine   Since 64 = 4 × 16 and 16 + 64 = 80, a number less than 16 will produce a sum less than 80. Thus, the answer is correct.

2 **Tina wants to earn at least $13 this week. Her father has agreed to pay her $5 to mow the lawn and $2 an hour to weed the garden. Suppose Tina mows the lawn. What is the minimum number of hours she will have to spend weeding the garden in order to earn at least $13?**

Explore   Let n = the number of hours Tina must spend weeding the garden. *Define a variable.*
                                                            Write an inequality.

Plan $5 plus $2 times the number of hours weeding is at least $13.

                5 + 2 × n $\geq$ 13

Solve $5 + 2n \geq 13$ *Solve the inequality.*
                  $2n \geq 8$
                  $n \geq 4$

Tina must spend a minimum of 4 hours weeding.

Examine   Since $2 an hour for 4 hours is $8, and $5 + $8 = $13, Tina earns less than $13 if she works less than 4 hours. Thus, the answer is correct.

Example

3 Wendi is making a rectangular tablecloth that must be 60 cm longer than its width. She has 1000 cm of tape with which to make a border for the tablecloth. What are the greatest possible dimensions of the rectangle if the perimeter can be no greater than 1000 cm?

The perimeter is no greater than 1000 cm means it is less than or equal to 1000 cm.

Explore   Let x = the width of the rectangle.   *Define a variable.*
            Then, $x + 60$ = the length of the rectangle.

Make a drawing.

Recall that $P = 2\ell + 2w$.
$P = 2(x + 60) + 2x$
$P = 2x + 120 + 2x$
$P = 4x + 120$

x cm

$(x + 60)$ cm

Plan   The perimeter is no greater than 1000 cm.   *Write an inequality.*

            $4x + 120$ $\leq$ 1000

Solve $4x + 120 \leq 1000$ *Solve the inequality.*
                    $4x \leq 880$
                    $x \leq 220$
                $x + 60 \leq 280$

The greatest possible dimensions are 220 cm by 280 cm.

Examine   Since the perimeter of this rectangle is $2(220) + 2(280)$ or 1000 cm, dimensions less than 220 cm by 280 cm produce a perimeter less than 1000 cm. Thus, the answer is correct.

Exercises

Exploratory   For each of the following exercises, define a variable and state an inequality that describes the situation.

1. If 5 times a number is increased by 4, the result is at least 19. Find the least possible number that satisfies these conditions.

2. The sum of two numbers is at least 28. One number is 3 times the other number. Find the least numbers that satisfy these conditions.

3. Bob and Jill earned at most a total of $260 last year. Jill earned 3 times as much as Bob. What is the greatest amount each person earned?

4. The length of a rectangle is 4 times its width. Its perimeter is no greater than 170 cm. Find the greatest possible dimensions of the rectangle.

5. The length of a rectangle is 6 cm less than 5 times its width. The perimeter is a minimum of 84 cm. What are the least the dimensions can be?

7. The sum of two consecutive odd integers is at most 20. What are the greatest possible integers?

6. In their new business venture, Seth and Jane expect to earn a combined total of at least $85,000. Their agreement states that Jane will make $5000 less than Seth. What is the least each can earn?

8. The sum of two consecutive even integers is not less than 3 times the lesser integer decreased by 16. What are the least possible integers?

Written **Solve the inequality in each of the following Exploratory Exercises.**

1. Exercise 1

3. Exercise 3

5. Exercise 5

7. Exercise 7

2. Exercise 2

4. Exercise 4

6. Exercise 6

8. Exercise 8

For each of the following, define a variable. Then write and solve an inequality.

9. Oakdale Municipal Garage charges $1.50 for the first hour and $0.50 for each additional hour. How many hours can you park your car if you wish to pay no more than $4.50?

11. Suppose the maximum weight of the crate in exercise 10 must be 80 lb. What is the greatest number of books that can be packed in the crate?

10. An empty book crate weighs 30 lb. What is the greatest number of books weighing 1.5 lb each that can be packed in the crate if the maximum weight must be 60 lb?

12. What is the greatest number of books that can be packed in the crate in exercise 10 if the maximum weight is 50 lb?

Look at the labels in the photograph at the right to solve each of the following.

13. If you order a 4-oz hamburger without a roll, how many french fries can you eat so that your meal contains less than 800 calories?

15. You want to eat 20 french fries and a hamburger with a roll. What is the most the hamburger can weigh so that the entire meal contains 1200 calories or less?

14. If you order a 4-oz hamburger with a roll, how many french fries can you eat so that your meal contains less than 800 calories?

16. If you order a 3-oz hamburger with a roll, how many french fries must you eat so that your meal contains at least 1000 calories?

4.7   Compound Inequalities

In the Olympics, a welterweight wrestler weighs more than 68 kilograms but not more than 74 kilograms. Suppose w is the number of kilograms of a wrestler's weight. Then the inequalities $w > 68$ and $w \le 74$ are both true for every welterweight.

Journal

What new things have you learned so far in this chapter? Be sure to give examples.

An open sentence is satisfied when the variable has been replaced and the resulting sentence is true.

Logic can be used in describing this situation. We write a compound sentence using the connective *and*. The sentence is a conjunction of the inequalities $w > 68$ and $w \le 74$. Instead of saying $w > 68$, we say $68 < w$. We call the sentence a **compound inequality.**

$$(68 < w) \wedge (w \le 74)$$

This conjunction is equivalent to the sentence $68 < w \le 74$. This sentence is read *w is greater than 68 and less than or equal to 74.*

The solution set of a conjunction must contain all the replacements for w that *satisfy* both conjuncts. The graph of the solution set is the set of points that are common to the graphs of both conjuncts.

Example

1 Graph the solution set of $68 < w \leq 74$.

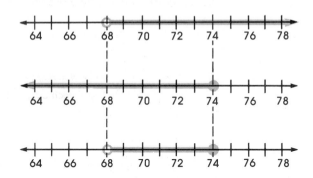

Graph the solution set of $68 < w$.

Graph the solution set of $w \leq 74$.

The graph of the solution set of $68 < w \leq 74$ is the set of points common to both graphs.

Suppose an Olympic wrestler is *not* a welterweight. Then the compound sentence $(w \leq 68) \lor (w > 74)$ describes his weight. This sentence is another kind of compound inequality, a disjunction. Recall that for a disjunction to be true at least one disjunct must be true. The solution set must contain all the replacements for w that satisfy at least one of the disjuncts.

Example

2 Graph the solution set of $(w \leq 68) \lor (w > 74)$.

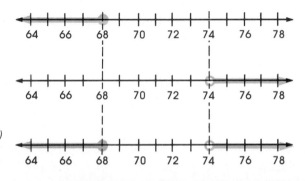

Graph the solution set of $w \leq 68$.

Graph the solution set of $w > 74$.

The graph of the solution set of $(w \leq 68) \lor (w > 74)$ is the set of points of at least one of the graphs.

Many compound inequalities can be solved using the methods you already have learned for solving an inequality.

Examples

3 Solve $3 < 2x - 1 < 9$. Then graph the solution set.

$3 < 2x - 1 < 9$
$(3 < 2x - 1) \wedge (2x - 1 < 9)$ *Write the sentence as a conjunction.*
$4 < 2x \qquad \wedge \qquad 2x < 10$ *Solve each conjunct.*
$2 < x \qquad \wedge \qquad x < 5$

The solution set is $\{x: 2 < x < 5\}$.

Graph $x > 2$.

Graph $x < 5$.

The solution set is the set of points common to both graphs.

4 Solve $(3y - 1 < 11) \vee (2y + 5 \geq 19)$. Then graph the solution set.

$(3y - 1 < 11) \vee (2y + 5 \geq 19)$
$3y < 12 \ \vee \qquad 2y \geq 14$ *Solve each disjunct.*
$y < 4 \ \vee \qquad y \geq 7$

The solution set is $\{y: (y < 4) \vee (y \geq 7)\}$.

Graph $y < 4$.

Graph $y \geq 7$.

The graph of the solution set is the points of each of the graphs.

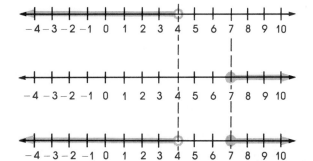

Exercises

Exploratory Suppose the domain of x is the set of integers, Z. State the solution set of each of the following.

1. $4 < x < 9$ **2.** $-3 < x < 2$ **3.** $-4 \leq x < 0$ **4.** $-12 \leq x \leq -10$

Suppose the domain of x is the set $\{-4, -3, -2, -1, 0, 1, 2, 3\}$. State the solution set of each of the following.

5. $(x < -1) \lor (x > 1)$ **6.** $(x \leq -3) \lor (x > 0)$
7. $(x > -2) \lor (x > 1)$ **8.** $(x \leq 2) \lor (x > -3)$

Write the compound sentence whose solution set is graphed in each of the following.

9.

10.

11.
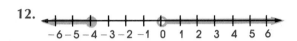

12.

13.

14.

Written Graph the solution set of each of the following.

1. $1 < x < 6$ **2.** $-5 < x \leq 3$ **3.** $0 \leq y < 2$
4. $0 \leq m \leq 2$ **5.** $(y < -2) \lor (y > 3)$ **6.** $(y \leq -1) \lor (y > 2)$
7. $(x \geq 2) \lor (x \leq -2)$ **8.** $(x > 3) \lor (x < -1)$ **9.** $(y < 2) \lor (y < 4)$
10. $(y < 2) \land (y < 4)$ **11.** $(x > 3) \land (x > 6)$ **12.** $(x < 5) \lor (x > 1)$

Solve each of the following compound inequalities. Then graph the solution set.

13. $2 < y + 4 < 11$ **14.** $1 < x - 2 < 7$
15. $(x - 4 < 1) \land (x + 2 > 1)$ **16.** $(y + 6 > -1) \land (y - 2 < 4)$
17. $2x + 1 \leq -9$ **18.** $1 - 4y \geq 9$
19. $(y + 2 < 3) \lor (y - 1 > 5)$ **20.** $(x + 4 < 2) \lor (x - 2 > 1)$
21. $4 < 2x - 2 < 10$ **22.** $-1 < 3m + 2 < 14$
23. $0 < 3y - 6 \leq 6$ **24.** $-3 - x < 2x < 3 + x$
25. $-5 < 4 - 3x < 13$ **26.** $(9 - 2m > 11) \land (5m < 2m + 9)$

Challenge Solve each of the following. Then graph the solution set.

1. $x - 2 < (2x + 5) < x + 6$ **2.** $y - 16 \leq 2(y - 4) < y + 2$
3. $k + 7 < (2k + 2) < 0$ **4.** $2r + 6 < (3r - 3) \leq 2(r + 3)$
5. $(4z + 8 \geq z + 6) \lor (7z - 14 \geq 2z - 4)$
6. $(5x + 7 > 2x + 4) \lor (3x + 3 < 24 - 4x)$

4.8   Sentences Containing Fractions

Suppose an equation or inequality contains coefficients which are fractions or decimals. Such an open sentence may be solved by finding an equivalent sentence that does not contain fractional coefficients. Consider the following examples.

Examples

1 Solve $\frac{2}{3}x + \frac{1}{4}x = 22$.

An equivalent equation may be found by multiplying both sides by the least common denominator (LCD) of $\frac{2}{3}$ and $\frac{1}{4}$.

$$\frac{2}{3}x + \frac{1}{4}x = 22$$

$$12\left(\frac{2}{3}x + \frac{1}{4}x\right) = 12 \cdot 22 \qquad \textit{Multiply by 12, the LCD of } \frac{2}{3} \textit{ and } \frac{1}{4}.$$

$$12 \cdot \frac{2}{3}x + 12 \cdot \frac{1}{4}x = 264 \qquad \textit{Use the distributive property.}$$

$$8x + 3x = 264 \qquad \textit{This equation has no fractional coefficients.}$$

$$11x = 264 \qquad \textit{Combine like terms.}$$

$$x = 24 \qquad \textit{Divide both sides by 11.}$$

2 Solve $\frac{2x}{5} \geq 9 - \frac{x}{2}$.

$$\frac{2x}{5} \geq 9 - \frac{x}{2} \qquad \textit{The LCD of the fractions is 10.}$$

$$10\left(\frac{2x}{5}\right) \geq 10\left(9 - \frac{x}{2}\right) \qquad \textit{Multiply both sides by 10.}$$

$$10\left(\frac{2x}{5}\right) \geq 10 \cdot 9 - 10\left(\frac{x}{2}\right) \qquad \textit{Use the distributive property.}$$

$$4x \geq 90 - 5x$$

$$9x \geq 90 \qquad \textit{Add 5x to both sides.}$$

$$x \geq 10 \qquad \textit{Divide both sides by 9.}$$

A sentence with decimal coefficients can be solved using this method. Recall that any decimal is a fraction. For example, 0.03 is $\frac{3}{100}$. The LCD of any two decimals is 10, 100, or some other power of 10.

Example

3 Solve $0.08x + 0.40(x + 10) = 13.6$.

$$0.08x + 0.40(x + 10) = 13.6 \qquad \text{\textit{The LCD of these decimals is 100.}}$$
$$100[0.08x + 0.40(x + 10)] = 100 \cdot 13.6 \qquad \text{\textit{Multiply both sides by 100.}}$$
$$100(0.08x) + 100 \cdot 0.40(x + 10) = 1360 \qquad \text{\textit{Use the distributive property.}}$$
$$8x + 40(x + 10) = 1360$$
$$8x + 40x + 400 = 1360$$
$$48x = 960$$
$$x = 20$$

Exercises

Exploratory Find each of the following products. State in simplest form.

1. $\frac{5}{6} \cdot \frac{3}{8}$ **2.** $\frac{2}{3} \cdot \frac{x}{4}$ **3.** $6 \cdot \frac{5}{12}$ **4.** $10 \cdot \frac{2y}{5}$ **5.** $15 \cdot \frac{2x}{3}$

6. $24 \cdot \frac{3m}{8}$ **7.** $8\left(\frac{x}{2} + \frac{1}{4}\right)$ **8.** $12\left(\frac{x}{6} - \frac{2}{3}\right)$ **9.** $30\left(\frac{2x}{5} + \frac{5}{6}\right)$ **10.** $36\left(\frac{4}{9} - \frac{5y}{12}\right)$

State the LCD of the fractions in each of the following exercises.

11. $\frac{1}{2}$ and $\frac{2}{3}$ **12.** $\frac{2}{7}$ and $\frac{1}{3}$ **13.** $\frac{3}{4}$ and $\frac{3}{8}$ **14.** 0.3 and 0.03 **15.** 0.05 and 0.27

16. $\frac{4}{5}$ and $\frac{5}{12}$ **17.** $\frac{2}{25}$ and $\frac{2}{9}$ **18.** $\frac{1}{54}$ and $\frac{5}{9}$ **19.** 0.12 and 0.3 **20.** 0.009 and 0.99

Written Solve each of the following.

1. $\frac{y}{5} = 3$ **2.** $\frac{1}{4}m = -6$ **3.** $\frac{2x}{5} > 8$ **4.** $\frac{3y}{4} < 6$

5. $\frac{y + 3}{5} = 7$ **6.** $\frac{x - 6}{3} = 4$ **7.** $\frac{2x - 1}{5} = 1$ **8.** $\frac{3m + 2}{2} = 7$

9. $\dfrac{4y - 6}{2} = -13$

10. $\dfrac{2x}{5} \geq \dfrac{5}{3}$

11. $\dfrac{y - 5}{45} = \dfrac{4}{9}$

12. $\dfrac{y}{5} + \dfrac{y}{7} = 12$

13. $\dfrac{x}{2} - \dfrac{x}{5} = 3$

14. $\dfrac{3y}{4} - \dfrac{y}{3} = -2$

15. $\dfrac{3x + 1}{8} = \dfrac{10x + 4}{12}$

16. $\dfrac{3x - 3}{5} < \dfrac{4x - 2}{6}$

17. $\dfrac{x}{2} + \dfrac{x}{3} = 5$

18. $\dfrac{3x}{4} - \dfrac{5x}{8} = -\dfrac{1}{2}$

19. $\dfrac{7x}{10} - \dfrac{x}{5} < \dfrac{3}{2}$

20. $\dfrac{y - 6}{6} + \dfrac{y + 3}{3} = -\dfrac{1}{3}$

21. $\dfrac{x + 11}{9} - \dfrac{x - 1}{3} > 2$

22. $\dfrac{x + 3}{4} + \dfrac{x + 2}{5} \leq \dfrac{5}{2}$

23. $\dfrac{10x - 3}{10} - \dfrac{3x + 3}{5} = \dfrac{1}{10}$

24. $\dfrac{y}{2} - \dfrac{y}{4} + \dfrac{y}{3} < 7$

25. $\dfrac{3x + 6}{4} - \dfrac{2x + 2}{3} = -1$

26. $\dfrac{4x + 5}{6} - \dfrac{x + 7}{4} = \dfrac{1}{2}$

27. $\dfrac{3y - 4}{5} = 1 - \dfrac{4 - 3y}{2}$

28. $\dfrac{x}{6} - 1 = \dfrac{x - 20}{8}$

29. $\dfrac{x + 8}{16} - 1 > \dfrac{4 - x}{12}$

30. $3.8 - 0.5x = 0.3$

31. $1.2y + 6.7 = 13.9$

32. $0.8z + 0.35 = 7.55$

33. $0.25y + 3.4 = 1.6$

34. $0.05(x - 8) = 0.07$

35. $0.04(y - 2) = 0.12y + 0.02$

36. $0.06x + 0.05(x + 2) = 0.1x$

37. $0.05(c - 1) + 3c = 0.25c + 1$

38. $\dfrac{1}{2}(3y - 1) - \dfrac{1}{3}(2y - 5) = 6 - y$

39. $\dfrac{1}{3}(2y - 2) + \dfrac{1}{6}(9y + 23) = 3y - 1$

Challenge   Solve each of the following.

1. $2 < \dfrac{1 - 5x}{4} - \dfrac{2x - 7}{3} \leq 8$

2. $3 < \dfrac{2x - 1}{5} - \dfrac{3x + 4}{2} \leq 5$

Solve each of the following for x. Give the values of a and b for which the solution is undefined.

3. $\dfrac{4x}{2b} + \dfrac{3}{b} = 2a$

4. $\dfrac{6x}{5b} + \dfrac{4}{b} = \dfrac{6}{a}$

5. $\dfrac{x + a}{a} + \dfrac{x + b}{b} = c$

Mixed Review

Find each of the following and simplify.

1. $2.9 + (-4.6)$ **2.** $17.2 - (-8.8)$ **3.** $(9.75)(-0.16)$ **4.** $8.12 - (-0.7)$

Solve and check each equation.

5. $23 - 8x = 17$

6. $3(2m - 7) = 11m - 6$

Let p represent "$-6y + 2 \leq 14$" and q represent "$(y < -1) \leftrightarrow (y - 6 \geq -5)$." If $y = -2$, determine the truth value of each statement.

7. $p \rightarrow q$ **8.** $p \wedge q$ **9.** $p \wedge \sim q$ **10.** $\sim p \vee q$

4.9 More Problem Solving

Sketches are often a convenient aid for solving verbal problems.

Examples

1 Two airplanes take off at the same time from an airport. One flies west at a constant speed of 800 km/h. The other flies east at a constant speed of 1000 km/h. In how many hours will they be 9000 km apart?

Explore Let h = the number of hours until the planes are 9000 km apart.   *Make a sketch.*

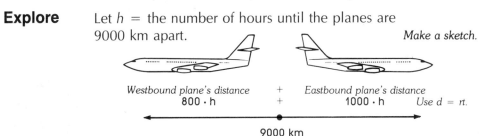

Westbound plane's distance + Eastbound plane's distance
800 · h   +   1000 · h   *Use d = rt.*

← 9000 km →

Plan
The westbound   the eastbound   the total
plane's distance *plus* plane's distance *is* distance.

$$800h \quad + \quad 1000h \quad = 9000$$
$$1800h = 9000$$

Solve
$$h = 5$$

In 5 hours, the planes will be 9000 km apart.

Examine Since the planes will be $5 \cdot 800 + 5 \cdot 1000$ or 9000 km apart after 5 hours, the answer is correct.

2 Almonds cost $1.60 per pound. Walnuts cost $1.10 per pound. How many pounds of each should be mixed to produce a 40-pound mixture that costs $1.25 per pound?

Explore Let n = the number of pounds of almonds.
Then, $40 - n$ = the number of pounds of walnuts.

Plan
The cost of   the cost of   the cost of
almonds *plus*   walnuts   *is* the mixture.

$$1.60n \quad + \quad 1.10(40 - n) = 1.25\,(40)$$
$$160n + 4400 - 110n = 5000$$

Solve
$$50n = 600$$
$$n = 12$$

The mixture contains 12 lb of almonds and $40 - 12$ or 28 lb of walnuts.   *Examine this solution.*

Example

3 Ace's Car Rental rate is $17 a day plus $0.09 a mile. Zip's Car Rental rate is $13 a day plus $0.11 a mile. Suppose you rent a car for one day. How many miles must you drive for Ace's Car Rental plan to be a better buy than Zip's Car Rental plan?

Explore   Let n = the number of miles driven in one day.
Then, $17 + 0.09n$ = the cost of Ace's plan
and $13 + 0.11n$ = the cost of Zip's plan.

Plan   *The cost of Ace's plan* *is less than* *the cost of Zip's plan.*

$$17 + 0.09n \quad < \quad 13 + 0.11n$$

Solve
$$4 + 0.09n < 0.11n \qquad \text{Add} - 13 \text{ to both sides.}$$
$$400 + 9n < 11n \qquad \text{Multiply both sides by 100.}$$
$$400 < 2n \qquad \text{Add } -9n \text{ to both sides.}$$
$$200 < n$$

You must drive more than 200 miles in one day for the Ace's Car Rental plan to be a better buy.   *Examine this solution.*

Exercises

Exploratory   Use $d = rt$ to answer each of the following.

1. Jean is driving at 80 km/h. How far will she travel in 3 hours? In x hours? In $x + 1$ hours?

2. John is biking at 8 km/h. How far will he ride in 3 hours? In h hours? In $8 - h$ hours?

3. Rick traveled 600 km. What is his rate if he made the trip in 12 hours? In 10 hours? In x hours?

4. Susie ran 26 miles. What is her rate if she ran 5 hours? h hours? $3h$ hours?

5. Jan traveled 400 km. How long did the trip take her if her rate was 60 km/h? 75 km/h?

6. Jim drove 350 miles. How long did the trip take him if his rate was 40 mph? 50 mph? r mph?

Write an expression that represents the cost of each of the following.

7. a mixture of 40 lb of coffee at $3.25 per pound and 25 lb of coffee at $3.80 per pound

8. a mixture of x gal of paint thinner at $1.50 per gallon and $(150 - x)$ gal of paint at $15 per gallon

Written   Solve each problem.

1. Two planes leave an airport at the same time. One plane flies west at 600 km/h. The other flies east at 340 km/h. In how many hours will they be 2820 km apart?

2. Two cars leave Salem and travel in opposite directions. Their rates are 50 km/h and 60 km/h. In how many hours will they be 440 km apart?

3. Two trains start toward each other at the same time from stations 1035 miles apart. One train travels at 40 mph and the other at 50 mph. In how many hours will they pass each other?

4. Two trains leave Central Station and travel in opposite directions. After 11 hours they are 1265 miles apart. The rate of one train is 15 mph greater than the other. Find both rates.

5. How much candy costing $1.20 per pound should be mixed with candy costing $1.90 per pound to produce a 70-pound mixture that costs $1.50 per pound?

6. Grade A seeds cost 80¢ per pound and Grade B seeds cost 50¢ per pound. How many pounds of each should be mixed to produce 30 pounds that cost 75¢ per pound?

7. How many pounds of coffee costing $3.20 per pound must be mixed with 18 lb of coffee costing $2.80 per pound to produce a mixture that costs $2.96 per pound?

8. How many gallons of punch costing 70¢ per gallon must be mixed with 1200 gal of punch costing 50¢ per gallon to produce a mixture that costs 65¢ per gallon?

9. City Bank charges $1.75 a month plus 8¢ per check. Bank One charges $2.50 a month plus 6¢ per check. When is an account at City Bank a better buy than an account at Bank One?

10. The XYZ Car Rental plan is $18.75 a day plus 16¢ a mile. The ABC Car Rental plan is $15.95 a day plus 19¢ a mile. How many miles would you have to drive in a day for the XYZ plan to be the better buy?

11. Ajax Printers charges $18 for setting the presses plus $1.85 for the first 100 cards plus 3¢ for each additional card over 100. How many cards can be purchased for less than $25?

12. Print Center charges $15 for setting the presses plus $1.60 for the first 75 cards plus 4¢ for each additional card over 75. How many cards can be purchased for less than $25?

13. Look at exercises 11 and 12. How many cards must be purchased from Ajax Printers to be a better buy than cards purchased from Print Center?

Challenge   Solve each problem.

1. Twice as many nickels as dimes are in a coin bank. If their total value is at least $2.95, find the least possible number of coins.

2. Is it possible for a total of 12 nickels and quarters to have a value of $4.25? Explain.

Mathematical Excursions

The absolute value of a number is the number of units it is from 0 on the number line. An open sentence containing absolute value can be interpreted as follows.

$|x| = 3$ x is 3 units from 0.

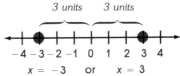

$x = -3$ or $x = 3$

$|x| = 3$ is equivalent to the disjunction $x = -3 \lor x = 3$.
Likewise, $|x - 4| = 6$ is equivalent to $(x - 4 = -6) \lor (x - 4 = 6)$.
                     Solve for x. $x = -2 \lor$ $x = 10$

$|x| > 3$ x is greater than 3 units from 0.

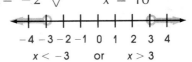

$x < -3$ or $x > 3$

$|x| > 3$ is equivalent to the disjunction $x < -3 \lor x > 3$.
Likewise, $|x - 4| > 1$ is equivalent to $(x - 4 < -1) \lor (x - 4 > 1)$.
                Solve for x. $x < 3$ $\lor$ $x > 5$
                The solution set is $\{x: (x < 3) \lor (x > 5)\}$.

$|x| < 3$ x is less than 3 units from 0.

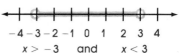

$x > -3$ and $x < 3$

$|x| < 3$ is equivalent to the conjunction $x > -3 \land x < 3$.
Likewise, $|x + 4| < 2$ is equivalent to $(x + 4 > -2) \land (x + 4 < 2)$.
                Solve for x. $x > -6 \land$ $x < -2$
                The solution set is $\{x: -6 < x < -2\}$.

Example   Solve $|2x + 1| \leq 7$. Then, graph the solution set.

$|2x + 1| \leq 7$ is equivalent to $(2x + 1 \geq -7) \land (2x + 1 \leq 7)$.
        $2x + 1 \geq -7 \land 2x + 1 \leq 7$
        $2x \geq -8 \land$ $2x \leq 6$
        $x \geq -4 \land$ $x \leq 3$

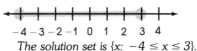

The solution set is $\{x: -4 \leq x \leq 3\}$.

Exercises Write an equivalent conjunction or disjunction for each of the following exercises. Then, solve each exercise and graph the solution set.

1. $|y| = 6$,   **2.** $|x| = 10$ **3.** $|y| > 7$ **4.** $|y| < 12$
5. $|x + 2| = 3$ **6.** $|y - 2| = 6$ **7.** $17 = |12 - d|$ **8.** $11 = |-8 + m|$
9. $|y + 2| > 4$ **10.** $|x - 5| \geq 2$ **11.** $|2 + x| < 5$ **12.** $|y + 5| \leq 2$

Portfolio Suggestion

Select an item from this chapter that you feel shows your best work and place it in your portfolio. Explain why you selected it.

Performance Assessment

Solve $ry + s = tx - m$ for y. Explain each step in your solution. Would there be any limitations for the value of each variable? If so, explain the limitation.

Chapter Summary

1. A mathematical sentence that contains any of the symbols $<$, $>$, $\leq$, $\geq$, or $\neq$ is called an **inequality.** (113)
2. **Addition Property for Inequalities:** Suppose a, b, and c are any numbers. If $a < b$, then $a + c < b + c$. If $a > b$, then $a + c > b + c$. (116)
3. **Multiplication Property for Inequalities:** Suppose a and b are any numbers. If $c > 0$ and $a < b$, then $ac < bc$. If $c > 0$ and $a > b$, then $ac > bc$. If $c < 0$ and $a < b$, then $ac > bc$. If $c < 0$ and $a > b$, then $ac < bc$. (118)

Chapter Review

4.1 **Solve each of the following.**

1. If $C = a + 4b$, find C when $a = -3$ and $b = 4$.

2. If $r = s + 3t$, find s when $r = 18$ and $t = 2$.

3. The formula for the area of a rectangle is $A = \ell w$. If the length ℓ is doubled and the width w is doubled, how does the area change?

4. Write a formula that can be used to find the amount A, in cents, of x nickels plus y dimes.

4.2 **Solve each of the following equations for x.**

5. $x + 2b = a$

6. $\dfrac{x}{a} = b$

7. $ax + b = c$

8. For what value of a is the formula $x = \dfrac{b}{5 - a}$ undefined?

4.3 **Graph each of the following solution sets.**

9. $y > -4$

10. $x \leq -2$

11. $y \neq -1$

Write an inequality for each of the following graphs.

12.

13.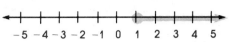

4.4 **Solve each of the following inequalities. Then, graph the solution set.**

14. $y - 7 < 1$ **15.** $5 + m > -2$ **16.** $3x + 4 - 2x \geq 3$

4.5 **Solve each of the following inequalities. Then, graph the solution set.**

17. $3x < 27$ **18.** $-2x > 8$ **19.** $2y + 1 \leq -7$
20. $3(y - 2) \geq 9$ **21.** $2x > 7(x - 5)$ **22.** $2y + 7(1 - y) < -8$

4.6 **Solve each of the following.**

23. Seven more than 5 times a whole number is at least 47. Find the minimum value of the number.

24. The length of a rectangle is 8 cm less than 3 times its width. The perimeter is at greatest 80 cm. What is the greatest possible length of the rectangle?

4.7 **Graph each of the following.**

25. $-3 < x < 2$
27. $(x > 4) \wedge (x > 5)$
29. $(y \geq 3) \vee (y \leq -1)$

26. $-1 < x < 3$
28. $(x < -1) \vee (x > 2)$
30. $(y \leq 2) \vee (y > -1)$

Solve each of the following. Then, graph the solution sets.

31. $1 < 3y - 5 < 13$

32. $(2x - 1 < -5) \vee (3x + 2 \geq 5)$

4.8 **Solve each of the following. Then, graph the solution sets of the inequalities.**

33. $\dfrac{y}{5} + \dfrac{y}{6} = -11$

34. $\dfrac{3y - 1}{3} - \dfrac{4y - 2}{5} < \dfrac{2}{3}$

35. $0.04(2x + 3) > 0.16$

36. $0.2x + 0.01(18 - x) = 0.01x$

4.9 **Solve each of the following.**

37. A train left Troy at 8 A.M. and traveled west at 45 mph. At 10 A.M. a train left the station traveling west at 54 mph. At what time did the second train overtake the first train?

38. Candy costing $1 a pound is to be mixed with candy costing $2.25 a pound to make 25 lb of candy costing $2 a pound. How much of each should be mixed?

39. A box of seeds weighs 1.5 lb. An empty carton weighs 3.5 lb. How many boxes of seeds can be packed in a carton if it can weigh no greater than 35 lb?

40. UniBank charges $2 a month plus 7¢ per check. Pine Trust charges $2.25 a month plus 6¢ per check. How many checks must you write to pay less at Pine Trust?

Chapter Test

Solve each of the following.

1. In the formula $A = \dfrac{h}{2}(a + b)$, find b if $A = 80$, $a = 6$, and $h = 8$.

2. Solve the formula $a(x - b) = c$ for x in terms of a, b, and c. State the values of a for which your answer is undefined.

3. A jogger ran 15 miles in $x + 2$ hours. Write an expression which represents the jogger's rate.

4. If the length of each side of a square is multiplied by 3, how is its area changed?

5. Write an inequality for the graph shown at the right.

Solve each of the following. Then, graph the solution set.

6. $5y \le -30$

7. $-7y < 21$

8. $4x - 1 > 27$

9. $10y < 6(2y + 4)$

10. $-3 - x < 2x < 3 + x$

11. $\dfrac{2}{3}x - 2 \ge \dfrac{5}{6}x + 2$

Solve each of the following.

12. The length of a rectangle is 3 times its width. Find the greatest possible width of the rectangle, if its perimeter is not greater than 850 cm.

13. June earns $3 an hour for the first 40 hours, then $4.50 an hour for overtime. How many hours must she work in order to make at least $165?

14. Theater tickets cost $3.50 for children and $6.50 for adults. Sabrina bought 8 tickets and spent $37.00. How many adult tickets did she buy?

15. Pine City Bank charges $4.25 a month plus $0.11 per check. Pine City Savings charges $5.85 a month plus $0.08 per check. How many checks must you write each month to pay less at Pine City Bank?

Aspects of Geometry

Application in Architecture

Architects use **geometry** to create buildings. They must create buildings that are both functional and pleasing to the eye.

One of the most prestigious architects of our time is Kenzo Tange from Japan. He combines traditional Japanese design with more angular forms often used in the western hemisphere. The photo at the right shows one of his buildings. What geometric shapes did Kenzo Tange use to design this building?

Individual Project:
Foreign Languages

If you are studying another language, choose a country that uses that language as its primary language. If you are not studying another language at this time, choose any country. Find pictures of architecture from the country that you have chosen and make a bulletin board using these pictures.

5.1   Points, Lines, and Planes

We see the shapes of geometry in the structures in which people live and work.

Geometric shapes are found in nature as well as in art.

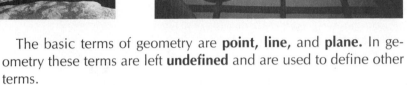

Undefined terms are words whose meaning is readily understood.

The basic terms of geometry are **point, line,** and **plane.** In geometry these terms are left **undefined** and are used to define other terms.

B • point B

A
•
point A

C
•
point C

An exact location or a pinhole suggests the idea of a point. Points have no size. Points are represented by dots and named by capital letters.

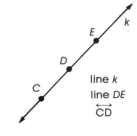

A line, in geometry, means a straight line.

line *k*
line *DE*
$\overleftrightarrow{CD}$

A never-ending straight path suggests the idea of a **line.** Lines extend indefinitely and have no thickness or width. Lines are represented by double arrows and uppercase letters or by a lowercase letter. A line also can be named by any two points of the line.

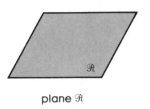

plane $\mathcal{R}$

A never-ending flat surface suggests the idea of a **plane.** A plane extends indefinitely in all directions and has no thickness. Planes are represented by four-sided figures and named by capital script letters.

The word *between* is also an undefined term. Note how it is used in the definition of a **line segment.**

Definition of Line Segment

> A line segment is part of a line consisting of two points and all the points between them. The two points are called the **endpoints** of the segment.

Line segments are named by the letters for their endpoints with a bar on the top.

line segment *AB*
$\overline{AB}$

Two lines, or lines and planes, **intersect** if they have points in common. For example, the two lines below intersect at point *E*.

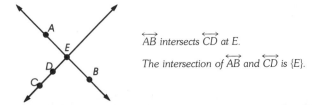

$\overleftrightarrow{AB}$ intersects $\overleftrightarrow{CD}$ at *E*.

The intersection of $\overleftrightarrow{AB}$ and $\overleftrightarrow{CD}$ is {*E*}.

Definition of Intersection

> The set of points which two sets of points have in common is the intersection of the two sets.

Example

1 Find the intersection of plane $\mathcal{A}$ and plane $\mathcal{B}$.

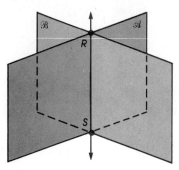

Plane $\mathcal{A}$ and plane $\mathcal{B}$ have $\overleftrightarrow{RS}$ in common. Their intersection is $\overleftrightarrow{RS}$.

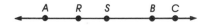

The symbol ∩ is used to represent an intersection. For example, $\overline{AS} \cap \overline{RB} = \overline{RS}$ and $\overleftrightarrow{AB} \cap \overline{SC} = \overline{SC}$ indicate intersections on line AB.

Definition of Union

The set of points in either or both of two sets of points is called the **union** of the two sets.

The symbol ∪ is used to represent a union. For example, $\overline{AS} \cup \overline{RB} = \overline{AB}$ and $\overleftrightarrow{AB} \cup \overline{SC} = \overleftrightarrow{AB}$ indicate unions on the line AB shown above.

Exercises

Exploratory State whether each of the following suggests a *point*, a *line*, or a *plane*.

1. corner of a box
2. side of a box
3. edge of a box
4. wall of a room
5. straw
6. grain of salt
7. telephone wire
8. clothesline
9. ceiling of a room
10. star in the sky
11. straight highway on a map
12. city on a map

Written   For each of the following, use the figure below.

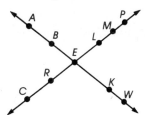

1. Write another name for $\overleftrightarrow{AB}$.
2. Write another name for line ℓ.
3. Name the intersection of $\overleftrightarrow{AD}$ and line n.
4. Name the union of $\overline{DH}$ and $\overline{CH}$.
5. Name two points that lie on $\overleftrightarrow{AD}$.
6. Name two points that lie on line p.
7. Write three other names for line n.
8. Write three other names for $\overleftrightarrow{BC}$.

Determine whether each of the following statements is *true* or *false*.

9. A line is a set of points.
10. A line is part of a line segment.
11. A line segment is part of a line.
12. No line can contain more than two segments.
13. There cannot be more than one line in a plane.
14. The intersection of two segments may be a segment.

Draw and label a diagram to show each of the following.

15. Plane $\mathscr{L}$ intersects plane $\mathscr{M}$ at $\overleftrightarrow{PQ}$.
16. Plane $\mathscr{A}$, plane $\mathscr{B}$, and plane $\mathscr{C}$ intersect at $\overleftrightarrow{ST}$.
17. $\overleftrightarrow{RS}$ and plane $\mathscr{M}$ do *not* intersect.
18. $\overleftrightarrow{PQ}$, $\overleftrightarrow{RS}$, and plane $\mathscr{N}$ intersect at T.

Use the figure below to find the simplest name for each of the following.

19. $\overline{EL} \cup \overline{LP}$
20. $\overline{RE} \cup \overline{EL}$
21. $\overleftrightarrow{BK} \cap \overleftrightarrow{KW}$
22. $\overleftrightarrow{BK} \cup \overleftrightarrow{KW}$
23. $\overleftrightarrow{BK} \cup \overleftrightarrow{AW}$
24. $\overleftrightarrow{BK} \cap \overleftrightarrow{AW}$
25. $\overleftrightarrow{AK} \cap \overleftrightarrow{RM}$
26. $\overline{RL} \cup \overline{LM}$
27. $\overline{RE} \cap \overline{EL}$
28. $\overline{EL} \cap \overline{MP}$
29. $\overline{AB} \cap \overleftrightarrow{KW}$
30. $\{B\} \cup \overline{AE}$

If $A = \{1, 2, 3, 4, 5\}$, $B = \{2, 4, 6, 8\}$, and $C = \{1, 8, \text{Akron}\}$, find each of the following.

Sample: $B \cap C = \{8\}$ $A \cup C = \{1, 2, 3, 4, 5, 8, \text{Akron}\}$

31. $A \cup B$
32. $A \cap B$
33. $A \cup (B \cup C)$
34. $(A \cap B) \cup (A \cap C)$

Use the figure at the right for each of the following.

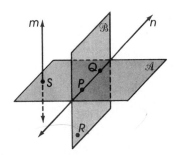

35. Name a point not on line n.
36. Name a point not in plane $\mathscr{B}$.
37. Name a point not in plane $\mathscr{A}$.
38. Name all points common to line m and plane $\mathscr{A}$.
39. Name all points common to plane $\mathscr{A}$ and plane $\mathscr{B}$.
40. Name all points common to line m and line n.

5.2 Rays and Angles

A never-ending straight path in one direction suggests the idea of a **ray.** A ray is part of a line. In contrast to a line segment, it has only one endpoint. Rays are named by two letters. The first is the endpoint of the ray, and the second is any other point on the ray.

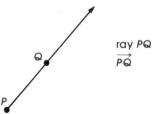

ray PQ
$\overrightarrow{PQ}$

Definition of Ray

> A ray is part of a line consisting of one endpoint and all the points on the line on one side of the endpoint.

Note that the definition of a ray uses the undefined terms *point* and *line* as does the following definition.

Definition of Collinear Points

> Two or more points are **collinear** if and only if they lie on the same line.

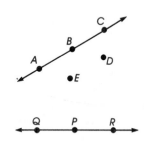

For example, A, B, and C are collinear points and A, D, and E are *not* collinear points.

Any point on a line determines two rays on that line. The point is the endpoint of the two rays that head in opposite directions. These rays are called **opposite rays.** $\overrightarrow{PQ}$ and $\overrightarrow{PR}$ are opposite rays if and only if P, Q, *and* R are collinear and P is between Q and R.

Definition of Angle

> An **angle** is the union of two rays that have the same endpoint. The rays are called the **sides** of the angle. The common endpoint is called the **vertex** of the angle.

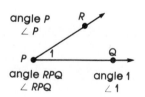

angle P
$\angle P$

angle RPQ
$\angle RPQ$

angle 1
$\angle 1$

Angles can be named in several ways, with letters or numbers. When three letters are used to name an angle, the letter naming the vertex is between the other two letters.

An angle separates a plane into three parts. The parts are the *interior*, the *exterior*, and the *angle itself*. Any point in the red part of the plane is in the interior of ∠P. Any point in the gray part of the plane is in the exterior of ∠ P

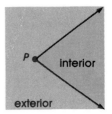

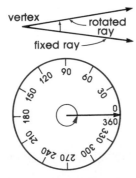

Another way of thinking of an angle is as a rotation where one of the rays is fixed and the other is rotated about the vertex.

Suppose a complete rotation is divided into 360 equal parts. This would produce 360 small angles. Each small angle would be $\frac{1}{360}$ of a complete rotation.

Definition of Degree

A **degree** is a unit of angle measure. One degree is symbolized by 1°. It is $\frac{1}{360}$ of a complete rotation of a ray.

The symbol $m\angle KLM$ represents the degree measure of ∠KLM. If ∠KLM measures 36° we write $m\angle KLM = 36$.

A protractor may be used to find the degree measure of a given angle. Place the center of the protractor at the vertex of the angle. Line up the 0 mark on the inner scale with one ray of the angle. Read where the other ray falls on the same scale.

The two scales make it convenient to measure angles in different positions.

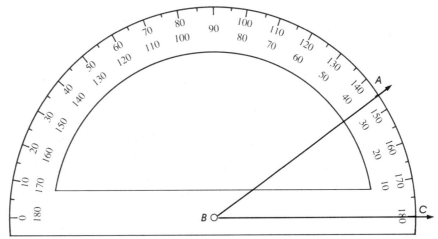

The degree measure of angle *ABC* is 35.
$$m\angle ABC = 35$$

Example

1 **Find the angle measurement for each sector of the circle graph below.**

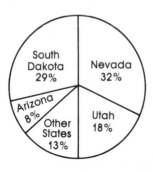

South Dakota 29%
Nevada 32%
Arizona 8%
Other States 13%
Utah 18%

Measure each angle with a protractor. The center of the circle is the vertex of each angle.

Nevada .... 115°
South Dakota .... 104°
Utah .... 65°
Arizona .... 29°
Other States .... 47°

Angles can be classified according to their measure.

Type of Angle	Degree Measure of Angle
Acute	less than 90
Right	90
Obtuse	between 90 and 180
Straight	180
Reflex	greater than 180

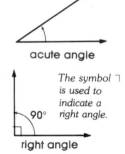

acute angle

The symbol ⌐ is used to indicate a right angle.

90°

right angle

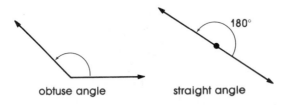

obtuse angle

180°

straight angle

reflex angle

Definition of Perpendicular

Two lines which intersect to form right angles are **perpendicular** to each other.

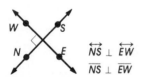

$\overleftrightarrow{NS} \perp \overleftrightarrow{EW}$
$\overline{NS} \perp \overline{EW}$

The symbol ⊥ means *is perpendicular to*. In the figure at the left, $\overleftrightarrow{NS}$ is perpendicular to $\overleftrightarrow{EW}$.

Parts of lines, such as line segments or rays, are perpendicular to each other if the lines containing them are perpendicular. For example, $\overline{NS} \perp \overline{EW}$.

Exercises

Exploratory   For each of the following use the figure below. State *yes* or *no*.

1. $\overrightarrow{GV}$ and $\overrightarrow{GC}$ name the same ray.
2. $\overrightarrow{AV}$ is another name for $\overrightarrow{VA}$.
3. E is the endpoint of $\overrightarrow{EF}$.
4. $\overrightarrow{AB}$ and $\overrightarrow{BA}$ have the same endpoint.
5. C lies on $\overrightarrow{VG}$.
6. $\overrightarrow{EV}$ and $\overrightarrow{VE}$ are opposite rays.
7. $\overline{AC} \perp \overleftrightarrow{GC}$
8. $m\angle 3 = m\angle 4$
9. $\angle GVA$ is a right angle.
10. $\overrightarrow{AE} \perp \overleftrightarrow{FV}$
11. $m\angle 1 + m\angle 2 = 90$
12. $m\angle 4 = m\angle 1 + m\angle 2$

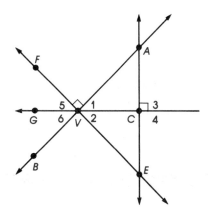

Use the figure below to find the degree measure for each of the following.

13. $\angle QVZ$
14. $\angle RVZ$
15. $\angle SVZ$
16. $\angle TVZ$
17. $\angle WVZ$
18. $\angle XVZ$
19. $\angle YVZ$
20. $\angle QVP$
21. $\angle SVP$
22. $\angle TVP$
23. $\angle WVP$
24. $\angle YVP$

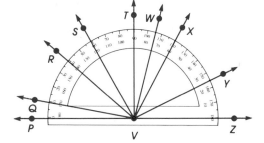

Written   Classify angles with each of the following measurements as *acute, right, obtuse, straight,* or *reflex*.

1. $85°$
2. $115°$
3. $180°$
4. $185°$
5. $90°$
6. $0.015°$
7. $90.05°$
8. $355°$

Use a protractor to find the measurement of each of the following angles.

9.
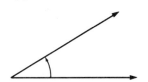

10.

11.

Use a protractor to draw angles having the following measurements.
12. $70°$ 13. $150°$ 14. $13°$

Use the figure below to answer each of the following.

15. ∠1 and ∠PQS name the same angle. Write
    yes or no.
16. Name the vertex of ∠TQR.
17. Name the sides of ∠2.
18. Name the common side of ∠1 and ∠2.
19. Name all angles with $\overrightarrow{QV}$ as a side.
20. Name a point that lies on ∠4.
21. Name a point in the interior of ∠TQR.
22. Name a point in the exterior of ∠2.

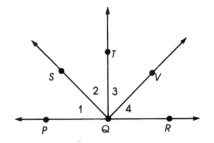

Find the measurement of the angle which corresponds to each of the following parts of a complete rotation of 360°.

23. $\frac{1}{20}$ 24. $\frac{5}{18}$ 25. $\frac{4}{9}$ 26. 25%   27. $16\frac{2}{3}$%

What part of a complete rotation is an angle having each of the following measurements?

28. 90°   29. 279°   30. 60°   31. 40°   32. $x°$

For each of the following, draw two angles to satisfy the given conditions.

33. The angles intersect in a point.   34. The angles intersect in two points.
35. The angles intersect in three points.   36. The angles intersect in four points.

Determine whether each of the following statements is *true* or *false*.

37. The intersection of two rays must be a   38. The intersection of two rays cannot be
    point.   a point.
39. The intersection of two line segments   40. The intersection of a line segment and
    may be a ray.   a ray must be a line segment.

Mathematical Excursions

Radian Measure

One unit of angle measure is the degree. Another common unit of angle measure is the radian. A radian is defined using a circle.

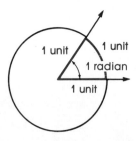

Cut a piece of string that is the same length as the radius of a circle. Stretch the string along the edge of the circle. Mark the beginning point and endpoint of the string on the circle. The angle formed by drawing two rays from the center, one through each point, measures 1 radian. If a circle has a 1 unit radius, then the measurement around the circle is 2π units.

5.3   Special Angles

When three or more rays in the same plane have a common endpoint, various angles may be formed. Two angles are **adjacent** when they have a common side, the same vertex, and *no* interior points in common.

In the figure at the right, ∠PQR and ∠RQS are adjacent. The angles PQS and RQS are *not* adjacent.   *Why?*

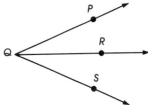

A ray is a **bisector of an angle** if and only if it is the common side of two adjacent angles having the same measure.

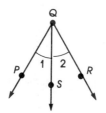

In the figure at the left, $m\angle 1 = m\angle 2$. Because $\overrightarrow{QS}$ separates ∠PQR into two angles having the same measure, it is the bisector of ∠PQR.

Two angles can be related by the sum of their measures.

| **Definition of Supplementary Angles** | Two angles are **supplementary** if and only if the sum of their degree measures is 180. |

Each angle is called a supplement of the other.

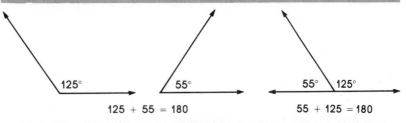

125 + 55 = 180   55 + 125 = 180

| **Definition of Complementary Angles** | Two angles are **complementary** if and only if the sum of their degree measures is 90. |

Each angle is called a complement of the other.

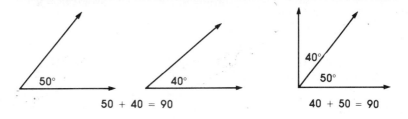

50 + 40 = 90   40 + 50 = 90

Examples

1 **Angles *A* and *B* are complementary. If *m*∠*A* = 73, find *m*∠*B*.**

Since ∠*A* and ∠*B* are complementary, the following is true.

$$m\angle A + m\angle B = 90$$
$$73 + m\angle B = 90$$
$$m\angle B = 17 \qquad 73 + 17 = 90 \ \checkmark$$

2 **Two angles are supplementary. If one angle measures 20 less than 3 times the other, find the degree measure of each angle.**

Let *x* represent the degree measure of one angle. Then the degree measure of the other must be $3x - 20$.

$$x + (3x - 20) = 180$$
$$4x - 20 = 180$$
$$4x = 200 \text{ or } x = 50$$

The degree measure of one angle is 50. The degree measure of the other is $3(50) - 20$ or 130. $130 + 50 = 180 \ \checkmark$

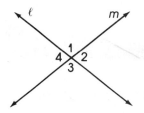

Vertical angles are two nonadjacent angles formed by two intersecting lines. In the figure, lines ℓ and *m* intersect, and ∠2 and ∠4 are vertical angles as are ∠1 and ∠3. Notice that the vertical angles appear to have the same degree measure.

Theorem 5-1

Vertical angles have the same measure.

Mathematical statements which can be proved are called **theorems.** The following is an *informal proof* of Theorem 5-1.

Suppose that lines ℓ and *m* intersect to form angles 1, 2, 3, and 4 as shown above. Angles 1 and 2 form a straight angle and are therefore supplementary. The same is true of angles 2 and 3.

$m\angle 1 + m\angle 2 = 180$	$m\angle 2 + m\angle 3 = 180$
$m\angle 1 = 180 - m\angle 2$	$m\angle 3 = 180 - m\angle 2$

Subtract m∠2 from both sides of each equation.

Thus, $m\angle 1 = m\angle 3$.   *Why?*

Examples

3 If $m\angle CED$ is 42, find the degree measure of the other angles.

$\angle BEC$ and $\angle CED$ are supplementary. Thus, $m\angle BEC = 180 - 42$ or 138. Since Theorem 5-1 says vertical angles have the same measure, $m\angle BEA = m\angle CED$ or 42 and $m\angle AED = m\angle BEC$ or 138.

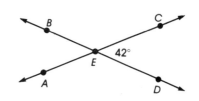

4 Find $m\angle PQR$ and $m\angle TQS$.

Since $\angle PQR$ and $\angle TQS$ are vertical angles, they have the same degree measure.
$$3x - 40 = 2x + 10$$
$$x - 40 = 10$$
$$x = 50$$
$$m\angle PQR = 2(50) + 10 \text{ or } 110$$
$$m\angle TQS = 3(50) - 40 \text{ or } 110$$

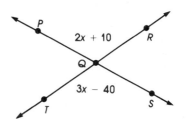

Exercises

Exploratory   Angles A and B are complementary. Find $m\angle A$ when $\angle B$ has each of the following degree measures.

1. 30	**2.** 15	**3.** 87	**4.** 22.29
5. $4\frac{1}{2}$	**6.** $63\frac{3}{8}$	**7.** 6.05	**8.** K

Angles M and P are supplementary. Find $m\angle M$ when $\angle P$ has each of the following degree measures.

9. 170	**10.** 15	**11.** 90	**12.** 55.5
13. $27\frac{2}{3}$	**14.** $132\frac{3}{5}$	**15.** 161.3	**16.** x

Written   Draw and label a diagram to show each of the following.

1. $\overrightarrow{CR}$ is the bisector of $\angle ACE$ and $m\angle ACE = 80$.

2. Angles ABC and CBD are adjacent angles.

3. Angles XYZ and ZYW are adjacent angles. $m\angle XYZ = 50$ and $m\angle ZYW = 25$.

4. Angles RST and TSW are adjacent angles. $m\angle RST = 40$ and $m\angle TSW = 20$. $\overrightarrow{SK}$ is the bisector of $\angle RSW$.

Each of the following pairs of angles is either complementary or supplementary. Find the degree measure of each angle.

5.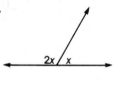
2x / x

6.
x + 10 / x

7.
y
4y

8.
x + 12
3x

9.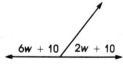
6w + 10 / 2w + 10

10.
8x + 25
5x

11.
5x − 5 \ 7x + 29

12.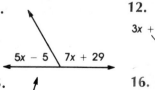
3x + 3 7x − 3

13.
x + 7
2x + 14

14.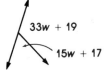
3y − 10
y + 8

15.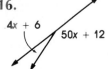
33w + 19
15w + 17

16.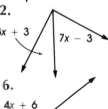
4x + 6
50x + 12

In each of the following, ∠R and ∠T are complementary. Find m∠R and m∠T.

17. m∠R is twice m∠T
18. m∠R is 10 more than m∠T
19. m∠R is 5 times m∠T
20. m∠R is 4 times m∠T
21. m∠R is 40 less than m∠T
22. m∠R is 5 more than m∠T
23. m∠R equals m∠T
24. m∠R is 21 more than twice m∠T
25. m∠R is 10 less than 3 times m∠T
26. m∠R is 20 more than 4 times m∠T

Find m∠R and m∠T for exercises 17-26 if ∠R and ∠T are supplementary.

27. Exercise 17
28. Exercise 18
29. Exercise 19
30. Exercise 20
31. Exercise 21
32. Exercise 22
33. Exercise 23
34. Exercise 24
35. Exercise 25
36. Exercise 26

Find the value of x in each of the following.

37.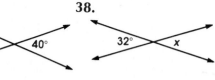
x 40°

38.
32° x

39.
50°
x

40.
x

41.
46°
x

42.
20°
80°
x

43.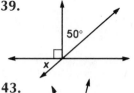
x 60°
80°

44.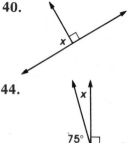
x
75°

45.
57° 2x − 3°

46.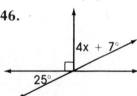
4x + 7°
25°

47.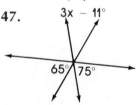
3x − 11°
65° 75°

48.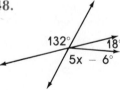
132° 18°
5x − 6°

Solve each problem.

49. Two angles are complementary. If one angle measures 12 more than the other, find the degree measure of each angle.

50. Two angles are complementary. If one angle measures 3 less than half the other, find the degree measure of each angle.

51. Two angles are supplementary. If one angle measures 4 times the other, find the degree measure of each angle.

52. Two angles are supplementary. If one angle measures 9 more than twice the other, find the degree measure of each angle.

53. If $x + 17$ represents the degree measure of an angle, write the expressions that represent the complement and the supplement of the angle.

54. If $15 - 4x$ represents the degree measure of an angle, write the expressions that represent the complement and the supplement of the angle.

Challenge Find $m\angle A$, $m\angle B$, and $m\angle C$ if their sum is 180 and the following conditions are true.

1. $m\angle A = 2(m\angle B)$; $m\angle B = m\angle C + 20$ **2.** $m\angle A = 3(m\angle C)$; $m\angle A - 30 = m\angle B$

3. $m\angle A = \frac{1}{5}(m\angle B)$; $m\angle B = \frac{1}{3}(m\angle C + 30)$ **4.** $m\angle A = m\angle B + 20$; $m\angle B = 3(m\angle C) - 5$

5. Find the measure of the obtuse angle formed by the hands of a clock at 4:40 p.m.

━━━━━━━ Mixed Review ━━━━━━━

Solve each of the following.

1. $17 - 6x = -25$ **2.** $5y + 12 = 10 + 3y$ **3.** $2a + 7 \le 13$

4. $\dfrac{4b + 17}{9} > \dfrac{9b - 8}{11}$ **5.** $\dfrac{10m - 3}{7} = \dfrac{41 - m}{3}$ **6.** $\dfrac{2n - 5}{3} - \dfrac{5n - 2}{16} = -\dfrac{3}{4}$

7. $-11 < 5 - 4p \le 13$ **8.** $5(2r + 3) - 3(2 - 5r) = r - 39$

9. $(3c + 11 < 23) \lor (1 - 8c < -47)$ **10.** $(d - 15 > 6d) \land (7d - 1 < 23 - d)$

Graph each of the following. Then identify each graph as a line, line segment, or ray.

11. $x \ge 2$ **12.** $(x \le 4) \lor (x \ge -1)$

13. $(x \le 3) \lor (x \le -2)$ **14.** $-5 \le x \le 1$

15. $(x \ge 0) \land (x \ge 6)$ **16.** $(-2x \le 6) \land (5x \le -5)$

17. The measure of the supplement of an angle is 20 more than twice the measure of its complement. Find the measure of the angle.

18. Leann plans to build a rectangular pen for her ducks. The length of the pen must be 3 m more than twice its width. What are the greatest possible dimensions of the pen if the perimeter can be no greater than 120 m?

5.4 Parallel Lines, Transversals, and Angles

When you connect two points with a line you assume there is exactly one way of doing so. However you cannot *prove* that there is exactly one line. A statement which is accepted as true *without proof* is called an **axiom.**

Axiom 5-1

For every two distinct points, there exists one and only one line containing both of them.

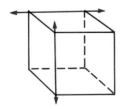

The rails on a railroad track never meet. Lines in the same plane that never meet are called **parallel lines.**

Definition of Parallel Lines

Two lines are parallel if and only if they lie in the same plane and they do *not* intersect.

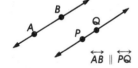

$$\overleftrightarrow{AB} \parallel \overleftrightarrow{PQ}$$

The symbol $\parallel$ means *is parallel to*. In the figure at the left, $\overleftrightarrow{AB}$ is parallel to $\overleftrightarrow{PQ}$.

Parts of lines, such as line segments or rays, are parallel to each other if the lines containing them are parallel. For example, $\overrightarrow{AB} \parallel \overline{PQ}$. Every line is considered parallel to itself.

Some lines do *not* intersect and yet are *not* parallel. For example, the two lines indicated on the cube will never meet and are *not* parallel. These are called **skew lines.**

Another statement which cannot be proved is given below.

Axiom 5-2

Given a line and a point *not* on the line, then there is one and only one line through the point which is parallel to the given line.

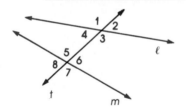

In a plane, a **transversal** is a line that intersects two other lines in two different points. For example, the transversal t intersects, or cuts, lines ℓ and m.

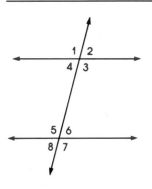

Eight angles are formed by the intersection of a transversal and two lines.

Alternate Interior Angles	∠4 and ∠6, ∠3 and ∠5
Corresponding Angles	∠1 and ∠5, ∠2 and ∠6
	∠3 and ∠7, ∠4 and ∠8

What happens if two parallel lines are cut by a transversal? Measure the sets of corresponding angles shown at the left. You might conclude that the corresponding angles have the same measure.

Axiom 5-3

If two parallel lines are cut by a transversal, then the corresponding angles have the same measure.

Example

1 **Parallel lines ℓ and m are cut by transversal t and form ∠1 and ∠2 which measure 70 and 110 respectively. Find m∠3 and m∠4.**

∠1 and ∠3 are corresponding angles as are ∠2 and ∠4. Thus, each pair of angles has the same measure.

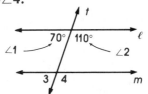

$m\angle 3 = m\angle 1 = 70$ and $m\angle 4 = m\angle 2 = 110$

The following theorem states another fact about the angles formed by parallel lines and a transversal.

Theorem 5-2

If two parallel lines are cut by a transversal, then the alternate interior angles have the same measure.

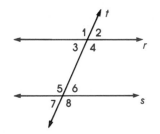

An informal proof of Theorem 5-2 is given below.

Suppose that parallel lines r and s are cut by transversal t. Angles 2 and 3 have the same measure because they are vertical angles. But ∠2 and ∠6 are corresponding. So they also have the same measure. Since ∠3 and ∠6 both have the same measure as ∠2, we can conclude $m\angle 3 = m\angle 6$.

Example

2 Find the degree measures of the eight angles formed when transversal n cuts parallel lines k and ℓ.

The angles whose degree measures are given are supplementary.

$$(x + 30) + 2x = 180$$
$$3x + 30 = 180$$
$$3x = 150$$
$$x = 50$$

Thus, $x + 30 = 80$ and $2x = 100$. Now, use your knowledge of vertical, corresponding, and alternate interior angles to find the degree measures of the other angles. They are shown in the figure at the right.

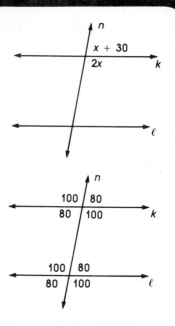

Exercises

Exploratory State whether each of the following suggests *intersecting, parallel,* or *skew lines.*

1. rungs on a ladder
2. railroad crossing sign
3. airline flight paths
4. lines on a football field
5. airport runways
6. electric power lines
7. guitar strings
8. marks on a ruler
9. artist's T-square
10. rows of corn in a field

Use the figure below. Suppose $\ell \parallel m$. Name each of the following.

11. Four pairs of corresponding angles
12. Four pairs of alternate interior angles

Suppose $t \parallel s$. Name each of the following.

13. Four pairs of corresponding angles
14. Four pairs of alternate interior angles

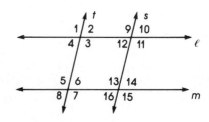

Written   Draw a diagram to illustrate each of the following.

1. $\overleftrightarrow{AB} \parallel \overleftrightarrow{CD}$

2. ℓ and m are skew.

3. k is a transversal to ℓ and m

4. $\overleftrightarrow{RS} \parallel \overleftrightarrow{TV}$, both lines are cut by $\overleftrightarrow{AB}$

In each of the following, ℓ and m are parallel. Find the degree measures of the numbered angles.

5.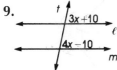

6.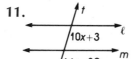

7.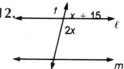

8.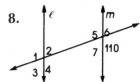

In each of the following, ℓ and m are parallel. Find the measures of each of the eight angles formed when ℓ and m are cut by t.

9.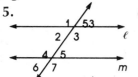

10.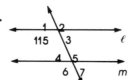

11.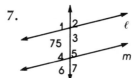

12.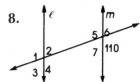

13.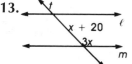

14.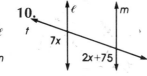

15.

16.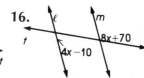

State whether each of the following is *always, sometimes,* or *never* true.

17. If ℓ and m are lines in the same plane, then they are parallel.

18. If ℓ and m are lines in the same plane, then they intersect in exactly two points.

19. If ℓ, m, and n are lines in one plane, then they are all parallel.

20. If ℓ and m are skew lines, and n is parallel to m, then n and ℓ are skew.

21. If ℓ is a line in the plane and D is a point in the same plane *not* on ℓ, then there is exactly one line containing D and parallel to ℓ.

22. If two parallel lines are cut by a transversal, then the alternate interior angles formed are supplementary.

23. If x and y are alternate interior angles, then they must be on the same side of the transversal.

24. If ℓ is a line and D is a point not on ℓ, then there exists only one plane containing ℓ and D.

Challenge   Use an informal proof to show that each of the following is true.

1. Angles on alternate sides of a transversal and outside the parallel lines are alternate exterior angles. If two parallel lines are cut by a transversal, then the alternate exterior angles have the same measure.

2. Assume that Theorem 5-2 is true without proving it. Then show that Axiom 5-3 is true using this assumption.

5.5 Classifying Polygons

Many-sided figures can be found in nature, art, and everyday life. You can trace the following figures without lifting your pencil from the paper. You can return to the starting point without tracing any point other than the starting point more than once. These figures are called **simple closed curves.**

Simple closed curves lie entirely in one plane.

The three figures above on the right are **polygons.** A polygon is a simple closed curve composed entirely of line segments called **sides.**

Number of Sides	Polygon
3	triangle
4	quadrilateral
5	pentagon
6	hexagon
8	octagon
10	decagon
12	dodecagon

Polygons can be classified by their number of sides. The chart at the left gives some common names.

In general, a polygon with n sides is called an *n-gon.* Thus, an octagon also is called an 8-gon. A polygon with 14 sides is called a 14-gon.

A polygon in which all sides have the same measure and all angles have the same measure is called a **regular polygon.**

The marks show sides and angles that have the same measure.

 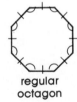

regular
octagon

regular
pentagon

regular
hexagon

Example

1 Classify the following polygons by the number of sides and as regular or not regular.

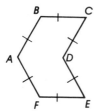

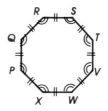

Polygon *ABCDEF* has 6 sides. It is a hexagon.
The angles do *not* have the same measure.
Thus, it is *not* regular.

Polygon *PQRSTVWX* has 8 sides. It is an octagon.
All sides have the same measure as do the angles.
Thus, it is regular.

Triangles can be classified even further. They can be classified by the number of sides that have the same measure.

Lab Activity

You can discover more about isosceles triangles in Lab 4 on page A7.

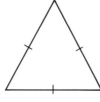

Equilateral
3 sides that have the same measure

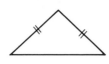

Isosceles
at least 2 sides that have the same measure

Scalene
no sides that have the same measure

Triangles also can be classified by angles. All triangles have at least two acute angles. Use the third angle to classify the triangle.

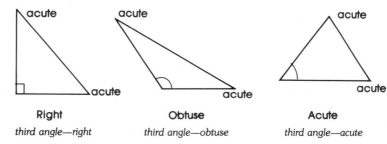

Right
third angle—right

Obtuse
third angle—obtuse

Acute
third angle—acute

The following table presents a classification of quadrilaterals.

Defining Characteristic	Name	Figure
a quadrilateral with exactly one pair of parallel sides	trapezoid	
a quadrilateral with two pairs of parallel sides	parallelogram	
a parallelogram with all sides having same measure	rhombus	
a parallelogram with four right angles	rectangle	
a rhombus with four right angles or a rectangle with all sides having same measure	square	

Exercises

Exploratory   Tell whether each figure is a simple closed curve. State *yes* or *no*.

1.   2.   3.   4.

5.   6.   7.   8.

Tell whether each figure in exercises 1-8 is a polygon. State *yes* or *no*.

9. Exercise 1   **10.** Exercise 2   **11.** Exercise 3   **12.** Exercise 4
13. Exercise 5   **14.** Exercise 6   **15.** Exercise 7   **16.** Exercise 8

Written   Classify each polygon by the number of sides and as regular or *not* regular.

1.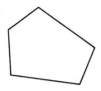

2.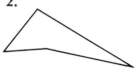

3.

4.

5.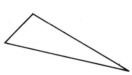

6.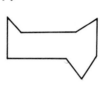

7.

8.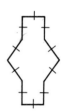

Classify each triangle by its sides and then by its angles.

9.

10.

11.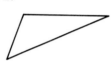

12.

13.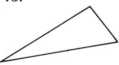

14.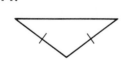

15.

16.

State if each of the following statements is *always, sometimes,* or *never* true.

17. A right triangle is isosceles.

18. An obtuse triangle is isosceles.

19. An obtuse triangle has a right angle.

20. An acute triangle has exactly one acute angle.

21. An acute triangle has exactly two acute angles.

22. An obtuse triangle has exactly two acute angles.

23. If a quadrilateral is a rectangle, then it is a parallelogram.

24. If a quadrilateral is a rectangle, then it is a square.

25. If a quadrilateral is a rhombus, then it is a square.

26. If a quadrilateral is a trapezoid, then it is a parallelogram.

27. If a quadrilateral is a square, then it is a parallelogram.

28. If a quadrilateral is a square, then it is a rhombus.

List all possible types of quadrilaterals with the following characteristics.

29. two pairs of parallel sides

30. exactly one pair of parallel sides

31. all sides have same measure

32. all angles have same measure

5.6   Angles of a Triangle

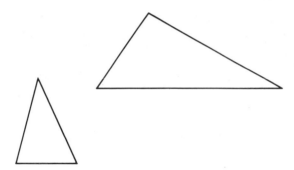

Suppose you measure the angles of a triangle. What do you think the sum of these measures would be? Use a protractor to measure the angles of the two triangles shown at the left. Add the measures of the angles of each triangle. If you have measured carefully, you should see that the sum of the degree measures of the angles is 180.

Theorem 5-3

> The sum of the degree measures of the angles of a triangle is 180.

An informal proof of Theorem 5-3 is given below.

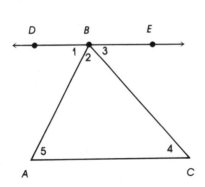

Draw $\overleftrightarrow{DE}$ through B and parallel to $\overline{AC}$ using Axiom 5-2. Then by the definition of transversal, $\overline{BC}$ and $\overline{AB}$ are transversals which cut $\overline{DE}$ and $\overline{AC}$. Theorem 5-2 says that $m\angle 3 = m\angle 4$ and $m\angle 1 = m\angle 5$. The sum of $m\angle 1$, $m\angle 2$, and $m\angle 3$ must be 180 since they form a straight angle. That is, $m\angle 1 + m\angle 2 + m\angle 3 = 180$. Finally, replace $m\angle 1$ by $m\angle 5$ and $m\angle 3$ by $m\angle 4$.
$$m\angle 5 + m\angle 2 + m\angle 4 = 180.$$

Example

1   The degree measures of two angles of a triangle are 34 and 72. Find the degree measure of the third angle.

The sum of the degree measures of the angles is 180.
$$34 + 72 + x = 180$$
$$106 + x = 180$$
$$x = 180 - 106 \text{ or } 74$$

The degree measure of the third angle is 74. $34 + 72 + 74 = 180$ √

Example

2 In △KLM, m∠K is two times m∠L and m∠L is 4 more than m∠M. Find m∠K, m∠L, and m∠M.

Let x represent m∠M. Then x + 4 represents m∠L and 2(x + 4) represents m∠K.

$$m∠M + m∠L + m∠K = 180 \qquad \text{Why?}$$
$$x + (x + 4) + 2(x + 4) = 180$$
$$4x + 12 = 180$$
$$x = 42$$

m∠M = x or 42 \qquad m∠L = x + 4 or 46 \qquad m∠K = 2(x + 4) or 92

$$42 + 46 + 92 = 180 \quad \checkmark$$

Exercises

Exploratory   Use the figure below to solve each of the following.
1. Find m∠1 if m∠2 = 40 and m∠3 = 55.
2. Find m∠1 if m∠2 = 60 and m∠3 = 60.
3. Find m∠1 if m∠2 = 81 and m∠3 = 74.
4. Find m∠2 if m∠1 = 45 and m∠4 = 105.
5. Find m∠2 if m∠1 = 47 and m∠4 = 132.
6. Find m∠2 if m∠1 = 44 and m∠4 = 121.
7. Find m∠3 if m∠1 = 45 and m∠5 = 98.
8. Find m∠3 if m∠1 = 60 and m∠5 = 114.

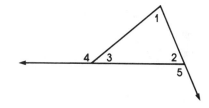

Written   Find the degree measures of the angles in each of the following.

1.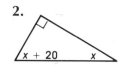
2.
3.
4.

5.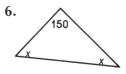
6.
7.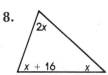
8.

9.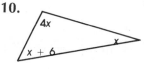
10.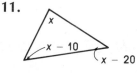
11.
12.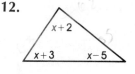

Solve each problem.

13. In $\triangle KLM$, $m\angle K = 120$ and $m\angle L = m\angle M$. Find $m\angle L$.

14. In $\triangle ABC$, $m\angle A = k$ and $m\angle B = 2k$. If $m\angle C = 54$, find $m\angle A$ and $m\angle B$.

15. In $\triangle PRS$, $m\angle P$ is 3 less than $m\angle R$, and $m\angle S$ is 21 more than four times $m\angle P$. Find $m\angle P$, $m\angle R$, and $m\angle S$.

16. In $\triangle DEF$, $m\angle D$ is 20 less than twice $m\angle E$, and $m\angle E$ equals $m\angle F$. Find $m\angle D$, $m\angle E$, and $m\angle F$.

State if each of the following statements is *always, sometimes,* or *never* true.

17. The sum of the measures of the angles of a scalene triangle is greater than 180.

18. A line exists containing any vertex of a triangle and parallel to the opposite side.

Challenge The measure of an exterior angle of a triangle is equal to the measure of the two remote interior angles. That is, $m\angle x + m\angle y = m\angle z$. Find the degree measure of angle y in each of the following.

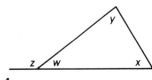

1.

2.

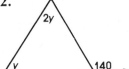

3.

4.

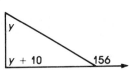

═══════Mixed Review═══════

Solve each formula for the indicated variable.

1. $p = mq + b$ for q

2. $s = \dfrac{Y - y}{X - x}$ for X

3. $t = \dfrac{-g}{a - g}$ for g

Solve each of the following.

4. $\dfrac{4x - 23}{6} + \dfrac{11 - 5x}{4} = 1 - x$

5. $-5 - 2y \le 3y < 15 + 2y$

Find $m\angle 1$ and $m\angle 2$ if the following conditions are true.

6. $\angle 1$ and $\angle 2$ are alternate interior angles formed when parallel lines ℓ and m are cut by transversal t; $m\angle 1 = 12x + 11$; $m\angle 2 = 7x + 36$

7. $\angle 1$ and $\angle 2$ are corresponding angles formed when parallel lines ℓ and m are cut by transversal t; $m\angle 1 = 91 - 5x$; $m\angle 2 = 8x + 65$

8. $\angle 1$ and $\angle 2$ are vertical angles; $m\angle 1 = 67 - 2x$; $m\angle 2 = 6x + 11$

9. $\angle 1$ and $\angle 2$ are complementary angles; $m\angle 1 = 67 - 2x$; $m\angle 2 = 6x + 11$

10. $\angle 1$ and $\angle 2$ are supplementary angles; $m\angle 1 = 67 - 2x$; $m\angle 2 = 6x + 11$

11. If $\angle A$ is an acute angle and $m\angle A = 5x - 70$, find all possible values for x.

5.7   Quadrilaterals

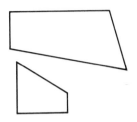

Suppose you measure the angles of a quadrilateral. What do you think the sum of these measures would be? Use a protractor to measure the angles of the two quadrilaterals shown at the left. Add the measures of the angles of each quadrilateral. If you have measured carefully, you should see that the sum of the degree measures of the angles is 360.

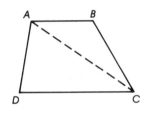

Theorem 5-4

> The sum of the degree measures of the angles of a quadrilateral is 360.

This theorem can be proved informally as follows.

Let *ABCD* be any quadrilateral. There exists a line through *A* and *C* according to Axiom 5-1. This line separates quadrilateral *ABCD* into two triangles. But, the sum of the degree measures of each triangle is 180. Thus, the total of the degree measures of the angles of both triangles is 360. We conclude that the sum of the degree measures of the angles in quadrilateral *ABCD* is 360.

Example

1 Find the degree measures of the angles in the quadrilateral.

The sum of the degree measures of the angles is 360.

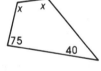

$$75 + 40 + x + x = 360$$
$$115 + 2x = 360$$
$$2x = 245$$
$$x = 122\tfrac{1}{2} \qquad 75 + 40 + 122\tfrac{1}{2} + 122\tfrac{1}{2} = 360 \quad \checkmark$$

The degree measures of the angles are 75, 40, $122\tfrac{1}{2}$, and $122\tfrac{1}{2}$.

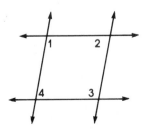

A parallelogram is formed when two pairs of parallel lines intersect as shown. Any two angles which share a side of the parallelogram are called **consecutive angles.** For example, in the figure at the left, ∠2 and ∠3 are consecutive angles. Two angles of a parallelogram which are *not* consecutive are called **opposite angles** of the parallelogram. For example, ∠2 and ∠4 are opposite angles.

Theorem 5-5	The sum of the degree measures of any two consecutive angles of a parallelogram is 180.

Theorem 5-6	Any two opposite angles of a parallelogram have the same measure.

Example

2 Find the degree measure of each angle in the parallelogram shown at the right.

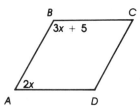

$2x + (3x + 5) = 180$ *Why?*
$5x + 5 = 180$
$5x = 175$ or $x = 35$

$m\angle A = 2x$ or 70 and $m\angle B = 3x + 5$ or 110. $70 + 110 = 180$ ✓
Using Theorem 5-6, $m\angle A = m\angle C$ and $m\angle B = m\angle D$.
Thus, the degree measures of the angles A, B, C, and D
are 70, 110, 70, and 110, respectively.

Exercises

Exploratory Use the parallelogram below to solve each problem.

1. Find $m\angle C$ if $m\angle A = 50$.   **2.** Find $m\angle D$ if $m\angle B = 65$.
3. Find $m\angle B$ if $m\angle C = 93$.   **4.** Find $m\angle A$ if $m\angle D = 88$.
5. Find $m\angle D$ and $m\angle B$ if $m\angle A = 105$.
6. Find $m\angle A$ and $m\angle C$ if $m\angle B = 39$.

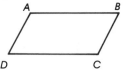

Written Find the degree measures of the angles in each of the following.

1.

80
120
x 100

2.
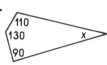
110
130
90   x

3.

x
125
x
35

4.

$x-15$
$x+10$
$x-30$ $x-5$

5.

$x+10$
$x+20$
$x-10$ $x-20$

6.

$x-20$
$x+50$
$x-70$ $x-40$

7.

$x+10$
$x+35$
80   $x-65$

8.

$3x-10$ $2x$
$x+35$
$x-15$

9.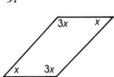

10.

11.

12.

13. The degree measure of one angle of a parallelogram is 30 more than that of a consecutive angle. Find the degree measure of each angle of the parallelogram.

14. The degree measure of one angle of a parallelogram is two times that of another angle. The angles are consecutive. Find the degree measure of each angle of the parallelogram.

15. Write an informal proof of Theorem 5-5.

16. Supply the reasons for each step in the proof of Theorem 5-6. That is, if polygon $ABCD$ is a parallelogram, then $m\angle A = m\angle C$ and $m\angle B = m\angle D$.

a. $\overline{AD} \parallel \overline{BC}$ and $\overline{AB} \parallel \overline{DC}$

b. $m\angle DAB = m\angle CBF$ and $m\angle ABC = m\angle EAD$

c. $m\angle CBF = m\angle BCD$ and $m\angle EAD = m\angle ADC$

d. $m\angle DAB = m\angle BCD$ (or $m\angle A = m\angle C$)
$m\angle ABC = m\angle ADC$ (or $m\angle B = m\angle D$)

a. _____

b. _____

c. _____

d. _____

Challenge A line segment joining two nonconsecutive vertices of a polygon is called a *diagonal* of the polygon. In each case below, the polygon is separated into triangles by the diagonals. **Find the sum of the degree measures of the angles of each polygon.**

1.

2.

3.

4. Copy and complete the following table.

Polygon	Number of Sides	Number of Triangles	Sum of Degree Measures of Angles
triangle	3	1	$1 \cdot 180 = 180$
quadrilateral	4	2	$2 \cdot 180 = 360$
pentagon	_____	3	_____
hexagon	_____	_____	_____
octagon	_____	_____	_____
$\vdots$	$\vdots$	$\vdots$	$\vdots$
n-gon	n	_____	_____

Problem Solving Application: Flow Proofs

In this chapter, we have used informal proofs to show why a theorem is true. Another type of proof is a *flow proof*. In a flow proof, statements are logically organized, starting with given statements which are always underlined. Arrows are drawn between statements to show the order that they should follow. A number is placed above each arrow to refer to the reasons that allow the statement to be made. The reasons are listed below the statements.

Examples

1 **Write a flow proof for the statement "if $3x + 7 = 10$, then $x = 1$."**

First, write the given and the prove statements.

Given: $3x + 7 = 10$ **Prove:** $x = 1$

The next step is to build the proof, connecting each step with the next. Make sure you have a valid reason for each step.

$$\underline{3x + 7 = 10} \xrightarrow{1} 3x + 7 + (-7) = 10 + (-7) \xrightarrow{2} 3x + 0 = 3 \xrightarrow{3}$$

$$3x = 3 \xrightarrow{4} \frac{1}{3} \cdot 3x = \frac{1}{3} \cdot 3 \xrightarrow{5} 1 \cdot x = 1 \xrightarrow{6} x = 1$$

1. Addition Property for Equations
2. Associative Property for Addition, Additive Inverse Property, Substitution
3. Additive Identity Property, Substitution
4. Multiplication Property for Equations
5. Associative Property for Multiplication, Multiplicative Inverse Property, Substitution
6. Multiplicative Identity Property, Substitution

Note that the given is always underlined and that each step may involve more than one reason.

2 **Write a flow proof for the following statement.**
If $\overline{BD}$ is a diagonal of parallelogram $ABCD$ and $m\angle ABD = m\angle ADB$, then $\overline{DB}$ bisects $\angle ADC$.

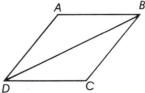

Given: parallelogram $ABCD$
        $\overline{BD}$ is a diagonal.
        $m\angle ABD = m\angle ADB$
Prove: $\overline{DB}$ bisects $\angle ADC$.

Parallelogram $ABCD$
$\overline{BD}$ is a diagonal.
$m\angle ABD = m\angle ADB$ $\left.\right\} \overset{1}{\rightarrow} \overline{BA} \parallel \overline{CD} \overset{2}{\rightarrow} \overline{BD}$ is a transversal cutting $\overline{BA}$ and $\overline{CD}. \overset{3}{\rightarrow}$

$\angle CDB$ and $\angle ABD$ are alternate interior angles. $\overset{4}{\rightarrow} m\angle CDB = m\angle ABD \overset{5}{\rightarrow}$
$m\angle CDB = m\angle ADB \overset{6}{\rightarrow} \overline{DB}$ bisects $\angle ADC$.

1. Definition of parallelogram
2. Definition of transversal
3. Definition of alternate interior angles
4. Theorem 5-2: If two parallel lines are cut by a transversal, then the alternate interior angles have the same measure.
5. Substitition
6. Definition of bisector of an angle

Exercises

Exploratory   **Identify the given statement and the prove statement of each conditional.**

1. If $2y - 3 < 7$ then $y < 5$.

2. If an angle is a right angle, then it is not an acute angle.

3. A student will be suspended if the student leaves campus without permission.

4. Two intersecting lines contain at least three points.

Written   **Prove each statement using a flow proof.**

1. If $-4x + 3 = 7$, then $x = -1$.

2. If $3 - 2x > -5$, then $x < 4$.

In the figure at the right, $m\angle GZE = 45$ and $\overrightarrow{ZD}$ bisects $\angle EZH$. Find the measure of each of the following angles. Then write a flow proof to justify your answer.

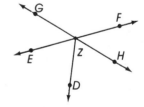

3. $\angle FZH$ 4. $\angle GZF$ 5. $\angle EZD$

In the figure at the right, $\overleftrightarrow{AB} \parallel \overleftrightarrow{CD}$ and $m\angle ARQ = m\angle BRS$. Prove each statement using a flow proof.

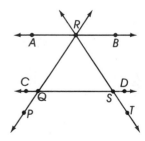

6. $m\angle CQP = m\angle RSQ$ 7. $\angle ARS$ is supplementary to $\angle TSD$.

8. If $m\angle ARQ = 40$, find $m\angle QRS$. Write a flow proof to justify your answer.

Portfolio Suggestion

Review the items in your portfolio. Make a table of contents of the items, noting why each item was chosen. Replace any items that are no longer appropriate.

Performance Assessment

Rosa is constructing a stained-glass window. The design she is creating is shown at the right.

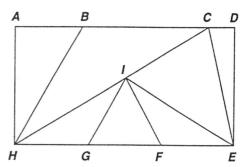

1. In your own words, define each of the following and name an example in the window.
   a. scalene triangle
   b. equilateral triangle
   c. obtuse triangle
2. Can a triangle have both an obtuse angle and a right angle? Explain. *Yes because 90 + obtuse anything about 130*
3. Name two triangles in the window that appear to be congruent.

∠HIG and ∠FIL

Chapter Summary

1. **Points** have no size. (140)
2. **Lines** extend indefinitely and have no thickness or width. (141)
3. A **plane** extends indefinitely in all directions and has no thickness. (141)
4. A **line segment** consists of two points and the points on the line between them. The two points are called the **endpoints** of the segment. (141)
5. The set of points which two sets of points have in common is the **intersection** of the two sets. (141)

6. The set of points in either or both of two sets of points is called the **union** of the two sets. (142)

7. A **ray** is part of a line consisting of one endpoint and all the points on one side of the endpoint. (144)

8. Two or more points are **collinear** if and only if they lie on the same line. (144)

9. An **angle** is the union of two rays that have the same endpoint. The rays are called the **sides** of the angle. The common endpoint is called the **vertex** of the angle. (144)

10. The **degree** (1°) is a unit of angle measure. It is $\frac{1}{360}$ of a complete rotation of a ray. (145)

11. Two lines which intersect to form right angles are **perpendicular** to each other. (146)

12. Two angles are **supplementary** if and only if the sum of their degree measures is 180. (149)

13. Two angles are **complementary** if and only if the sum of their degree measures is 90. (149)

14. **Vertical angles** are two nonadjacent angles formed by two intersecting lines. (150)

15. Theorem 5-1: Vertical angles have the same measure. (150)

16. Axiom 5-1: For every two distinct points, there exists one and only one line containing both of them. (154)

17. Two lines are **parallel** if and only if they lie in the same plane and they do not intersect. (154)

18. Axiom 5-2: Given a line and a point not on the line, then there is one and only one line through the point which is parallel to the given line. (154)

19. Axiom 5-3: If two parallel lines are cut by a transversal, then the corresponding angles have the same measure. (155)

20. Theorem 5-2: If two parallel lines are cut by a transversal, then the alternate interior angles have the same measure. (155)

21. A **polygon** is a simple closed curve composed entirely of line segments called **sides.** (158)

22. A polygon in which all sides have the same measure and all angles have the same measure is called a **regular polygon.** (158)

23. Theorem 5-3: The sum of the degree measures of the angles of a triangle is 180. (162)

24. Theorem 5-4: The sum of the degree measures of the angles of a quadrilateral is 360. (165)

25. Theorem 5-5: The sum of the degree measures of any two consecutive angles of a parallelogram is 180. (166)

26. Theorem 5-6: Any two opposite angles of a parallelogram have the same measure. (166)

⧄ Chapter Review

5.1 For each of the following use the figure on the right.
1. Write another name for $\overleftrightarrow{AB}$.
2. Write another name for k.
3. What points do $\overleftrightarrow{ED}$ and $\overleftrightarrow{BF}$ have in common?
4. Find $\overleftrightarrow{BF} \cap \overleftrightarrow{EB}$. 5. Find $\overline{BF} \cap \overleftrightarrow{DC}$.
6. Find $\overline{DC} \cup \overleftrightarrow{ED}$. 7. Find $\overline{AB} \cup \overline{BC}$.

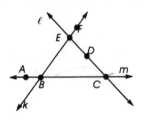

5.2 Classify angles with each of the following measurements as *acute, right, obtuse, straight,* or *reflex.*

8. 72°	**9.** 114°	**10.** 3°	**11.** 60°
12. 340°	**13.** 270°	**14.** 183.7°	**15.** 40°

For each of the following, use the figure above.

16. Name the vertex of $\angle EBC$. **17.** Name the sides of $\angle ACE$.
18. Use a protractor to find the measurement of $\angle CEF$. **19.** Name a point in the interior of $\angle EBC$.

5.3 Each of the following pairs of angles is either complementary or supplementary and have the indicated measures. Find the degree measure of each angle.

20. 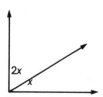 **21.**   **22.** 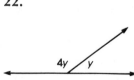 **23.**

Find $m\angle 1$ and $m\angle 2$ if $\angle 1$ and $\angle 2$ are vertical angles and the following conditions are true.

24. $m\angle 1 = x + 30$ **25.** $m\angle 1 = 5x - 60$ **26.** $m\angle 1 = 6x$
$\quad\ m\angle 2 = 3x$ $\qquad m\angle 2 = x$ $\qquad m\angle 2 = 3x + 9$

5.4 Assume lines k and ℓ are parallel. Find the degree measures of each of the eight angles formed when the transversal m cuts lines k and ℓ.

27. 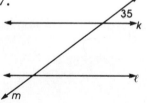 **28.** 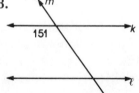 **29.**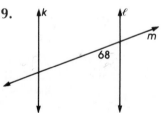

30. 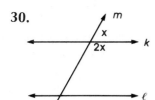   **31.** 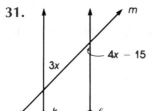   **32.**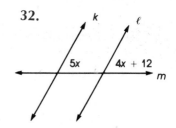

5.5 **Classify each polygon by the number of sides and as regular or *not* regular.**

33. 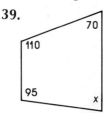   **34.** 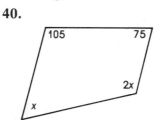   **35.** 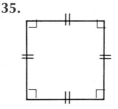   **36.**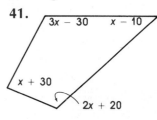

5.6 **Solve each problem.**

37. In △DEF, the degree measure of ∠D is 10 less than that of ∠E, and ∠F is a right angle. Find the degree measure of each angle.

38. In △GHI, the degree measure of ∠G is 20 more than three times that of ∠H. The degree measure of ∠I equals that of ∠G. Find the degree measure of each angle.

5.7 **Find the degree measures of the angles in each of the following.**

39.    **40.**    **41.**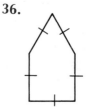

⚖️ Chapter Test

For each of the following, use the figure on the right.

1. Write another name for $\overleftrightarrow{DB}$.
2. Write another name for $\overrightarrow{AB}$.
3. Write another name for $\angle FBG$.
4. Find the intersection of $\overline{AC}$ and $\overline{DH}$.
5. Find $\overline{FE} \cap \overline{FB}$.   6. Find $\overleftrightarrow{DB} \cup \overleftrightarrow{HG}$.
7. Name the vertex of $\angle ABG$.
8. Use a protractor to find the measurement of $\angle EBC$.

Classify each of the following angles in the figure above as *acute, right, obtuse,* or *straight*.

9. $\angle ABC$ 10. $\angle CBF$ 11. $\angle CBG$ 12. $\angle DBE$

For each of the following, use the figure above. State whether each statement is *true* or *false*.

13. $m\angle EBD = m\angle FBA$
14. $m\angle CBD = 90$
15. $\angle ABE$ and $\angle EBC$ are supplementary.
16. $\angle CBE$ and $\angle EBD$ are vertical angles.

Find the degree measures of supplementary angles 1 and 2 given the following conditions.

17. $m\angle 1$ is 14 less than $m\angle 2$.
18. $m\angle 1$ equals $m\angle 2$

In the figure at the right, $\ell \parallel m \parallel n$.

19. Find $m\angle 11$.   20. Find $m\angle 10$.
21. Find $m\angle 7$.   22. Find $m\angle 5$.
23. Are $\angle 1$ and $\angle 2$ complementary angles?
24. Are $\angle 1$ and $\angle 7$ corresponding angles?

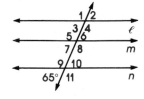

Classify each polygon by the number of sides and as regular or *not* regular.

25.   26.   27.   28.

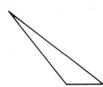

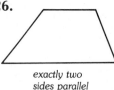
exactly two
sides parallel

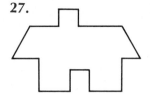

Find the degree measures of the angles in each of the following.

29.   30.   31.   32.

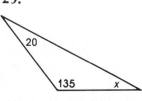

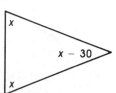

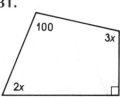

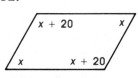

Geometric Relationships | 6

Application in Automobile Design

Automotive engineers try to design cars that are aerodynamic, fuel efficient, safe, and pleasing to the eye. Usually a team of engineers and technicians work together for many months to design a car. During this time, they make numerous **scale drawings** and scale models of the proposed car and its various parts.

Maggie Greene is making a scale drawing of part of a car that she is helping to design. Suppose one length on the proposed automobile is 20 centimeters and is represented by 2 centimeters on the drawing. Find the length that corresponds to 9 centimeters on the drawing.

Individual Project: *Drafting*

Make a scale drawing of your mathematics classroom or room at home. Write a paragraph that describes your drawing and how you made it.

6.1 Ratio

A certain bill passed the Senate. For every 3 *yes* votes, there were 2 *no* votes. We say that the bill passed by a ratio of 3 to 2. A **ratio** is a comparison of two numbers by division.

A ratio of 3 to 2 can be expressed in the following ways.

$$\frac{3}{2} \qquad 3:2 \qquad 3 \div 2 \qquad 1.5$$

In general, the ratio of *a* to *b* is written $\frac{a}{b}$ or *a* : *b*. The numbers *a* and *b* are called **terms of the ratio.**

Example

1 **Write a ratio that compares 14 to 100.**

The ratio of 14 to 100 is $\frac{14}{100}$. This ratio could also be written as 14 : 100, 0.14, or 14 ÷ 100.

Calculator Hint

A calculator can be used to determine if

$\frac{14}{100}$ and $\frac{56}{400}$ are

equivalent ratios.
14 ÷ 100 = 0.14
56 ÷ 400 = 0.14
Since 0.14 = 0.14,

$$\frac{14}{100} \qquad \frac{56}{400}$$

Since 7 and 50 have no common divisor except 1, they are relatively prime.

The ratio $\frac{14}{100}$ can be simplified to $\frac{7}{50}$. The ratios $\frac{14}{100}$ and $\frac{7}{50}$ name the same number. They are **equivalent ratios.**

You can find other ratios equivalent to these by using fractions equivalent to 1 as in the following examples.

$$\frac{7}{50} \cdot \frac{10}{10} = \frac{70}{500} \qquad\qquad \frac{14}{100} \cdot \frac{4}{4} = \frac{56}{400}$$

If *a* is any number and *b* and *k* are any numbers except zero, then $\frac{a}{b}$ and $\frac{ak}{bk}$ are equivalent ratios.

When you simplify the ratio $\frac{14}{100}$ to $\frac{7}{50}$, you know that no whole number other than 1 divides both 7 and 50 evenly. The ratio is said to be in **simplest form.**

Simplest Form of a Ratio

A ratio is in simplest form if no whole number other than 1 divides both its terms evenly.

Example

2 Compare the measures of the two segments in the figure at the right by writing the ratio in simplest form.

$AB = 30$ The measure of $\overline{AB}$ is symbolized AB.

$CD = 50$ The measure of $\overline{CD}$ is symbolized CD.

The ratio of AB to CD is $\dfrac{30}{50}$.

$$\frac{30}{50} = \frac{3 \cdot 10}{5 \cdot 10}$$

$$= \frac{3 \cdot 10 \div 10}{5 \cdot 10 \div 10}$$ Simplify the ratio by dividing the numerator and denominator by their greatest common factor, 10.

$$= \frac{3}{5}$$

The ratio in simplest form is $\dfrac{3}{5}$.

Figure: A $\bullet$—30 units—$\bullet$ B ; C $\bullet$—50 units—$\bullet$ D

Exercises

Exploratory   Write each of the following as a ratio in two different ways.

1. 3 to 5   **2.** 5 to 3   **3.** 8 to 5   **4.** 7 to 10
5. 4 to 1   **6.** 9 to 3   **7.** 13 : 100   **8.** 4 : 6
9. 40 cm to 25 cm   **10.** 25 g to 100 g   **11.** 100 m to 100 m   **12.** 36° to 200°

Find an equivalent fraction for 1 in the form $\dfrac{k}{k}$ to use in changing each of the following to $\dfrac{100}{200}$.

13. $\dfrac{1}{2}$ **14.** $\dfrac{4}{8}$ **15.** $\dfrac{5}{10}$ **16.** $\dfrac{50}{100}$

Written   Find the value of x so that each of the following is equivalent to $\dfrac{1}{3}$.

1. $\dfrac{x}{6}$ **2.** $\dfrac{x}{30}$ **3.** $\dfrac{200}{x}$ **4.** $\dfrac{1.5}{x}$ **5.** $\dfrac{3.7}{x}$

Write each of the following ratios in simplest form.

6. $\dfrac{6}{10}$ **7.** $\dfrac{10}{6}$ **8.** $\dfrac{26}{13}$ **9.** $\dfrac{300}{200}$ **10.** $\dfrac{55}{90}$

11. $\dfrac{10}{15}$ **12.** $\dfrac{65}{40}$ **13.** $\dfrac{12}{32}$ **14.** $\dfrac{14}{18}$ **15.** $\dfrac{56}{72}$

16. $\dfrac{48}{64}$ **17.** $\dfrac{34}{1700}$ **18.** $\dfrac{150}{1000}$ **19.** $\dfrac{500}{350}$ **20.** $\dfrac{20}{150}$

Write the ratio of *AB* to *CD* in simplest form for each of the following.

21. $AB = 2,\ CD = 10$ **22.** $AB = 14,\ CD = 4$ **23.** $AB = 18,\ CD = 27$

24. $AB = 37,\ CD = 1$ **25.** $AB = \dfrac{1}{2},\ CD = \dfrac{3}{4}$ **26.** $AB = 3.5,\ CD = 10.5$

Write each of the following as ratios in simplest form.

27. 3 cm to 5 cm   **28.** 3 m to 9 m   **29.** 3 cm to 1 m

30. 3 mm to 1 cm   **31.** 1 cm to 1 m   **32.** 3.2 cm to 0.8 m

Challenge   **Solve each of the following problems.**

1. The Senate passed a bill by a ratio of 8 to 1. If 90 senators voted all together, how many voted for and how many voted against the bill?

2. A committee has 15 members. The ratio of women to men on the committee is 2 : 1. How many men and how many women are on the committee?

3. The measures of the angles of a triangle are in the ratio 4 : 5 : 9. Find the measure of each angle of the triangle.

4. The lengths of the sides of a triangle are in the ratio 3 : 5 : 6. If the perimeter of the triangle is 42 cm, find the length of each side.

5. The ratio of the length to the width of a rectangle is 2 : 1. If the perimeter of the rectangle is 90 m, find its dimensions.

Mixed Review

Solve each of the following.

1. $4x + 11 \le -15$ **2.** $7(y - 3) > 2y + 9$ **3.** $-3x < 5x - 8 < 3x$

4. Find $m\angle 1$ and $m\angle 2$ if $\angle 1$ and $\angle 2$ are vertical angles, $m\angle 1 = 6x - 31$, and $m\angle 2 = 2x + 25$.

5. Find $m\angle 1$ and $m\angle 2$ if $\angle 1$ and $\angle 2$ are complementary angles, $m\angle 1 = 6x - 31$, and $m\angle 2 = 2x + 125$.

6. Find $m\angle 1$ and $m\angle 2$ if $\angle 1$ and $\angle 2$ are supplementary angles, $m\angle 1 = 6x - 31$, and $m\angle 2 = 2x + 25$.

7. In $\triangle DEF$, $m\angle D$ is 32 more than half $m\angle E$ and $m\angle F$ is 5 less than 3 times $m\angle E$. Find $m\angle D$, $m\angle E$, and $m\angle F$.

8. In quadrilateral *EFGH*, $m\angle E = 2x - 1$, $m\angle F = x - 26$, $m\angle G = 3x + 4$, and $m\angle H = 6x - 13$. Find $m\angle E$, $m\angle F$, $m\angle G$, and $m\angle H$.

6.2   Proportion

The Flashes and Pioneers tied this year each having won half their games. The Flashes won 20 games out of 40, and the Pioneers, 22 of 44. The ratios of games won to games played are equivalent. These ratios may be written as a **proportion.**

$$\frac{20}{40} = \frac{22}{44} \qquad \textit{The two ratios are equivalent.}$$

Definition of Proportion

A proportion is an equation of the form $\frac{a}{b} = \frac{c}{d}$, which states that two ratios are equivalent.

$a : b = c : d$
means
extremes

Every proportion has four terms; a, b, c, and d. The terms are said to be *in proportion* or *proportional*. The terms a and d are called the **extremes** and b and c are the **means.**

Proportions involving algebraic fractions often are used to solve problems. A useful property can be found by simplifying the fractions in the following proportion.

$$\frac{a}{b} = \frac{c}{d}$$

$$\frac{a}{b}(bd) = \frac{c}{d}(bd) \qquad \textit{Multiply both sides of the equation by bd.}$$

$$ad = bc$$

Property of Proportions

In a proportion, the product of the means is equal to the product of the extremes.

If $\frac{a}{b} = \frac{c}{d}$, then $ad = bc$.

The process that uses this property is called **cross multiplication.** If $\frac{a}{b} \bowtie \frac{c}{d}$, then $ad = bc$.

Examples

1 Solve the proportion $\dfrac{x}{2} = \dfrac{65}{26}$.

$$\frac{x}{2} = \frac{65}{26}$$
$$26x = 2 \cdot 65 \qquad \textit{Cross multiply.}$$
$$26x = 130$$
$$x = 5$$

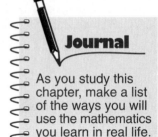

Journal

As you study this chapter, make a list of the ways you will use the mathematics you learn in real life.

2 Solve the proportion $\dfrac{5}{x + 3} = \dfrac{3}{2x - 8}$.

$$\frac{5}{x + 3} = \frac{3}{2x - 8}$$
$$5(2x - 8) = (x + 3)3 \qquad \textit{Cross multiply.}$$
$$10x - 40 = 3x + 9 \qquad \textit{Use the distributive property.}$$
$$7x = 49 \text{ or } x = 7$$

3 Find two numbers that are in the ratio 3 : 4 if one is 17 more than the other.

Define a variable.   Let x represent the lesser number.
Then $x + 17$ represents the greater number.

Write a proportion. $\dfrac{x}{x + 17} = \dfrac{3}{4}$

Solve the proportion. $4x = 3(x + 17) \qquad \textit{Cross multiply.}$
$4x = 3x + 51 \qquad \textit{Use the distributive property.}$
$x = 51$

Since $x = 51$, $x + 17 = 68$. Therefore, the two numbers are 51 and 68.

4 The Pine City Art Club has 110 members. The ratio of women to men is 5 : 6. How many members are women and how many are men?

Define a variable.   Let x represent the number of women.
Then $110 - x$ represents the number of men.

Write a proportion. $\dfrac{x}{110 - x} = \dfrac{5}{6}$

Solve the proportion. $6x = (110 - x)5 \qquad \textit{Cross multiply.}$
$6x = 550 - 5x \qquad \textit{Use the distributive property.}$
$11x = 550 \text{ or } x = 50$

Since $x = 50$, $110 - x = 60$. Thus, the group has 50 women and 60 men.

Exercises

Exploratory Name the product of the means and the product of the extremes in each of the following proportions.

1. $\dfrac{x}{6} = \dfrac{2}{3}$

2. $\dfrac{x}{3} = \dfrac{40}{60}$

3. $\dfrac{y}{3} = \dfrac{4}{8}$

4. $\dfrac{x}{10} = \dfrac{9}{30}$

5. $\dfrac{10}{k} = \dfrac{8}{6}$

6. $\dfrac{3}{4} = \dfrac{3}{w}$

7. $\dfrac{x}{5} = \dfrac{15}{25}$

8. $\dfrac{x}{1} = \dfrac{2}{3}$

Written Solve each of the following proportions.

1. $\dfrac{x}{8} = \dfrac{3}{4}$

2. $\dfrac{x}{8} = \dfrac{20}{32}$

3. $\dfrac{10}{a} = \dfrac{20}{28}$

4. $\dfrac{13}{2} = \dfrac{78}{y}$

5. $\dfrac{5}{6} = \dfrac{x}{72}$

6. $\dfrac{k}{8} = \dfrac{2}{3}$

7. $\dfrac{k}{14} = \dfrac{3}{4}$

8. $\dfrac{x}{25} = \dfrac{3}{4}$

9. $\dfrac{2x}{10} = \dfrac{13}{20}$

10. $\dfrac{3}{5} = \dfrac{6}{2w}$

11. $\dfrac{x+1}{8} = \dfrac{3}{4}$

12. $\dfrac{7}{y+9} = \dfrac{1}{2}$

13. $\dfrac{x-1}{x+1} = \dfrac{6}{10}$

14. $\dfrac{3}{7} = \dfrac{x-2}{x+2}$

15. $\dfrac{x}{0.02} = \dfrac{1}{2}$

16. $\dfrac{1.32}{x} = \dfrac{4}{5}$

Solve each of the following problems.

17. The measures of two segments are in the ratio 2 : 3. If the longer segment has a measure of 12, what is the measure of the shorter segment?

18. Two numbers are in the ratio $\dfrac{2}{3}$. The greater number is 11 more than the other number. Find the numbers.

19. The sum of two numbers is 14. They are in the ratio $\dfrac{3}{4}$. Find the numbers.

20. A committee has 40 members. The ratio of men to women is 3 to 5. How many men and how many women are on the committee?

21. Suppose that 75 Senators voted on a certain bill. How many Senators voted for and how many against the bill if it passed by a ratio of 14 to 1?

22. Two angles are complementary. Their measures are in the ratio of 8 : 1. Find the measures.

23. Two angles are supplementary. Find their measures if they are in the ratio $\dfrac{1}{5}$.

24. The scale on a map is 2 cm to 5 km. Pine City and Maplewood are 15.75 km apart. How far apart are they on the map?

Challenge Solve each of the following problems.

1. Two numbers are in the ratio 1 : 4. If 1 is added to the lesser number, the ratio of the two numbers becomes 1 to 3. Find the numbers.

2. Two numbers are in the ratio 3 : 10. If 12 is added to each number, the ratio of the two numbers becomes 1 to 2. Find the numbers.

Many problems in mathematics and science require computations involving measurements. Solutions to these problems contain the correct measure as well the correct unit of measure. One method for performing computations with measurements is called **dimensional analysis.** This method is based on the idea that units of measure, or dimensions, can be treated as factors. When units of measure are common factors, they may be "divided out" or "cancelled."

Example The distance an object moves when traveling at a constant speed is given by the formula $d = rt$ where r represents the constant speed (or rate) and t represents time. If Josie is driving a car at a constant speed of 72 kilometers per hour, how far will she travel in 3 hours?

$$d = rt$$
$$d = 72 \, \frac{\text{km}}{\text{h}} \times 3 \, \text{h}$$
$$d = 216 \text{ km}$$

Josie will travel 216 kilometers in 3 hours.

We use conversion factors to convert from one unit of measure to another. Conversion factors are ratios with a value equal to one and are constructed from equalities.

Equality	Conversion Factor	
$24 \text{ h} = 1 \text{ day}$	$\dfrac{24 \text{ h}}{1 \text{ day}}$ or $\dfrac{1 \text{ day}}{24 \text{ h}}$	$\dfrac{24 \text{ h}}{1 \text{ day}}$ is read "24 hours per day."
$5280 \text{ ft} = 1 \text{ mi}$	$\dfrac{5280 \text{ ft}}{1 \text{ mi}}$ or $\dfrac{1 \text{ mi}}{5280 \text{ ft}}$	$\dfrac{5280 \text{ ft}}{1 \text{ mi}}$ is read "5280 feet per mile."
$144 \text{ in}^2 = 1 \text{ ft}^2$	$\dfrac{144 \text{ in}^2}{1 \text{ ft}^2}$ or $\dfrac{1 \text{ ft}^2}{144 \text{ in}^2}$	$\dfrac{144 \text{ in}^2}{1 \text{ ft}^2}$ is read "144 square inches per square foot."

When a measurement is multiplied by a conversion factor, the measure and unit of measure change, but the value of the measurement does not. Thus, we can convert units of measure by multiplying with the appropriate conversion factors.

Example Convert 55 miles per hour (mph) to feet per second.

In order to convert 55 mph to feet per second, conversion factors for miles to feet and for hours to seconds must be used.

$$60 \text{ s} = 1 \text{ min}$$
$$3600 \text{ s} = 60 \text{ min} \qquad \textit{Multiply each side by 60.}$$
$$3600 \text{ s} = 1 \text{ h} \qquad \textit{Substitute 1 h for 60 min.}$$

Conversion factors for hours to seconds: $\dfrac{3600 \text{ s}}{1 \text{ h}}$ or $\dfrac{1 \text{ h}}{3600 \text{ s}}$

The number of feet per second equivalent to 55 mph can now be found by multiplying 55 mph by the appropriate conversion factors for miles to feet and hours to seconds.

$$55 \frac{\text{mi}}{\text{h}} = \square \frac{\text{ft}}{\text{s}}$$

$$55 \frac{\text{mi}}{\text{h}} = 55 \frac{\cancel{\text{mi}}}{\cancel{\text{h}}} \times \frac{5280 \text{ ft}}{1 \cancel{\text{mi}}} \times \frac{1 \cancel{\text{h}}}{3600 \text{ s}}$$

The conversion factor for miles to feet should have feet in the numerator. The conversion factor for hours to seconds should have seconds in the denominator.

$$55 \frac{\text{mi}}{\text{h}} = 80\frac{2}{3} \frac{\text{ft}}{\text{s}}$$

Thus, 55 mph is equivalent to $80\frac{2}{3}$ feet per second.

Exercises Make the following conversions.

1. 1 week to minutes

2. 1 square mile to square feet

3. 1 cubic yard to cubic inches

4. 10 mph to mi/s

5. 0.2 m/s to m/hr

6. 8800 yd/min to mph

7. 48 km/h to m/s

8. 48 mph to ft/s

9. 90 mi/(h²) to ft/(s²)

10. 9.8 m/(s²) to km/(h²)

Use $d = rt$ to answer each question.

11. Janet drove for 4 hours at 50 miles per hour. What was her distance traveled in miles and in feet?

12. Jerrod traveled at a rate of 88 km/h for 396 kilometers. What was his travel time in hours and in min?

13. Jansen traveled 411 kilometers in 5 hours. What was his speed in km/h and in m/s? At that rate, how many minutes would it take him to travel 137 kilometers?

When an object is dropped, the distance it falls is given by the formula $d = \frac{1}{2}gt^2$ where g represents the acceleration due to gravity and t represents time. Suppose a rock is dropped from the top of a 490-meter cliff. ($g = 9.8$ meters per second squared)

14. How far will the rock fall in 4 seconds?

15. How long, to the nearest second, will it take the rock to hit the ground?

6.3   Percents in Proportions

Percent problems can be solved by using the following proportion.

Percent Proportion

$$\frac{\text{Percentage}}{\text{Base}} = \text{Rate} \quad \text{or} \quad \frac{P}{B} = \frac{r}{100}$$

The **percentage** is a number which is compared to another number called the **base.** The **rate** is a **percent.**

Definition of Percent

A percent is the ratio of a number to 100.

For example, 14% means $\frac{14}{100}$ or 0.14.

Examples

1 **Express 3.5% as a decimal.**

3.5% means $\frac{3.5}{100}$.

$$\frac{3.5}{100} \cdot \frac{10}{10} = \frac{35}{1000} \qquad \textit{Multiplying by a number equivalent to 1 eliminates the decimal value in the numerator.}$$

3.5% is the decimal 0.035.

2 **What number is 15% of 620?**

$$\frac{P}{B} = \frac{r}{100} \qquad \textit{Use the percent proportion.}$$

$$\frac{P}{620} = \frac{15}{100} \qquad \begin{array}{l}\textit{15\% tells you that r = 15.}\\ \textit{Some number is compared to 620, so B = 620.}\end{array}$$

$$100P = 9300 \qquad \textit{Cross multiply.}$$

$$P = 93$$

Therefore, 93 is 15% of 620.

Example

3 A stereo system is on sale with $90 off the $600 list price. The discount, or amount the price is reduced, is what percent of the list price?

$$\frac{P}{B} = \frac{r}{100}$$

$$\frac{90}{600} = \frac{r}{100}$$ *The discount is compared to the list price.*
       P = 90 and B = 600.

$$600r = 9000$$ *Cross multiply.*

$$r = 15$$ The discount is 15% of the list price.

Exercises

Exploratory   Express each percent as a decimal and each decimal as a percent.

1. $7\frac{1}{2}\%$ **2.** $24\frac{1}{2}\%$ **3.** 0.365   **4.** $88\frac{3}{4}\%$ **5.** $16\frac{1}{4}\%$
6. 0.512   **7.** 1.73   **8.** 0.08   **9.** 125%   **10.** 0.0025

Written   Use a proportion to solve each of the following problems.

1. 30% of 30 is what number?
2. 40% of what number is 80?
3. 20 is what percent of 60?
4. What percent of 50 is 35?
5. Find x if 60 is 75% of x.
6. What number is 75% of 80?
7. 20% of 30 is what number?
8. 5% of 80 is what number?
9. 1.8 is 60% of what number?
10. 6 is what percent of 48?
11. 34 is what percent of 17?
12. 23 is 25% of what number?
13. Janice scored 85% on a test. She answered 34 questions correctly. How many questions were on the test?
14. Eric paid $0.30 sales tax on a $7.50 purchase. What is the percent of sales tax?
15. Kelvin plays basketball for a university team. He made 122 field goals, or 40% of the total he attempted. How many field goals did he attempt?
16. Alice plays basketball for a university team. She attempted 25 free throws and made 80% of them. How many free throws did she make?
17. A coat was reduced in price $6. The percent of discount was 10%. What was the list price of the coat?
18. June earns $125 per week in salary and 8% commission on all sales. How much must she sell to earn $200 per week?

19. If a theater is filled to 75% of its capacity, how many of the theater's 720 seats are filled?

20. If 8% of a 150-kilogram alloy is copper, how many grams of copper are in the alloy?

21. A raincoat that regularly costs $110.25 is on sale for $73.50. What percent of the regular price is the discount on the raincoat?

22. On January 1, the price of a gallon of gasoline was $1.20. On June 1, the price of a gallon of this gasoline was $1.29. What is the percent of increase in this time period?

23. Coaxial Cablevision increased the monthly service charge for their basic cable package to $27.95. If this price represents a 7.5% increase, what was the original price?

24. Pine City Appliance advertises a stereo system for $663.00. This price represents a discount of 22% off the original price. What was the original price of the stereo system?

25. Candidate Block received 45% of all votes cast for mayor in the last election. Only 52% of the registered voters turned out for this election. If there were 8000 registered voters, how many voted for candidate Block?

Challenge   **Use a proportion to solve each problem.**

1. This week all dresses at Di's Fashions are on sale at a 25% discount. Every Friday, all items in the store are reduced an additional 15%. Find the regular price of a dress that costs $33.15 on Friday.

2. The list price of each item at Al's Sports Shop is increased to make a 10% profit on each purchase. On Monday's, there is a 20% off sale on tennis racquets. Find the list price of a tennis racket that costs $47.52 on Monday.

━━━━━━━━━━━━━━**Mixed Review**━━━━━━━━━━━━━━

Solve each equation.

1. $11a - 23 = 31 - 7a$

2. $3(4 - x) - 5(2x + 9) = 15 - x$

Solve each proportion.

3. $\dfrac{b}{5} = \dfrac{6}{15}$

4. $\dfrac{44}{y} = \dfrac{24}{25}$

5. $\dfrac{k - 1}{15} = \dfrac{45}{27}$

6. $\dfrac{x - 4}{x + 5} = \dfrac{5}{8}$

Let *p*, *q*, and *r* represent the following statements about lines *ℓ*, *m* and *n*.
  p: If *ℓ* and *m* are parallel, and *m* and *n* are parallel, then *ℓ* and *n* must be parallel.
  q: If *ℓ* and *m* are skew, and *m* and *n* are skew, then *ℓ* and *n* must be skew.
  r: If *ℓ* and *m* are parallel, and *m* and *n* are skew, then *ℓ* and *n* must be skew.
Determine whether each of the statements represented by *p*, *q*, and *r* is true or false. Then determine the truth value of each of the following.

7. $q \to (r \wedge p)$

8. $(\sim p \leftrightarrow q) \vee r$

9. $\sim p \leftrightarrow (q \vee r)$

10. Two angles are supplementary. Find their measures if they are in the ratio 11 : 4.

6.4   Similar Figures

A photograph of an object is the same shape as the actual object but usually a different size. In mathematics, figures that have the same shape but not necessarily the same size are called **similar figures.**

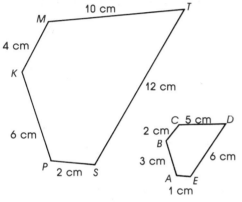

For example, the two figures shown at the left are **similar polygons.**

We write $PKMTS \sim ABCDE$ and we say *polygon PKMTS is similar to polygon ABCDE.* The vertices correspond in the order in which they are named. This means it is *not* correct to write $PKMTS \sim EABCD$.

These similar polygons are positioned so that it is easy to see **corresponding parts.**

Compare the measures of the corresponding parts.

The measures of corresponding angles are equal.

$$m\angle P = m\angle A \qquad m\angle K = m\angle B \qquad m\angle M = m\angle C$$
$$m\angle T = m\angle D \qquad m\angle S = m\angle E$$

The ratios of the measures of corresponding sides are equivalent. That is, the measures of corresponding sides are in proportion.

$$\frac{PK}{AB} = \frac{6}{3} = 2 \qquad \frac{KM}{BC} = \frac{4}{2} = 2 \qquad \frac{MT}{CD} = \frac{10}{5} = 2$$

$$\frac{TS}{DE} = \frac{12}{6} = 2 \qquad \frac{PS}{AE} = \frac{2}{1} = 2$$

Definition of Similar Polygons

Two polygons are similar if their corresponding angles have the same measure and the measures of their corresponding sides are in proportion.

The converse of this definition is also true.

If two polygons are similar, then their corresponding angles have the same measure and the measures of their corresponding sides are in proportion.

Examples

1 The two quadrilaterals at the right are similar. Name the pairs of corresponding sides.

The figures are *not* positioned on the page so that it is easy to see corresponding parts. We must think of the way to position the figures so that we can determine which parts correspond.

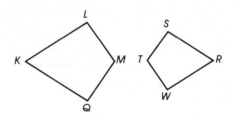

$\angle K$ corresponds to $\angle R$
$\angle L$ corresponds to $\angle S$
$\angle M$ corresponds to $\angle T$
$\angle Q$ corresponds to $\angle W$

$\overline{KL}$ corresponds to $\overline{RS}$
$\overline{LM}$ corresponds to $\overline{ST}$
$\overline{MQ}$ corresponds to $\overline{TW}$
$\overline{QK}$ corresponds to $\overline{WR}$

2 Triangle *ABC* is similar to triangle *RST*. What is the value of *k*?

Comparing $\triangle ABC$ to $\triangle RST$, we see the ratio of measures of corresponding sides is 3 : 6 or 1 : 2.

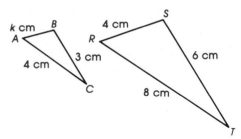

$$\frac{k}{4} = \frac{1}{2} \qquad \textit{Write a proportion}$$
$$2k = 4 \qquad \textit{Cross multiply.}$$
$$k = 2 \qquad \textit{Solve for k.}$$

3 What is the value of *k* if $\triangle LMN \sim \triangle VWX$?

Here, $\overline{LM}$ corresponds to $\overline{VW}$ and $\overline{MN}$ corresponds to $\overline{WX}$.

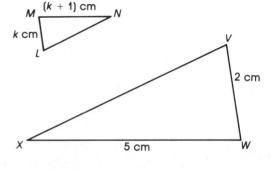

$$\frac{k}{2} = \frac{k+1}{5} \qquad \textit{Write a proportion.}$$
$$5k = 2(k+1)$$
$$5k = 2k + 2$$
$$3k = 2 \qquad \textit{Subtract 2k.}$$
$$k = \frac{2}{3} \qquad \textit{Solve for k.}$$

Exercises

Exploratory   Answer each of the following.

1. Give some examples, other than photographs and maps, of similarity outside the classroom.

2. If two polygons are similar, what do you know about the measures of corresponding angles?

3. What do you know about the measures of corresponding sides of similar polygons?

Written   The pair of figures in each of the following exercises are similar. Name the pairs of corresponding angles and corresponding sides.

1.

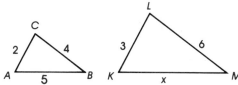

2.

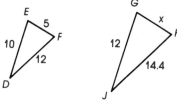

3.

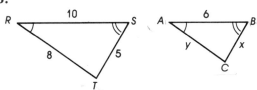

4.

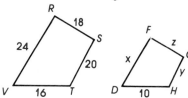

5.

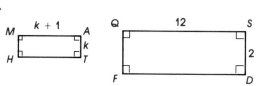

6.

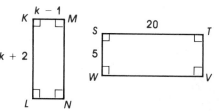

7.

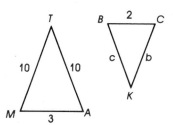

8.

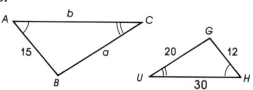

Find the values of the variables given in the figures in each of the following. The measurements given are in centimeters.

9. Exercise 1

10. Exercise 2

11. Exercise 3

12. Exercise 4

13. Exercise 5

14. Exercise 6

15. Exercise 7

16. Exercise 8

By completing each of the following steps in order, you will discover an important fact about similar triangles.

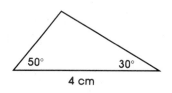

1. Examine the triangle at the left. Check the indicated measurements with your ruler and protractor.
2. Use these tools to draw another triangle on your paper with a base 8 centimeters and acute angles of 50° and 30°. What must be the measurements of the third angles of both triangles?

3. Measure the remaining sides of both triangles. Check the ratios of measures of sides of your triangle to those of corresponding sides of the triangle given above. Are they equivalent to 2 : 1?
4. What can you say about the two triangles?
5. Repeat steps 2 through 4 with a base of 12 centimeters.
6. Repeat steps 2 through 4 with a 16-centimeter base.
7. Explain why your investigation indicates that the following statement is true.

Two triangles are similar if two angles of one triangle have the same measures as the corresponding angles of the other.

Exercises Use the statement given above to answer the following.

1. Tell whether the following statement is true or false. Two right triangles are similar if an acute angle of one has the same measure as an acute angle of the other.

State whether the two triangles in each of the following are similar. Explain your answer.

2.

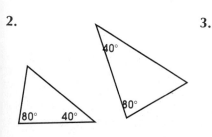

3.

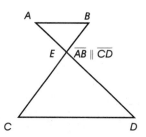

$\overline{AB} \parallel \overline{CD}$

4.

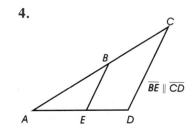

$\overline{BE} \parallel \overline{CD}$

6.5   Congruence and SAS

Certainly all three figures in the drawing shown above are similar since they have the same shape. The two figures on the left have the same shape and the same size. These two figures are **congruent figures.**

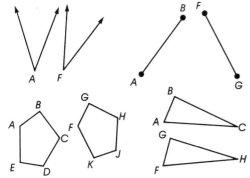

The figures in each pair at the left are also congruent.

Two angles with the same measure are **congruent angles.**

Two line segments with the same measure are **congruent segments.**

Two polygons with the sides and angles of one congruent to the corresponding sides and angles of the other are **congruent polygons.**

We write $\angle A \cong \angle F$, $\overline{AB} \cong \overline{FG}$, $\triangle ABC \cong \triangle FGH$, and so on. We say *angle A is congruent to angle F, segment AB is congruent to segment FG, and triangle ABC is congruent to triangle FGH.*

Note that two triangles are congruent if six parts of each triangle—three sides and three angles—are congruent to the corresponding parts of the other triangle.

It is *not* necessary to show all six pairs of parts are congruent in order to conclude that the triangles are congruent.

Suppose that a triangle has one side 3 cm long and another 5 cm long. The **included angle** for the sides is a 25° angle. Draw a triangle on your paper with these measurements. Start with a 5-cm segment. With one endpoint of the segment as vertex, draw an angle of 25° using your protractor.

An included angle between two segments means the segments are on the sides of the angle. The vertex of the angle is a common endpoint of the segments.

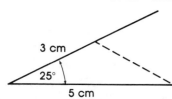

Extend the other side of the angle so that you can draw a 3 cm segment on it. Draw the third side of the triangle by connecting the endpoints.

Can you draw another triangle that fulfills these requirements but is not congruent to the first triangle? You will find that you *cannot.*

This and many other examples lead us to accept the following principle as an axiom.

Axiom 6-1
SAS

If two sides and the included angle of one triangle are congruent to the corresponding sides and included angle of another triangle, then the triangles are congruent. The abbreviation SAS stands for Side—Angle—Side.

It is important that the known angle is included between the two known sides. In the figure given below, the angle in each triangle is *not* included between the two sides. In this case, the two triangles are *not* congruent.

Congruent parts are two pairs of sides and a pair of angles not included between the sides.

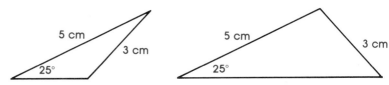

We often can show that two triangles are congruent using only three pairs of parts. Then, we can conclude that the other three pairs are congruent.

Example

1 **In isosceles triangle *ABC* at the right, show that angle *A* is congruent to angle *C*.**

Let $\overline{BD}$ be the bisector of $\angle ABC$. This means that $\angle 1 \cong \angle 2$. Also, $\overline{AB} \cong \overline{BC}$ because $\triangle ABC$ is isosceles. And, surely, $\overline{BD} \cong \overline{BD}$. Since $\angle 1$ is included between $\overline{AB}$ and $\overline{BD}$ and $\angle 2$ is between $\overline{BD}$ and $\overline{BC}$, we conclude that $\triangle ABD \cong \triangle CBD$ by SAS. Since congruent triangles have corresponding parts congruent, $\angle A \cong \angle C$.

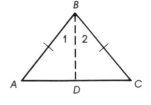

An informal argument of the following theorem was given in example 1.

Theorem 6-1

If two sides of a triangle are congruent, then the angles opposite those sides are congruent.

Exercises

Exploratory   **Explain what it means for pairs of the following geometric figures to be congruent.**

1. line segments   **2.** angles   **3.** triangles   **4.** parallelograms

Complete each of the following exercises.

5. Draw $\triangle ABC$ so that $\overline{AB}$ is 3 cm long, $\overline{BC}$ is 2 cm, and $\angle B$ has 30°. Compare your drawing with your classmates' to see how many are congruent.

6. Draw two triangles that are *not* congruent which *each* have sides of 4 cm and 7 cm and an angle of 40°.

Written   **Name the corresponding parts of each pair of congruent triangles.**

1.

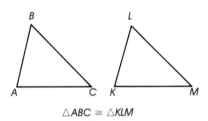

$\triangle ABC \cong \triangle KLM$

2.

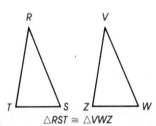

$\triangle RST \cong \triangle VWZ$

3.

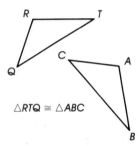

$\triangle RTQ \cong \triangle ABC$

Use the figure in exercise 1. Suppose you know the following information. Write what you need to know to show that the triangles are congruent.

4. $\overline{AB} \cong \overline{KL}$
$\overline{BC} \cong \overline{LM}$

5. $\overline{AB} \cong \overline{KL}$
$\angle A \cong \angle K$

6. $\overline{AC} \cong \overline{KM}$
$\overline{BC} \cong \overline{LM}$

Use the figure in the exercise named in each of the following. List the information needed to show that the triangles are congruent by SAS.

7. Exercise 2   **8.** Exercise 3

Use the figure at the right. Triangle *QRS* is isosceles. Find the measure of every angle given the measures in each of the following.

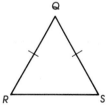

9. $m\angle Q = 40$ **10.** $m\angle Q = 25$

11. $m\angle S = 70$ **12.** $m\angle R = 25$

13. $m\angle Q = x, m\angle R = 2x$ **14.** $m\angle Q = x - 45, m\angle S = x$

15. $m\angle R = x, m\angle Q = x + 15$ **16.** $m\angle S = x + 21, m\angle Q = 6x + 8$

17. $m\angle Q = 7x, m\angle R = x$ **18.** $m\angle Q = x, m\angle R = x$

6.6 The ASA Axiom

Can two triangles of different shapes have two pairs of congruent angles with congruent sides included between the angles?

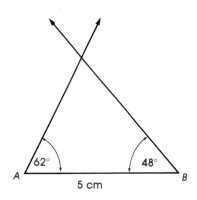

Try to draw two different triangles with angles of 62° and 48° and the side included 5 centimeters long.

First, draw a line segment of 5 centimeters on your paper. Then, draw angles of 62° and 48° at the endpoints of the segment. Extend the sides until· they meet forming a triangle.

Can a different triangle be drawn with the same measurements? Remember that the 5-centimeter side must be included between the two angles. Would you get a different size or shape triangle if you drew a 62° angle at *B* and a 48° angle at *A*?

Actually there is only one triangular shape that has the given angles and side measurements. We accept the principle based on this idea as an axiom.

Axiom 6-2
ASA

> If two angles and the included side of one triangle are congruent to corresponding parts of another triangle, then the triangles are congruent. The abbreviation ASA stands for Angle—Side—Angle.

This axiom is used to help prove a useful theorem about parallelograms.

Theorem 6-2

> A diagonal of a parallelogram separates the parallelogram into two congruent triangles.

Lab Activity

You can discover whether Side-Side-Angle is a test for triangle congruence in Lab 5 on page A8.

Thus far we have used informal arguments to show why a theorem is true. To *prove* this theorem we shall give a more **formal proof** in *two-column* or *statement-reason* form.

It is important to plan the proof logically step-by-step. Statements are listed in the left column. Reasons are listed in the right column. Each statement has a reason. The reasons can be given information, definitions, axioms, or theorems that have already been proved.

Example

1 Write a two-column proof of Theorem 6-2: A diagonal of a parallelogram separates the parallelogram into two congruent triangles.

Given: Parallelogram *ABCD*
$\overline{AC}$ is a diagonal

Prove: △*ABC* ≅ △*CDA*

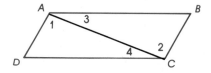

Proof:

STATEMENTS	REASONS
1. *ABCD* is a parallelogram	**1.** Given
2. $\overline{AB} \parallel \overline{DC}$ and $\overline{AD} \parallel \overline{BC}$	**2.** Definition of parallelogram
3. $\overline{AC}$ is a transversal cutting $\overline{AB}$ and $\overline{DC}$ and also cutting $\overline{AD}$ and $\overline{BC}$.	**3.** Definition of transversal
4. ∠3 and ∠4 are alternate interior angles. ∠1 and ∠2 are alternate interior angles.	**4.** Definition of alternate interior angles
5. $m\angle 3 = m\angle 4$ $m\angle 1 = m\angle 2$	**5.** Theorem 5-2: If two parallel lines are cut by a transversal, then the alternate interior angles have the same measure.
6. ∠3 ≅ ∠4, ∠1 ≅ ∠2	**6.** Two angles with the same measure are congruent.
7. $\overline{AC} \cong \overline{AC}$	**7.** Any line segment is congruent to itself.
8. △*ABC* ≅ △*CDA*	**8.** ASA Axiom

All pairs of corresponding sides of congruent triangles are congruent. Therefore, from the theorem just proved, the opposite sides of a parallelogram are congruent.

Here is a summary of facts about a parallelogram.

1. Opposite sides are parallel.
2. A diagonal separates it into two congruent triangles.
3. Opposite sides are congruent.
4. Opposite angles are congruent.
5. Consecutive angles are supplementary.

▰▰▰ Exercises ▰▰▰

Exploratory Write what you need to know to prove $\triangle ABC \cong \triangle KLM$ given the information in each of the following.

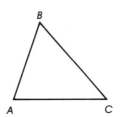
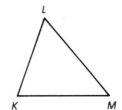

1. $\overline{AB} \cong \overline{LK}$

   $\angle B \cong \angle L$

2. $\angle B \cong \angle L$

   $\angle C \cong \angle M$

3. $\angle A \cong \angle K$

   $\overline{AC} \cong \overline{KM}$

4. $\angle C \cong \angle M$

   $\overline{AC} \cong \overline{KM}$

Answer each of the following.

5. Is there an SSA principle for establishing congruence of two triangles? Why or why not? Illustrate your answer.

6. List the information needed to show that $\triangle DEF \cong \triangle GHK$.

Written State whether each of the following is *always, sometimes,* or *never* true.

1. If two sides and an angle of one triangle are congruent to two sides and an angle of another, the triangles are congruent.

2. If three angles of one triangle are congruent to three angles of another triangle, then the triangles are similar.

3. A diagonal of a parallelogram separates the parallelogram into two congruent triangles.

4. A diagonal of a parallelogram bisects the angles whose vertices are endpoints of the diagonal.

5. If $m\angle A$ in $\square ABCD$ is 60, then the angle opposite $\angle A$ also measures 60.

6. The measures of opposite sides of a parallelogram are in the ratio 1 : 2.

7. In a parallelogram, two consecutive angles have the same measure.

8. All the sides of a parallelogram are the same in measure.

Use the figure at the right to answer exercises 9-12. The measurements given are in centimeters. Suppose that $\triangle ABC \cong \triangle KLM$.

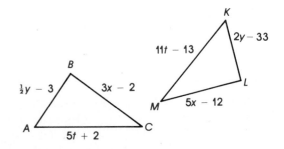

9. Name all corresponding parts.

Find each of the following measures.

10. AB 11. BC 12. AC

Complete each of the following proofs.

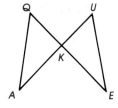

13. Given: $QK = KU$
$\angle Q \cong \angle U$

Prove: $\triangle QKA \cong \triangle UKE$

Proof:

STATEMENTS	REASONS
1. $QK = KU$, $\angle Q \cong \angle U$	**1.** _____
2. $\overline{QK} \cong \overline{KU}$	**2.** _____
3. $\angle QKA$ and $\angle UKE$ are vertical angles.	**3.** _____
4. $m\angle QKA = m\angle UKE$	**4.** _____
5. $\angle QKA \cong \angle UKE$	**5.** _____
6. $\triangle QKA \cong \triangle UKE$	**6.** _____

14. Given: $\overline{QR} \parallel \overline{ST}$
$PS = RW$
$\angle P \cong \angle W$

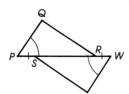

Prove: $\triangle QRP \cong \triangle TSW$

Proof:

STATEMENTS	REASONS
1. $\overline{QR} \parallel \overline{ST}$	**1.** _____
2. $\overline{PW}$ is a transversal cutting $\overline{QR}$ and $\overline{ST}$.	**2.** _____
3. $\angle QRP$ and $\angle TSW$ are alternate interior angles.	**3.** _____
4. $m\angle QRP = m\angle TSW$	**4.** _____
5. $\angle QRP \cong \angle TSW$	**5.** _____
6. $PS = RW$	**6.** _____
7. $SR = SR$	**7.** _____
8. $PS + SR = RW + SR$	**8.** Addition of equals
9. $PR = SW$	**9.** Substitution
10. $\overline{PR} \cong \overline{SW}$	**10.** _____
11. $\angle P \cong \angle W$	**11.** _____
12. $\triangle QRP \cong \triangle TSW$	**12.** _____

Challenge   **Answer each of the following.**

1. Write the converse of Theorem 6-1: If two sides of a triangle are congruent, then the angles opposite those sides are congruent.

2. Investigate a proof of the converse of Theorem 6-1. What information is needed to complete the proof?

3. Write an AAS axiom.

4. How could an AAS axiom be used to prove the converse of Theorem 6-1?

6.7   Constructions

Constructions differ from drawings.

In **constructions,** you may use only a straightedge and compass to draw geometric figures. In the **drawings** of figures in previous lessons, you could use other tools, including measuring devices such as rulers and protractors.

CONSTRUCTION 1:   **Construct a segment congruent to a given segment.**

Given: $\overline{AB}$ with measure k

Method:

1. Using a straightedge, draw a segment longer than $\overline{AB}$.
2. Choose any point on the segment. Label it C.
3. Place the metal point of the compass on A and pencil point on B.   *The measure of the distance between the compass points is k. The measure may be "copied" or transferred by keeping the compass set.*
4. Place the metal point on C and draw a pencil mark through the segment. This mark is called an arc. Point C is the center of the arc. Label the intersection D. Then $\overline{CD} \cong \overline{AB}$.

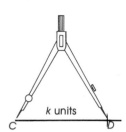

CONSTRUCTION 2:   **Construct a segment twice as long as a given segment.**

Given: $\overline{AB}$ with measure k

Method:

1. Proceed as in construction **1.**
2. With the compass set at measure k and center at D, draw a second arc at E. $\overline{CE}$ has the measure $2k$.

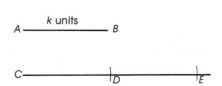

CONSTRUCTION 3:   **Construct a triangle with sides congruent to 3 given segments.**

Given: Segments with measures k, ℓ, and m

Method:

1. Construct $\overline{AB}$ with measure k.
2. With compass set at measure ℓ and center at A, draw an arc as shown.
3. With compass set to match measure m and center at B, draw an arc intersecting the arc in step **2.** Label the intersection C.
4. Draw $\overline{AC}$ and $\overline{BC}$. $\triangle ABC$ has sides congruent to the given segments.

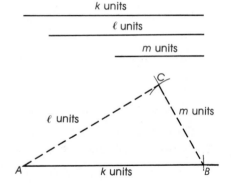

In the preceding construction, there is another possible intersection point D on the other side of $\overline{AB}$. You should check to see that $\triangle ABC$ and $\triangle ABD$ are congruent.

Apparently, only one size and shape of triangle can be obtained from the segments given. The construction does *not* prove this idea. We accept the principle as an axiom.

Axiom 6-3
SSS

> If three sides of one triangle are congruent to three sides of another triangle, the triangles are congruent. The abbreviation SSS stands for Side—Side—Side.

CONSTRUCTION 4: Construct a line perpendicular to a given line at a given point on the line.

Given: $\overleftrightarrow{AC}$ with point B on $\overleftrightarrow{AC}$

Method:

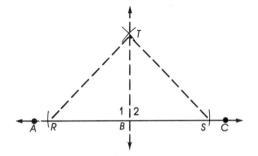

1. With center at B and a convenient compass setting, draw arcs on both sides of B intersecting $\overleftrightarrow{AC}$. Label the intersections R and S.
2. Set the compass at a measure greater than BR. With center first at R, then at S, draw intersecting arcs. Label the intersection T.
3. Draw $\overleftrightarrow{BT}$. This line is perpendicular to $\overleftrightarrow{AC}$ at B.

This line is also the perpendicular bisector of $\overline{RS}$.

Exercises

Written **Draw a segment with measure k. Construct each of the following.**

1. a segment with measure k
2. a segment with measure $4k$
3. an equilateral triangle with sides of measure k
4. a triangle with sides of measure $3k$, $3k$, and $4k$
5. a triangle with sides of measure $2k$, $5k$, and $6k$
6. a parallelogram with sides of measure $2k$ and $4k$ and diagonal of measure $3k$

7. Draw a segment. Construct a line perpendicular to it at an endpoint.
8. Draw a segment. Construct the perpendicular bisector of the segment.

Challenge **Copy the figure at the right. Complete each construction.**

1. Construct a line through A perpendicular to ℓ.
2. Construct a line through B perpendicular to ℓ.
3. Construct a line through B parallel to ℓ.

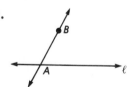

Problem Solving Application: Use a Diagram

Many problems can be solved more easily if a picture or diagram is drawn to represent the situation. Sometimes a diagram can help you decide how to work a problem. At other times, the diagram will show the answer to the problem.

Example

1 Find the height of a transmitter tower if it casts a shadow 72 feet long at the same time a 6-foot 3-inch pole near the tower casts a shadow 3 feet 9 inches long.

Explore The problem asks for the height of the tower.
Let x = the measure of the height of the tower.
Now, draw a diagram to represent the situation.

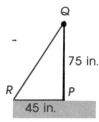

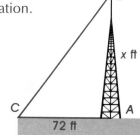

$\overline{AB}$ measures x ft.
$\overline{AC}$ measures 72 ft.
$\overline{PQ}$ measures 6 ft 3 in. or 75 in.
$\overline{PR}$ measures 3 ft 9 in. or 45 in.

Plan Assume that the tower and the pole both form right angles with the ground. Also, $\angle ABC$ and $\angle PQR$ have the same measure since they represent the angles formed by the beam of light from the sun. Thus, $\triangle ABC \sim \triangle PQR$. *Why?*

Solve

$$\frac{AB}{PQ} = \frac{AC}{PR} \qquad \textit{Write a proportion.}$$
$$\frac{x}{75} = \frac{72}{45}$$
$$45x = 5400 \qquad \textit{Cross multiply.}$$
$$x = 120 \qquad \textit{Solve for x.}$$

The height of the transmitter tower is 120 feet.

Examine The ratio of the height of the tower to the length of its shadow is 120 : 72 or 5 : 3. The ratio of the height of the pole to the length of its shadow is 75 : 45 or 5 : 3. Since these ratios are the same, the solution is correct.

Example

2 In a maximum circuit network, there is a connection, or circuit, between each pair of terminals in the network. How many circuits are required in a maximum circuit network with 4 terminals?

Explore   The network has 4 terminals. There must be a circuit between each pair of terminals.

Plan   Each terminal can be represented by a point, and each circuit represented by a line segment. Then the number of circuits in the network is same as the number of line segments. Draw a diagram to represent the situation.

Solve   Count the number of line segments. Since there are 6 segments in the diagram, there must be 6 circuits in the network.

Examine this solution.

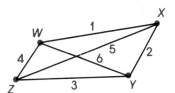

Exercises

Written   **Solve each problem.**

1. Sid was standing next to a pine tree with a shadow of 30 feet. If Sid is 5 feet 8 inches tall with a shadow of 8 feet, find the height of the tree.

2. How many circuits are required in a maximum circuit network with 5 terminals? with 6 terminals?

3. You can cut a pizza into 7 pieces with only 3 straight cuts. What is the greatest number of pieces you can make with 5 straight cuts?

4. An ant is climbing a 30-ft pole. Each day it climbs 7 feet. Each night it slips back 4 feet. How many days will it take the ant to reach the top of the pole?

5. How many games will occur in a 6-team baseball tournament if the teams play each other exactly once?

6. A staircase with 3 steps can be built with 6 blocks. How many blocks are needed to build a 7-step staircase?

7. In a minimum circuit network, there is a path of one or more circuits between each pair of terminals in the network and as few circuits as possible are used. How many circuits are required for a minimum circuit network with 4 terminals? with 5 terminals?

8. How many gallons of punch costing $1.05 per gallon must be mixed with 900 gallons of punch costing $1.13 per gallon to produce a mixture that costs $1.10 per gallon?

Portfolio Suggestion

Select some of your work from this chapter that shows how you used a calculator or computer. Place it in your portfolio.

Performance Assessment

At a certain time of the day, Lisa's shadow is 8 feet long. Lisa is 5 feet 6 inches tall, and the tree in her yard is 44 feet tall. Lisa stands with her back to the tree so that her shadow extends beyond the tree's shadow. Then she backs up until the end of her shadow coincides with the edge of the tree's shadow. How far from the tree is Lisa standing? Include a drawing and explanation of your solution.

Chapter Summary

1. A **ratio** is in **simplest form** if no whole number other than 1 divides both its terms evenly. (176)

2. A **proportion** is an equation of the form $\frac{a}{b} = \frac{c}{d}$, which states that two ratios are **equivalent.** In a proportion, the product of the **means** is equal to the product of the **extremes.** That is, $ad = bc$. (179)

3. A **percent** is the ratio of a number to 100. $\frac{\text{Percentage}}{\text{Base}}$ = Rate or $\frac{P}{B} = \frac{r}{100}$. (184)

4. Two polygons are **similar** if their **corresponding angles** have the same measure and the measures of their **corresponding sides** are in proportion. (187)

5. Axiom 6-1, **SAS:** If two sides and the included angle of one triangle are **congruent** to the corresponding sides and included angle of another triangle, then the triangles are congruent. (192)

6. Theorem 6-1: If two sides of a triangle are congruent, then the angles opposite those sides are congruent. (192)

7. Axiom 6-2, **ASA:** If two angles and the included side of one triangle are **congruent** to corresponding parts of another triangle, then the triangles are congruent. (194)

8. Theorem 6-2: A **diagonal** of a parallelogram separates the parallelogram into two **congruent** triangles. (194)

9. Axiom 6-3, **SSS:** If three sides of a triangles are congruent to three sides of another **triangle, the** triangles are congruent. (199)

Chapter Review

6.1 Write each of the following as a ratio in simplest form and also as a decimal.

1. 70 to 100   **2.** 11 : 55   **3.** $\dfrac{15}{80}$ **4.** 1.5 to 60

6.2 Solve each of the following proportions.

5. $\dfrac{11}{4} = \dfrac{t}{28}$ **6.** $\dfrac{w}{3} = \dfrac{2}{5}$ **7.** $\dfrac{3}{4} = \dfrac{y+1}{24}$ **8.** $\dfrac{2x+4}{3x-4} = \dfrac{6}{7}$

Solve each problem.

9. Two numbers are in a ratio of 5 to 2. Find the numbers if the greater is 21 more than the other.

10. Two complementary angles are in the ratio 19 : 11. Find the measure of each angle.

6.3 Solve each problem.

11. Express $40\dfrac{1}{2}\%$ as a decimal.

12. What is the price reduction of a coat at 15% discount if its list price was $125?

6.4 In the figure at the right, △*DEF* ~ △*KLM*. Use the figure to complete each of the following.

13. ∠*D* corresponds to __.
14. $\overline{EF}$ corresponds to __.
15. __ corresponds to $\overline{KM}$.
16. *m∠F* = *m∠* __
17. The measures of corresponding sides are __.
18. The measures of corresponding angles are __.

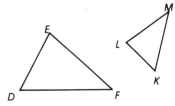

6.5 and 6.6 State whether each of the following is *always, sometimes,* or *never* true.

19. If two sides and an angle of a triangle are congruent to corresponding parts of another triangle, the triangles are congruent.

20. In an isosceles triangle, the measures of the angles opposite the congruent sides are equal.

21. The diagonals of a parallelogram are congruent.

22. Consecutive angles of a parallelogram are supplementary.

23. Opposite angles of a parallelogram are supplementary.

24. The diagonals of a parallelogram are perpendicular.

6.7 **25.** Draw a segment with measure *k*. Then construct a triangle with sides of measure 2*k*, 2*k*, and *k*.

Chapter Test

Write each of the following ratios as a decimal.

1. $\dfrac{30}{100}$

2. 7.14%

3. 3 to 5

4. 3.5 : 7

Solve each proportion.

5. $\dfrac{k}{5} = \dfrac{3}{4}$

6. $\dfrac{x + 2}{18} = \dfrac{1}{2}$

7. $\dfrac{z + 1}{2z - 1} = \dfrac{3}{5}$

Solve each problem.

8. A committee has 20 members. The ratio of Republicans to Democrats is 4 : 1. Find the number of Democrats.

9. Suppose $\triangle ABC \sim \triangle DEF$. The measures of their sides are in the ratio 4 : 7. Find AB if DE is 9 more than AB.

Tell if the information given in each exercise is sufficient to prove that $\triangle DEF$ is congruent to $\triangle SRT$. Write *yes* or *no*.

10. $\overline{EF} \cong \overline{RT}$, $\angle D \cong \angle S$, $\overline{DF} \cong \overline{ST}$
11. $\angle D \cong \angle S$, $\angle E \cong \angle R$, $\angle F \cong \angle T$
12. $\overline{DE} \cong \overline{SR}$, $\overline{EF} \cong \overline{RT}$, $\overline{DF} \cong \overline{ST}$
13. $\overline{DE} \cong \overline{SR}$, $\angle D \cong \angle S$, $\angle E \cong \angle R$

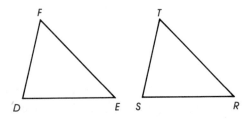

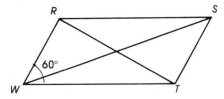

In the figure at the left, *RSTW* is a parallelogram and $m\angle W = 60$. Tell if each of the following must be true. Write *yes* or *no*.

14. $m\angle T = 120$ **15.** $\triangle RWT \cong \triangle TSR$
16. $m\angle RWS = 30$ **17.** $\overline{RT} \perp \overline{WS}$

18. In the figure at the right, $\overline{SA}$ and $\overline{NK}$ bisect each other. Show that $SK = NA$.

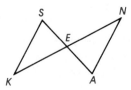

Use the figure at the left. Construct segments of each of the following measures.

19. $2k$ **20.** $3k$ **21.** $4k$

22. Construct a triangle with sides of measures $2k$, $3k$, and $4k$.

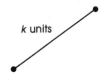

k units

The Real Numbers

Application in Engineering

Many people like to ride roller coasters. In order to make the rides safe, people who design these types of rides must have a thorough understanding of scientific and mathematical principles.

To keep passengers from falling out of the coaster when it goes upside down, designers must plan for it to be going a certain speed when it makes the loops. The **formula** for the minimum speed is $v = \sqrt{32r}$ where v is the speed in feet per second and r is the radius of the loop in feet. How fast must the coaster be going if the radius is 45 feet?

Individual Project: *Softball*

Find out the distance from one base to the next in a softball diamond. Choose five locations on the third base line including third base itself and calculate how far a player would have to throw the ball from each location to first base. Include drawings or models with your calculations.

7.1 Rational Numbers as Decimals

Between any two rational numbers on the number line there are an unlimited number of rational numbers.
Consider the graph of 0 and 1 shown below.

One way to find a rational number between 0 and 1 is to take the average of 0 and 1. You will obtain $\frac{1}{2}$. If you take the average of $\frac{1}{2}$ and 1 you obtain $\frac{3}{4}$. The average of $\frac{3}{4}$ and 1 is $\frac{7}{8}$, and so on. You can continue doing this forever. Thus, there are an infinite number of rational numbers between any two rational numbers. This property is called the **density property** of the rational numbers.

▰Example▰

1 Find a rational number between $\frac{2}{7}$ and $\frac{3}{7}$. Show the graph of the numbers on a number line.

Find the average of $\frac{2}{7}$ and $\frac{3}{7}$.

$\frac{2}{7} + \frac{3}{7} = \frac{5}{7}$ *Find the sum.*

$\frac{5}{7} \div 2 = \frac{5}{14}$ *Divide by the number of addends.*

A rational number between $\frac{2}{7}$ and $\frac{3}{7}$ is $\frac{5}{14}$.

A rational number can be written in several ways. For example, $\frac{1}{2}$, 50%, and 0.5 each represent the same rational number.

Example

2 Change the fractions $\frac{2}{11}$ and $\frac{17}{3}$ to decimals.

In each case, divide the numerator by the denominator.

```
       0.1818 . . .              5.666 . . .
   11)2.0000                  3)17.000
     1 1                        15
      ───                       ──
       90                        2 0
       88                        1 8
       ──                        ───
        20                        20
        11                        18
        ──                        ──
         90                        20
         88                        18
         ──                        ──
          2   and so on            2   and so on
```

Notice that the same remainders keep repeating.

The decimal form for $\frac{2}{11}$ is 0.1818 . . . and for $\frac{17}{3}$ is 5.666 . . . Notice that the division process could continue forever. However, a pattern is established. The digit or group of digits which form the pattern are indicated by a bar over the digits. Therefore, $\frac{2}{11}$ can be written as $0.\overline{18}$ and $\frac{17}{3}$ can be written as $5.\overline{6}$.

Decimals in which a digit or a group of digits repeats are called **repeating decimals.** The digit or group of repeating digits is called the **repetend.**

Example

3 Change the fractions $\frac{3}{4}$ and $\frac{6}{5}$ to decimals.

Divide 3 by 4 and 6 by 5.

$$
\begin{array}{r}
0.7500\ldots \\
4)\overline{3.0000} \\
2\,8 \\
\hline
20 \\
20 \\
\hline
0 \\
0 \\
\hline
0
\end{array}
\qquad
\begin{array}{r}
1.200\ldots \\
5)\overline{6.000} \\
5 \\
\hline
10 \\
10 \\
\hline
0 \\
0 \\
\hline
0
\end{array}
$$

The decimal form for $\dfrac{3}{4}$ is $0.75\overline{0}$ and for $\dfrac{6}{5}$ is $1.2\overline{0}$.

Decimals in which only a 0 repeats are written without the repeating 0's. Therefore $\dfrac{3}{4}$ can be written as 0.75 and $\dfrac{6}{5}$ can be written as 1.2.

Calculator Hint

Many calculators display up to 10 digits, which makes determining whether a decimal repeats or terminates easier than manual calculations.

Decimals like $0.75\overline{0}$ and $1.2\overline{0}$ in which only a 0 repeats are a special type of repeating decimal called **terminating decimals.**

Every rational number can be represented by a repeating decimal.

Exercises

Exploratory Show that each of the following is a rational number.

1. 3%   **2.** 30%   **3.** -0.6666 **4.** 140.39

5. $-0.\overline{6}$ **6.** $17.\overline{2}$%   **7.** $\dfrac{\frac{5}{8}}{\frac{2}{3}}$ **8.** $\dfrac{0.\overline{4}}{0.\overline{5}}$

State the numbers, in order, from least to greatest.

9. $\frac{1}{2}, \frac{1}{4}, \frac{1}{3}$ **10.** $-4.2, -4.02, -4.22$ **11.** $0.01, 0.10, 0.0013$

12. $\frac{3}{4}, \frac{6}{7}, \frac{4}{5}$ **13.** $-\frac{2}{15}, -\frac{1}{6}, -\frac{1}{5}$ **14.** $-1.7, -\frac{5}{4}, -1.07$

Express each of the following as a decimal.

15. $\frac{1}{2}$ **16.** $\frac{1}{4}$ **17.** $\frac{3}{4}$ **18.** $\frac{4}{5}$ **19.** $\frac{6}{5}$

Written   Find the average of each pair of numbers. Then, graph the three numbers on a number line.

1. 60, 61   **2.** 8.6, 8.7   **3.** 4.2, 4.3   **4.** 5.83, 5.834

5. $\frac{1}{5}, \frac{1}{4}$ **6.** $-\frac{1}{3}, -\frac{1}{2}$ **7.** $-\frac{1}{4}, -\frac{1}{5}$ **8.** $-\frac{3}{7}, -\frac{1}{4}$

9. 140.01, 140.02   **10.** $-2.6418, -2.641$

Express each of the following as a decimal.

11. $\frac{2}{3}$ **12.** $\frac{7}{8}$ **13.** $\frac{4}{9}$ **14.** $\frac{7}{4}$ **15.** $\frac{5}{9}$ **16.** $\frac{3}{8}$

17. $\frac{5}{12}$ **18.** $\frac{11}{7}$ **19.** $\frac{7}{11}$ **20.** $\frac{3}{20}$ **21.** $\frac{8}{25}$ **22.** $-\frac{2}{9}$

23. $-\frac{11}{5}$ **24.** $\frac{9}{16}$ **25.** $\frac{41}{200}$ **26.** $\frac{3}{16}$ **27.** $\frac{16}{5}$ **28.** $-\frac{16}{125}$

Find three rational numbers between each of the following pair of numbers.

29. 48, 49   **30.** $-4.1, -4.12$ **31.** $-\frac{2}{5}, -\frac{3}{5}$ **32.** 3.25, 3.251

Challenge   Solve each problem.

1. Find a rational number which is $\frac{1}{3}$ of the way between 1.2 and -0.05.

2. Suppose $\frac{1}{4}$ is $\frac{1}{6}$ of the way between $\frac{1}{8}$ and n. Find n where $n > \frac{1}{4}$.

━━━━━━━━━━━━━━━━━━━ **Mixed Review** ━━━━━━━━━━━━━━━━━━━

Choose the best answer.

1. Which of the following is *not* sufficient to show that $\triangle DEF$ is congruent to $\triangle RST$?

   a. $\overline{DE} \cong \overline{RS}, \angle D \cong \angle R, \angle E \cong \angle S$
        b. $\overline{DE} \cong \overline{RS}, \angle D \cong \angle R, \overline{DF} \cong \overline{RT}$
   c. $\overline{DE} \cong \overline{RS}, \angle D \cong \angle R, \overline{EF} \cong \overline{ST}$
        d. $\overline{DE} \cong \overline{RS}, \overline{DF} \cong \overline{RT}, \overline{EF} \cong \overline{ST}$

2. Suppose $\angle A$ and $\angle B$ are supplementary angles. If $m\angle A = 6x - 17$ and $m\angle B = 4x + 7$, what is the value of x?

   a. $x = 12$ **b.** $x = 19$ **c.** $x = 10$ **d.** $x = 17$

3. The scale on a map is 3 cm to 10 km. If Maple City and Pinewood are 202.5 km apart, how far apart are they on the map?

   a. 60.75 cm   **b.** 675 cm   **c.** 6.75 cm   **d.** 6.075 cm

4. A suit jacket was on sale for $108.54 at Magnum's Mens Wear. If this price represents a $33\frac{1}{3}\%$ discount, what was the original price of the jacket?

   a. $144.72   **b.** $162.81   **c.** $72.36   **d.** $325.62

5. Let p represent "the ratio of x to $x + 4$ is 4 : 5" and q represent "$3x - 17 \leq 30$." Which statement is true when $x = 16$?

   a. $p \rightarrow q$
        b. $\sim p \vee q$
   c. the contrapositive of $p \rightarrow q$
        d. the inverse of $p \rightarrow q$

7.2 Repeating Decimals for Rational Numbers

Numbers expressed as repeating decimals can be represented as rational numbers in the form $\frac{a}{b}$, where a and b are integers, and $b \neq 0$. Terminating decimals can be changed to fractions by the following method.

Example

1 Express 2.15 and 0.112 in the form $\frac{a}{b}$.

$$2.15 = \frac{215}{100} \text{ or } \frac{43}{20}$$

The denominator in each case is the same as the place value of the last digit in the decimal.

$$0.112 = \frac{112}{1000} \text{ or } \frac{14}{125}$$

You can express other repeating decimals as fractions using the following method.

Example

2 Express $0.\overline{54}$ in the form $\frac{a}{b}$.

Let $n = 0.\overline{54}$.
Therefore $100n = 54.\overline{54}$.

Multiply each side by 10^2 or 100 because the repetend has two digits.

Subtract the first equation from the second.

$$
\begin{array}{r}
100n = 54.\overline{54} \\
n = 0.\overline{54} \\
\hline
99n = 54
\end{array}
$$

Solve for n.

$$n = \frac{54}{99} \text{ or } \frac{6}{11} \qquad \text{Thus, } 0.\overline{54} = \frac{6}{11}.$$

Examples

3 **Express $0.4\overline{3}$ in the form $\frac{a}{b}$.**

Let $n = 0.4\overline{3}$.
Therefore, $10n = 4.\overline{3}$.   *Multiply both sides by 10.*

$$\begin{array}{r} 10n = 4.3\overline{3} \\ n = 0.4\overline{3} \\ \hline 9n = 3.9 \end{array}$$ *$4.\overline{3}$ is written as $4.3\overline{3}$ to make the subtraction easier.*

$n = \dfrac{3.9}{9}$ or $\dfrac{13}{30}$ Thus, $0.4\overline{3} = \dfrac{13}{30}$.

4 **Express $0.\overline{132}$ in the form $\frac{a}{b}$.**

Let $n = 0.\overline{132}$.
Therefore, $1000n = 132.\overline{132}$.   *Multiply both sides by 1000.*

$$\begin{array}{r} 1000n = 132.\overline{132} \\ n = 0.\overline{132} \\ \hline 999n = 132 \end{array}$$ *Subtract.*

$n = \dfrac{132}{999}$ or $\dfrac{44}{333}$ Thus, $0.\overline{132} = \dfrac{44}{333}$.

5 **Express $5.\overline{9}$ in the form $\frac{a}{b}$.**

Let $n = 5.\overline{9}$.

$$\begin{array}{r} 10n = 59.\overline{9} \\ n = 5.\overline{9} \\ \hline 9n = 54 \end{array}$$ *Multiply both sides by 10.*
    Subtract.

$n = \dfrac{54}{9}$ or 6 *Note that the fraction also names a whole number.*

Thus, $5.\overline{9} = 6$.

Journal

Do you think decimals or fractions are used more often in the real world? Explain your answer.

The method used in the previous examples can be used to change any repeating decimal to the form $\frac{a}{b}$ where a and b are integers and $b \neq 0$. Therefore, the following statement can be concluded.

Every repeating decimal represents a rational number.

This statement and its converse, given on page 208, are true. Using your knowledge of logic, you can write the following statement.

A number is rational if and only if it can be represented by a repeating decimal.

Exercises

Exploratory Express each of the following in the form $\dfrac{a}{b}$ where a and b are integers and $b \neq 0$.

1. 0.8	**2.** 0.48	**3.** $0.63\overline{0}$	**4.** -0.427
5. 3.52	**6.** -14.19	**7.** $0.55\overline{5}$	**8.** 0.55
9. $0.\overline{4}$	**10.** $0.\overline{5}$	**11.** $0.\overline{37}$	**12.** $0.75\overline{75}$
13. $0.\overline{6}$	**14.** $0.\overline{3}$	**15.** $2.\overline{3}$	**16.** $25.\overline{27}$

Written Express each of the following in the form $\dfrac{a}{b}$ where a and b are integers and $b \neq 0$.

1. 2.4	**2.** 4.2	**3.** $2.\overline{75}$	**4.** $2.\overline{135}$
5. $2.51\overline{3}$	**6.** -1.215	**7.** $0.\overline{15}$	**8.** $0.6\overline{4}$
9. $-0.4\overline{8}$	**10.** 2.5713	**11.** $-19.1\overline{3}$	**12.** $0.2\overline{53}$
13. $5.12\overline{877}$	**14.** $-2.0\overline{14}$	**15.** $8.10\overline{56}$	**16.** $0.133\overline{2123}$
17. $2.3\overline{9}$	**18.** $62.\overline{8}$	**19.** $4.1\overline{9}$	**20.** $0.261\overline{261}$
21. $5.26\overline{9}$	**22.** $0.4\overline{4}$	**23.** $2.74\overline{9}$	**24.** $0.\overline{9}$

Express each of the following as an equivalent repeating decimal.

25. 7	**26.** 4	**27.** 3.4	**28.** 6.8	**29.** 10.75	**30.** 1

Challenge Find a rational number which is between each pair of rational numbers.

1. $0.\overline{9}$ and $0.\overline{8}$

2. 0.37 and $0.\overline{37}$

3. 0.3333 and $0.\overline{3}$

4. $7.\overline{9}$ and $7.\overline{1}$

5. 0.99 and $0.\overline{9}$

6. 0.1818 and $\dfrac{1}{11}$

Find the sum of the following pairs of decimals.

7. 0.2 and $0.\overline{2}$

8. 0.33 and $0.0\overline{9}$

9. $0.\overline{7}$ and $0.\overline{2}$

10. $-0.\overline{3}$ and $0.\overline{3}$

11. $0.\overline{27}$ and $0.\overline{72}$

12. $0.\overline{37}$ and $0.\overline{86}$

7.3   The Real Numbers

In the decimal 0.75775777577775 . . ., after the decimal point, there are first one 7 and one 5. Then there are two 7's and one 5, then three 7's and one 5 and so on. There is a pattern, but there is no repetend. This decimal is an example of a **nonrepeating decimal.**

**Definition of an
Irrational Number**

> A number which is represented by a nonrepeating decimal is an irrational number.

A well-known number is π (pi).

$$\pi = 3.14159 \ldots$$

In the decimal representation of π, there is no systematic pattern to the infinite listing of digits. Thus, it is an irrational number. Every decimal represents either a rational number or an irrational number.

**Definition of a
Real Number**

> The set consisting of all rational numbers and all irrational numbers is called the set of real numbers.

The symbol $\mathscr{R}$ represents the set of real numbers.

The point corresponding to 0.757757775 . . . is between 0.7 and 0.8 on a number line. It is also between 0.75 and 0.76.

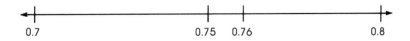

To be even more precise, 0.75775777 . . . is between 0.757 and 0.758. Continuing this procedure, you close in from both sides to some point P. Point P corresponds to the irrational number 0.75775777 . . . .

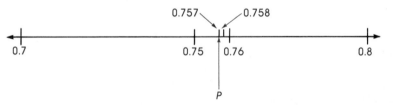

For every real number, rational or irrational, there is a corresponding point on the number line.

**Completeness
Property of the Real
Numbers**

Each point on a number line corresponds to a real number, and each real number corresponds to a point on the number line.

All the properties of the rational numbers hold for the set of real numbers.

Example

1 **Solve and graph the solution set of $2x + 3 < 7$ in the domain of real numbers.**

$$2x + 3 < 7$$
$$2x < 4 \qquad \text{\textit{Subtract 3.}}$$
$$x < 2 \qquad \text{\textit{Divide by 2.}}$$

The solution is the set which includes all real numbers less than 2.

$\{x : x \in \mathcal{R} \text{ and } x < 2\}$ *Note that the solution set contains irrational numbers such as 1.62662666 . . .*

Graph the solution set.

When the domain of x is $\mathcal{Q}$, the rational numbers, it is understood that irrational numbers are not included in the graph of the solution set.

Exercises

Exploratory   State whether each decimal is repeating or nonrepeating.

1. 0.47575757 . . .
2. 0.47557555755557 . . .
3. −2.23611561156 . . .
4. 8.0101101110 . . .

Explain whether a number may satisfy the following conditions.

5. rational but not real
6. irrational but not real
7. real but not irrational
8. real but not rational

Written   State whether the following numbers are rational or irrational.

1. 6
2. 0.4343 . . .
3. 0.161161116 . . .
4. π
5. $\dfrac{17}{17}$
6. $\dfrac{-5}{6}$
7. 3.1323334353 . . .
8. $2.\overline{9}$

State whether each statement is *true* or *false*.

9. Every rational number is a real number.
10. Every number is a rational number.
11. The number 0 is real.
12. The number π is real.
13. The number −4.2 is irrational.
14. Every point on the number line corresponds to a rational number.
15. Every repeating decimal represents a rational number.

Simplify each of the following.

16. $0.\overline{14} + 0.\overline{25}$
17. 0.1414414441 . . . + 0.5252252225 . . .
18. $0.4 + 0.\overline{4}$
19. −0.30330333 . . . + 0.30330333 . . .
20. $0.6 + 0.\overline{6}$
21. $0.9 + 0.\overline{9}$
22. $0.\overline{3} + 0.\overline{6}$
23. $1.\overline{4} + 2.12122122212222$ . . .
24. $6.\overline{2} + 0.81881888$ . . .
25. 0.0101101110 . . . + 0.101001000 . . .
26. $0.\overline{4} + 0.\overline{6}$
27. 1.313313331 . . . + 0.4224222422224 . . .
28. $3.5 + 0.\overline{4}$
29. 2.343343334 . . . + 5.212212221 . . .

Solve each of the following in the domain of real numbers. In the case of inequalities, graph the solution set on the real number line.

30. $3x + 2 < 6$
31. $4x − 5 < 3$
32. $4x − 1 < x + 2$
33. $16 \le x + 9$
34. $−7 + (−2x) \le −8x$
35. $6x > 5x + 6$
36. $9x + 7 = 8x − 2$
37. $−3 + 14y = 15y$
38. $14 + 7x > 8x$
39. $9y − 6 > 10y$
40. $5x + 4(3 − x) = −17x$
41. $\frac{1}{2}y − \frac{2}{3} = \frac{3}{4}y + \frac{1}{9}$
42. $−10x + 2(x − 5) = 13x − 7$
43. $−4 − 7(2t − 3) \ge 10t − 5(t + 1)$
44. $8r + 3(r − 4) < 5r + 12$
45. $10p − 5(3p − 1) \le 6(2p − 5) + 11p$

7.4 Square Roots

Squaring a number means using that number as a factor two times.

To find the area of a square whose side has a length of 4 units, find the square of 4.

$$4 \times 4 = 16$$

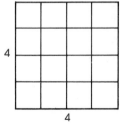

The area of the square is 16 square units.

The inverse of squaring is finding the **square root.**

To find the length of a side of a square whose area is 49 square units, find two equal factors whose product is 49.

$$x^2 = x \cdot x = 49$$

Since $7^2 = 49$, the length of a side of the square is 7 units. The square root of 49 is 7.

Since $(-7) \times (-7)$ also equals 49, another square root of 49 is -7.

Area = 49 square units

Definition of a Square Root

A number x is called a square root of a number y if $x^2 = y$.

The symbol $\sqrt{}$, called a **radical sign,** is used to denote the positive square root of a number. The expression under the radical sign is called the **radicand.**

radical sign $\sqrt{81}$ radicand

$\sqrt{49} = 7$ $\sqrt{49}$ indicates the positive or **principal** square root of 49.

$-\sqrt{49} = -7$ $-\sqrt{49}$ indicates the negative square root of 49.

$\pm\sqrt{49} = \pm7$ $\pm\sqrt{49}$ indicates either one or both square roots of 49.

Examples

1 Find $\sqrt{144}$.

$\sqrt{144} = 12$ since $12 \cdot 12 = 144$.

2 Find $-\sqrt{81}$.

$-\sqrt{81} = -9$ since $\sqrt{81} = 9$.

3 Solve the equation $x^2 = \dfrac{36}{121}$ for x where $x \in \mathcal{R}$.

By the definition of square root, $x = \sqrt{\dfrac{36}{121}}$ or $x = -\sqrt{\dfrac{36}{121}}$

$$x = \frac{6}{11} \quad \text{or} \quad x = -\frac{6}{11}$$

The solution set is $\left\{ \dfrac{6}{11}, -\dfrac{6}{11} \right\}$.

Rational numbers like 144, 81, and $\dfrac{36}{121}$, whose square roots are rational numbers are called **perfect squares.** In general the square root of a positive integer which is *not* the square of an integer, such as $\sqrt{2}$, $\sqrt{3}$, and $\sqrt{7}$, is an irrational number.

You should be careful not to confuse the negative square root of a number with the square root of a negative number. For example, note the differences in the meanings of $-\sqrt{9}$ and $\sqrt{-9}$. Since $\sqrt{9} = 3$, $-\sqrt{9} = -3$. But there are *no* two equal real factors whose product is -9. Therefore, $\sqrt{-9}$ *cannot* be a real number. If x is a negative number, then $\sqrt{x}$ is *not* a real number.

Exercises

Exploratory   State the principal square root of each number.

1. 36   **2.** 144   **3.** $\frac{4}{9}$ **4.** 0.81

5. 0.01   **6.** 0.09   **7.** $\dfrac{100}{121}$ **8.** $\dfrac{25}{49}$

Express each number in simplified form.

9. $\sqrt{25}$ **10.** $-\sqrt{36}$ **11.** $\sqrt{144}$ **12.** $\sqrt{121}$

13. $-\sqrt{64}$ **14.** $-\sqrt{\dfrac{4}{9}}$ **15.** $-\sqrt{49}$ **16.** $\sqrt{\dfrac{25}{36}}$

Written   **Express each number in simplified form.**

1. $\sqrt{400}$ **2.** $\sqrt{0}$ **3.** $-\sqrt{0}$ **4.** $\sqrt{484}$

5. $\sqrt{\dfrac{4}{25}}$ **6.** $-\sqrt{324}$ **7.** $\dfrac{\sqrt{49}}{\sqrt{225}}$ **8.** $\dfrac{\sqrt{100}}{\sqrt{81}}$

9. $-\sqrt{\dfrac{784}{25}}$ **10.** $\sqrt{\dfrac{441}{361}}$ **11.** $\dfrac{\sqrt{0.16}}{\sqrt{0.36}}$ **12.** $\sqrt{1\dfrac{7}{9}}$

13. Find all integers between 1 and 100 whose square roots are integers.   **14.** Find all integers between 100 and 200 whose square roots are integers.

15. If x represents a real number, under what circumstances will $\sqrt{x}$ also represent a real number?

State whether each of the following represents a real number. If it does, state whether it is *rational* or *irrational*.

16. $\sqrt{4}$ **17.** $-\sqrt{36}$ **18.** $-\sqrt{5}$ **19.** $\sqrt{-5}$

20. $\sqrt{(-6)^2}$ **21.** $\sqrt{-10}$ **22.** $(\sqrt{2})^2$ **23.** $2 + \sqrt{2}$

24. $5\sqrt{-3}$ **25.** $\sqrt{4} \times \sqrt{9}$ **26.** $-\sqrt{5} + \sqrt{5}$ **27.** $(\sqrt{5,243,896})^2$

Find the solution set of each equation in the domain $\mathcal{R}$.

28. $x^2 = 49$ **29.** $y^2 = 36$ **30.** $m^2 + 9 = 25$

31. $a^2 + 64 = 100$ **32.** $x^2 + x^2 = 50$ **33.** $y^2 + 64 = 289$

Simplify each of the following.

34. $\sqrt{5^2}$ **35.** $\sqrt{3^2}$ **36.** $\sqrt{(19,000)^2}$ **37.** $\sqrt{0^2}$

38. $\sqrt{(-2)^2}$ **39.** $\sqrt{(-17)^2}$ **40.** $\sqrt{(1.896)^2}$ **41.** $\sqrt{(-38)^2}$

42. $\sqrt{(-13.5)^2}$ **43.** $-\sqrt{3^2}$ **44.** $-\sqrt{(-2)^2}$ **45.** $-\sqrt{(15.6)^2}$

46. $\pm\sqrt{\dfrac{4}{9}}$ **47.** $-\sqrt{\dfrac{16}{25}}$ **48.** $\pm\sqrt{\dfrac{36}{49}}$ **49.** $\pm\sqrt{\left(\dfrac{15}{16}\right)^2}$

50. $\pm\sqrt{(36)^2}$ **51.** $\pm\sqrt{(3.1)^2}$ **52.** $\pm\sqrt{(13,225)^2}$ **53.** $-\sqrt{x^2}$

Challenge   **Answer each of the following.**

1. Does $\sqrt{x^2}$ always equal x? Explain your answer.

2. Does $\sqrt{x^2}$ always equal $|x|$? Explain your answer.

The problem of trying to find a rational number whose square is 2 dates back to the ancient Greeks. They were able to demonstrate that the search for such a rational number is a hopeless task!

Consider the prime factorization of 24 and $(24)^2$.

$24 = 2 \cdot 2 \cdot 2 \cdot 3$ *three factors of 2 and one factor of 3*
$(24)^2 = 2 \cdot 2 \cdot 2 \cdot 2 \cdot 2 \cdot 2 \cdot 3 \cdot 3$ *six factors of 2 and two factors of 3*

Consider a number x whose prime factorization is given below.

$x = 2 \cdot 2 \cdot 3 \cdot 5 \cdot 5 \cdot 5 \cdot 7 \cdot 7 \cdot 7 \cdot 7 \cdot 7 \cdot 7 \cdot 7 \cdot 7 \cdot 7$

There are two factors of 2, one factor of 3, three factors of 5, and nine factors of 7. Thus, x^2 will have twice these factors; that is, four factors of 2, two factors of 3, six factors of 5, and eighteen factors of 7. In each case, if x has n factors of a prime number k, then x^2 has $2n$ factors of k. That is, x^2 must have an even number of factors of any prime number.

To try to find a rational number whose square is 2, we use a form of argument known as indirect reasoning. Consider the following statements p and $\sim p$.

p: There is no rational number whose square is 2.
$\sim p$: There is a rational number whose square is 2.

The plan is to assume $\sim p$ is true. Then, show that this assumption leads to a contradiction. If $\sim p$ leads to a contradiction, then $\sim p$ is false and p must be true.

Let $\dfrac{a}{b}$ be a rational number whose square is 2. Then,

$$\left(\frac{a}{b}\right)^2 = 2 \quad \text{or} \quad \frac{a^2}{b^2} = 2.$$

Multiply each side of the equation by b^2.

$$a^2 = 2b^2$$

a^2 must have an even number of factors of 2.
b^2 must have an even number of factors of 2.

But $2b^2$ has an extra factor of 2. Therefore, $2b^2$ has an odd number of factors of 2 if a has an even number of factors of 2. Since $a^2 = 2b^2$, they are the same number. This means that a^2 has an even number of factors of 2 and also an odd number of factors of 2. Of course, this is impossible!

The assumption, $\sim p$, that there is a rational number whose square is 2, has led to a contradiction. Therefore, we can conclude that $\sim p$ is false, and p is true. That is, there is no rational number whose square is 2.

Exercises Prove there is no rational number whose square is each of the following.

1. 3   2. 5   3. 6   4. 7   5. 8

7.5 Approximating Square Roots

The area of the carpet at the left is 30 square yards. The length of one side is $\sqrt{30}$ yards. Since 30 is not a perfect square of some integer, $\sqrt{30}$ is an irrational number.

You can get a rough approximation of $\sqrt{30}$ by first finding the two consecutive integers between which $\sqrt{30}$ lies. For example, since $5^2 = 25$ and $6^2 = 36$, $\sqrt{30}$ is between 5 and 6.

$$5 < \sqrt{30} < 6$$

Example

1 **Find two consecutive integers between which $\sqrt{51}$ lies.**

$7^2 = 49$ and $8^2 = 64$. 51 is between 49 and 64. Therefore, $7 < \sqrt{51} < 8$.

One method for approximating square roots is called the **divide-and-average method.** This method is used below to approximate $\sqrt{30}$ to the nearest hundredth.

Step 1

Choose a first approximation for $\sqrt{30}$. Choose 5.5 because $\sqrt{30}$ is approximately midway between $\sqrt{25}$ and $\sqrt{36}$.

$$25 < \ 30 \ < 36$$
$$5^2 < \ 30 \ < 6^2$$
$$5 < \sqrt{30} < 6$$

Step 2

Divide 30 by 5.5.
Carry the quotient to one more decimal place than the divisor.

$$
\begin{array}{r}
5.45 \\
5.5\overline{\smash{)}30.0\,00} \\
\underline{27\ 5} \\
2\ 5\ 0 \\
\underline{2\ 2\ 0} \\
3\ 00 \\
\underline{2\ 75} \\
25
\end{array}
$$

Step 3

Average the quotient, 5.45, and the divisor.

$$\frac{5.45 + 5.5}{2} = \frac{10.95}{2}$$
$$\approx 5.47$$

Step 4

Use the average, 5.47, as the new divisor. Repeat Steps 2 and 3.

$$5.47_\wedge\overline{)30.00_\wedge000}$$ *Divide.*

quotient: 5.484

$$\frac{5.47 + 5.484}{2} \approx 5.477$$ *Average.*

An approximation for $\sqrt{30}$ is 5.477. Rounding to the nearest hundredth $\sqrt{30} \approx 5.48$.

Example

2 **Approximate $\sqrt{0.176}$. Use the divide-and-average method.**

$$0.16 < 0.176 < 0.25$$
$$(0.4)^2 < 0.176 < (0.5)^2$$
$$0.4 < \sqrt{0.176} < 0.5$$

Since 0.176 is much closer to 0.16, choose 0.4 as the first approximation.

$$0.4_\wedge\overline{)0.1_\wedge76}$$ *Divide.*   quotient: 0.44

$$\frac{0.44 + 0.4}{2} = \frac{0.84}{2}$$ *Average.*
$$\approx 0.42$$

Use 0.42 as the second approximation.

$$0.42_\wedge\overline{)0.17_\wedge600}$$ *Divide.*   quotient: 0.419

$$\frac{0.419 + 0.42}{2} = \frac{0.839}{2}$$ *Average.*
$$\approx 0.4195$$

An approximation for $\sqrt{0.176}$ is 0.4195.

Another method of finding rational approximations for positive square roots is to use a table of square roots. Such a table can be found on page 482 of this book.

Examples

3 **Use the table of square roots to find an approximate value for $\sqrt{7}$.**

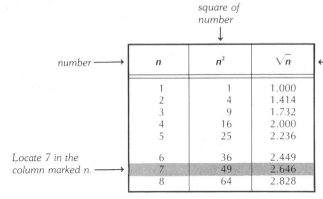

square of number ↓

number →

square root of number ←

Locate 7 in the column marked n. →

Locate the value for $\sqrt{7}$ in the column marked $\sqrt{n}$.

n	n²	√n
1	1	1.000
2	4	1.414
3	9	1.732
4	16	2.000
5	25	2.236
6	36	2.449
7	49	2.646
8	64	2.828

An approximate value for $\sqrt{7}$ is 2.646.

4 **Use the table of square roots to find an approximate value for $\sqrt{43}$.**

n	n²	√n
42	1764	6.481
43	1849	6.557
44	1936	6.633

An approximate value for $\sqrt{43}$ is 6.557.

Some calculators have a key marked $\sqrt{x}$ which can be used to obtain square root approximations. To approximate $\sqrt{30}$, first enter the number 30. Then, depress the $\sqrt{x}$ key. The number appearing in the display is an approximation of $\sqrt{30}$.

Exercises

Exploratory **State two consecutive integers between which each of the following lies.**

1. $\sqrt{3}$ 2. $\sqrt{5}$ 3. $\sqrt{12}$ 4. $-\sqrt{18}$ 5. $\sqrt{47}$

6. $\sqrt{51}$ 7. $\sqrt{68}$ 8. $-\sqrt{75}$ 9. $\sqrt{143}$ 10. $-\sqrt{\frac{1}{2}}$

11. $\sqrt{3.6}$ 12. $-\sqrt{8.9}$ 13. $\sqrt{17.8}$ 14. $-\sqrt{86.42}$ 15. $\sqrt{0.6}$

Written State two consecutive integers between which each of the following lies.

1. $\sqrt{7}$ **2.** $\sqrt{11}$ **3.** $-\sqrt{17}$ **4.** $-\sqrt{23}$

5. $\sqrt{154}$ **6.** $-\sqrt{0.5}$ **7.** $-\sqrt{81.2}$ **8.** $\sqrt{101.5}$

Use the divide-and-average method to find a rational approximation correct to the nearest tenth.

9. $\sqrt{47}$ **10.** $\sqrt{75}$ **11.** $\sqrt{19}$ **12.** $\sqrt{23}$

13. $\sqrt{86.3}$ **14.** $\sqrt{92.8}$ **15.** $\sqrt{103}$ **16.** $\sqrt{146.2}$

17. $\sqrt{0.3}$ **18.** $\sqrt{412}$ **19.** $\sqrt{586}$ **20.** $\sqrt{631}$

21. $\sqrt{83.24}$ **22.** $\sqrt{341.62}$ **23.** $\sqrt{248.26}$ **24.** $\sqrt{1482.6}$

Using the approximations in the table on page 482, or a calculator, find each of the following to the nearest tenth.

25. $\sqrt{56}$ **26.** $\sqrt{30}$ **27.** $\sqrt{25}$ **28.** $\sqrt{13}$

29. $\sqrt{21}$ **30.** $\sqrt{110}$ **31.** $\sqrt{3} \times \sqrt{7}$ **32.** $\sqrt{11} \times \sqrt{10}$

33. $\sqrt{5} \times \sqrt{6}$ **34.** $\sqrt{7} \times \sqrt{8}$ **35.** $\sqrt{4} + \sqrt{9}$ **36.** $\sqrt{16} + \sqrt{9}$

Using the approximations in the table on page 482, or a calculator, find the perimeter, to the nearest tenth, of each figure. The measurements are given in centimeters.

37. 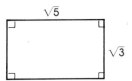   **38.**    **39.** 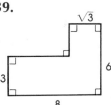   **40.**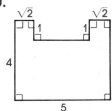

Using the approximations in the table on page 482, or a calculator, compute the area, to the nearest tenth, of each figure in exercises 37–40.

41. Exercise 37   **42.** Exercise 38   **43.** Exercise 39   **44.** Exercise 40

Mixed Review

Solve each of the following. For inequalities, graph the solution set on the real number line.

1. $3(2 - y) + 11y = 5y - 13$ **2.** $2(3n - 1) = 5(2n + 1) - (14 - 5n)$

3. $25 + x^2 = 169$ **4.** $2p^2 - 13 = 85$

5. $7a - 13 > -5a + 10(2a - 1)$ **6.** $b - 3(11b - 10) \geq 5(5b - 3) + 6b$

7. $\dfrac{r}{35} = \dfrac{7}{10}$ **8.** $\dfrac{45}{16} = \dfrac{36}{t}$ **9.** $\dfrac{60}{2k - 3} = \dfrac{15}{16}$ **10.** $\dfrac{z + 5}{6 - z} = \dfrac{3}{8}$

7.6   Simplifying Radicals

To find the square root of a perfect square such as 196, you can use its prime factorization.

$$\sqrt{196} = \sqrt{2 \cdot 2 \cdot 7 \cdot 7}$$
$$= \sqrt{2^2 \cdot 7^2}$$
$$= \sqrt{2^2} \cdot \sqrt{7^2}$$
$$= 2 \cdot 7 \quad \text{or} \quad 14 \qquad \textit{Check: } 14^2 = 196 \quad \checkmark$$

A property of square roots was used to simplify $\sqrt{196}$.

Product Property of Square Roots

For any nonnegative numbers a and b,
$$\sqrt{a \cdot b} = \sqrt{a} \cdot \sqrt{b}.$$

Examples

1 **Simplify $\sqrt{200}$.**   *A square root is in simplest form when its radicand does not contain any perfect square factors other than 1.*

$$\sqrt{200} = \sqrt{5 \cdot 5 \cdot 2 \cdot 2 \cdot 2} \qquad \textit{Prime factorization of 200}$$
$$= \sqrt{5^2 \cdot 2^2 \cdot 2}$$
$$= \sqrt{5^2} \cdot \sqrt{2^2} \cdot \sqrt{2} \qquad \textit{Use the product property of square roots.}$$
$$= 5 \cdot 2\sqrt{2} \text{ or } 10\sqrt{2} \qquad \textit{Check: } (10\sqrt{2})^2 = 100 \cdot 2 \text{ or } 200 \quad \checkmark$$

2 **Simplify $\frac{1}{6}\sqrt{27}$.**

$$\frac{1}{6}\sqrt{27} = \frac{1}{6}\sqrt{9 \cdot 3}$$
$$= \frac{1}{6}\sqrt{3^2 \cdot 3}$$
$$= \frac{1}{6} \cdot 3\sqrt{3} \text{ or } \frac{\sqrt{3}}{2}$$

3 **Simplify $\sqrt{50y^3}$.**

$$\sqrt{50y^3} = \sqrt{5 \cdot 5 \cdot 2 \cdot y \cdot y \cdot y} \qquad \textit{y cannot represent a negative number in}$$
$$= \sqrt{5^2 \cdot 2 \cdot y^2 \cdot y} \qquad \textit{this example. Why?}$$
$$= \sqrt{5^2} \cdot \sqrt{2} \cdot \sqrt{y^2} \cdot \sqrt{y}$$
$$= 5 \cdot y \cdot \sqrt{2} \cdot \sqrt{y} \text{ or } 5y\sqrt{2y}$$

Find the area of the rectangle at the right.

$$2\sqrt{3} \cdot 3\sqrt{2} = 2 \cdot 3 \cdot \sqrt{3} \cdot \sqrt{2}$$
$$= 6\sqrt{6}$$

$3\sqrt{2}$
units

$2\sqrt{3}$
units

The area of the rectangle is $6\sqrt{6}$ square units.

Example

4 **Simplify $(4\sqrt{6})(7\sqrt{2})$.**

$$
\begin{aligned}
(4\sqrt{6})(7\sqrt{2}) &= 4 \cdot 7 \cdot \sqrt{6 \cdot 2} \\
&= 28\sqrt{12} \\
&= 28\sqrt{2 \cdot 2 \cdot 3} \\
&= 28 \cdot \sqrt{2^2} \cdot \sqrt{3} \\
&= 28 \cdot 2\sqrt{3} \text{ or } 56\sqrt{3}
\end{aligned}
$$

Exercises

Exploratory **Simplify.**

1. $\sqrt{8}$ 2. $\sqrt{12}$ 3. $\sqrt{18}$ 4. $\sqrt{32}$
5. $\sqrt{40}$ 6. $\sqrt{48}$ 7. $\sqrt{45}$ 8. $\frac{1}{2}\sqrt{20}$

Written **Simplify.**

1. $\sqrt{28}$ 2. $\sqrt{72}$ 3. $\sqrt{54}$ 4. $\sqrt{80}$
5. $\sqrt{63}$ 6. $\sqrt{90}$ 7. $\sqrt{99}$ 8. $\sqrt{108}$
9. $\sqrt{242}$ 10. $\sqrt{128}$ 11. $5\sqrt{18}$ 12. $3\sqrt{8}$
13. $4\sqrt{200}$ 14. $6\sqrt{45}$ 15. $\frac{1}{2}\sqrt{80}$ 16. $\frac{2}{3}\sqrt{12}$
17. $\dfrac{\sqrt{18}}{3}$ 18. $\dfrac{\sqrt{24}}{2}$ 19. $\dfrac{\sqrt{48}}{4}$ 20. $\frac{3}{5}\sqrt{300}$
21. $\sqrt{x^3}$ 22. $\sqrt{m^4}$ 23. $\sqrt{9y^4}$ 24. $3x\sqrt{20x^2}$
25. $(\sqrt{2})(\sqrt{6})$ 26. $(\sqrt{2})(\sqrt{10})$ 27. $(3\sqrt{6})(2\sqrt{3})$
28. $(5\sqrt{2})(\sqrt{8})$ 29. $(5\sqrt{6})(2\sqrt{6})$ 30. $(4\sqrt{2})^2$
31. $(2\sqrt{3})^2$ 32. $(6\sqrt{3})(2\sqrt{15})$ 33. $(5\sqrt{8})(3\sqrt{3})$

Find the area, in simplest form, for each of the following rectangles. The measurements are given in centimeters.

34.

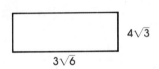

$3\sqrt{6}$ $4\sqrt{3}$

35.

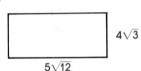

$5\sqrt{12}$ $4\sqrt{3}$

36.

$8\sqrt{12}$ $2\sqrt{3}$

37.

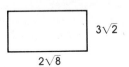

$2\sqrt{8}$ $3\sqrt{2}$

38.

$4\sqrt{12}$ $5\sqrt{5}$

39.

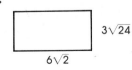

$6\sqrt{2}$ $3\sqrt{24}$

Solve each of the following in the domain $\mathscr{R}$.

40. $x^2 = 28$ **41.** $y^2 = 72$ **42.** $2x^2 = 12$ **43.** $26 + y^2 = 74$
44. $4x^2 = 32$ **45.** $y^2 - 13 = 35$ **46.** $2x^2 - 3 = 11$ **47.** $y^2 = 108$

Solve each equation. Then use the approximations in the table on page 482, or a calculator, to approximate each solution to the nearest tenth.

48. $x^2 - 12 = 25$ **49.** $2y^2 + 23 = 39$ **50.** $3r^2 - 13 = r^2 + 23$
51. $2(x^2 + 29) = 7x^2 - 62$ **52.** $102 - 6a^2 = 5(2a^2 - 34)$
53. $y^2 - 8y + 14 = 4(y^2 - 2y - 19)$ **54.** $3(8x - 3x^2) = 4(x^2 + 6x - 195)$

Challenge   Answer each of the following.

1. If $a = 16$ and $b = 9$, does $\sqrt{a + b} = \sqrt{a} + \sqrt{b}$?   **2.** Does $\sqrt{a + b} = \sqrt{a} + \sqrt{b}$ for any nonnegative values given to a and b?

Mathematical Excursions

Mathematics in Space

Spacecraft attempting to leave a planet must have a minimum escape velocity. The escape velocity, v, is calculated by the following formula.

$$v = \sqrt{\frac{2GM}{r}}$$

G is a gravitational constant.
M is the mass of the planet.
r is the radius of the planet.

Notice that the formula requires the use of a square root. Ability to use and compute square roots is often a necessary tool in science and engineering.

7.7   Adding and Subtracting Radicals

To find the perimeter of the triangle shown at the right, add the lengths of each side.

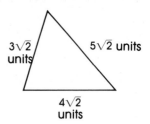

$3\sqrt{2} + 4\sqrt{2} + 5\sqrt{2} = (3 + 4 + 5)\sqrt{2}$ *Use the distributive property. $\sqrt{2}$ is a common factor of each term.*
$$= 12\sqrt{2}$$

The perimeter of the triangle is $12\sqrt{2}$ units.

You can add or subtract square roots in the same way you add or subtract like terms. Notice how the following additions and subtractions are similar.   *What is the common factor for each addition and each subtraction?*

$$\left.\begin{array}{r} 4x + 2x = (4 + 2)x \text{ or } 6x \\ 4\sqrt{3} + 2\sqrt{3} = (4 + 2)\sqrt{3} \text{ or } 6\sqrt{3} \end{array}\right\} \quad additions$$

$$\left.\begin{array}{r} 6y - 2y = (6 - 2)y \text{ or } 4y \\ 6\sqrt{7} - 2\sqrt{7} = (6 - 2)\sqrt{7} \text{ or } 4\sqrt{7} \end{array}\right\} \quad subtractions$$

Example

1 **Simplify $8\sqrt{3} - 5\sqrt{3} + 2\sqrt{3}$.**

$8\sqrt{3} - 5\sqrt{3} + 2\sqrt{3} = (8 - 5 + 2)\sqrt{3}$
$$= 5\sqrt{3}$$

Expressions like $2\sqrt{3} + \sqrt{5}$ do not have a common factor other than 1. Thus, $2\sqrt{3} + \sqrt{5}$ *cannot* be expressed as a single term. Sometimes an expression which does not appear to have a common factor other than 1 can be simplified and then added or subtracted.

Examples

2 **Simplify $5\sqrt{3} + \sqrt{27}$.**

First, simplify $\sqrt{27}$.

$\sqrt{27} = \sqrt{9 \cdot 3}$ or $3\sqrt{3}$

$5\sqrt{3} + \sqrt{27} = 5\sqrt{3} + 3\sqrt{3}$ *Replace $\sqrt{27}$ by $3\sqrt{3}$.*

$= (5 + 3)\sqrt{3}$

$= 8\sqrt{3}$

3 **Simplify $\sqrt{18} - \sqrt{8} + 2\sqrt{3}$.**

Simplify $\sqrt{18}$.

$\sqrt{18} = \sqrt{9 \cdot 2}$ or $3\sqrt{2}$

Simplify $\sqrt{8}$.

$\sqrt{8} = \sqrt{4 \cdot 2}$ or $2\sqrt{2}$

$2\sqrt{3}$ cannot be simplified.

$\sqrt{18} - \sqrt{8} + 2\sqrt{3} = 3\sqrt{2} - 2\sqrt{2} + 2\sqrt{3}$

$= (3 - 2)\sqrt{2} + 2\sqrt{3}$

$= \sqrt{2} + 2\sqrt{3}$

4 **Simplify $2\sqrt{3} (4\sqrt{2} - 2\sqrt{5})$.**

$2\sqrt{3}(4\sqrt{2} - 2\sqrt{5}) = (2\sqrt{3})(4\sqrt{2}) - (2\sqrt{3})(2\sqrt{5})$ *Use the distributive property to simplify.*

$= 8\sqrt{6} - 4\sqrt{15}$

Exercises

Exploratory Express each of the following as a single term.

1. $4\sqrt{3} + 2\sqrt{3}$ 2. $4\sqrt{5} + 3\sqrt{5}$ 3. $2\sqrt{6} + 4\sqrt{6} + \sqrt{6}$
4. $8\sqrt{3} - 2\sqrt{3}$ 5. $11\sqrt{7} - \sqrt{7}$ 6. $15\sqrt{2} - 5\sqrt{2} - 4\sqrt{2}$
7. $5\sqrt{2} + 3\sqrt{2} - 8\sqrt{2}$ 8. $16\sqrt{5} - 4\sqrt{5} + 8\sqrt{5}$ 9. $4\sqrt{3} + 7\sqrt{3} - 2\sqrt{3}$

State whether each of the following statements is *true* or *false*.

10. $\sqrt{9} + \sqrt{4} = \sqrt{13}$ 11. $\sqrt{9} \cdot \sqrt{4} = \sqrt{36}$ 12. $\frac{1}{9} > \sqrt{\frac{1}{9}}$

Written Simplify. Assume all variables represent positive numbers.

1. $2\sqrt{5} + 3\sqrt{5}$ 2. $6\sqrt{2} - 4\sqrt{2}$ 3. $5\sqrt{3} - 2\sqrt{3}$

4. $8\sqrt{7} + 15\sqrt{7}$ **5.** $10\sqrt{6} + 4\sqrt{6}$ **6.** $4\sqrt{2} - \sqrt{2}$

7. $7\sqrt{10} - \sqrt{10}$ **8.** $\sqrt{11} - 5\sqrt{11}$ **9.** $8\sqrt{x} + \sqrt{x}$

10. $5\sqrt{6} - 11\sqrt{6} + 8\sqrt{6}$ **11.** $2\sqrt{m} - 4\sqrt{m}$ **12.** $8\sqrt{2} + 3\sqrt{2} + 4\sqrt{2}$

13. $6\sqrt{3} - 2\sqrt{3} + 4\sqrt{3}$ **14.** $3\sqrt{2y} + 5\sqrt{2y} - 12\sqrt{2y}$ **15.** $\sqrt{3} + \sqrt{27}$

16. $3\sqrt{45} - \sqrt{5}$ **17.** $\sqrt{8} - \sqrt{2}$ **18.** $\sqrt{50} + 3\sqrt{2}$

19. $2\sqrt{12} - \sqrt{48}$ **20.** $5\sqrt{32} - 4\sqrt{8}$ **21.** $4\sqrt{20} + 2\sqrt{45}$

22. $\sqrt{27} + \sqrt{12}$ **23.** $2\sqrt{27} + 5\sqrt{63}$ **24.** $4\sqrt{3} + 2\sqrt{27} + 4\sqrt{18}$

25. $3\sqrt{8} + 5\sqrt{2} - \sqrt{32}$ **26.** $4\sqrt{2} + 6\sqrt{5} - 3\sqrt{2}$

27. $\sqrt{48} - \sqrt{12} + \sqrt{300}$ **28.** $8\sqrt{72} + 2\sqrt{20} - 3\sqrt{5}$

29. $\sqrt{6x} - 3\sqrt{24x}$ **30.** $\sqrt{27x^2} + \sqrt{12x^2}$

For each of the following, find an equivalent expression which is in simplified form that does not contain parentheses.

31. $3\sqrt{2}(\sqrt{2} - 1)$ **32.** $\sqrt{3}(3 + \sqrt{3})$

33. $\sqrt{5}(\sqrt{5} + \sqrt{10})$ **34.** $3\sqrt{3}(2\sqrt{5} - 4\sqrt{20} - 2\sqrt{45})$

35. $2\sqrt{5}(\sqrt{8} - \sqrt{2})$ **36.** $4\sqrt{3}(\sqrt{12} - 4\sqrt{3} + \sqrt{27})$

37. $8\sqrt{6}(4\sqrt{3} + 13\sqrt{6})$ **38.** $3\sqrt{2}(\sqrt{8} - 5\sqrt{2} + 11)$

39. $10(\sqrt{5} - \sqrt{2})$ **40.** $3\sqrt{2}(6\sqrt{7} - 2\sqrt{28} + 5\sqrt{63})$

41. $3(3\sqrt{21} - 3\sqrt{12})$ **42.** $5\sqrt{18}(4\sqrt{2} + 6\sqrt{10} - 4\sqrt{6})$

Using the approximations in the table on page 482, or a calculator, find the perimeter, to the nearest tenth, of each figure. The measurements are given in centimeters.

43.   **44.**   **45.**

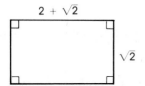

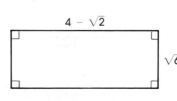

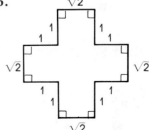

Using the approximations in the table on page 482, or a calculator, compute the area, to the nearest tenth, of each figure in exercises 43-45.

46. Exercise 43   **47.** Exercise 44   **48.** Exercise 45

Challenge   State whether each of the following statements is *true* or *false*.

1. $(\sqrt{5})^2$ is a rational number.   **2.** If $x > 1$, then $\sqrt{x} > x$.

3. If $x < 0$, then $x^2 > x$.   **4.** If $0 < x < 1$, then $\sqrt{x} > x$.

5. If $a^2 = b^2$, then $a = b$.   **6.** $3 + \sqrt{5}$ is an irrational number.

7. $\sqrt{2}$ can be expressed as a repeating decimal.   **8.** The sum of two irrational numbers is an irrational number.

7.8 The Pythagorean Theorem

A painter has a 15-foot ladder. If the bottom of the ladder is 7 feet from the wall, how high on the wall does the ladder reach? The ladder against the wall forms a right triangle.

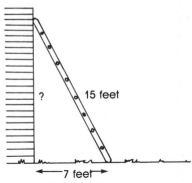

The side opposite the right angle in a right triangle is called the **hypotenuse.** The other two sides are called the **legs** of the triangle.

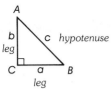

To find the length of the hypotenuse, given the lengths of the legs, you can use a formula generalized by the Greek mathematician Pythagoras.

The Pythagorean Theorem

In a right triangle, if a and b represent the measures of the legs and c represents the measure of the hypotenuse, then $c^2 = a^2 + b^2$.

The Pythagorean Theorem can be illustrated geometrically as follows.

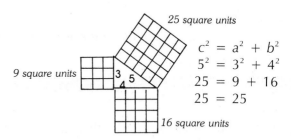

$$c^2 = a^2 + b^2$$
$$5^2 = 3^2 + 4^2$$
$$25 = 9 + 16$$
$$25 = 25$$

For the reach of the ladder, let $a = 7$ and $c = 15$. Find the length of the leg whose measure is b.

$$c^2 = a^2 + b^2$$
$$15^2 = 7^2 + b^2$$
$$225 = 49 + b^2$$
$$176 = b^2 \qquad \text{Subtract 49 from both sides.}$$
$$\sqrt{176} = b \qquad \text{Why is } -\sqrt{176} \text{ not a possible solution?}$$

An approximation of $\sqrt{176}$ is 13.3. So, the ladder can reach a height of about 13.3 feet on the wall.

Example

1 Find the measure of the hypotenuse of a right triangle with $a = 15$ and $b = 8$.

$$c^2 = a^2 + b^2 \qquad \textit{Use the Pythagorean Theorem.}$$
$$c^2 = 15^2 + 8^2 \qquad \textit{Replace a by 15 and b by 8.}$$
$$c^2 = 225 + 64$$
$$c^2 = 289$$
$$c = \sqrt{289} \text{ or } 17$$

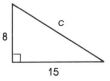

The measure of the hypotenuse is 17.

It can be shown that the converse of the Pythagorean Theorem is also true.

Converse of the Pythagorean Theorem

If the sum of the squares of the measures of two sides of a triangle is equal to the square of the measure of the third side, then the triangle is a right triangle.

From your knowledge of logic, the following biconditional can be stated.

A triangle is a right triangle if and only if $c^2 = a^2 + b^2$.

Example

2 Determine whether the following could be the measures of the sides of a right triangle. $a = 5$, $b = 7$, and $c = 9$

If a, b, and c are the measures of the sides of a right triangle, then they satisfy the Pythagorean Theorem.

$$c^2 = a^2 + b^2$$
$$9^2 \overset{?}{=} 5^2 + 7^2$$
$$81 \overset{?}{=} 25 + 49$$
$$81 \neq 74$$

The measures do *not* satisfy the Pythagorean Theorem. Therefore, they are *not* the measures of the sides of a right triangle.

Exercises

Exploratory State the value of *a, b,* or *c.*

1. $3^2 + 4^2 = c^2$ **2.** $6^2 + 8^2 = c^2$ **3.** $8^2 + 15^2 = c^2$

4. $6^2 + b^2 = 10^2$ **5.** $12^2 + b^2 = 20^2$ **6.** $5^2 + b^2 = 13^2$

7. $a^2 + 15^2 = 17^2$ **8.** $a^2 + 24^2 = 25^2$ **9.** $a^2 + 16^2 = 20^2$

Written Find the missing measure in each right triangle.

1.   **2.**    **3.**   **4.**

5.   **6.**   **7.**   **8.**

In each of the following, *c* is the measure of the hypotenuse of a right triangle. Find the measure of the missing side to the nearest tenth.

9. $a = 5, b = 12$ **10.** $a = 12, b = 15$ **11.** $a = 5, c = 7$

12. $a = 4, c = 9$ **13.** $a = \sqrt{3}, c = \sqrt{15}$ **14.** $b = 4, c = 8$

15. $a = 5, b = 7$ **16.** $a = 2, b = 5$ **17.** $a = 4, c = 9\frac{1}{2}$

18. $a = \sqrt{5}, b = \sqrt{2}$ **19.** $a = 3\sqrt{2}, c = 12$ **20.** $b = 3\sqrt{5}, c = 4\sqrt{7}$

Find the length of a diagonal of a rectangle whose sides are given.

21. 8′ and 15′   **22.** 10″ and 24″   **23.** 15 yd and 20 yd   **24.** 2.1″ and 2″

The measures of three sides of a triangle are given in each of the following. Determine whether each triangle is a right triangle.

25. 9, 16, 20   **26.** 4, 5, 7   **27.** 6, 8, 10

28. 9, 40, 41   **29.** 11, 12, 16   **30.** 45, 60, 75

For each problem, make a drawing. Then write and solve an equation. Round your answers to the nearest tenth using the table on page 482 or a calculator.

31. A baseball diamond is a square. The distance from home plate to first base is 90 feet. Approximate, to the nearest tenth of a foot, the distance from home plate to second base.

32. Clara decides to take a short cut to school by walking diagonally across a square field which is 120 yards on a side. How many yards shorter is the shortcut than the usual route?

33. A telephone pole is 26 feet high. A wire is stretched from the top of the pole to a point on the ground which is 15 feet from the bottom of the pole. How long is the wire?

34. Jessica hikes 7 miles due east and then 3 miles due north. How far is she from the starting point?

7.9   Dividing Square Roots

In order to simplify the quotient of two square roots such as $\frac{\sqrt{4}}{\sqrt{9}}$, it is important to understand the following property.

Quotient Property of Square Roots

If a is a nonnegative real number and b is a positive real number, then $\sqrt{\frac{a}{b}} = \frac{\sqrt{a}}{\sqrt{b}}$.

Examples

1 **Simplify** $\sqrt{\frac{144}{169}}$.

$$\sqrt{\frac{144}{169}} = \frac{\sqrt{144}}{\sqrt{169}}$$

$$= \frac{12}{13}$$

2 **Find a rational approximation for** $\frac{1}{\sqrt{2}}$.

First write $\frac{1}{\sqrt{2}}$ as a fraction which does not have an irrational denominator.

$$\frac{1}{\sqrt{2}} = \frac{1}{\sqrt{2}} \cdot \frac{\sqrt{2}}{\sqrt{2}}$$

Since $\frac{\sqrt{2}}{\sqrt{2}} = 1$, multiplying $\frac{1}{\sqrt{2}}$ by $\frac{\sqrt{2}}{\sqrt{2}}$ does not change its value.

$$= \frac{1 \cdot \sqrt{2}}{\sqrt{2} \cdot \sqrt{2}} \text{ or } \frac{\sqrt{2}}{2}$$

Now find the rational approximation for $\frac{\sqrt{2}}{2}$.

$$\frac{\sqrt{2}}{2} \approx \frac{1.414}{2} \text{ or } 0.707$$

The approximate value of $\sqrt{2}$ can be found on page 482 or by using a calculator.

The method used to simplify $\frac{1}{\sqrt{2}}$ is called **rationalizing the denominator.** Notice that the denominator becomes a rational number.

An expression containing square roots is said to be simplified when the following conditions are met.

1. The radicand does not contain any perfect square factors other than 1.
2. The radicand does not contain a fraction.
3. A radical does not appear in a denominator.

Examples

3 Simplify $\dfrac{6}{\sqrt{3}}$ and $\sqrt{\dfrac{2}{3}}$.

In each case rationalize the denominator.

$$\frac{6}{\sqrt{3}} = \frac{6}{\sqrt{3}} \cdot \frac{\sqrt{3}}{\sqrt{3}} = \frac{6\sqrt{3}}{3} = 2\sqrt{3}$$

$$\sqrt{\frac{2}{3}} = \frac{\sqrt{2}}{\sqrt{3}} = \frac{\sqrt{2}}{\sqrt{3}} \cdot \frac{\sqrt{3}}{\sqrt{3}} = \frac{\sqrt{6}}{3}$$

4 Simplify $\dfrac{\sqrt{5}}{\sqrt{8}}$.

Rationalize the denominator.

$$\frac{\sqrt{5}}{\sqrt{8}} \cdot \frac{\sqrt{8}}{\sqrt{8}} = \frac{\sqrt{40}}{8} = \frac{\sqrt{4} \cdot \sqrt{10}}{8} = \frac{2\sqrt{10}}{8} = \frac{1}{4}\sqrt{10}$$

Another way to rationalize the denominator is to multiply by $\dfrac{\sqrt{2}}{\sqrt{2}}$.

$$\frac{\sqrt{5}}{\sqrt{8}} \cdot \frac{\sqrt{2}}{\sqrt{2}} = \frac{\sqrt{10}}{\sqrt{16}} = \frac{\sqrt{10}}{4} \text{ or } \frac{1}{4}\sqrt{10}$$

Exercises

Exploratory Simplify.

1. $\dfrac{\sqrt{42}}{\sqrt{7}}$ 2. $\dfrac{\sqrt{18}}{\sqrt{2}}$ 3. $\dfrac{\sqrt{54}}{\sqrt{6}}$ 4. $\dfrac{\sqrt{26}}{\sqrt{13}}$

5. $\dfrac{\sqrt{27}}{\sqrt{3}}$ 6. $\dfrac{2}{\sqrt{3}}$ 7. $\dfrac{1}{\sqrt{5}}$ 8. $\dfrac{4}{\sqrt{6}}$

Written   Simplify. Assume all variables represent positive numbers.

1. $\dfrac{\sqrt{20}}{\sqrt{5}}$

2. $\sqrt{\dfrac{5}{9}}$

3. $\dfrac{\sqrt{42}}{\sqrt{6}}$

4. $\sqrt{\dfrac{3}{4}}$

5. $\dfrac{\sqrt{8}}{4}$

6. $\sqrt{\dfrac{5}{16}}$

7. $\sqrt{\dfrac{2}{9}}$

8. $\dfrac{\sqrt{15}}{\sqrt{3}}$

9. $\dfrac{3}{\sqrt{5}}$

10. $\dfrac{3}{\sqrt{3}}$

11. $\dfrac{5}{\sqrt{5}}$

12. $\dfrac{2}{\sqrt{6}}$

13. $\sqrt{\dfrac{4}{5}}$

14. $\sqrt{\dfrac{4}{7}}$

15. $\sqrt{\dfrac{9}{10}}$

16. $\sqrt{\dfrac{2}{5}}$

17. $5\sqrt{\dfrac{3}{25}}$

18. $18\sqrt{\dfrac{3}{2}}$

19. $\sqrt{2\dfrac{1}{4}}$

20. $2\sqrt{\dfrac{27}{4}}$

21. $\dfrac{3\sqrt{2}}{2\sqrt{6}}$

22. $\dfrac{2}{5}\sqrt{\dfrac{1}{2}}$

23. $\dfrac{4\sqrt{7}}{3\sqrt{2}}$

24. $\left(\sqrt{\dfrac{1}{2}}\right)\left(\sqrt{\dfrac{3}{2}}\right)$

25. $\sqrt{\dfrac{3b^2}{4}}$

26. $3y\sqrt{\dfrac{x}{y}}$

27. $\dfrac{2+\sqrt{3}}{\sqrt{3}}$

28. $\dfrac{4\sqrt{2}-6}{\sqrt{2}}$

29. $\dfrac{2\sqrt{3}+3}{\sqrt{3}}$

30. $\dfrac{3\sqrt{2}+8}{\sqrt{6}}$

31. $\dfrac{\sqrt{27}-\sqrt{48}}{\sqrt{3}}$

32. $\dfrac{4\sqrt{7}-10\sqrt{2}}{\sqrt{2}}$

33. $\sqrt{\dfrac{m^4}{7}}$

34. $\sqrt{\dfrac{27}{b}}$

35. $\sqrt{\dfrac{54}{r}}$

36. $\sqrt{\dfrac{5n^3}{4m^2}}$

37. $\dfrac{5\sqrt{21}}{\sqrt{5}}$

38. $\sqrt{\dfrac{0}{3}}$

39. $\sqrt{\dfrac{b}{6}}$

40. $\sqrt{\dfrac{a^2}{5}}$

41. $\sqrt{\dfrac{2}{5}}$

42. $\sqrt{\dfrac{2}{3}}\sqrt{\dfrac{5}{2}}$

43. $\sqrt{\dfrac{3}{7}}\sqrt{\dfrac{7}{2}}$

44. $\dfrac{2\sqrt{18}}{\sqrt{3}}$

45. $\dfrac{6\sqrt{5}}{\sqrt{3}}$

46. $\dfrac{3\sqrt{7}}{\sqrt{5}}$

47. $\dfrac{2\sqrt{8}}{\sqrt{7}}$

48. $\dfrac{4\sqrt{12}}{\sqrt{10}}$

49. $\dfrac{3\sqrt{5}+6}{\sqrt{8}}$

50. $\dfrac{5\sqrt{7}+2\sqrt{7}}{\sqrt{7}}$

51. $\dfrac{2\sqrt{7}-5\sqrt{3}}{\sqrt{2}}$

52. $\sqrt{\dfrac{27y}{2x}}$

Find the value of the expression $x^2 - 2x + 3$ if x has the following values. Express the result in simplest form.

53. $\sqrt{2}$

54. $\sqrt{7}$

55. $4\sqrt{5}$

56. $\sqrt{\dfrac{1}{2}}$

57. $\sqrt{\dfrac{2}{3}}$

58. $\dfrac{1}{2}\sqrt{\dfrac{3}{4}}$

Challenge   Find the solution set, in simplest form, of each of the following equations if the domain is $\mathcal{R}$.

1. $2x + \sqrt{7} = 9$

2. $x\sqrt{3} - \sqrt{5} = 7$

3. $-x\sqrt{2} + 3 = -7$

4. $x\sqrt{3} + 9 = 6 + \sqrt{3}$

5. $2x\sqrt{7} - 3\sqrt{5} = 27$

6. $5\sqrt{3} - 7x\sqrt{10} = \sqrt{10}$

Problem Solving Application: Look for a Pattern

Many problems can be solved more easily by studying the information given and **looking for a pattern.** When using this problem-solving strategy, it is important to organize the information in the problem. Sometimes a pattern may be obvious, but in many instances, you will need to "play" with the information in order to find a pattern.

Examples

1 **Find the sum of the first 100 odd integers.**

Explore   The 100th odd integer is 199.   *Why?*
Thus, we want to find the sum $1 + 3 + 5 + \cdots + 199$.

Plan   You could compute this sum directly, but even using a calculator, the computation would be time consuming. To find the sum of the first 100 odd integers, examine the sums of the first few odd integers and *look for a pattern.*

Solve
Sum of the first 1 odd integer: 1 $= 1$
Sum of the first 2 odd integers: $1 + 3$ $= 4$
Sum of the first 3 odd integers: $1 + 3 + 5$ $= 9$
Sum of the first 4 odd integers: $1 + 3 + 5 + 7$ $= 16$
Sum of the first 5 odd integers: $1 + 3 + 5 + 7 + 9 = 25$

Notice that each sum is a *perfect square,* and that it is the square of the number of addends. Generalizing this pattern, the sum of the first 100 odd integers will be 100^2 or 10,000.
You can use a calculator to check this result.

2 **Complete the chart at the right.**

x	1	2	3	4	5	6
y	17	21	25			

Explore   To complete this chart, we must find a relationship between the variables x and y.

Plan   To find the relationship, look for a pattern by finding the differences between successive values of x and successive values of y.

Solve   Notice that the changes in *x* are 1, while the changes in *y* are 4.

Continue this pattern to complete the table.

$$+1 \quad +1$$

x	1	2	3	4	5	6
y	17	21	25			

$$+4 \quad +4$$

x	1	2	3	4	5	6
y	17	21	25	29	33	37

Exercises

Written   **Copy and complete each table.**

1.

a	1	2	3	4	5	6
b	5	7	9			

2.

x	1	2	3			6
y	8	16		32		

Solve each problem.

3. Find the sum of the first 100 even integers.

4. Find the following sum.
$$1 + 2 + 4 + 8 + \cdots + 2^{100}$$

5. There were 10 people at a party. Each person shook hands with each of the other people exactly once. How many handshakes occurred?

6. How many squares are shown at the right?   *Hint: There are more than 16.*

7. How many rectangles are shown below?

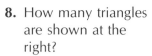

8. How many triangles are shown at the right?

9. Study the figures at the right. If the pattern of dots continues, how many dots will be in the figure with ten dots in its bottom row?

10. Imagine 25 lockers, all closed, and 25 people. Suppose the first person opens every locker. Then the second person closes every second locker. Next the third person changes the state of every third locker. (If it's open, she closes it. If it's closed, she opens it.) Suppose this procedure is continued until the 25th person changes the state of the 25th locker. Which lockers will be open at the end of the procedure?

Portfolio Suggestion

Select a homework item or class notes that shows something new you learned in this chapter and place it in your portfolio.

Performance Assessment

Mrs. Raines is having a sprinkler system installed in her flower bed, shown at the right. The sprinklers are at points *A, B, C,* and *D.* How much pipe would be required to hook each sprinkler directly to the water source, *S*? Include a drawing and explanation of your solution.

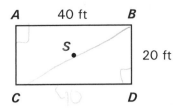

 # Chapter Summary

1. A number is rational if and only if it can be represented by a **repeating decimal.** A repeating decimal is a decimal in which a digit or a group of digits repeats. (207, 212)
2. An **irrational number** is a number which cannot be represented by a repeating decimal. (213)
3. The set consisting of all rational numbers and all irrational numbers is called the set of **real numbers,** represented by the symbol $\mathcal{R}$. (213)
4. **Completeness property of the real numbers:** Each point on a number line corresponds to a real number, and each real number corresponds to a point on the number line. (214)
5. A number x is called a **square root** of a number y if $x^2 = y$. (216)
6. The symbol $\sqrt{}$, called a **radical sign,** is used to denote the positive square root of a number. The expression under the radical sign is called the **radicand.** (216)
7. **Product property of square roots:** For any nonnegative numbers a and b, $\sqrt{a \cdot b} = \sqrt{a} \cdot \sqrt{b}$. (224)
8. The side opposite the right angle in a right triangle is called the **hypotenuse.** The other two sides are called the **legs** of the triangle. (230)
9. The **Pythagorean Theorem:** In a right triangle, if a and b represent the measures of the legs and c represents the measure of the hypotenuse, then $c^2 = a^2 + b^2$. (230)

10. **Converse of the Pythagorean Theorem:** If the sum of the squares of the measures of two sides of a triangle is equal to the square of the measure of the third side, then the triangle is a right triangle. (231)

11. **Quotient property of square roots:** If a is a nonnegative number and b is a positive real number, then $\dfrac{\sqrt{a}}{\sqrt{b}} = \sqrt{\dfrac{a}{b}}$. (233)

 # Chapter Review

7.1   Express each of the following as a decimal.

1. $\dfrac{7}{12}$ 2. $\dfrac{5}{9}$ 3. $\dfrac{67}{200}$ 4. $\dfrac{311}{999}$ 5. $\dfrac{17}{5}$

7.2   Express each of the following in the form $\dfrac{a}{b}$ where a and b are integers and $b \neq 0$.

6. -0.75 7. 2.18 8. $3.6\overline{851}$ 9. $5.80\overline{9}$ 10. 0.33333

7.3   Solve each of the following in the domain of real numbers. Graph the solution set on the real number line.

11. $-4 - 7(2x - 3) \geq 10x - 5(x + 1)$ 12. $\frac{1}{2}y - \frac{2}{3} \geq \frac{3}{4}y + \frac{1}{9}$

7.4   Express each of the following in simpler form.

13. $\sqrt{81}$ 14. $-\sqrt{625}$ 15. $\sqrt{(1.386)^2}$ 16. $-\sqrt{(48)^2}$

7.5   Use the divide-and-average method to find a rational approximation correct to the nearest tenth for each of the following.

17. $\sqrt{65}$ 18. $\sqrt{38}$ 19. $\sqrt{88}$ 20. $\sqrt{220}$

7.6   Simplify each of the following.

21. $(\sqrt{3})(\sqrt{8})$ 22. $(2\sqrt{6})(5\sqrt{3})$ 23. $(4\sqrt{2})^2$ 24. $\left(\dfrac{\sqrt{48}}{4}\right)\left(\dfrac{\sqrt{24}}{2}\right)$

7.7   Simplify each of the following. Assume all variables represent positive numbers.

25. $2\sqrt{7} + 8\sqrt{7}$ 26. $3\sqrt{54} + 4\sqrt{6} + 8\sqrt{18}$ 27. $\sqrt{72y^2}$

7.8   In each of the following, c is the measure of the hypotenuse of a right triangle. Find the missing measure to the nearest tenth.

28. $a = 5, c = 18$ 29. $a = \sqrt{5}, b = \sqrt{2}$ 30. $b = 2, c = 8\frac{3}{4}$

7.9   Simplify each of the following.

31. $\dfrac{5}{\sqrt{7}}$ 32. $\sqrt{\dfrac{4}{7}}$ 33. $\dfrac{7\sqrt{5}}{3\sqrt{3}}$ 34. $\dfrac{4\sqrt{2} - 8}{\sqrt{2}}$ 35. $\sqrt{\dfrac{3}{8}} + \sqrt{72} - \sqrt{\dfrac{8}{3}}$

 **Chapter Test**

Tell whether each of the following is *true* or *false*.

1. Each rational number corresponds to a point on the number line.

2. Each point on the number line corresponds to a rational or irrational number.

3. The sum of two rational numbers is always a rational number.

4. The sum of two irrational numbers is always an irrational number.

5. The number 0.171171117 . . . is a repeating decimal.

6. The number 0.71384238423842 . . . is a repeating decimal.

7. Every integer is a real number.

8. Every rational number is a real number.

9. Not every irrational number is a real number.

10. Not every real number is an irrational number.

Express each of the following as a decimal.

11. $\dfrac{3}{16}$

12. $\dfrac{8}{3}$

13. $\dfrac{17}{40}$

14. $\dfrac{22}{99}$

Express each of the following in the form $\dfrac{a}{b}$ where a and b are integers and $b \neq 0$.

15. 0.8

16. 63.6

17. $3.49\overline{9}$

18. $0.52\overline{0}$

Find the sum of each of the following.

19. $\dfrac{2}{9} + \dfrac{37}{99}$

20. $0.72\overline{3} + 0.\overline{3}$

21. $0.232232223 \ldots + 1.\overline{1}$

Express each of the following in simplest form. Assume all variables represent positive numbers.

22. $\sqrt{49}$

23. $-\sqrt{144}$

24. $\sqrt{(3.86)^2}$

25. $\sqrt{72}$

26. $\sqrt{\dfrac{45}{24}}$

27. $\dfrac{\sqrt{44}}{\sqrt{x}}$

28. $\dfrac{9\sqrt{5}}{3\sqrt{8}}$

29. $\sqrt{48y^2}$

30. $(\sqrt{6})(3\sqrt{10})$

31. $4\sqrt{8} + 2\sqrt{7} - 5\sqrt{2} + 8\sqrt{28}$

32. Find the length of a diagonal of a rectangle whose sides measure 0.9 cm and 4.0 cm.

33. Find the solution set of $x^2 = 99$ for $x \in \mathcal{R}$.

34. The measures of the sides of a triangle are 6, 8, and 10. Is the triangle a right triangle?

35. Jerome hikes 15 km due west and then 20 km due south. Janice starts at the same place and hikes 10 km due east and then 24 km due north. Is Jerome or Janice closer to the starting point?

Measurement and Geometry

Application in Agriculture

U.S. farmers produce a large amount of the world's food supply. Over 30% of the world's supply of corn and over 40% of the world's supply of soybeans are produced in the United States. In order for U.S. farmers to produce so much food, they must be knowledgeable about many things including **area** and **volume**.

Because of the paths of two streams, one of Alfred Taylor's fields is in the shape of a trapezoid. He needs to know the area of the field in order to determine how much fertilizer he should use for that field. The length of the two parallel sides of the field are 2000 feet and 1400 feet. One side of the trapezoid is perpendicular to the parallel sides, and it is 1100 feet long. What is the area of the field?

Group Project: *Sports*

Choose five sports and draw a sketch of the playing field for each sport. Label all of the distances on the sketch. Find the area of each playing field and area of any special parts of the playing field. For example, if you choose soccer, find the area of the field and the area of the goalie boxes.

8.1 Distance in the Metric System

In a measurement system, a **standard unit** is a basic measurement that everyone agrees upon. The **meter** (m) is the standard unit of length in the metric system. When the metric system was developed, it was agreed that the meter was one ten-millionth of the distance from the equator to the north pole.

Other commonly used units of distance are the centimeter (cm), millimeter (mm), and kilometer (km).

A door is about 1 m wide.

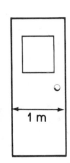

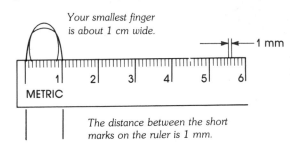

Your smallest finger is about 1 cm wide.

1 mm

The distance between the short marks on the ruler is 1 mm.

Five city blocks are about 1 km long.

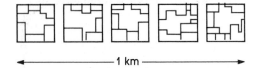

← 1 km →

Long distances, such as the distances between cities, are measured in kilometers. Shorter distances, such as the height of a mountain, are measured in meters.

Objects such as books or photographs are measured in centimeters. Millimeters are used for very short distances. Photographic film and tiny mechanical parts are measured in millimeters.

Example

1 **Find the length of $\overline{AB}$ in centimeters and millimeters.**

A   B

Use a metric ruler.
The length of $\overline{AB}$ is 4.5 cm. *To find this length in millimeters, count the short marks on the ruler. Or, since there are 10 mm in 1 cm, multiply 4.5 by 10.*

The length of $\overline{AB}$ is 45 mm.

You may know that 1 meter is 100 centimeters and 1 kilometer is 1000 meters.

As we saw in example 1, there are 10 millimeters in 1 centimeter. This fact illustrates an important feature of the metric system. Every unit is equivalent to some other unit multiplied, or divided, by 10. Thus, the system is closely related to the decimal place-value system.

Journal

What experiences have you had that used the metric system?

Place Values

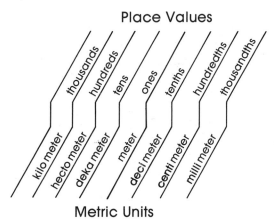

Metric Units

Example

2 **Find the number of millimeters in 1 meter.**

Look at the chart at the right. Any unit is equivalent to the unit at its right multiplied by 10.

Millimeter is 3 places to the right of meter. Thus, 10 is a factor 3 times in the following product.

1 meter = 10 · 10 · 10 millimeters
        = 1000 millimeters   There are 1000 millimeters in 1 meter.

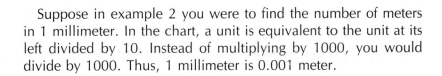

Suppose in example 2 you were to find the number of meters in 1 millimeter. In the chart, a unit is equivalent to the unit at its left divided by 10. Instead of multiplying by 1000, you would divide by 1000. Thus, 1 millimeter is 0.001 meter.

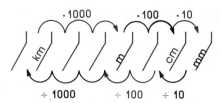

The chart at the left shows how the most commonly used units are related.

Examples

3 **Find the number of centimeters in 3.5 meters.**

$\cdot 100$

3.5 m = ___?___ cm   *Multiply by 100 since there are 100 centimeters in 1 meter.*
3.5 · 100 = 350
    3.5 m = 350 cm

There are 350 centimeters in 3.5 meters.

4 **Find the number of kilometers in 625 meters.**

625 m = ___?___ km   *Divide by 1000 since there are 1000 meters in 1 kilometer.*
÷ 1000
625 ÷ 1000 = 0.625
    625 m = 0.625 km

There is 0.625 kilometer in 625 meters.

Exercises

Exploratory   **Find the metric unit that completes each of the following. Use km, m, cm, or mm.**

1. A ski is 2 _____ long.

2. A sheet of theme paper is 21.5 _____ wide.

3. A quilt is 1.5 _____ wide.

4. A nickel is 2 _____ thick.

5. An Olympic race is 0.8 _____ long.

6. A record album is 30 _____ wide.

7. A paper clip is 8 _____ wide.

8. A window is 1.34 _____ wide.

9. 1 m = 100 _____

10. 1 cm = 0.01 _____

11. 1 cm = 10 _____

12. 1 km = 1000 _____

13. 1 mm = 0.1 _____

14. 1 m = 0.001 _____

15. 30 cm = 300 _____

16. 2 m = 2000 _____

17. 7000 m = 7 _____

18. 65 mm = 6.5 _____

19. 5.5 km = 5500 _____

20. 17 m = 1700 _____

Written Find the number of centimeters in each of the following lengths.

1. 2m	**2.** $\frac{1}{2}$ m	**3.** 80 m	**4.** 48 m
5. 1 km	**6.** 5 km	**7.** 3 m	**8.** 18 km
9. 1200 mm	**10.** 750 mm	**11.** 3 mm	**12.** 8 mm
13. 2.735 km	**14.** 0.5 km	**15.** 17.381 m	**16.** 13.8 mm

Find the number of meters in each of the following lengths.

17. 2 km	**18.** 30 km	**19.** 700 cm	**20.** 35 cm
21. 3350 cm	**22.** 1400 mm	**23.** 347 cm	**24.** 1.3 km
25. 42 mm	**26.** 613 mm	**27.** 8.3 km	**28.** 13.5 cm
29. 0.1 mm	**30.** 41.5 mm	**31.** 1200 km	**32.** 3148.1 km

Measure the length of each of the following segments. Express each length in centimeters, millimeters, and meters.

33.

34.

35.

36.

37.

38.

39.

40.

41.

42.

Copy and complete each of the following.

43. 3 cm = _____ mm	**44.** 8 km = _____ m	**45.** 2 m = _____ cm
46. 4200 mm = _____ m	**47.** 80 mm = _____ cm	**48.** 317 cm = _____ m
49. 3.2 cm = _____ m	**50.** 3.4 cm = _____ mm	**51.** 8.3 m = _____ mm
52. 2.4 km = _____ m	**53.** 30,000 m = _____ km	**54.** 80 m = _____ km
55. 1200 mm = _____ cm	**56.** 2.3 km = _____ cm	**57.** 3001 cm = _____ m
58. 0.3 km = _____ m	**59.** 803 m = _____ km	**60.** 0.0728 cm = _____ m
61. 2.4 cm = _____ mm	**62.** 2.4 m = _____ cm	**63.** 1 mm = _____ km
64. 1.78 km = _____ cm	**65.** 0.83 cm = _____ km	**66.** 0.0034 km = _____ mm

Challenge Suppose the metric prefixes were used in our money system. State the value in dollars of each of the following units if such a unit existed.

1. decidollar	**2.** centidollar	**3.** kilodollar
4. millidollar	**5.** hectodollar	**6.** dekadollar

Find the equivalent measurement of each of the following in grams or liters.

7. 1000 milligrams	**8.** 1 kilogram	**9.** 20 deciliters
10. 10 centiliters	**11.** 2000 centigrams	**12.** 10 kiloliters

8.2 Areas of Rectangles and Parallelograms

In order to determine how much soil nutrient to apply to a field, a farmer must know the **area** of the field.

The area of a polygon is the measurement of the region formed by the polygon and its interior. In the metric system, some measurements of area are given in square meters (m²). If the side of a square is 1 meter long, we say it has an area of 1 square meter. How would you define 1 square centimeter (cm²)?

The area is 1 m².

Jodi has a rectangular garden with dimensions of 5 meters by 3 meters. A drawing of Jodi's garden is shown at the left. Each small square represents 1 square meter. There are 5 · 3 or 15 squares. The area is 15 square meters.

In the same way, a rectangle with dimensions of 0.8 centimeter by 0.3 centimeter has an area of 0.24 square centimeter.

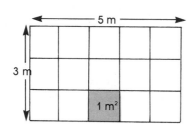

The area is 15 m².

These and other similar examples suggest the following statement.

Area of a Rectangle

If a rectangle has an area of A square units, a length of ℓ units, and a width of w units, then A = ℓw.

Examples

1 **Find the area of a square meter in square centimeters.**

$A = \ell w$ *Use the formula for area of a rectangle.*

$A = s \cdot s$ or s^2 *In a square $\ell = w$. So, let $s =$ the length of each side.*

$A = 100^2$ *Since 1 m = 100 cm, replace s by 100.*

$A = 10{,}000$

The area of a square meter is 10,000 square centimeters.
There are 10,000 square centimeters in 1 square meter.

2 **Find a rational approximation to the nearest hundredth for the area of a rectangle with a length of $\sqrt{10}$ cm and width of $\sqrt{5}$ cm.**

$A = \ell w$ *Use the formula.*

$A = \sqrt{10} \cdot \sqrt{5}$ *Replace ℓ by $\sqrt{10}$ and w by $\sqrt{5}$.*

$ = \sqrt{50}$

$ = 5\sqrt{2}$ *The area of the rectangle is exactly $5\sqrt{2}$ cm².*

$A \approx 5(1.414) = 7.07$ *$\sqrt{2} \approx 1.414$*

The area is approximately 7.07 square centimeters.

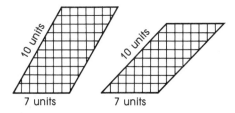

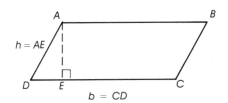

Suppose a parallelogram has sides of 10 units and 7 units. Is its area 10 • 7 square units? Look at the two parallelograms at the left. You can see that their areas differ though they both have sides of 10 units and 7 units.

Consider $\square ABCD$ shown at the left. Any side may be called the **base,** with length of b units. In this case, call $\overline{CD}$ the base. Draw $\overline{AE}$ perpendicular to $\overline{CD}$. In a parallelogram, a segment perpendicular to the base with endpoints on the base and the line on which the opposite side lies is called a **height** or **altitude** of the parallelogram. We say the height is h units long. Thus, $\overline{AE}$ is a height of $\square ABCD$ that corresponds to base $\overline{CD}$.

Then, for $\square ABCD$, $b = CD$ and $h = AE$. We can show that $A = bh$.

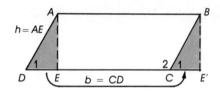

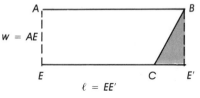

$A = \ell w$ or $A = EE' \cdot AE$

Suppose that $\triangle AED$ in $\square ABCD$ could be moved so that $\overline{AD}$ and $\overline{BC}$ lie on a line as shown at the left. Point E is on E'. Since $\angle 1$ and $\angle 2$ are consecutive angles of the parallelogram, the sum of their measures is 180. This means that $\angle ECE'$ is a straight angle.

The new figure, $ABE'E$, is a rectangle. This rectangle has the same area as $\square ABCD$. But the measure of the area of $\square ABE'E$ is the product, $EE' \cdot AE$.

You can see that $EE' = CD$ or b. Also, $AE = h$. Thus, we have shown that for $\square ABCD$, $A = bh$.

Area of a Parallelogram

If a parallelogram has an area of A square units, a base of b units, and a corresponding height of h units, then $A = bh$.

Example

3 Find the exact area of $\square PQRS$ shown at the right.

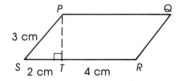

Let $b = SR$
$$SR = ST + TR$$
$$= 2 + 4$$
$$= 6 \qquad \textit{Thus, } b = 6.$$

Then, find h.

Since $\overline{PT} \perp \overline{SR}$, $h = PT$ and $\triangle PTS$ is a right triangle.

$ST^2 + PT^2 = PS^2$	*Use the Pythagorean Theorem.*
$2^2 + PT^2 = 3^2$	*Replace ST by 2 and PS by 3.*
$4 + PT^2 = 9$	
$PT^2 = 5$	*Subtract 4 from both sides.*
$PT = \sqrt{5}$	*Reject $-\sqrt{5}$ since distance is positive. Thus, $h = \sqrt{5}$.*
$A = bh$	*Use the formula for the area of a parallelogram.*
$A = 6\sqrt{5}$	*Replace b by 6 and h by $\sqrt{5}$.*

The area of $\square PQRS$ is exactly $6\sqrt{5}$ square centimeters.

Exercises

Exploratory Find the area of a rectangle with the dimensions given in each of the following.

1. 3 cm, 2 cm   **2.** 10 cm, 4 cm   **3.** 3 m, 1.2 m   **4.** 10 m, 4.2 m
5. 20 ft, 15 ft   **6.** 15 in., 7 in.   **7.** 1.1 m, 4.9 m   **8.** 300 cm, 25 cm

Find the perimeter of the rectangle in each of the following.

Sample: The rectangle has dimensions of 6 cm by 1 cm. Use $P = 2\ell + 2w$.
$P = 2 \cdot 6 + 2 \cdot 1$ or 14 cm. The perimeter of the rectangle is 14 cm.

9. Exercise 1   **10.** Exercise 2   **11.** Exercise 3   **12.** Exercise 4
13. Exercise 5   **14.** Exercise 6   **15.** Exercise 7   **16.** Exercise 8

Complete each of the following.

17. 1 m² = ___ cm²   **18.** 1 cm² = ___ mm²   **19.** 1 km² = ___ m²   **20.** 1 ft² = ___ in.²

Written Find the area of a rectangle with the dimensions given in each of the following. Express your answer in simplest form with the most appropriate square units.

1. 1.04 cm, 3.02 cm   **2.** 1.08 m, 0.6 m   **3.** 3 mm, 0.5 m
4. 0.06 km, 1.14 m   **5.** 8 ft, 1 ft   **6.** 3 yd, 2 ft
7. $\sqrt{2}$ cm, $\sqrt{7}$ cm   **8.** $3\sqrt{15}$ m, $\sqrt{3}$ m   **9.** $2\sqrt{5}$ cm, $3\sqrt{5}$ cm
10. $\frac{1}{2}\sqrt{20}$ m, $\sqrt{5}$ m   **11.** $\frac{1}{3}\sqrt{6}$ km, $\sqrt{21}$ km   **12.** $\sqrt{14}$ m, $\frac{1}{6}\sqrt{42}$ m

The top, bottom, and 4 sides of each cardboard box shown below are to be covered with colored paper. For each box, what is the total area to be covered?

13.
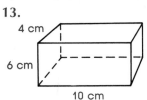
4 cm
6 cm
10 cm

14.
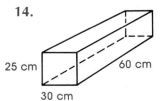
25 cm
60 cm
30 cm

15.
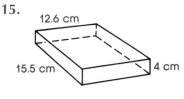
12.6 cm
15.5 cm
4 cm

In each of the following, the measures of a parallelogram are given. Find a rational approximation for A to the nearest hundredth. Use the table on page 482 or a calculator.

16. $b = 144, h = 80$ **17.** $b = 122, h = 36$ **18.** $b = 2.23, h = 0.95$
19. $b = 2\sqrt{6}, h = 3\sqrt{3}$ **20.** $b = 4\sqrt{5}, h = \sqrt{5}$ **21.** $b = 3\sqrt{42}, h = 4\sqrt{21}$

In each of the following, the measures of the area and a base of a parallelogram are given. Find h, expressed in simplest form.

22. $A = 20, b = 5$ **23.** $A = 6.5, b = 5$ **24.** $A = 200, b = 14$
25. $A = 36\sqrt{2}, b = 2\sqrt{2}$ **26.** $A = 24, b = 2\sqrt{3}$ **27.** $A = 75\sqrt{7}, b = 5\sqrt{5}$

Find the area of each of the following parallelograms or write *insufficient information given*.

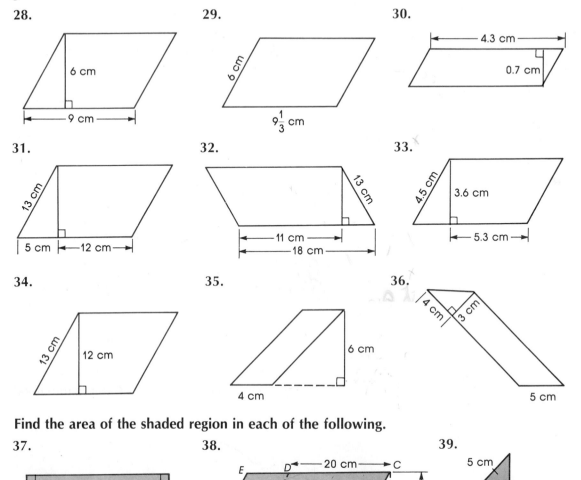

28.

6 cm

9 cm

29.

6 cm

$9\frac{1}{3}$ cm

30.

4.3 cm

0.7 cm

31.

13 cm

5 cm | 12 cm

32.

13 cm

11 cm

18 cm

33.

4.5 cm

3.6 cm

5.3 cm

34.

13 cm

12 cm

35.

6 cm

4 cm

36.

4 cm | 3 cm

5 cm

Find the area of the shaded region in each of the following.

37.

A   B
15 cm
5 cm
D   C

28 cm

ABCD is a parallelogram.

38.

E   D ← 20 cm → C

15 cm   15 cm

A   B

ABCD and DEFG are parallelograms.

20 cm

10 cm

F   8 cm   G

39.

5 cm

Congruent segments are indicated.

Challenge   Answer each of the following.

1. Which is smaller, 1000 square meters or 1 square kilometer?

2. Which is smaller, 0.01 square meters or 10 square centimeters?

3. Write a two-column proof to show that the area of a parallelogram is the product of the lengths of its base and its height.

8.3 The Area of a Triangle

Next season the Pine City Zoo plans to resurface the seal pool, which is triangular in shape. To decide how much surfacing material is needed, the contractor must know the area of the triangle.

Any side of a triangle may be called the **base.** The corresponding height is perpendicular to the line on which the base lies. Its endpoints are on the line and at the opposite vertex. The area of a triangle is found by using the formula which follows.

Area of a Triangle

If a triangle has an area of *A* square units, a base of *b* units, and a corresponding height of *h* units, then

$$A = \frac{1}{2}bh.$$

We can show that this formula is true by noticing that the area of a triangle is related to the area of a parallelogram.

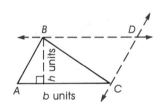

Look at $\triangle ABC$ shown at the left. The line through *B* parallel to $\overline{AC}$ is $\overleftrightarrow{BD}$. The line through *C* parallel to $\overline{AB}$ is $\overleftrightarrow{CD}$. Quadrilateral *ABDC* is a parallelogram by definition.

Diagonal *BC* separates $\square ABDC$ into two congruent triangles. Since congruent triangles have the same area, the area of $\triangle ABC$ must be one-half the area of $\square ABDC$. Since for $\square ABDC$, $A = bh$, then for $\triangle ABC$, $A = \frac{1}{2}bh$.

Example

1 **Find the area of $\triangle ABC$.**

$A = \frac{1}{2}bh$ *Use the formula.*
$A = \frac{1}{2} \cdot 8 \cdot 5$ *Replace b by 8 and h by 5.*
$A = 20$

The area of $\triangle ABC$ is 20 square units.

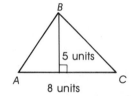

Examples

2 **Find the area of △DEF.**

The height corresponding to a base need not be in the interior of the triangle. In the case shown at the right, base EF must be extended as shown. But b remains the measure of the original base EF.

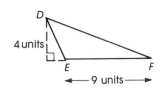

$A = \frac{1}{2}bh$ *Use the formula.*
$A = \frac{1}{2} \cdot 9 \cdot 4$ *Replace b by 9 and h by 4.*
$A = 18$

The area of △DEF is 18 square units.

3 **Find the area of right triangle RST.**

Let $RT = b$
Then, $SR = h$
$A = \frac{1}{2}bh$
$A = \frac{1}{2} \cdot 12 \cdot 5$
$A = 30$

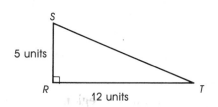

The area of △RST is 30 square units.

Lab Activity

You can learn more about the area of a triangle in Lab 6 on page A9.

In example 3, $\overline{SR}$ also might be called the base. Then $\overline{RT}$ is the corresponding height. This means that *for any right triangle, the area is one-half the product of the lengths of the legs.*

Exercises

Exploratory In each of the following, the measures of a triangle are given. Express *A* in simplest form.

1. $b = 3, h = 4$ **2.** $b = 7, h = 5$ **3.** $b = 1.2, h = 80$ **4.** $b = 1.5, h = 3$
5. $b = \frac{1}{2}, h = \frac{1}{4}$ **6.** $b = 3\frac{1}{2}, h = 7$ **7.** $b = 2\sqrt{2}, h = \sqrt{6}$ **8.** $b = \sqrt{3}, h = 3\sqrt{3}$

Written   Find the exact area of each of the following triangles.

1.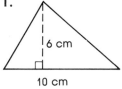
6 cm
10 cm

2.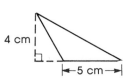
4 cm
|← 5 cm →|

3.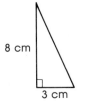
8 cm
3 cm

4.
15 cm
9 cm
4 cm

5.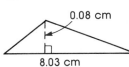
0.08 cm
8.03 cm

6.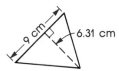
11 cm
16 cm

7.
9 cm
6.31 cm

8.
3 cm
4 cm

9.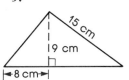
15 cm
9 cm
|← 8 cm →|

10.
3 cm
45°

11.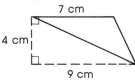
7 cm
4 cm
9 cm

12.
4 cm

Find the area of the shaded region in each of the following.

13.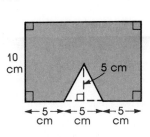
10 cm
5 cm
←5→←5→←5→
cm cm cm

14.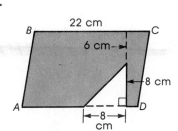
22 cm
6 cm→
8 cm
|←8→|
cm
ABCD is a parallelogram.

15.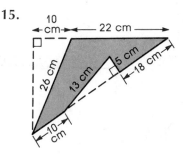
10 cm 22 cm
26 cm
13 cm
5 cm
18 cm
10 cm

State if each of the following is *always, sometimes,* or *never* true.

16. Two congruent triangles have the same areas.

17. Two triangles with the same area are congruent.

18. The height of a triangle is not in the interior of the triangle.

19. A parallelogram is separated by a diagonal into two triangles with the same area.

20. Two similar triangles have the same areas.

21. Two heights of a triangle have the same length.

Challenge   Write a two-column proof for each of the following.

1. The area of a triangle is one-half the product of the lengths of the base and the height.

2. The area of a right triangle is one-half the product of the lengths of the legs.

8.4   The Area of a Trapezoid

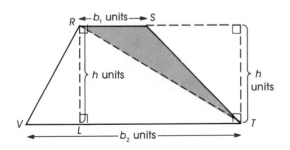

The parallel sides of a trapezoid are called the **bases** of the trapezoid. A **height** or **altitude** of the trapezoid is a segment perpendicular to both bases. Its endpoints are on the bases. At the left, $\overline{RS}$ and $\overline{VT}$ are the bases and $\overline{RL}$ is a height of trapezoid $RSTV$.

A trapezoid is separated by a diagonal into two triangles. Notice that $RL = h$ for both $\triangle RST$ and $\triangle RTV$ in the figure shown at the left.   *Why?*

You can find the formula for the area of a trapezoid by finding the sum of the areas of the triangles.

$\underbrace{\text{The area of trapezoid } RSTV}$ $\underbrace{\text{is}}$ $\underbrace{\text{the area of } \triangle RST}$ $\underbrace{\text{plus}}$ $\underbrace{\text{the area of } \triangle RTV.}$

$$A \qquad = \qquad \frac{1}{2}b_1h \qquad + \qquad \frac{1}{2}b_2h$$

$$A = \frac{1}{2}h(b_1 + b_2) \qquad \textit{Use the distributive law.}$$

Area of a Trapezoid

> If a trapezoid has an area of A square units, bases of b_1 units and b_2 units, and a height of h units, then $A = \frac{1}{2}h(b_1 + b_2)$.

Example

1 **Suppose the area of trapezoid $ABCD$ is 115 square units. Find b_2.**

$A = \frac{1}{2}h(b_1 + b_2)$ *Use the formula.*

$115 = \frac{1}{2} \cdot 10(8 + b_2)$ *Replace A by 115, h by 10, and b_1 by 8.*

$115 = 5(8 + b_2)$

$115 = 40 + 5b_2$ *Use the distributive law.*

$75 = 5b_2$

$15 = b_2$

Exercises

Exploratory State if each of the following is *true* or *false*.

1. If a figure is a trapezoid, then it is a polygon.
2. If a figure is a parallelogram, then it is a trapezoid.
3. If a figure is a polygon, then it is a parallelogram.
4. If a figure is not a rectangle, then it is not a parallelogram.

State the converse of the statements in each of the following. Then state if the converse is *true* or *false*.

5. Exercise 1　　6. Exercise 2　　7. Exercise 3　　8. Exercise 4

State the inverse of the statements in each of the following. Then state if the inverse is *true* or *false*.

9. Exercise 1　　10. Exercise 2　　11. Exercise 3　　12. Exercise 4

Written Find the missing measure of the trapezoid in each of the following.

	A	h	b_1	b_2		A	h	b_1	b_2
1.	?	10	10	20	**2.**	?	7	6	8
3.	?	0.8	0.8	1.8	**4.**	?	1.7	1.3	2.5
5.	?	13	18	27	**6.**	?	226	150	152
7.	?	0.5	1.8	0.4	**8.**	8	2	?	5
9.	47.5	5	4	?	**10.**	105	?	3	3.5
11.	34	?	8	8.5	**12.**	0.08	0.2	?	0.3

Find the area of each of the following regions.

13.

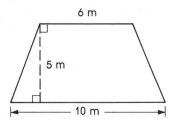

14.

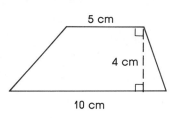

15.

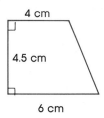

16.

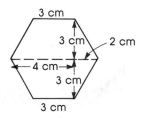

17.

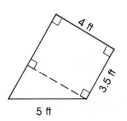

18.

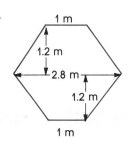

Mathematical Excursions

When two polygons are similar, the measures of their corresponding sides are in proportion. What about their perimeters and areas? Are the measures of these also in proportion? If so, how do these ratios relate to the ratio of the measures of the corresponding sides?

In the figure at the right, $\triangle ABC \sim \triangle XYZ$. Comparing $\triangle ABC$ to $\triangle XYZ$, we see that the ratio of the measures of their corresponding sides is $20 : 30$ or $2 : 3$. Now find the ratio of their perimeters and the ratio of their areas.

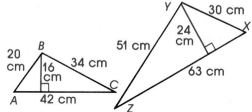

$$\frac{\text{perimeter of } \triangle ABC}{\text{perimeter of } \triangle XYZ} = \frac{20 \text{ cm} + 34 \text{ cm} + 42 \text{ cm}}{30 \text{ cm} + 51 \text{ cm} + 63 \text{ cm}}$$

$$= \frac{96 \text{ cm}}{144 \text{ cm}}$$

$$= \frac{2}{3}$$

$$\frac{\text{area of } \triangle ABC}{\text{area of } \triangle XYZ} = \frac{\frac{1}{2} (42 \text{ cm})(16 \text{ cm})}{\frac{1}{2} (63 \text{ cm})(24 \text{ cm})}$$

$$= \frac{336 \text{ cm}^2}{756 \text{ cm}^2}$$

$$= \frac{4}{9} \text{ or } \left(\frac{2}{3}\right)^2$$

These examples suggest the following property.

Perimeters and Areas of Similar Polygons	If two polygons are similar, then the ratio of their perimeters is equivalent to the ratio of the measures of their corresponding sides and the ratio of their areas is equivalent to the square of the ratio of the measures of their corresponding sides.

Exercises Complete the table below if $\triangle ABC \sim \triangle XYZ$.

	Perimeter of $\triangle ABC$	Perimeter of $\triangle XYZ$	Area of $\triangle ABC$	Area of $\triangle XYZ$	Ratio of Sides	Ratio of Perimeters	Ratio of Areas
1.	21	14	18	_____	_____	_____	_____
2.	_____	75	27	75	_____	_____	_____
3.	10	_____	7	_____	$\frac{1}{2}$	_____	_____
4.	_____	$7\frac{1}{2}$	8	_____	_____	$\frac{4}{3}$	_____
5.	12	_____	_____	$8\frac{3}{4}$	_____	_____	$\frac{16}{25}$

8.5   Circles and Circumference

In constructions, you use a compass to "copy" or transfer a certain distance such as the length of a line segment.

As you know, a compass also is used to draw a **circle.** The metal point is held at a point on the paper. Then, with the distance fixed between the point and the pencil, the circle is drawn around the point.

Definition of a Circle

> A circle is the set of all points in a plane a given distance from a point in the plane called the **center** of the circle.

The symbol ⊙ means circle. A circle usually is named by its center. Thus, the circle at the left is called ⊙O.

The figure also shows segments that are related to the circle.

Any segment whose endpoints are the center and a point of the circle is a **radius** of the circle. From the definition of a circle, we can conclude that all radii of a circle are congruent.

A segment whose endpoints are points of the circle is called a **chord** of the circle.

A chord that contains the center of the circle is a **diameter** of a circle. All diameters of a circle are congruent. Note that a diameter of a circle is twice as long as a radius.

Two circles that have congruent radii are **congruent.**

The circle separates a plane into three parts. The parts are the circle *itself,* its *interior,* and its *exterior.* Consider points *E, F,* and *G* in the figure at the left. Only point *F* is a point of the circle. Point *E* is in the *interior* of the circle and point *G* is in the *exterior.*

The **circumference** of a circle is the distance around the circle. Thus, the circumference of the circle in the photograph above is the distance the pencil moved to complete the circle.

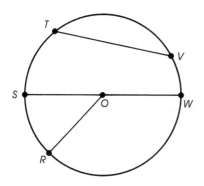

$\overline{OR}$ is a radius.
$\overline{TV}$ and $\overline{SW}$ are chords.
$\overline{SW}$ is also a diameter.
$\overline{OS}$ and $\overline{OW}$ are also radii.

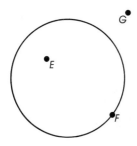

Suppose C, the measure of the circumference of a circle, is compared to d, the measure of the diameter.

The ratio $\dfrac{C}{d}$ can be found for different circles. Perform the following experiment. Find an approximate measure of the diameter of the top of a can by measuring the top at its greatest width. Then, measure the distance around the can with a tape measure (or a string and ruler) to find the measure of the circumference. Record the measures d and C and the ratio $\dfrac{C}{d}$ in a chart like the one shown at the left.

Repeat this experiment several times with cans of different sizes. The ratio always appears to have a value slightly greater than 3. In fact, the ratio $\dfrac{C}{d}$ has the same value for every circle.

The value of $\dfrac{C}{d}$, denoted by the Greek letter π, is an irrational number. The exact value cannot be given in fractional or decimal form. The following express the most commonly used rational approximations.

Circle	Diameter, d	Circumference, C	$\dfrac{C}{d}$
1			
2			
3			
4			

$$\pi \approx 3.14 \qquad \pi \approx \frac{22}{7}$$

Calculator Hint

The π key on a calculator uses an approximation greater than 3.14.

Example

1 To the nearest tenth, what is the value of 5π? Use $\pi \approx 3.14$.

$5\pi \approx 5(3.14)$ or 15.7

Since $\dfrac{C}{d} = \pi$, then $C = \pi d$. But if r is the measure of the radius, then $d = 2r$. Thus, $C = 2\pi r$.

Circumference of a Circle

> If a circle has a circumference of C units and a radius of r units, then $C = 2\pi r$.

Examples

2 **What is the exact circumference of a circle with a diameter of 2 cm? Also, find an approximate measurement to the nearest tenth.**

$C = \pi d$ *Use the formula for the circumference.*
$C = \pi \cdot 2$ *Replace d by 2.*
$C \approx 2(3.14)$ *Recall that $\pi \approx 3.14$.*
$C \approx 6.28$

The circumference is exactly 2π cm and approximately 6.3 cm.

3 **The circumference of a circle is 125.6 meters. Find approximate measurements of a radius and diameter of the circle.**

$C = 2\pi r$ *Use the formula for the circumference.*
$125.6 \approx 2 \cdot 3.14r$ *Replace C by 125.6 and π by 3.14.*
$125.6 \approx 6.28r$
$20 \approx r$ *Divide both sides by 6.28.*
$d \approx 2 \cdot 20$ or 40 $d = 2r$

The circle has a radius of about 20 meters and a diameter of about 40 meters.

Exercises

Exploratory   Use the figure at the right to answer each of the following.

1. Name the center of the circle.
2. Name the circle.
3. Name three radii of the circle.
4. Name two chords of the circle.
5. Name a diameter of the circle.
6. Name two points in the interior of the circle.
7. Name a point in the exterior of the circle.
8. Name 5 points of the circle.

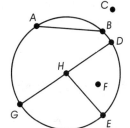

State the exact circumference of a circle with each of the following measurements.

9. diameter, 10 km	10. diameter, 2 cm	11. diameter, 1 km
12. diameter, 35 m	13. diameter, 12.5 cm	14. diameter, 0.01 m
15. radius, 7 in.	16. radius, 20 ft	17. radius, 3 in.
18. radius, 6 in.	19. radius, 0.7 cm	20. radius, 0.042 m

Written   Find a rational approximation to the nearest tenth for the circumference in each of the following circles in Exploratory Exercises. Use $\pi \approx 3.14$.

1. Exercise 9

2. Exercise 10

3. Exercise 11

4. Exercise 12

5. Exercise 13

6. Exercise 14

7. Exercise 15

8. Exercise 16

9. Exercise 17

10. Exercise 18

11. Exercise 19

12. Exercise 20

In each of the following, the circumference of a circle is given. Find the approximate measurements to the nearest tenth of the radius and diameter of the circle.

13. 12π cm

14. 2π km

15. 10π ft

16. 4π in.

17. 3.14 cm

18. 12.56 m

19. 10 cm

20. 100 m

State whether each of the following is *always, sometimes,* or *never* true.

21. A diameter of a circle is a chord of the circle.

22. A radius of a circle is a chord of the circle.

23. Two radii of a circle form a diameter of the circle.

24. A chord of a circle is a diameter of the circle.

25. The line segment connecting a point interior to a circle to a point exterior to a circle intersects the circle in exactly one point.

26. A line connecting two points exterior to a circle intersects the circle in two points.

Find the exact distance for each of the following.

27. one-half the distance around a circle with a radius of 5 cm

28. one-third of the distance around a circle with a radius of 13 cm

29. one-half the circumference of a circle with a radius of π units

30. twice the circumference of a circle with a radius of w units

Answer each of the following.

31. If you double the length of the radius of a circle, how does the circumference change?

32. If you triple the length of the radius of a circle, how does the circumference change?

33. If you multiply the length of the radius of a circle by k, how does the circumference change?

Challenge   Given: Circle O with diameters AE and BD. Show that each of the following is true.

1. $\triangle AOB \cong \triangle DOE$

2. $AB = DE$

3. $\angle AED \cong \angle ABD$

Suppose you slice straight through a rubber ball.

4. What shape will be made on each part of the ball by the slice?

5. How can you slice the ball to obtain the largest such shape?

8.6   The Area of a Circle

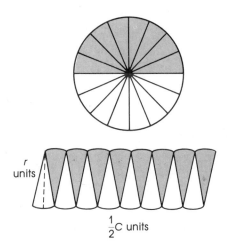

r units

$\frac{1}{2}C$ units

The **area** of a circle is the measurement of the region formed by the circle and its interior.

Suppose we separate a circle into parts as shown in the top figure at the left. Then, to help to find the area, we can arrange these parts in a figure that looks like a parallelogram.

$$A = bh \qquad \text{\textit{Use the formula for area of a parallelogram.}}$$
$$A = \frac{1}{2}Cr \qquad \text{\textit{Replace b by } } \frac{1}{2}C \text{ \textit{ and h by r. Why?}}$$
$$A = \frac{1}{2}(2\pi r)r \qquad \text{\textit{Replace C by } } 2\pi r.$$
$$A = \pi r^2$$

The preceding is not a formal derivation, but it may help you to understand the following statement.

Area of a Circle

If a circle has an area of *A* square units and a radius of *r* units, then $A = \pi r^2$.

Examples

1 **What is the exact area of a circle with a radius of 12 units?**

$$A = \pi r^2 \qquad \text{\textit{Use the formula for the area of a circle.}}$$
$$A = \pi \cdot 12^2 \qquad \text{\textit{Replace r by 12.}}$$
$$A = 144\pi$$

The exact area of the circle is 144π square units.

2 **The area of a circle is 36π square units. Find the length of the radius of the circle.**

$$A = \pi r^2 \qquad \text{\textit{Use the formula.}}$$
$$36\pi = \pi r^2 \qquad \text{\textit{Replace A by } } 36\pi.$$
$$36 = r^2 \qquad \text{\textit{Divide both sides by } } \pi.$$
$$\pm 6 = r \qquad \text{\textit{Reject } } -6 \text{ \textit{ since length is positive.}}$$

The length of the radius is 6 units.

Example

3 **If the length of the radius of a circle is doubled, how is the area changed?**

Let x = the measure of the radius of the original circle.
Then, $2x$ = the measure of the radius of the new circle.
For the original circle, $A = \pi x^2$.   *Use the formula $A = \pi r^2$.*
For the new circle, $A = \pi(2x)^2$ or $4\pi x^2$.

The area of the new circle is 4 times the area of the original circle.

Exercises

Exploratory   In each of the following, the length of the radius of a circle is given. Find the exact area of the circle.

1. 1 unit	**2.** 3 units	**3.** 7 cm	**4.** 10 cm	**5.** 1.4 km
6. 12.3 ft	**7.** $\frac{1}{2}$ in.	**8.** $3\frac{1}{2}$ in.	**9.** $2\sqrt{21}$ mm	**10.** $9\sqrt{14}$ m

Written   In each of the following, the length of the radius of a circle is given. Find a rational approximation to the nearest tenth for the area of the circle. Use $\pi \approx 3.14$.

1. 10 cm	**2.** 2 in.	**3.** 8 m	**4.** 5 mm	**5.** 25 in.
6. 3 km	**7.** 6 ft	**8.** 18 cm	**9.** 140 m	**10.** 200 km
11. 1.1 mm	**12.** 1.5 m	**13.** 15.6 km	**14.** t units	**15.** $3r$ cm

In each of the following, the area of a circle is given. Find the exact length of the radius of the circle.

16. 16π cm²	**17.** 36π cm²	**18.** 100π m²	**19.** 1.21π m²	**20.** 24π cm²
21. πs^2 m²	**22.** 0.04 cm²	**23.** 15 ft²	**24.** 507 km²	**25.** B cm²

In each of the following, the area of a circle is given. Find a rational approximation to the nearest tenth for the circumference of the circle. Use $\pi \approx 3.14$ and the table on page 482 or a calculator.

26. 25π m²	**27.** 400π in.²	**28.** 0.09π cm²	**29.** 361π yd²	**30.** 256π cm²
31. 81π m²	**32.** 75π cm²	**33.** 88π in.²	**34.** 32π ft²	**35.** 245π km²

Find how the area of a circle is changed if the length of the radius is multiplied by each of the following numbers.

36. 3	**37.** 4	**38.** 5	**39.** 10	**40.** 0.25	**41.** 1.5

In the figure at the right, the two circles have the same center, O, but have radii of different lengths. Find the area of the shaded region if the radii of the circles have the following lengths.

42. 5 cm, 2 cm

43. 2 in., $\sqrt{2}$ in.

44. $\sqrt{10}$ cm, $\sqrt{6}$ cm

45. $\sqrt{3}$ ft, 2 in.

Challenge   The four vertices of a square lie on a circle. Find the area of the circle if the square has the following perimeter.

1. 16 m   **2.** 32 in.   **3.** $16\sqrt{2}$ cm   **4.** $24\sqrt{3}$ ft

Find the exact area of the shaded portion in each figure.

5.   **6.**   **7.** 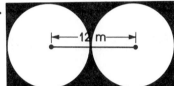   **8.**

Mathematical Excursions

Centers of Circles

The following steps can be used to find the center of a circle.

Place a sheet of paper over the circle so that one corner touches the circle.	The paper intersects the circle at two other points, A and B. Draw the chord, $\overline{AB}$.	Place the paper so that it touches the circle at a different point. Repeat step 2, and draw $\overline{EF}$.

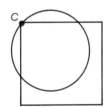   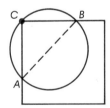   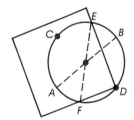

The intersection of $\overline{AB}$ and $\overline{EF}$ is the center of the circle.

Exercises   On a sheet of paper, trace around each of the following circular objects. Find the center and then the length of the radius to the nearest millimeter.

1. a half-dollar   **2.** a dinner plate   **3.** a soda can   **4.** a coffee mug

8.7 Measuring Volume

Empty cracker boxes and soup cans enclose regions of space. They are examples of **geometric solids.**

A can has a curved surface. A box has flat surfaces. These flat surfaces form rectangles which are parts of intersecting planes.

A geometric solid formed by parts of planes is a **polyhedron.** A box is a model of a special kind of polyhedron called a **rectangular solid.**

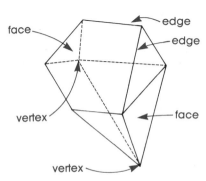

The parts of planes which form polyhedra are polygons. These polygons and their interiors are called **faces.** The intersection of pairs of faces are line segments called **edges.** The points where edges intersect are called **vertices.**

A typical polyhedron is shown at the left.

Polyhedra are classified according to the number of their faces. The names of some polyhedra are given in the table at the right.

Polyhedron	Number of Faces
Tetrahedron	4
Hexahedron	6
Octahedron	8
Dodecahedron	12
Icosahedron	20

Example

1 **Name the faces, edges, and vertices of the tetrahedron at the right.**

The 4 faces are $\triangle ABC$, $\triangle CBD$, $\triangle CDA$, and $\triangle ABD$. The 6 edges are $\overline{AB}$, $\overline{AD}$, $\overline{BD}$, $\overline{AC}$, $\overline{BC}$, and $\overline{DC}$. The 4 vertices are A, B, C, and D.

Prisms are an important subset of the set of polyhedra.

**Definitions of Prism
and Right Prism**

> A prism is a polyhedron that has two parallel faces, called **bases,** and three or more other faces, called **lateral faces,** that are parallelograms. A prism with lateral faces that are rectangles is a **right prism.**

Prisms are classified by the shape of their bases. Thus, there are triangular prisms, pentagonal prisms, and so on. The following are right prisms except for the one on the left. *Rectangular prism* is another name for a rectangular solid since all its faces are rectangles.

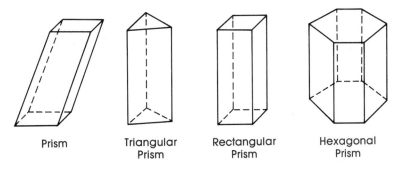

Prism   Triangular   Rectangular   Hexagonal
          Prism   Prism   Prism

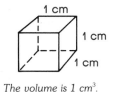

The volume is 1 cm³.

The volume of a solid is the measurement of the space enclosed by the solid. In the metric system, some measurements of volume are given in cubic centimeters (cm³). If each edge of a cube is 1 centimeter long, we say that the volume of the cube is 1 cubic centimeter.

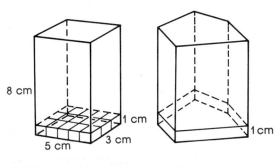

Figure 1   Figure 2

To find the volume of a rectangular solid, consider a slice one centimeter deep at the bottom of the solid as shown in figure 1 at the left. The number of cubic centimeters the slice contains is the number of square centimeters in the area of its base. In this case, that is 5 • 3 or 15. The volume of the slice is 15 cm³. Since the solid is 8 cm high, there are 8 such slices. The total volume is 8 • 15 or 120 cm³. A similar argument holds for any right prism, as shown in figure 2.

Volume of a Right Prism

> If a right prism has a volume of V cubic units, a base with an area of B square units, and a height of h units, then $V = Bh$.

We can use this formula to find formulas for the volumes of other special right prisms. Suppose a rectangular solid has a base with length of ℓ units and width of w units. Then, $B = \ell w$. The solid has a height of h units. The formula for the volume of a rectangular solid is $V = \ell wh$. We use $V = s^3$ for a cube with an edge of s units.

In $V = Bh$, replace B by ℓw.

Example

2 Find the volume of the rectangular solid shown at the right.

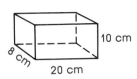

10 cm
8 cm
20 cm

$V = \ell wh$

$V = 20 \cdot 8 \cdot 10$ *Replace ℓ by 20, w by 8, and h by 10.*

$V = 1600$ The volume is 1600 cm^3.

The Great Pyramid has a base that is a square.

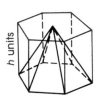

h units

The area of the base is B square units.

A **pyramid** is a polyhedron formed by joining the vertices of a polygon to a point not in the plane of the polygon. The polygon is called the **base** of the pyramid and the point is called the **vertex**.

In the figure at the left, the volume of the pyramid is smaller than the volume of the prism. If a pyramid and a prism have the same base and height, then the volume of the pyramid is one-third the volume of the prism.

Suppose the area of the base of the prism shown in the figure at the left is 300 cm^2 and the prism has a height of 10 cm. The volume of the prism is $300 \cdot 10$ or 3000 cm^3. Then, the volume of the pyramid shown is $\frac{1}{3} \cdot 3000$ or 1000 cm^3.

Volume of a Pyramid

> If a pyramid has a volume of V cubic units, a base with an area of B square units, and a height of h units, then $V = \frac{1}{3}Bh$.

Exercises

Exploratory   Name the faces, edges, and vertices of each of the following polyhedra.

1.

2.

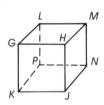

3.

Find the volume if an edge of a cube has the length in each of the following.

4. 2 cm   **5.** 3 cm   **6.** 8 cm   **7.** 1 m   **8.** 2.5 mm   **9.** 1.3 cm

Written   Find the missing measure of the rectangular solid in each of the following.

	V	ℓ	w	h		V	ℓ	w	h
1.	?	2	2	3	**2.**	?	4	8	3
3.	?	2	$2\frac{1}{2}$	$3\frac{1}{2}$	**4.**	?	$\frac{1}{2}$	2	$2\frac{1}{4}$
5.	90	?	5	6	**6.**	5760	24	?	10

Find the volume of the right triangular prism with each of the following measures.

7. In $\triangle EFG$, $A = 25$, $FK = 12$
8. In $\triangle EFG$, $A = 70$, $GL = 45$
9. $EG = 10$, $FD = 6$, $FK = 15$
10. $EG = 12$, $FD = 5$, $GL = 14$
11. $EG = 100$, $FD = 40$, $FK = 600$
12. $ED = 5$, $DG = 15$, $EF = 13$, $FK = 50$
13. $ED = 9$, $EG = 50$, $EF = 41$, $FK = 200$

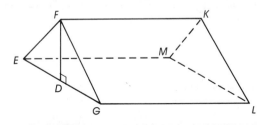

Find the volume of the pyramid that is described in each of the following.

14. It has a base with an area of 180 cm²
and a height of 10 cm.
15. It has a base with an area of 320 cm²
and a height of 12 cm.
16. Its base is a rectangle 20 cm long and 30
cm wide. The pyramid is 15 cm high.

Challenge   Find the volume of the Great
Pyramid in Egypt when it was first com-
pleted. It had a height of 146.6 m and a
square base with a side of 230 m.

8.8 Solids With Curved Surfaces

Right Prism   **Right Cylinder**

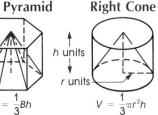

$V = Bh$ $V = \pi r^2 h$

In this book, the word "cylinder" refers to a right circular cylinder.

A **right circular cylinder** has a curved lateral surface such that any segment entirely on it is perpendicular to the two circular bases of the cylinder. A right cylinder may be compared with a right prism. The volume of a right prism is found by thinking of this solid as a layer of slices each one unit high. The volume of a right cylinder is found the same way. The area of the base is found by using $A = \pi r^2$.

Volume of a Right Circular Cylinder

If a right circular cylinder has a volume of V cubic units and a height of h units, and its base has a radius of r units, then $V = \pi r^2 h$.

Right Pyramid   **Right Cone**

$V = \frac{1}{3}Bh$ $V = \frac{1}{3}\pi r^2 h$

In this book, the word "cone" refers to a right circular cone.

A **right circular cone** is another solid with a curved surface. Like a cylinder, it has a circular base; but it has a vertex like that of a pyramid. The volume of a cone is less than the volume of a cylinder with the same base and height. Compare the volume of a cone with the volume of a right pyramid, as shown at the left.

Volume of a Right Circular Cone

If a right circular cone has a volume of V cubic units and a height of h units, and its base has a radius of r units, then $V = \frac{1}{3}\pi r^2 h$.

Sphere

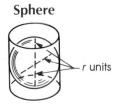

r units

A **sphere** has an entirely curved surface. Every point on the sphere is the same distance from the *center* of the sphere. A segment with endpoints on the sphere and the center is a *radius*.

Suppose a height of a cylinder is the same as a diameter of a sphere. A radius of the cylinder's base is the same as a radius of the sphere. The volume of this cylinder is found by using $V = \pi r^2(2r)$ or $2\pi r^3$. Compare this volume to the volume of the sphere, as shown at the left. The volume of the sphere is less than the volume of the cylinder.

Volume of a Sphere

If a sphere has a volume of V cubic units and a radius of r units, then $V = \frac{4}{3}\pi r^3$.

Examples

1 **Find the exact volume of a cylinder with a height of 10 cm and a base with a radius of 4 cm.**

$V = \pi r^2 h$ *Use the formula for the volume of a cylinder.*
$V = \pi \cdot 4^2 \cdot 10$ *Replace r by 4 and h by 10.*
$V = 160\pi$ The volume is 160π cm^3.

2 **Find the exact volume of a cone with a height of 12 cm and its base has a radius of 5 cm.**

$V = \dfrac{1}{3}\pi r^2 h$ *Use the formula for the volume of a cone.*

$V = \dfrac{1}{3}\pi \cdot 5^2 \cdot 12$ *Replace r by 5 and h by 12.*

$V = 100\pi$ The volume is 100π cm^3.

3 **Find an approximation to the nearest tenth for the volume of a sphere with a radius of 6 cm.**

Lab Activity

You can learn about the surface area of a sphere in Lab 7 on page A10.

$V = \dfrac{4}{3}\pi r^3$ *Use the formula for the volume of a sphere.*

$V = \dfrac{4}{3}\pi \cdot 6^3$ *Replace r by 6.*

$V = \dfrac{4}{3}\pi \cdot 216$ or 288π *Use $\pi \approx 3.14$.*

$V \approx 288 \cdot 3.14$ or 904.32 The volume is approximately 904.3 cm^3.

4 **Find the exact volume of the figure at the right.**

The lower portion of the figure is a cylinder. Find its volume.
$V = \pi r^2 h$ *Use the formula.*
$V = \pi \cdot 4^2 \cdot 3$ or 48π

The top of the figure is a cone.

$V = \dfrac{1}{3}\pi r^2 h$ *Use the formula.*

$V = \dfrac{1}{3}\pi \cdot 3^2 \cdot 8$ or 24π

8 cm

3 cm

3 cm

4 cm

The volume of the figure is 48π cm^3 + 24π cm^3 or 72π cm^3.

Exercises

Exploratory In each of the following, the measures of a cylinder are given. Find the exact volumes if the measurements are in centimeters.

1. $r = 1, h = 3$ **2.** $r = 2, h = 5$ **3.** $r = 3, h = 3$ **4.** $r = 2, h = 3$
5. $r = 5, h = 20$ **6.** $r = 8, h = 50$ **7.** $r = 7, h = 1.1$ **8.** $r = 9, h = 3.3$

In each of the following, the measures of a cone are given. Find the exact volume if the measurements are in meters.

9. $r = 10, h = 3$ **10.** $r = 2, h = 6$ **11.** $r = 3, h = 12$ **12.** $r = 6, h = 3$
13. $r = 5, h = 2$ **14.** $r = 4, h = 1$ **15.** $r = 20, h = 15$ **16.** $r = 12\sqrt{2}, h = 7\sqrt{2}$

Written In each of the following the measures of a cylinder are given. Find the exact volumes if the measurements are in centimeters.

1. $h = 10, r = 5$ **2.** $h = 12, r = 4$ **3.** $h = 10, r = \sqrt{5}$ **4.** $h = 8, r = 2\sqrt{3}$
5. $h = 3.5, r = 2$ **6.** $h = 5, r = 2$ **7.** $h = 16, r = 12$ **8.** $h = 41, r = 22$

Find a rational approximation to the nearest tenth of the volume of a cone with the measurements given in each of the following. Use $\pi \approx 3.14$.

9. h, 2 cm; r, 3 cm   **10.** h, 3 m; r, 6 m   **11.** h, 4 cm; r, 15 cm
12. h, 12 cm; r, 5 cm   **13.** h, 1.1 m; r, 2 cm   **14.** h, 3.4 cm; r, 6 mm
15. h, 5 m; r, $\sqrt{2}$ m   **16.** h, 1.8 cm; r, $2\sqrt{3}$ cm   **17.** h, 16 cm; r, $3\sqrt{3}$ mm

Find the exact volume of a sphere with a radius having each of the following lengths.

18. 1 m   **19.** 10 cm   **20.** 2 cm   **21.** 3 m   **22.** 5 ft   **23.** 3.5 m
24. $\frac{1}{2}$ mm   **25.** $\sqrt{3}$ m   **26.** $2\sqrt{2}$ cm   **27.** $\frac{1}{4}$ in.   **28.** $1\frac{3}{4}$ in.   **29.** $2r$ cm

State how the volume of a cylinder changes in each of the following situations.

30. The height is tripled.   **31.** The radius is doubled.

Find the exact volumes of the figures shown in each of the following.

32.
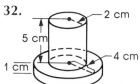
33.

a half sphere with a radius of 3 cm and a cone with a height of 6 cm

34.

Challenge   Solve each of the following.

1. Find the total volume of 2 cones, each of which has a height of r units and a base with a radius of r units.

2. Which is greater, the volume of a sphere with a radius of r units or the volume described in Challenge exercise 1?

The edges, radii, or height of a solid are called the *linear dimensions* of the solid. The corresponding linear dimensions of two solids are used to define *similar solids*.

> **Two solids are similar if the measures of the angles on their corresponding faces are equal and their corresponding linear dimensions are in proportion.**

In the figure below, the two rectangular prisms are similar. The ratio of their corresponding linear dimensions is $\dfrac{3 \text{ cm}}{2 \text{ cm}}$ or $\dfrac{3}{2}$.

For similar polygons, the ratio of the areas is related to the ratio of the measures of the corresponding sides. Does such a relationship exist between the ratio of the volumes of similar solids and the ratio of their corresponding linear dimensions?

Find the ratio of the volumes of the prisms below and compare it to the ratio of their corresponding linear dimensions.

$$\frac{\text{Volume of Prism 1}}{\text{Volume of Prism 2}} = \frac{(3 \text{ cm})(6 \text{ cm})(4\tfrac{1}{2} \text{ cm})}{(2 \text{ cm})(4 \text{ cm})(3 \text{ cm})}$$

$$= \frac{81 \text{ cm}^3}{24 \text{ cm}^3}$$

$$= \frac{27}{8} \text{ or } \left(\frac{3}{2}\right)^3$$

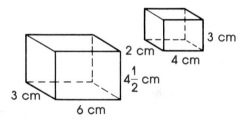

Volumes of Similar Solids

> If two solids are similar, then the ratio of their volumes is equivalent to the cube of the ratio of their corresponding linear dimensions.

If the ratio of the corresponding linear dimensions is $a : b$, then the ratio of the volumes is $a^3 : b^3$.

Exercises Solve each problem.

1. Laundry soap can be purchased in two different size boxes which are similar rectangular prisms. The height of the smaller box is 12 in., and the height of the larger box is 16 in. If the volume of the smaller box is 648 in³, what is the volume of the larger box?

2. A 4-inch tall can with a circular base has a volume of 25π in³. This can is similar to a can that has a volume of 43.2π in³. What is the height of the larger can?

8.9 Metric Units of Mass and Capacity

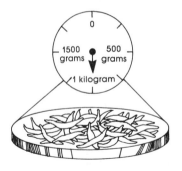

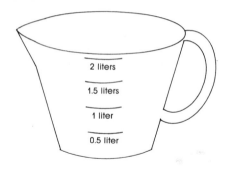

The mass of 1 cm³ of water is 1 g.

A basic unit of mass in the metric system is the **gram** (g). One gram is defined as the mass of 1 cubic centimeter of water at its maximum density. The mass of two paper clips is about one gram.

For many uses, a gram is too small to be practical. The kilogram (kg), which is 1000 grams, is used instead. The mass of this book is about 1 kilogram. An even larger unit is a metric ton (t), which is 1000 kilograms. The mass of a small car is about 1 metric ton.

A unit of capacity is the **liter** (L). It is used for volumes of liquids and gases. A liter is 1000 cubic centimeters. You can buy club soda by the liter.

A liter has the same volume as a cube with an edge of 10 centimeters.

Example

1 **At its maximum density, what is the capacity in liters of 1 kilogram of water?**

1 g is the mass of 1 cm³ of water.
1 kg is the mass of 1000 cm³ of water.   *1 kg = 1000 g*
1000 cm³ = 1 L

There is 1 liter in 1 kilogram of water.

The metric prefixes are used for units of mass and capacity.

<table>
<tr><td colspan="2">Units of Mass</td><td colspan="2">Units of Capacity</td></tr>
</table>

Units of Mass	Units of Capacity
1 kilogram = 10 hectograms	1 kiloliter = 10 hectoliters
1 hectogram = 10 decagrams	1 hectoliter = 10 decaliters
1 decagram = 10 grams	1 decaliter = 10 liters
1 gram = 10 decigrams	1 liter = 10 deciliters
1 decigram = 10 centigrams	1 deciliter = 10 centiliters
1 centigram = 10 milligrams	1 centiliter = 10 milliliters

Examples

2 **Express 1.42 grams in milligrams and kilograms.**

The chart at the right is related to the table of units of mass above and the decimal place-value system. A unit is equivalent to the unit at its right multiplied by 10.

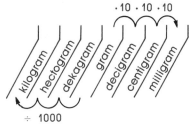

$\cdot 1000$
1.42 g = __?__ mg   *Multiply by 1000 since*
1.42 · 1000 = 1420 *1000 mg = 1 g.*
1.42 g = 1420 mg   There are 1420 milligrams in 1.42 grams.

In the chart, a unit is equivalent to the unit at its left divided by 10.

1.42 g = __?__ kg   *Divide by 1000 since there are 1000 g in 1 kg.*
÷ 1000
1.42 ÷ 1000 = 0.00142
1.42 g = 0.00142 kg   There is 0.00142 kilogram in 1.42 grams.

3 **Express 23 milliliters in liters and then in cubic centimeters.**

23 mL = __?__ L   *Divide by 1000 since there are 1000 mL in 1 L.*
÷ 1000

23 ÷ 1000 = 0.023   There is 0.023 liter in 23 milliliters.

$\cdot 1000$
0.023 L = ____ cm³   *Multiply by 1000 since there are 1000 cm³ in 1 L.*
0.023 · 1000 = 23
0.023 L = 23 cm³   There are 23 cubic centimeters in 23 milliliters.

Exercises

Exploratory State the metric unit you would use to measure each of the following. Use gram, kilogram, liter, or milliliter.

1. a bag of oranges
2. ground pepper
3. eyedrops
4. gasoline for a car
5. vanilla extract
6. an aspirin tablet
7. a wrestler
8. a hamster
9. liquid laundry detergent

State the metric unit of mass that completes each of the following.

10. The mass of a dime is 4 ___.
11. The mass of a box of cereal is 200 ___.
12. The mass of a football player is 110 ___.
13. The mass of a truck is 4 ___.
14. The mass of a pencil is 10 ___.
15. A sheet of paper has a mass of 0.2 ___.

State the metric unit of capacity that completes each of the following.

16. A ketchup bottle holds 0.5 ___.
17. A drinking glass holds 200 ___ of water.
18. A hot air balloon holds 25,000 ___ of air.
19. An aquarium holds 20 ___ of water.
20. A heart pumps about 7.8 ___ of blood a minute.
21. A mug holds 0.21 ___ of coffee.

Written Complete each of the following.

1. 5000 g = ___ kg
2. 8 kg = ___ g
3. 142 mg = ___ g
4. 1.6 g = ___ mg
5. 3500 mg = ___ g
6. 18.5 g = ___ mg
7. 3000 mL = ___ L
8. 2L = ___ mL
9. 4.6 L = ___ mL
10. 2100 mL = ___ L
11. 4200 mL = ___ L
12. 5.06 L = ___ mL
13. 3 L = ___ cm^3
14. 2000 cm^3 = ___ L
15. 3.08 L = ___ cm^3
16. 21 cm^3 = ___ L
17. 8.05 kg = ___ g
18. 6.71 t = ___ kg
19. 230 kg = ___ t
20. 105 mL = ___ L
21. 2.3 L = ___ cm^3
22. 250 cm^3 = ___ L
23. 23 g = ___ mg
24. 48 g = ___ mg
25. 948 cm^3 = ___ L
26. 9.7 L = ___ cm^3
27. 28 mL = ___ cm^3
28. 3.12 cm^3 = ___ mL
29. 2 mL = ___ cm^3
30. 1 m^3 = ___ L

Answer each of the following.

31. What is the capacity in liters of an aquarium that is 65 cm by 30 cm by 27 cm?

32. What is the mass of the water in the aquarium in exercise 31 when the aquarium is filled?

Challenge Solve each of the following.

1. In general, decreasing the temperature of matter causes it to become more dense. But ice floats, indicating that it is less dense than water in liquid form. In what form is water at its maximum density?

2. Weight is the pull that the earth exerts on an object. There is greater pull on an object at sea level than on a mountain. Does the mass of 1 cubic centimeter of water differ at sea level and on a mountain?

Problem Solving Application: Solve a Simpler Problem

An important strategy for solving complicated or unfamiliar problems is to *solve a simpler problem*. This strategy involves setting aside the original problem and solving a simpler or more familiar case of the problem. The same relationships and concepts that are used to solve the simpler problem can then be used to solve the original problem.

Examples

1 **Find the sum of the first 1000 natural numbers.**

Explore   The first 1000 natural numbers are the numbers 1 through 1000.
Thus, we want to find the sum $1 + 2 + 3 \cdots + 1000$.

Plan   This problem could be solved by actually adding all the numbers, but this computation would be very tedious even if a calculator is used.
Consider a simpler problem. Find the sum, S, of the first 10 natural numbers.

$$
\begin{aligned}
S &= 1 + 2 + 3 + \cdots + 9 + 10 \\
+ S &= 10 + 9 + 8 + \cdots + 2 + 1 \\
\hline
2S &= 11 + 11 + 11 + \cdots + 11 + 11 \\
2S &= 10(11) \\
S &= 5(11) \quad \text{or} \quad 55
\end{aligned}
$$

There are 10 addends in each sum, S.

Solve   Now extend this concept to solve the original problem.

$$
\begin{aligned}
S &= 1 + 2 + 3 + \cdots + 999 + 1000 \\
+ S &= 1000 + 999 + 998 + \cdots + 2 + 1 \\
\hline
2S &= 1001 + 1001 + 1001 + \cdots + 1001 + 1001 \\
2S &= 1000(1001) \\
S &= 500(1001) \quad \text{or} \quad 500,500
\end{aligned}
$$

There are 1000 addends in each sum, S.

The sum of the first 1000 natural numbers is 500,500.
You may wish to use a calculator to check this result.

2 **Find the number of line segments needed to connect 1001 different points such that a line segment connects each pair of points.**

We could try to solve the problem by first plotting 1001 different points. Then we could draw line segments connecting each pair of points and count how many line segments were drawn. An easier method would be to look at some simpler cases.

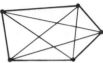

1 point	2 points	3 points	4 points	5 points
0 segments	1 segment	3 segments	6 segments	10 segments

In order to determine how many line segments are needed for 1001 points, we must now look for a pattern in the number of line segments for these cases.

2 points: 1   1 segment
3 points: 1 + 2   or   3 segments
4 points: 1 + 2 + 3   or   6 segments
5 points: 1 + 2 + 3 + 4 or   10 segments

Now extend this relationship to the original problem. The number of line segments needed for 1001 points is equal to the sum of the numbers 1 through 1000. Based on the results of Example 1, 500,500 line segments are needed.

Exercises

Written   Solve each problem.

1. A drain pipe is 750 cm long. A spider climbs up 100 cm during the day but falls back 80 cm during the night. If the spider begins at the bottom of the pipe, on what day will it get to the top?

2. Find the total number of squares in the checkerboard shown at the right.

3. Determine the number of lateral faces for a prism whose bases are 100-gons.

4. Determine the number of vertices for a prism whose bases are 100-gons.

5. Determine the number of edges for a prism whose bases are 100-gons.

6. Find the sum of the first n positive integers.

7. Find the number of line segments needed to connect n different points such that a line segment connects each pair of points.

8. A total of 3001 digits were used to print the page numbers of the Northern College annual. How many pages are in the annual?

Mixed Review

Solve each equation in the domain $\mathcal{R}$.

1. $x^2 = 40$ **2.** $4y^2 = 32$ **3.** $a^2 + 1 = 453$ **4.** $12 - 6n^2 = -636$

5. Find the area of a rectangle with a length of 15 cm and a diagonal that is 17 cm long.

Portfolio Suggestion

Select an item from your homework or classwork from this chapter that you feel shows your best work and place it in your portfolio. Explain why you selected it.

Performance Assessment

The city of Anderson is installing a new water tower.

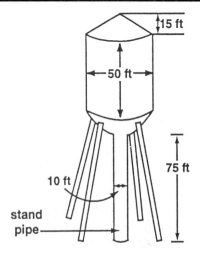

a. Give a plan for finding the volume of the water tank including the stand pipe and conical roof.

b. Then use your plan to find the volume of the water tank.

c. The average family uses about 40 cubic feet of water per day. Estimate the maximum population that one tankful might serve.

Chapter Summary

1. If a rectangle has an **area** of A square units, a length of ℓ units, and a width of w units, then $A = \ell w$. (246)
2. If a parallelogram has an area of A square units, a base of b units, and a corresponding height of h units, then $A = bh$. (248)
3. If a triangle has an area of A square units, a base of b units, and a corresponding height of h units, then $A = \frac{1}{2}bh$. (251)
4. If a trapezoid has an area of A square units, bases of b_1, units and b_2 units, and a height of h units, then $A = \frac{1}{2}h(b_1 + b_2)$. (254)
5. A **circle** is the set of all points in a plane a given distance from a point in the plane called the **center.** (257)
6. If a circle has a **circumference** of C units and a **radius** of r units, then $C = 2\pi r$. (258)
7. If a circle has an area of A square units and a radius of r units, then $A = \pi r^2$. (261)

8. A **prism** is a **polyhedron** that has two parallel faces, called **bases,** and three or more other **faces,** called **lateral faces,** that are parallelograms. (264)

9. A prism with lateral faces that are rectangles is a **right prism.** (265)

10. If a right prism has a volume of V cubic units, a base with an area of B square units, and a height of h units, then $V = Bh$. (266)

11. If a **pyramid** has a volume of V cubic units, a base with an area of B square units, and a height of h units, then $V = \frac{1}{3}Bh$. (266)

12. If a **right circular cylinder** has a volume of V cubic units and a height of h units, and its base has a radius of r units, then $V = \pi r^2 h$. (268)

13. If a **right circular cone** has a volume of V cubic units and a height of h units, and its base has a radius of r units, then $V = \frac{1}{3}\pi r^2 h$. (268)

14. If a **sphere** has a volume of V cubic units and a radius of r units, then $V = \frac{4}{3}\pi r^3$. (268)

⊿∑∇ Chapter Review

8.1 **Measure each of the following segments. Express the length in centimeters and millimeters.**

1.   2.

Complete each of the following.

3. 5 cm = _____ mm   4. 1275 mm = _____ m   5. 8 m = _____ cm
6. 325 cm = _____ m   7. 4500 m = _____ km   8. 9.1 km = _____ m
9. 25 mm = _____ m   10. 6625 cm = _____ km   11. 4.25 m = _____ mm

8.2 **Find the area of a rectangle with the dimensions given in each of the following. Express your answer in simplest form.**

12. 12 cm, 8 cm   13. 300 m, 20 m   14. 0.3 mm, 3.1 mm
15. $\sqrt{21}$ cm, $\sqrt{42}$ cm   16. $2\sqrt{7}$ km, $\sqrt{63}$ km   17. $3\sqrt{7}$ m, $4\sqrt{2}$ m

In each of the following the measures of a parallelogram are given. Find h.

18. $A = 60, b = 5$ 19. $A = 3.5, b = 0.7$ 20. $A = 325, b = 15$
21. $A = 72, b = 16$ 22. $A = 48\sqrt{3}, b = 3\sqrt{3}$ 23. $A = 129, b = 3\sqrt{3}$

8.3 and 8.4 **Find the area of each of the following figures.**

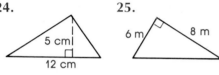

24.

25.

26.

27.

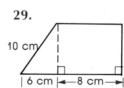

28. 6 cm

29.

30.

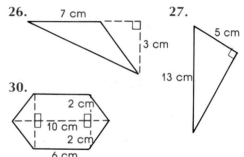

8.5 and 8.6 **Use ⊙K shown at the right to name each of the following.**

31. the center   **32.** 3 radii
33. 3 chords   **34.** 1 diameter

Find a rational approximation to the nearest tenth for each of the following. Use π ≈ 3.14.

35. the circumference of a circle with a radius of 12 cm

36. the area of a circle with a diameter of 30 m

37. the diameter of a circle with a circumference of 125.6 cm

38. the radius of a circle with an area of 314 ft²

8.7 and 8.8 **Find the volume of each of the following solids with the given measurements.**

39. a right prism with a height of 8 cm and a base with an area of 120 cm²

40. a cube with an edge of 10 m

41. a rectangular solid with a length of 14 cm, a width of 12 cm, and a height of 5 cm

42. a pyramid with a height of 30 cm and a rectangular base with a length of 8 cm and a width of 4 cm

43. a cylinder with a height of 12 cm and a base with a radius of 3 cm

44. a cone with a height of 10 cm and a base with a radius of 6 cm

45. a cylinder with a height of 9 m and a base with an area of 30 m²

46. a sphere with a radius of 5 cm

8.9 **Complete each of the following.**

47. 2400 g = _____ kg
48. 7.1 kg = _____ g
49. 1.8 g = _____ mg
50. 3.2 mg = _____ g
51. 4 L = _____ mL
52. 0.5 L = _____ mL
53. 2.8 L = _____ cm³
54. 5600 cm³ = _____ mL
55. 8000 cm³ = _____ L

Chapter Test

Complete each of the following.

1. 30 cm = _____ mm

2. 60 km = _____ m

3. 18 m = _____ cm

4. 2.1 mm = _____ cm

5. 1725 m = _____ km

6. 8.1 L = _____ cm³

7. 81 kg = _____ g

8. 247 mg = _____ g

9. 35 mL = _____ L

Find the area of each of the following figures. Express your answer in simplest form.

10.

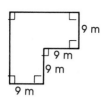

11.

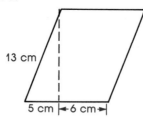

12.

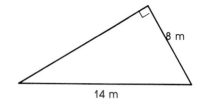

13.

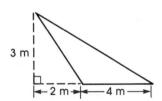

14.

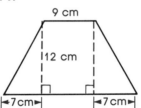

15.

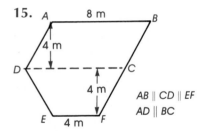

Solve each of the following.

16. Find the width of a rectangle with an area of 124 cm² and a length of 31 cm.

17. Find the height of a triangle with an area of 28 cm² and a base of 7 cm.

18. Find rational approximations to the nearest tenth for the circumference and area of a circle with a radius of 16 cm. Use $\pi \approx 3.14$.

19. Find the perimeter and area of a square with a diagonal that is 16 cm long.

Find the exact volume of each of the following.

20. a right prism with a height of 40 cm and a square base with sides of 2 cm

21. a pyramid with a height of 12 cm and a right triangle for a base with legs of 6 cm and 8 cm

22. a cube with an edge of 9 cm

23. a sphere with a radius of 4 mm

24. a cylinder with a height of 3 cm and a base with a radius of $3\sqrt{3}$ cm

25. a cone with a height of 10 cm and a base with a radius of 1.5 cm

Polynomials

Application in Wildlife Management

Park rangers are responsible for the protection and safety of the wildlife within the confines of the park. The rangers often provide herds of deer or other animals with food, medical care, shelter in winter, and protection from predators. These interventions can cause the size of the herd to grow out of control.

Without the controls of nature, an animal population develops polynomially. That is, if one female animal has x female offspring, then it will have an average of x^2 female grandoffspring, x^3 female great-grandoffspring, and so on. Write a **polynomial** that represents the total number of female descendants of one female after six generations.

Individual Project: *Biology*

Choose an endangered species to study. Write a report explaining how and why the population has come to be endangered. Then assume the species can be protected and write a polynomial that represents the female descendants of one animal. Assuming that 50% of the offspring are female, find the number of female descendants one female is likely to have after five generations.

9.1 Monomials

A constant names a specific number.

Expressions such as y, 15, $-7x^2$, and $\sqrt{11}x^3y^2$ are called **monomials** (mo-*no*-me-als). A monomial is a constant, a variable, or the product of a constant and one or more variables. Expressions like $2x + 1$, $\dfrac{3}{x^2}$, and $\sqrt{x}$ are *not* monomials.

Each of the following expressions for area or volume are examples of monomials.

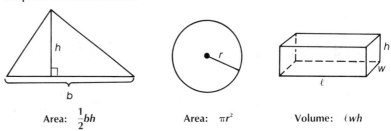

Area: $\dfrac{1}{2}bh$ Area: πr^2 Volume: ℓwh

The **degree** of a monomial is the sum of the exponents of its variables.

The exponent of y is 1. The degree of a nonzero constant is 0. The constant 0 has no degree.

Monomial	Variables	Exponents	Degree
y	y	1	1
$-7x^2$	x	2	2
$\sqrt{11}x^3y^2$	x, y	3, 2	$3 + 2 = 5$

The coefficient in a monomial is its numerical factor.

If two monomials are the same or differ only by their coefficients, they are *like monomials*. For example, $-7xy^2$ and $6xy^2$ are like monomials. The monomials $5y^2$ and $3y$ are *not* like monomials. *Why?*

The sum or difference of two or more like monomials can be simplified using the distributive property. *This is also referred to as combining like terms.*

Example

1 Simplify $7x + 11x$.

$$7x + 11x = (7 + 11)x \qquad \textit{Use the distributive property.}$$
$$= 18x$$

Examples

2 Simplify $-6xy + 8xy$.

$$-6xy + 8xy = (-6 + 8)xy$$
$$= 2xy$$

3 Simplify $4a^2b + 8a^2b - 11a^2b$.

$$4a^2b + 8a^2b - 11a^2b = (4 + 8 - 11)a^2b$$
$$= 1a^2b \text{ or } a^2b$$

Exercises

Exploratory State whether each of the following expressions is a monomial. If it is, then name its coefficient.

1. $7x$
2. y^2
3. -8
4. $-5xy$
5. $3xy + y$
6. $\frac{2}{3}x$
7. $\frac{11xy}{7}$
8. $\sqrt{xy}$
9. $-\frac{y}{11}$
10. $\frac{3}{x}$

State the degree of each of the following monomials.

11. $11m$
12. $-6y^3$
13. 12
14. $5xy$
15. $3x^2y$
16. $-\frac{2}{3}p^2qr^4$
17. -8
18. $\frac{y}{7}$
19. $-b$
20. $13x^3y^4z^5$

Written Simplify each of the following, if possible.

1. $-16y + 9y$
2. $13a - 18a$
3. $12x^2 + 4x^2$
4. $13r^3 - 15r^3$
5. $-6a^3 + 4a^3$
6. $y^4 + 2y^4$
7. $\frac{1}{4}x^2 + \frac{3}{4}x^2$
8. $2ab - 6ab$
9. $-6xy - 15xy$
10. $6a^2 + a$
11. $3mn - mn$
12. $7a^2b + 4a^2b$
13. $10xy^2 - xy^2$
14. $4ab^2 + a^2b$
15. $-6r^3s + 19r^3s$
16. $2x^3 + 3x^3 + 6x^3$
17. $5ab - 6ab + 10ab$
18. $7x^2y^3 - 4x^2y^3 - 8x^2y^3$
19. $st^2 + 6 - a$
20. $x^3 - \frac{1}{3}x^3 + \frac{3}{4}x^3$
21. $5y^2 - 12y^2 + y^2 - 3$

Express the measure of the perimeter of each figure as a monomial in simplest form.

22.

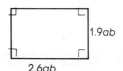

2.6ab, 1.9ab

23.

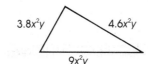

3.8x²y, 4.6x²y, 9x²y

24.

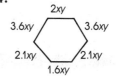

2xy, 3.6xy, 3.6xy, 2.1xy, 2.1xy, 1.6xy

9.2 Multiplying Monomials

The expression x^2 means $x \cdot x$. Recall that x is the base and 2 is the exponent.

Consider the product of the powers x^2 and x^4.

$$x^2 = x \cdot x \qquad \text{and} \qquad x^4 = x \cdot x \cdot x \cdot x$$

Thus, $x^2 \cdot x^4 = (x \cdot x) \cdot (x \cdot x \cdot x \cdot x)$.

$$\underbrace{x \cdot x}_{2} \cdot \underbrace{x \cdot x \cdot x \cdot x}_{4} = x^6$$
$$2 + 4 = 6$$
$$x^2 \cdot x^4 = x^{2+4} \text{ or } x^6$$

Likewise, $5^4 \cdot 5^7 = \underbrace{(5 \cdot 5 \cdot 5 \cdot 5)}_{4} \cdot \underbrace{(5 \cdot 5 \cdot 5 \cdot 5 \cdot 5 \cdot 5 \cdot 5)}_{7}$ or 5^{11}.

$$4 + 7 = 11$$

These and other similar examples suggest the following rule.

Product of Powers

For all numbers a and positive integers m and n,
$$a^m \cdot a^n = a^{m+n}.$$

Examples

1 Simplify $(2x)(4x^3)$.

$(2x)(4x^3) = (2 \cdot 4)(x \cdot x^3)$ *Recall $x = x^1$.*
$\qquad\qquad = 8 \cdot x^{1+3}$
$\qquad\qquad = 8x^4$

2 Simplify $(3x^2y^3)(-2xy^4)$.

$(3x^2y^3)(-2xy^4) = (3 \cdot -2)(x^2 \cdot x)(y^3 \cdot y^4)$ *Only exponents with the same base can*
$\qquad\qquad\qquad = -6 \cdot x^{2+1} \cdot y^{3+4}$ *be added.*
$\qquad\qquad\qquad = -6x^3y^7$

3 Simplify $(-2a^2bc)(-3ab^3c)(ab^4c)$.

$(-2a^2bc)(-3ab^3c)(ab^4c) = (-2)(-3)(a^2 \cdot a \cdot a)(b \cdot b^3 \cdot b^4)(c \cdot c \cdot c)$
$\qquad\qquad\qquad\qquad\quad = 6a^4b^8c^3$

Example

4 Express the area measure of the triangle in simplest form.

$A = \frac{1}{2}bh$

$\quad = \frac{1}{2}(6x^2y)(3xy)$

$\quad = (\frac{1}{2} \cdot 6 \cdot 3)(x^2 \cdot x)(y \cdot y)$

$\quad = 9x^3y^2$

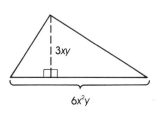

$3xy$

$6x^2y$

Exercises

Exploratory   State whether each of the following is *true* or *false*.

1. $2^3 \cdot 2^4 = 2^7$
2. $10^3 \cdot 10 = 10^4$
3. $3^2 \cdot 3^4 = 3^8$
4. $2^2 \cdot 2^3 = 6^5$

5. $8 \cdot 8^4 = 8^5$
6. $3^2 + 3^3 = 3^5$
7. $4^4 \cdot 4^2 = 8^6$
8. $6 \cdot 6 = 6^2$

9. $5^2 \cdot 5^5 = 5^7$
10. $6r \cdot r = 6r^2$
11. $y^2 \cdot y = y^2$
12. $ab^2 \cdot a^2b = 3ab$

Written   Find and simplify each product.

1. $x^2 \cdot x^3$
2. $y^4 \cdot y^7$
3. $b^3 \cdot b^5$
4. $c^2 \cdot c^2$

5. $r^3 \cdot r^6$
6. $y^3 \cdot y^3$
7. $d \cdot d^3$
8. $x^2 \cdot x \cdot x^3$

9. $y \cdot y \cdot y^3$
10. $y^{21} \cdot y^{18} \cdot y$
11. $(2x^2)(x)$
12. $3y \cdot 2y$

13. $(-2)(3xy)$
14. $4(-3)(2y)$
15. $(4ab)(-2c)$
16. $-3x \cdot 2x^2$

17. $(2a)(-4ab)$
18. $(-3y)(-2y^2)$
19. $(4x^2)(3xy)$
20. $2ab \cdot 3a^2$

21. $5mn^2 \cdot 2m$
22. $(x^3y^2)(xy^3)$
23. $a^2b \cdot a^3b^2$
24. $(r^4s^3)(rs^4)$

25. $(-2mn^2)(5m^2n)$
26. $(-6a^3b)(3a^2b^2)$
27. $(-5x^2y)(-2x^4y^7)$

28. $ab \cdot ac \cdot bc$
29. $(-4xy^2)(-8xz^3)$
30. $(-3axy^2)(-ax)$

31. $(mn)(np)(pr)$
32. $(\frac{1}{2}a^2b)(ab)(-\frac{2}{7}ac^2)$
33. $(-\frac{3}{4}ab)(\frac{4}{5}a^2b)$

34. $(3.1x^2)(4.3xy)$
35. $(\sqrt{3}xy^2)(\sqrt{8}x^2y^3)$
36. $(2\sqrt{2}a^2b)(\sqrt{2}ab)$

Express the measure of the area of each figure in simplest form.

37.

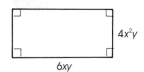

$4x^2y$

$6xy$

38.

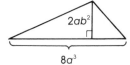

$2ab^2$

$8a^3$

39.

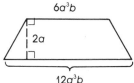

$6a^3b$

$2a$

$12a^3b$

Challenge   Find the value of *r* which makes each of the following sentences true.

1. $x^2 \cdot x^r = x^5$
2. $x^2 \cdot x^3 \cdot x^r = x^{12}$
3. $7 \cdot 7^r = 7$
4. $5 \cdot 5 \cdot 5^r = 5^9$

5. $x^{13} \cdot x^7 \cdot x^8 = x^{3r+7}$
6. $2^{r+5} = 2^{2r-1}$
7. $2^{2r+1} = 32$
8. $3^{r+1} = 3 \cdot 3^2$

9.3   Powers of Monomials

Suppose 4^2 is raised to the third power.

$$(4^2)^3 = 4^2 \cdot 4^2 \cdot 4^2$$
$$= 4^{2+2+2}$$
$$= 4^6$$

How can you obtain the exponent 6 from the exponents 2 and 3?

$$2 \cdot 3 = 6$$

Likewise, $(x^3)^4 = x^3 \cdot x^3 \cdot x^3 \cdot x^3$
$$= x^{12} \qquad \textit{Note that } 3 \cdot 4 = 12.$$

These and other similar examples suggest the following rule.

Power of a Power

> For all numbers a and positive integers m and n,
> $$(a^m)^n = a^{mn}.$$

Examples

1 **Simplify $(a^2)^5$.**

$(a^2)^5 = a^{2 \cdot 5}$ *Multiply the exponents 2 and 5.*
$ = a^{10}$

2 **Simplify $(x^2y)(x^3)^2$.**

$(x^2y)(x^3)^2 = (x^2y)(x^6)$ *Power of a power rule.*
$ = (x^2 \cdot x^6)(y)$
$ = x^8y$ *Product of a power rule.*

Suppose xy is raised to the third power.

$(xy)^3 = (xy)(xy)(xy)$
$ = (x \cdot x \cdot x)(y \cdot y \cdot y)$ *Commutative and associative properties*
$ = x^3y^3$

This and other similar examples suggest the following rule.

Power of a Product

> For all numbers a and b and all positive integers m,
> $$(ab)^m = a^m b^m.$$

Examples

3 Simplify $(2ab)^3$.

$$(2ab)^3 = 2^3 \cdot a^3 \cdot b^3 \qquad \text{\textit{Power of a product rule.}}$$
$$= 8a^3b^3$$

4 Simplify $(-4x)^2(xy)$.

$$(-4x)^2(xy) = (-4)^2(x^2)(xy)$$
$$= 16(x^2 \cdot x)(y)$$
$$= 16x^3y$$

5 Express the volume measure of the cube in simplest form.

$$V = (3x^2y^4)^3$$
$$= (3)^3 \cdot (x^2)^3 \cdot (y^4)^3$$
$$= 27x^6y^{12}$$

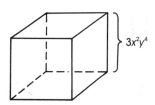

$3x^2y^4$

Exercises

Exploratory   State whether each of the following is *true* or *false*.

1. $(2^2)^3 = 2^6$
2. $3^3 \cdot 3^2 = 3^6$
3. $(5^3)^4 = 5^7$
4. $(5 \cdot 7)^2 = 5^2 \cdot 7^2$
5. $x^2 + x^2 = x^4$
6. $(3x^2)^3 = 3x^6$
7. $(m^2)^4 = m^8$
8. $(4x)^3 = 64x^3$
9. $(x^5y^6)^3 = x^8y^9$
10. $(-pq)^2 = p^2q^2$
11. $(n^3r^2)^4 = n^{12}r^8$
12. $(mn^3)^3 = mn^9$

Written   Simplify.

1. $(3^2)^3$
2. $(a^2)^4$
3. $(m^3)^3$
4. $(2y)^2$
5. $(3a)^3$
6. $(4b)^2$
7. $(ab)^2$
8. $(xy)^3$
9. $(2c)^5$
10. $(-2x)^2$
11. $(-3y)^3$
12. $(-5xy)^2$
13. $(\frac{1}{2}ab)^3$
14. $(3a^2)^2$
15. $(-2b^3)^4$
16. $(a^2b)^3$
17. $(4xy^2)^2$
18. $(-2xy^2z)^3$
19. $(-6x^3y)^2$
20. $(-2a^2b^3)^4$
21. $(-2x^2yz)^3$
22. $3y(2y)^3$
23. $4x(-3x)^3$
24. $2b^2(2ab)^3$

25. $-3x(4xy)^2$

26. $-3x(2y)^3$

27. $(-2x^2)^3(3xy)$

28. $(xy)^2(x^2)^3$

29. $(a^2)^3(ab^2)^2$

30. $3b^2(2bc)^3$

31. $-2xy^2(-3x)^2$

32. $(4mn^2)^2(m^2n^3)^4$

33. $(-2ab)^2(a^3b^4)^5$

34. $(-3x^2y)^2(3xy^4)^3$

35. $(-2xy^2z)^2(x^2y)^4$

36. $(5xy)^2(-2x^2y)^2$

37. $y(x^2y)^3$

38. $-7b^5(a^2b)^4$

39. $(2a^2b^3)(3a)^2$

40. $(\sqrt{3}x)^2$

41. $(\sqrt{5}x^2y)^2(2xy^3)^3$

42. $(2\sqrt{2}a^2b)^3(ab^2c)^2$

Express the volume measure of each cube in simplest form.

43.

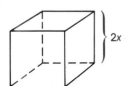

$2x$

44.

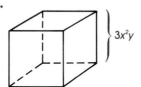

$3x^2y$

45.

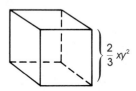

$\frac{2}{3}xy^2$

Challenge Find the value of r which makes each of the following sentences true.

1. $(x^r)^2 = x^8$

2. $(2^{10})^2 = 2^r$

3. $(x^2)^3 = x^r$

4. $(x^r)^3 = x^{12}$

5. $(3^5)^{2r} = 3^{10}$

6. $2^{r+30} = (2^r)^{16}$

7. $(3^{r+2})^2 = 3^{r+8}$

8. $(6^2)^{r+1} = 6^{14}$

9. $(x^3 \cdot x^r)^5 = x^{30}$

Write each of the following as a monomial in simplest form.

10. $(2x^2y)(3x^2y^3) + (xy)^4$

11. $(3y^2)(5y)^3 - (2y)^5$

12. $(-5m)^2(-2m)^3 - \left(\frac{1}{2}m\right)^2(2m)^3$

13. $(-4r)^3\left(\frac{1}{2}rs\right)^2 + \left(\frac{1}{2}r\right)^5(-2s)^2$

Mixed Review

Simplify each of the following. Assume all variables represent positive numbers.

1. $\sqrt{121}$

2. $-\sqrt{(64x)^2}$

3. $\sqrt{54}$

4. $\sqrt{500y^3}$

5. $(2\sqrt{15})(5\sqrt{24})$

6. $\sqrt{24} + 3\sqrt{6} - 5\sqrt{54}$

7. $5\sqrt{2}(\sqrt{18} + 3\sqrt{24})$

8. Find the area of a parallelogram with a base of 3 cm and a height of 7 cm.

9. Find the height of a triangle with an area of 72 cm² and a base of 16 cm.

10. Find the area of a trapezoid with bases of 5 m and 7 m and a height of 3 m.

11. Find the exact circumference of a circle with a diameter of 15 cm.

12. Find the exact length of the radius of a circle with an area of 80π mm².

13. Find the exact volume of a sphere with a radius of 7 in.

14. The lengths of the sides of a triangle are 4 cm, $2\sqrt{13}$ cm, and 6 cm. Is the triangle a right triangle?

15. The congruent sides of an isosceles triangle are each 26 cm long. Find the area of the triangle if its base is 24 cm long.

9.4   Polynomials

Every monomial is also a polynomial.

Expressions such as $5x + 1$, $x^2y + x$, and $5 - 3y + 6y^2$ are called **polynomials** (*pol-e-no-*me-als). A polynomial is the sum or difference of two or more monomials. The expression $x^2 + \dfrac{2}{x}$ is *not* a polynomial since $\dfrac{2}{x}$ is not a monomial.

Each monomial in a polynomial is called a term of the polynomial. For example, the polynomial $5 - 3y + 6y^2$ has three unlike terms; 5, $-3y$, and $6y^2$. A polynomial with two unlike terms is called a *binomial*. A polynomial with three unlike terms is called a *trinomial*.

A polynomial is in *simplest form* if it contains no like terms.

Examples

1 Simplify $2xy + 5 - 6xy$.

$$\begin{aligned} 2xy + 5 - 6xy &= 2xy - 6xy + 5 \qquad &\text{\textit{Commutative property for addition}} \\ &= (2 - 6)xy + 5 \qquad &\text{\textit{Use the distributive property.}} \\ &= -4xy + 5 \end{aligned}$$

2 Simplify $-3x^2y + xy + x^2y - 7xy$.

$$\begin{aligned} -3x^2y + xy + x^2y - 7xy &= -3x^2y + x^2y + xy - 7xy \\ &= -2x^2y - 6xy \end{aligned}$$

The degree of a polynomial is the same as that of the term of greatest degree.

Example

3 Find the degree of $5x^3y + 2xy - 2x^3y + 3$.

First combine like terms. $5x^3y + 2xy - 2x^3y + 3 = 3x^3y + 2xy + 3$

The degree of $3x^3y$ is 4. The degree of $2xy$ is 2. The degree of 3 is 0. Thus, the degree of $5x^3y + 2xy - 2x^3y + 3$ is 4.

Polynomials such as r^2, $s + 1$, and $3y + 2y^2 - 7$ are examples of *polynomials in one variable*. The terms of a polynomial in one variable are usually arranged so that the powers of the variable are in *descending* order. Such polynomials are said to be in **standard form.** The polynomial $3y^4 - 9y^3 + 2y - 6$ is in standard form. The polynomial $2a^2 + 8 - a$ is *not* in standard form. *Why?*

Exercises

Exploratory State whether each of the following is a polynomial. If it is, identify it as a *monomial, binomial,* or *trinomial.*

1. $5m$
2. $2x + 1$
3. $x^2 + 5$
4. $-5x^2y$

5. $r^2 + 6r + 1$
6. $x^3 - y^3$
7. $4 + 3y - 8y^3$
8. $\sqrt{x} - 5$

9. $\dfrac{3}{x}$
10. $\dfrac{y}{6}$
11. $\dfrac{3k}{2} + \dfrac{4k^2}{5}$
12. $\dfrac{a + 6}{b + 5}$

13. $9x^3y^4z^5 + \dfrac{1}{2}$
14. $\dfrac{6}{x + y}$
15. $\dfrac{4ab}{c} - \dfrac{2d}{x}$

16. -17
17. $5ab^2 - 2a + b^2$
18. $15n^2 + 3nt$

Written Write the degree of each polynomial.

1. $3y^2$
2. $-5x^3yz^2$
3. $2x + 1$
4. $14x + 3y$

5. $31y - y^4$
6. $4xy + 6x$
7. $4xy^2 - 3x^5$
8. $5abc + 2a^2b + 6ab^2$

Write each polynomial in standard form and then state its degree.

9. $2x + 5x^2$
10. $r^2 - 4 + 6r$
11. $-3 + y$

12. $-8 + m^2 - 3m$
13. $-6y - 2y^4 + 4 - 8y^2$
14. $-5 + 2x^4 + 3x$

15. $1 + 2x^2 + 3x^3 - x$
16. $3 - 6x + 7x^3 - 5x^2$
17. $17x^2 + 29x^3 + 5$

Simplify each polynomial, if possible, and then state its degree.

18. $3y + y + 5$
19. $7x + 2x - 5 + 3$
20. $6a + 2a + 3b + 5b$

21. $7m - 4m + 8n + 11n$
22. $6y + y + 4y^2$
23. $3y + 4y + 6y^2 - 3y^2$

24. $8xy - 6xy + 4xy$
25. $6a + b + 2a$
26. $x^2 + y + 4x^2$

27. $y^2 + y + 5y^2$
28. $4ab^2 - 2ab + 6ab^2$
29. $3c + 2b + 5c + 6b$

30. $4a + 2b - 5a + b$
31. $5b^2 - 3b^3 + 2b^2 + b^3$

32. $x^3 + y^2 + 6xy + 3x^3$
33. $-4xy + 2y + 4xy + 2y$

34. $2(x + 3y) + 3(2x + 4y)$
35. $-3(x^2 + y) + 4(2x^2 - y)$

Evaluate each polynomial if $a = -1$, $b = 2$, and $c = -3$.

36. $a^3 - 2ab$
37. $b^3 - 2ac$
38. $4a^5 - 2ab^2$

39. $2a^3 - 3a^2 + 7$
40. $-3a^7 - 4b^2 + c^3$
41. $4a^4b^4c + 3b^3 - 7c$

9.5   Adding and Subtracting Polynomials

You have used several properties to simplify polynomials. Likewise, you can use the same properties to add polynomials.

Examples

1 Add $(2x + 3y)$ and $(-6x + 5y)$.

$(2x + 3y) + (-6x + 5y) = (2x - 6x) + (3y + 5y)$ *Associative and commutative properties for addition*

$= (2 - 6)x + (3 + 5)y$ *Distributive property*

$= -4x + 8y$

2 Find the sum of $(2x^2 + 5x - 3)$ and $(x^2 - 7x - 6)$.

$(2x^2 + 5x - 3) + (x^2 - 7x - 6) = (2x^2 + x^2) + (5x - 7x) + (-3 - 6)$

$= (2 + 1)x^2 + (5 - 7)x + (-9)$

$= 3x^2 - 2x - 9$

3 Find the sum of $(3x^2y + 5x^2 + y^2)$ and $(3y^2 + 5x^2y - 7x^2)$.

Sometimes it is helpful to arrange the like terms in vertical columns and then add.

$$
\begin{array}{r}
3x^2y + 5x^2 + y^2 \\
5x^2y - 7x^2 + 3y^2 \\
\hline
8x^2y - 2x^2 + 4y^2
\end{array}
$$

To subtract a number from a number, you add its *additive inverse*.

$$\overbrace{}^{\text{additive inverses}}$$

$9 - 3 = 6 \qquad 9 + (-3) = 6$

$$\underbrace{}_{\text{same result}}$$

This same method is used to subtract polynomials.

You can find the additive inverse of a number by multiplying the number by -1. For example, the additive inverse of 3 is $(-1)3$ or -3. You can find the additive inverse of a polynomial in the same way.

Polynomial	Multiply by -1	Additive Inverse
$-y$	$-1(-y)$	y
$4x - 3$	$-1(4x - 3)$	$-4x + 3$
$2xy + 5y$	$-1(2xy + 5y)$	$-2xy - 5y$
$2x^2 - 3x - 5$	$-1(2x^2 - 3x - 5)$	$-2x^2 + 3x + 5$

Examples

4 Find $(3x - 5y) - (6x - 8y)$.

Add the additive inverse of $6x - 8y$.

$$
\begin{aligned}
(3x - 5y) - (6x - 8y) &= (3x - 5y) + (-1)(6x - 8y) \\
&= 3x - 5y - 6x + 8y \\
&= 3x - 6x - 5y + 8y \\
&= -3x + 3y
\end{aligned}
$$

5 Find $(3x^2 - 6xy + 4y^2) - (x^2 - 4xy)$.

$$
\begin{aligned}
(3x^2 - 6xy + 4y^2) - (x^2 - 4xy) &= (3x^2 - 6xy + 4y^2) + (-x^2 + 4xy) \\
&= 3x^2 - x^2 - 6xy + 4xy + 4y^2 \\
&= 2x^2 - 2xy + 4y^2
\end{aligned}
$$

6 Find $(7xy^2 - 10y^2 + 6x^3y - y) - (y^2 + 2xy^2 - 5y)$.

Subtraction also can be performed in column form. *Add the additive inverse of each term of the second polynomial.*

$$
\begin{array}{r}
7xy^2 - 10y^2 + 6x^3y - y \\
-(2xy^2 + y^2 - 5y) \\
\hline
\end{array}
\quad \Longrightarrow \quad
\begin{array}{r}
7xy^2 - 10y^2 + 6x^3y - y \\
-2xy^2 - y^2 + 5y \\
\hline
5xy^2 - 11y^2 + 6x^3y + 4y
\end{array}
$$

Exercises

Exploratory   State the additive inverse of each polynomial.

1. a

2. $-3x$

3. $7a + 6b$

4. $4x^2 - 5$

5. $-8m + 7n$

6. $-11x - 13y$

7. $x^2 - 2xy + y^2$

8. $-6x - 11 + 4y^2$

9. $ab - 5a^2 + b$

10. $-x^3 - x - 1$

11. $-6x^2 - 2xy + 12$

12. $-x^3 + 5xy^2 - \sqrt{3}$

Written   Find each sum.

1. $(5x - 7y) + (6x + 8y)$
2. $(7n + 2t) + (4n - 3t)$
3. $(7x - 2y) + (9x + 4y)$
4. $(17n - 5m) + (11n - 3m)$
5. $(11xy - 4yz) + (9xy + 4yz)$
6. $(-12y - 6y^2) + (-7y + 6y^2)$
7. $(-2y^2 - 4y + 7) + (2y^2 + 4y - 7)$
8. $(3y^2 + 5y - 12) + (-3y^2 - 5y + 12)$
9. $(3a^2 - 7a - 2) + (5a^2 - 3a - 17)$
10. $(5x^2 + 2x - 3) + (x^2 - 7)$
11. $(5a - c + 2b) + (-9a + 7b - 18c)$
12. $(7t + 4m + a) + (-3a - 7t + m)$
13. $(-3x + 5y - 10z) + (-6x + 8z - 6y)$
14. $(3a - 2b + 3c) + (-a - 3b - 4c)$
15. $-4(a + 2b) + 7(a + b)$
16. $4(a^2 - b^2) + 3(a^2 + b^2)$
17. $(x^2 - x - 2) + (2x - 7x^2 + 4) + (6x^2 + x - 7)$
18. $(2 - a^2 + 2a) + (-5 + 2a^2 - 3a) + (5a^2 - 3a)$
19. $3(x - 3 - 5c) + 2(x - a - 4c) + 4(a - 5c - 7)$

20. $2x + 3$
    $\underline{4x - 7}$

21. $5x + 5y$
    $\underline{6x - 8y}$

22. $4a + 5b - c$
    $\underline{8a - 6b + c}$

23. $-6b^2 + 13b - 5$
    $\underline{ 2b^2 - 10}$

24. $-21y^2 + 11y - 32$
    $\underline{18y^2 - 8y + 10}$

25. $-9ax^3 - 5ax^2 + 6ax$
    $\underline{-3ax^3 - 6ax^2 - 7ax}$

Find each difference.

26. $(3x + 5y) - (9x + y)$
27. $(7m + 3n) - (3m + 5n)$
28. $(-3x - 8y) - (8x + 3y)$
29. $(-5x - 6x^2) - (4x + 5x^2)$
30. $(-7x + 3y) - (-4x - 8y)$
31. $(-5r + 2s) - (-3r - 2s)$
32. $(5r - 3s) - (7r + 5s)$
33. $(8x - 7z) - (9x - 3z)$
34. $(11x^2 + 5x) - (7x^2 + 3)$
35. $(4a^2 + 17b) - (2a + 7b)$
36. $(3a^2 - 5d + 17) - (-a^2 + 5d - 3)$
37. $(15t^2 + 8t) - (19t^2 + 7)$
38. $(3z^2 + 5z - 4) - (3z^2 + 5z - 4)$
39. $(7h^2 - 3hk + 5k^2) - (7h^2 - 3hk + 5k^2)$
40. $(4x + 11) - (3x^2 + 7x - 3)$
41. $(10n^2 - 3nt + 4t^2) - (3n^2 + 5nt)$
42. $(16m^4n^2 - 3m^2n + 12) - (4m^4n^2 - 9)$
43. $(-1 + x^3y - 5x^2y^2) - (xy^3 - 5x^2y^2 - 7 + 4x^3y)$
44. $(x^3 - 3x^2y + 4xy^2 + y^3) - (7x^3 + x^2y - 9xy^2 + y^3)$

45. $5y^2 + 7y + 9$
    $\underline{2y^2 + 3y + 6}$

46. $7m^2 - 6m + 13$
    $\underline{7m^2 + 3m - 5}$

47. $a + 3b - c$
    $\underline{2a + b + 7c}$

48. $3a^2 + 5ab$
    $\underline{ - 2ab + b^2}$

49. $11m^2n^2 + 4mn - 6$
    $\underline{5m^2n^2 - 6mn + 17}$

50. $7z^2 + 4$
    $\underline{3z^2 + 2z - 6}$

Express the perimeter measure of each figure in simplest form.

51.
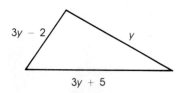

52.

53.
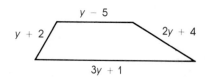

A rectangle has the given perimeter and width. Express the length as a polynomial in simplest form.

54. perimeter: $(12y - 4)$ cm
width: $(y + 2)$ cm

55. perimeter: $(16y - 2)$ m
width: $(2y - 4)$ m

56. perimeter: $(15x - 10)$ m
width: $(3x + 8)$ m

Challenge Solve each of the following.

1. $(2y + 4) - (y + 6) = 13$

2. $(-3x - 2) - (x + 5) = 9$

3. $(10n + 4) - (8n - 3) < 8$

4. $(-4t - 7) - (t + 1) \geq t$

Mathematical Excursions

The Pythagorean Theorem

The Pythagorean Theorem says that for any right triangle, the square of the measure of the hypotenuse equals the sum of the squares of the measures of the legs.

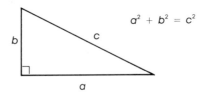

$$a^2 + b^2 = c^2$$

The following geometric proof can be given for the Pythagorean Theorem. Consider the two large congruent squares shown below.

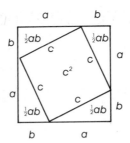

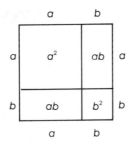

To find the area of each square, add the areas of the figures which compose it.

$$A = c^2 + \tfrac{1}{2}ab + \tfrac{1}{2}ab + \tfrac{1}{2}ab + \tfrac{1}{2}ab$$
$$= c^2 + 4(\tfrac{1}{2}ab)$$
$$= c^2 + 2ab$$

$$A = a^2 + ab + ab + b^2$$
$$= a^2 + 2ab + b^2$$

Since the squares are congruent, their areas are the same.

$$c^2 + 2ab = a^2 + 2ab + b^2$$
$$c^2 = a^2 + b^2 \qquad \text{Subtract } 2ab \text{ from both sides.}$$

Notice that this is the Pythagorean Theorem.

9.6   Multiplying a Polynomial by a Monomial

The measure of the area of the rectangle shown on the right is the product of $3x$ and $2x + 4$ or $3x(2x + 4)$.

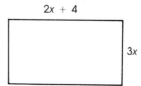

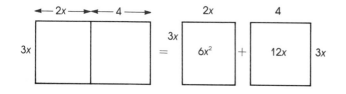

Suppose the original rectangle is separated into two smaller rectangles. The measure of the area of the large rectangle is equal to the sum of the measures of the areas of the two smaller rectangles.

$$3x(2x + 4) = (3x)(2x) + (3x)(4)$$
$$= 6x^2 + 12x$$

The same result can be obtained by using the distributive property.

$$3x(2x + 4) = 3x(2x) + 3x(4)$$
$$= 6x^2 + 12x$$

Examples

1 **Find $3y(2x + 6)$.**

$3y(2x + 6) = 3y(2x) + 3y(6)$ *Use the distributive property.*
$\qquad\qquad = 6xy + 18y$

2 **Find $-6r(r^2 - 3r + 2)$.**

$-6r(r^2 - 3r + 2) = -6r(r^2) - (-6r)(3r) + (-6r)(2)$
$\qquad\qquad\qquad = -6r^3 - (-18r^2) + (-12r)$
$\qquad\qquad\qquad = -6r^3 + 18r^2 - 12r$

3 **Find $2m^2n(5mn - 3m^3n^2 + 4mn^4)$.**

$2m^2n(5mn - 3m^3n^2 + 4mn^4) = 2m^2n(5mn) - 2m^2n(3m^3n^2) + 2m^2n(4mn^4)$
$\qquad\qquad\qquad\qquad\qquad = 10m^3n^2 - 6m^5n^3 + 8m^3n^5$

Lab Activity

You can learn how to use algebra tiles to multiply a polynomial by a monomial in Lab 8 on page A11.

Exercises

Exploratory   Find each product.

1. $3(4x)$
2. $-3x(12x^2)$
3. $(-10r^3)(-4r^5)$
4. $(a^2b)(-ab^2)$
5. $(-25m^4)(8am)$
6. $(7x^2y)(9x^3y^2)$
7. $2(y - 3)$
8. $3(4 + x)$
9. $-6(2a + 3b)$
10. $-2(5x + 2y)$
11. $-3(x - y)$
12. $-3(3x - 4w)$

Written   Find each product.

1. $4a(3a^2b)$
2. $4a(3a^2 + b)$
3. $x^2(x - 5)$
4. $2r(s - r)$
5. $4x^2(2x - 5xy)$
6. $-2x(3x - 5y)$
7. $-7a^2(a^3 - ab)$
8. $4f(gf^2 - 6h)$
9. $(3 - x)5x$
10. $11(-4d^2 + 6d + 12)$
11. $x(7x^3 + 21x - 13)$
12. $a^2(-4a^4 + 3a^2 - 9)$
13. $(x^2 - 3x - 4)(2x)$
14. $(y^3 - 4xy + y)(3y)$
15. $\frac{2}{3}(12x - 9)$
16. $3a^3b^2(-2ab^2 + 4a^2b - 7a)$
17. $-5mn^2(-3m^2n + 6m^3n - 3m^4n^4)$
18. $-4rs(2r^2s + 3s - r^5)$
19. $(3c^2 + 2cd^2 + 4d^2)10c^2d$
20. $4ax^2(-9a^3x^2 + 8a^2x^3 + 6a^4x^4)$
21. $17b^3d^2(-4b^2d^2 - 11b^3d^3 - 5bd^4)$

Express the area of each of the following using a polynomial in simplest form.

22. a rectangle with a length of $(4t + 7)$ units and a width of $3t$ units
23. a triangle with a base of $(x + 11)$ units and a height of x^2 units
24. a parallelogram with a base of $(4x^2 + 7x + 3)$ units and a height of $4x$ units
25. a trapezoid with bases of $(6x^2 + x)$ and $(2x^3 + 4x^2 + 5)$ units and a height of $3x^2$ units

Express the area of each of the four rectangles in each figure using a polynomial in simplest form.

26.

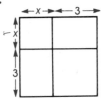

27.

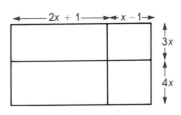

Solve each of the following.

28. $3t + 5(6 - t) = 4$
29. $2y - 1 - 3(y + 6) = 4$
30. $4(5 - 2y) - 3(6 + 2y) = 16$
31. $3(2x - 2) + 2(x - 5) = 4(x + 2)$
32. $x(x + 5) = x(x + 6) + 2$
33. $y(y + 3) = y(y - 6) + 3$

Solve each of the following. Then graph each solution set on a number line.

34. $2x - 3(x - 5) < 4x - 5$
35. $2(3z - 5) \geq 7(4z + 2)$
36. $2x(x - 6) \geq 2x(x - 7) - 2(x + 3)$

9.7   Multiplying Binomials

The measure of the area of the rectangle shown on the right is the product of $(x + 3)$ and $(x + 2)$ or $(x + 3)(x + 2)$.

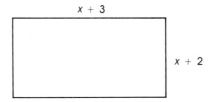

Suppose the original rectangle is separated into four smaller rectangles. The measure of the area of the original rectangle is equal to the sum of the measures of the areas of the four smaller rectangles.

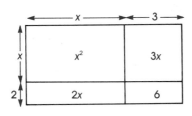

$$(x + 3)(x + 2) = x^2 + 3x + 2x + 6$$
$$= x^2 + 5x + 6 \quad \textit{Combine like terms.}$$

The same result can be obtained by using the distributive property.

$$(x + 3)(x + 2) = x(x + 2) + 3(x + 2)$$
$$= (x^2 + 2x) + (3x + 6)$$
$$= x^2 + 5x + 6$$

Lab Activity

You can learn how to use algebra tiles to multiply polynomials in Lab 9 on pages A12- A13.

Examples

1 Find $(x + 6)(x + 3)$.

$$\begin{aligned}(x + 6)(x + 3) &= x(x + 3) + 6(x + 3) & \textit{Use the distributive property.}\\ &= (x^2 + 3x) + (6x + 18) & \textit{Use the distributive property again.}\\ &= x^2 + 9x + 18 & \textit{Combine like terms.}\end{aligned}$$

2 Find $(y - 10)(y + 7)$.

$$\begin{aligned}(y - 10)(y + 7) &= y(y + 7) - 10(y + 7)\\ &= (y^2 + 7y) - (10y + 70)\\ &= y^2 + 7y - 10y - 70\\ &= y^2 - 3y - 70\end{aligned}$$

Consider the product of $(2x + 3)$ and $(x + 5)$.

$$
\begin{aligned}
(2x + 3)(x + 5) &= 2x(x + 5) + 3(x + 5) \\
&= 2x^2 + 10x + 3x + 15 \\
&= 2x^2 + 13x + 15
\end{aligned}
$$

The following shortcut can be used to multiply binomials.

First   *Last*

$$(2x + 3)(x + 5) = \underbrace{2x \cdot x}_{First} + \underbrace{2x \cdot 5}_{Outer} + \underbrace{3 \cdot x}_{Inner} + \underbrace{3 \cdot 5}_{Last}$$

Inner

Outer

F   Multiply the *first* terms of each binomial.
O   Multiply the *outer* terms.
I   Multiply the *inner* terms.
L   Multiply the *last* terms.

This process is called the **FOIL** (**F**irst, **O**uter, **I**nner, **L**ast) method for multiplying binomials.

Examples

3 Find $(x + 6)(x - 3)$.

$$
\begin{aligned}
&\qquad\quad\;\; \text{F} \qquad\; \text{O} \qquad\;\; \text{I} \qquad\; \text{L} \\
(x + 6)(x - 3) &= x \cdot x + x(-3) + 6 \cdot x + 6(-3) \\
&= x^2 - 3x + 6x - 18 \\
&= x^2 + 3x - 18
\end{aligned}
$$

4 Find $(3x - 1)(5x - 4)$.

$$
\begin{aligned}
&\qquad\quad\;\;\;\; \text{F} \qquad\;\; \text{O} \qquad\quad\; \text{I} \qquad\quad\; \text{L} \\
(3x - 1)(5x - 4) &= 3x \cdot 5x + 3x(-4) + (-1)(5x) + (-1)(-4) \\
&= 15x^2 - 12x - 5x + 4 \\
&= 15x^2 - 17x + 4
\end{aligned}
$$

5 Solve $(x + 2)(x + 3) = x^2 + 46$.

$$
\begin{aligned}
(x + 2)(x + 3) &= x^2 + 46 \\
x^2 + 3x + 2x + 6 &= x^2 + 46 \qquad && \textit{Multiply } (x + 2) \textit{ and } (x + 3). \\
x^2 + 5x + 6 &= x^2 + 46 \qquad && \textit{Combine like terms.} \\
5x + 6 &= 46 \qquad && \textit{Subtract } x^2 \textit{ from both sides.} \\
5x &= 40 \qquad && \textit{Subtract 6 from both sides.} \\
x &= 8 \qquad && \textit{Divide both sides by 5.}
\end{aligned}
$$

Exercises

Exploratory   Copy and complete each of the following.
1. $(x + 7)(x + 5) = $ ____$(x + 5) + $ ____$(x + 5)$
2. $(2y + 3)(y + 2) = $ ____$(y + 2) + $ ____$(y + 2)$
3. $(a + 6)(a - 7) = $ ____$(a - 7) + $ ____$(a - 7)$
4. $(x + 1)(3x - 4) = x($____$) + 1($____$)$
5. $(5y - 3)(y + 7) = 5y($____$) - 3($____$)$
6. $(7a - 9)(5a - 3) = 7a($____$) - 9($____$)$
7. $(2n + m)(n - 3m) = 2n($____$) + m($____$)$

Find the area measure of each of the following rectangles by adding the area measures of the smaller rectangles.

8.

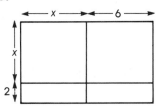

9.

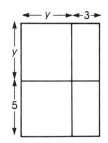

10.

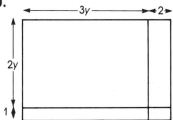

Written   Find each product.
1. $(x + 7)(x + 2)$
2. $(y + 2)(y + 3)$
3. $(m + 6)(m + 9)$
4. $(t - 5)(t + 4)$
5. $(m - 7)(m + 5)$
6. $(v - 4)(v - 7)$
7. $(x - 6)(x - 5)$
8. $(y + 9)(y - 7)$
9. $(2a + 1)(a + 3)$
10. $(2b + 3)(b + 1)$
11. $(3y - 1)(2y + 3)$
12. $(m - 6)(m - 6)$
13. $(a - 9)(a - 9)$
14. $(x - y)(x - y)$
15. $(2a - b)(2a - b)$
16. $(2x - 5)(2x + 1)$
17. $(5x + 3)(2x + 1)$
18. $(2y - 5)(3y - 1)$
19. $(3t - 5)(2t - 7)$
20. $(6a - 5)(7a - 9)$
21. $(r + 2s)(2r + 3s)$
22. $(2x + 3y)(3x - 5y)$
23. $(x + 3)(x - 3)$
24. $(x - 5)(x + 5)$
25. $(2x + 7)(2x - 7)$
26. $(x + y)(x - y)$
27. $(a + 10)(a + 10)$
28. $(n + 8)(n + 8)$
29. $(2x + y)(2x + y)$
30. $(x + y)(x + y)$
31. $(2 + \sqrt{3})(5 - \sqrt{3})$
32. $(\sqrt{2}a + 7)(a - 6)$
33. $(\sqrt{5} - x)(2\sqrt{5} + 3x)$
34. $(6y - \sqrt{2})(y - \sqrt{3})$
35. $(\sqrt{3}b - 2)(5b + \sqrt{3})$
36. $(3x - \sqrt{2})(\sqrt{2}x + 4)$

Solve each of the following.
37. $(s + 3)^2 - s^2 = 15$
38. $(x + 2)(x + 3) = x^2 + 41$
39. $(x - 5)(x - 1) = x^2 + 13$
40. $(t + 2)(t - 3) \geq t(t + 7)$
41. $m(m + 1) = (m - 6)(m + 4)$
42. $(6 + u)(4 - u) = 42 - (u^2 - 4u - 16)$
43. $(x + 5)^2 = x(x + 5)$
44. $(r + 2)^2 + 2r^2 < 3r(r - 4)$
45. $(y + 1)^2 < (y - 6)(y + 4)$
46. $(x - 2)^2 - (x - 2)^2 \geq 2$
47. $(x + 5)^2 - (x + 3)(x - 3) > 4$
48. $6x(x + 1) + 3x^2 + 4 \leq (3x - 4)^2$

9.8 Multiplying Polynomials

The distributive property is used to multiply binomials. In the same manner, it is used to multiply polynomials.

Examples

1 Find $(x + 3)(x^2 - 4x + 5)$.

$$
\begin{aligned}
(x + 3)(x^2 - 4x + 5) &= x(x^2 - 4x + 5) + 3(x^2 - 4x + 5) && \text{Distributive property} \\
&= (x^3 - 4x^2 + 5x) + (3x^2 - 12x + 15) \\
&= x^3 - 4x^2 + 5x + 3x^2 - 12x + 15 \\
&= x^3 - x^2 - 7x + 15
\end{aligned}
$$

2 Find $(2x - 4)(x^3 - 2x^2 - 3x + 1)$.

Multiplication can be performed in vertical form.

$$
\begin{array}{l}
x^3 - 2x^2 - 3x + 1 \\
2x - 4 \\
\hline
2x^4 - 4x^3 - 6x^2 + 2x \qquad \text{\textit{Multiply by 2x.}}\\
 - 4x^3 + 8x^2 + 12x - 4 \qquad \text{\textit{Then multiply by} } -4. \\
\hline
2x^4 - 8x^3 + 2x^2 + 14x - 4 \qquad \text{\textit{Then add vertically.}}
\end{array}
$$

3 Find $(a^3 + 2a^2 - 6)(4a - 3)$.

$$
\begin{array}{l}
a^3 + 2a^2 + 0a - 6 \qquad \text{\textit{Use 0a as a placeholder.}} \\
4a - 3 \\
\hline
4a^4 + 8a^3 + 0a^2 - 24a \\
 - 3a^3 - 6a^2 - 0a + 18 \qquad \text{\textit{How does the use of placeholders}} \\
\hline
4a^4 + 5a^3 - 6a^2 - 24a + 18 \qquad \text{\textit{help in this procedure?}}
\end{array}
$$

4 Solve $(x - 2)(x^2 - 3x + 1) = x(x - 2)(x - 3)$.

$$
\begin{aligned}
(x - 2)(x^2 - 3x + 1) &= x(x - 2)(x - 3) \\
x^3 - 3x^2 + x - 2x^2 + 6x - 2 &= x(x^2 - 5x + 6) && \text{\textit{Multiply.}} \\
x^3 - 5x^2 + 7x - 2 &= x^3 - 5x^2 + 6x && \text{\textit{Simplify on left. Multiply on right.}} \\
7x - 2 &= 6x && \text{\textit{Subtract }} (x^3 - 5x^2) \text{ \textit{from both sides.}} \\
-2 &= -x && \text{\textit{Subtract 7x from both sides.}} \\
2 &= x && \text{\textit{Multiply both sides by }} -1.
\end{aligned}
$$

▰▰▰ Exercises ▰▰▰

Exploratory　Find each product.

1. $5(2a^2 - 7)$
2. $-3(d^2 - 2d)$
3. $a(3a^2 + 2)$
4. $-b(b^3 + 3b^2)$
5. $6k(4k^2 + 7)$
6. $-3n(20n^4 - 6n^2)$
7. $7(a^2 - 2a + 3)$
8. $-9(2k^3 - 3k - 6)$
9. $d^2(12d^3 - 16d^2 + 12d)$
10. $2y(y^2 + 3y - 10)$
11. $3a^2(a^3 + 2a^2 - 6a)$
12. $10c^2d(3c^2 + 2cd^2 + 4d^2)$

Written　Find each product.

1. $(x + 3)(x^2 + 2x + 1)$
2. $(x + 2)(x^2 + 3x + 4)$
3. $(2x + 1)(x^2 + x + 1)$
4. $(2x + 1)^3$
5. $(y - 3)(y^3 + y - 2)$
6. $(x + 2)(x^2 + 3x + 2)$
7. $(y + 1)(y^2 + 2y + 3)$
8. $(m + 3)(m^2 + 2m + 5)$
9. $(x^2 - 5)(x^2 + 6)$
10. $(t - 4)(2t^2 + 3t - 6)$
11. $(2y - 4)(3y^2 - 7y + 5)$
12. $(a + b)(a^2 + 2ab + b^2)$
13. $(a - b)(a^2 - 2ab + b^2)$
14. $(a^2 + a + 1)(2a - 5)$
15. $(x^2 + 2x + 1)(x^2 + 2x + 1)$
16. $(x^4 - x^3 + x^2 + 1)(x + 1)$
17. $(x^4 + x^3 + x^2 + 1)(x - 1)$
18. $(3x^2 - 16x)(3x^2 - 2x)$
19. $(-2xy + 3y^2)(x - 6xy - 5y^2)$
20. $(3x + 2y)(9x^2 + 12xy + 4y^2)$

Simplify.

21. $(y - 3)^2 - 4y^2$
22. $4(2x - 3)(3x - 4) + 7x^2 - 5$
23. $(2y - 1)^3 + (y + 1)^2$
24. $y(2y^2 - 1)^2 - y^2(y + 1)^2$

Solve each of the following.

25. $(x + 2)(x^2 + x - 2) = x(x + 2)(x + 1)$
26. $(y - 4)(y^2 + 3y + 1) = y(y - 6)(y + 5)$
27. $(a + 3)(2a^2 - a + 2) = 2a(a + 1)^2 + a^2$
28. $(y - 3)(y^2 - 2y + 2) = y(y - 1)(y - 1) - 3y^2$
29. $(2x - 5)^2 - (2x - 7)^2 = -40$
30. $(x - 3)^3 - (x + 3)^3 = 9x(1 - 2x)$
31. $(7x - 3)(x + 2) - (3x + 1)(2x - 1) = x^2 + 1$
32. $(a - 6)(a^2 + a + 1) - a(a + 1)(a - 1) = a(2 - 5a)$

33. The product of two consecutive integers is 9 less than the square of the second integer. Find the two integers.

34. The square of an integer is 97 less than the square of a consecutive integer. Find the two integers.

35. The rectangular floor of the City Zoo is 8 meters longer than it is wide. If the length is increased by 4 meters and the width is decreased by 1 meter, the floor area remains the same. Find the dimensions of the floor.

36. The length and width of a rectangular prism are 4 cm longer and 3.2 cm shorter respectively, than an edge of a certain cube. The rectangular prism and the cube have the same height and the same surface areas. Find the length of an edge of the cube.

37. Find three consecutive integers such that the square of the second integer is one more than the product of the first and third integer.

38. Find two consecutive even integers such that when the square of the first integer is subtracted from the square of the second integer the result is 92.

39. The rectangular cafeteria at East High School is to be enlarged. Its new length will be 16 feet more than its present length and its new width will be 10 feet more than its present width. The floor area of the new cafeteria will contain 880 square feet more than the floor area of the present cafeteria. If the present cafeteria is twice as long as it is wide, find the dimensions of the new cafeteria.

40. Baseball and softball "diamonds" are actually squares. In baseball, the distance between bases is 30 feet longer than it is in softball. The area of the baseball diamond is 4500 square feet more than the area of the softball diamond. What is the distance between bases on a softball diamond?

41. The length of a rectangle is 3 cm longer than a side of a square. The width of the rectangle is 2 cm less than a side of the square. If the area of the square is the same as the area of the rectangle, find the dimensions of the rectangle.

42. The length of a rectangle is 3 cm longer than its width. If the length is increased by 2 cm, the area of the new rectangle is 10 sq cm more than the area of the original rectangle. Find the width and length of the original rectangle.

43. The product of two consecutive even integers is 1 less than the square of their average. Find the two integers.

44. A square picture is in a frame 1 inch wide. If the area of the frame is 52 square inches, find the dimensions of the picture.

Mixed Review

Choose the best answer.

1. Which of the following is the simplest form of $3\sqrt{12} + 4\sqrt{12} + \sqrt{18}$?
   a. $8\sqrt{42}$ **b.** $7\sqrt{12} + \sqrt{18}$ **c.** $14\sqrt{3} + 3\sqrt{2}$ **d.** $10\sqrt{5}$

2. What is the height of a trapezoid with an area of 132 cm^2 and bases of 7 cm and 4 cm?
   a. 24 cm   **b.** 12 cm   **c.** 33 cm   **d.** 6 cm

3. What is the volume of a cone with a height of 8 cm and a base with a radius of 15 cm?
   a. 2400π cm^3 **b.** 600π cm^3 **c.** 320π cm^3 **d.** 1800π cm^3

4. Which of the following is the simplest form of $(3ax^2)^2(2a^2x)^3$?
   a. $6a^8x^7$ **b.** $72a^7x^7$ **c.** $72a^8x^7$ **d.** $72a^{10}x^7$

5. Which of the following is the simplest form of $2(x + y) - 5(2x - 6y)$?
   a. $-8x - 28y$ **b.** $-8x + 32y$ **c.** $12x - 28y$ **d.** $-9x + 15y$

9.9   Dividing Monomials

If you add exponents when you multiply, then it seems reasonable to subtract exponents when you divide.

Remember that x ≠ 0 since division by 0 is undefined.

$$\frac{x^5}{x^3} = \frac{x \cdot x \cdot \cancel{x} \cdot \cancel{x} \cdot \cancel{x}}{\cancel{x} \cdot \cancel{x} \cdot \cancel{x}} \qquad \leftarrow 5 \text{ factors}$$
$$\qquad\qquad\qquad\qquad \leftarrow 3 \text{ factors}$$
$$= x \cdot x \qquad \leftarrow 2 \text{ or } (5 - 3) \text{ factors}$$
$$= x^{5-3} \text{ or } x^2$$

Now consider another example.

$$\frac{x^2}{x^6} = \frac{\cancel{x} \cdot \cancel{x}}{x \cdot x \cdot x \cdot x \cdot \cancel{x} \cdot \cancel{x}} \qquad \leftarrow 2 \text{ factors}$$
$$\qquad\qquad\qquad\qquad \leftarrow 6 \text{ factors}$$
$$= \frac{1}{x \cdot x \cdot x \cdot x} \qquad \leftarrow 4 \text{ or } (6 - 2) \text{ factors}$$
$$= \frac{1}{x^{6-2}} \text{ or } \frac{1}{x^4}$$

What happens if the exponents are the same? Try $\frac{3^2}{3^2}$ and $\frac{x^2}{x^2}$ for $x \neq 0$. The quotient in each case is 1.

These and other similar examples suggest the following rules.

Dividing Powers

> For positive integers m and n and nonzero number a,
>
> | if $m > n$ | if $m = n$ | if $n > m$. |
> |---|---|---|
> | $\dfrac{a^m}{a^n} = a^{m-n}$ | $\dfrac{a^m}{a^n} = 1$ | $\dfrac{a^m}{a^n} = \dfrac{1}{a^{n-m}}$ |

Examples

1 Simplify $\dfrac{a^{10}}{a^7}$.

$$\frac{a^{10}}{a^7} = a^{10-7} \qquad \textit{Use the dividing powers rule where } m > n.$$
$$= a^3$$

2 Simplify $\dfrac{y^2}{y^4}$.

$$\frac{y^2}{y^4} = \frac{1}{y^{4-2}} \qquad \textit{Use the dividing powers rule where } n > m.$$
$$= \frac{1}{y^2}$$

Example

3 Simplify $\dfrac{6x^5y^3}{2x^2y^7}$.

$$\frac{6x^5y^3}{2x^2y^7} = \left(\frac{6}{2}\right)\left(\frac{x^5}{x^2}\right)\left(\frac{y^3}{y^7}\right)$$

Note that $\dfrac{ab}{cd} = \dfrac{a \cdot b}{c \cdot d}$ *or* $\left(\dfrac{a}{c}\right)\left(\dfrac{b}{d}\right)$.

$$= 3 \cdot x^{5-2} \cdot \frac{1}{y^{7-3}}$$

$$= 3x^3 \cdot \frac{1}{y^4}$$

$$= \frac{3x^3}{y^4}$$

What is the meaning of a zero exponent? By the dividing powers rule, $\dfrac{x^2}{x^2} = 1$ if $x \neq 0$. Now if we simplify the quotient by subtracting exponents, we get $\dfrac{x^2}{x^2} = x^{2-2}$ or x^0. This example suggests the following definition. 0^0 *is not defined.*

Definition of a^0

> For every nonzero real number a, $a^0 = 1$.

What is the meaning of a negative exponent? By the dividing powers rule, $\dfrac{x^3}{x^7} = \dfrac{1}{x^{7-3}}$ or $\dfrac{1}{x^4}$. Now if we simplify the quotient by subtracting exponents, we get $\dfrac{x^3}{x^7} = x^{3-7}$ or x^{-4}. This example suggests the following definition.

Definition of a^{-n}

> For every nonzero real number a and every positive integer n,
>
> $$a^{-n} = \frac{1}{a^n}.$$

Exercises

Exploratory   Simplify each quotient.

1. $\dfrac{x^6}{x^2}$
2. $\dfrac{y^7}{y^2}$
3. $\dfrac{a^6}{a^4}$
4. $\dfrac{m^6}{m^5}$
5. $\dfrac{x^2}{x^3}$

6. $\dfrac{y^4}{y^5}$
7. $\dfrac{k^9}{k^4}$
8. $\dfrac{r^5}{r^2}$
9. $\dfrac{n^5}{n^5}$
10. $\dfrac{w^2}{w^9}$

Written   Find each quotient.

1. $\dfrac{x^9}{x^4}$
2. $\dfrac{y^8}{y^6}$
3. $\dfrac{an^6}{n^5}$
4. $\dfrac{xy^7}{y^4}$

5. $\dfrac{b^6c^5}{b^3c^2}$
6. $\dfrac{-a^4b^8}{a^4b^7}$
7. $\dfrac{-x^6y^6}{x^3y^4}$
8. $\dfrac{m^5n^2}{mn}$

9. $\dfrac{m^3}{m^7}$
10. $\dfrac{b^4}{b^8}$
11. $\dfrac{an^3}{n^5}$
12. $\dfrac{cm^2}{m^4}$

13. $\dfrac{x^3y^3}{x^3y^6}$
14. $\dfrac{a^4b^2}{a^6b^2}$
15. $\dfrac{48a^8}{12a}$
16. $\dfrac{15b^9}{3b^2}$

17. $\dfrac{4x^3}{28x}$
18. $\dfrac{12b^4}{60b}$
19. $\dfrac{20y^5}{40y^2}$
20. $\dfrac{10m^4}{30m}$

21. $\dfrac{16b^6c^5}{4b^4c^2}$
22. $\dfrac{20n^5m^9}{20nm^7}$
23. $\dfrac{-15r^5s^8}{5r^5s^2}$
24. $\dfrac{-27w^5t^7}{-3w^3t^2}$

25. $\dfrac{-2a^3b^6}{24a^2b^2}$
26. $\dfrac{-9c^4d^5}{-45c^3d^3}$
27. $\dfrac{4y^3t^5}{4y^4t^5}$
28. $\dfrac{a^2b^6}{-8a^3b^7}$

29. $\dfrac{-6x^2y^3z^3}{24x^2y^7z^3}$
30. $\dfrac{8a^3b^6c}{48a^3b^7c^4}$
31. $\dfrac{5ac}{8ab^5c^2}$
32. $\dfrac{2x^5y^3z^3}{8x^3y^7z}$

Express each of the following with positive exponents.

Sample: $2^{-3} \cdot 2^5 = \dfrac{1}{2^3} \cdot 2^5 = \dfrac{2^5}{2^3} = 2^{5-3}$ or 2^2

33. 4^{-6}
34. $\left(\dfrac{1}{y}\right)^{-5}$
35. $\dfrac{1}{r^{-3}}$
36. $\left(\dfrac{2}{b}\right)^{-7}$

37. $\left(\dfrac{y}{x}\right)^{-3}$
38. $(x^3y^2)^{-1}$
39. $(m^4n^5)^{-2}$
40. $\dfrac{5x}{y^{-1}}$

41. $(3.4)^{-2}$
42. $4^{-3} \cdot 4^2$
43. $(4^{-2})^3$
44. $(3^2 + 5^{-4})^0 + 4^{-2}$

Challenge   Solve each of the following for k.

1. $x^{k-12} = (x^2)^{-k-3}$

2. $m^k \cdot m^{-15} = (m^3)^{k+2}$

3. $x^{2k} \cdot x^{3k} = x^{15}$

4. $x^{2k+1} = (x^5)^{k-4}$

Mathematical Excursions

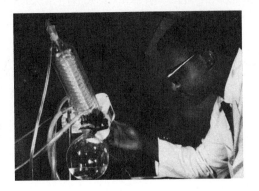

Scientists often work with very large or very small numbers. For example, the distance from the sun to the planet Neptune is about 2,790,000,000 miles, and the weight of one molecule of water is 0.00000000000000000000003 gram. These numbers are easier to compute with when expressed in **scientific notation.**

Definition of Scientific Notation

A number is expressed in scientific notation when it is in the following form.
$a \times 10^n$ where $1 \le a < 10$ and n is an integer

Example Express 2,790,000,000 and 0.00000000000000000000003 in scientific notation.

$$2{,}790{,}000{,}000 = 2.79 \times 10^9$$

9 places

$$0.00000000000000000000003 = 3.0 \times 10^{-23}$$

23 places

Numbers expressed in scientific notation can be multiplied using the commutative and associative properties for multiplication.

Example Find $(2 \times 10^{-5})(4 \times 10^9)$.

$$(2 \times 10^{-5})(4 \times 10^9) = (2)(4)(10^{-5})(10^9)$$
$$= 8 \times 10^4$$

Exercises Express each of the following in scientific notation.

1. 5800
2. 450,000
3. 0.0085
4. 0.00000045
5. 60,000
6. 86,000,000,000
7. 300,000,000
8. 0.00036
9. 0.004
10. 0.00087
11. 93,000,000
12. 0.000000000000166

Express each product in scientific notation.

13. $(2 \times 10^4)(4 \times 10^8)$
14. $(3 \times 10^4)(2 \times 10^6)$
15. $(6 \times 10^{-4})(5 \times 10^{-8})$
16. $(2.5 \times 10^{-7})(2.5 \times 10^{16})$
17. $(4.3 \times 10^3)(2.0 \times 10^{-5})$
18. $(3.4 \times 10^{-4})(5.6 \times 10^{10})$

9.10 Dividing Polynomials

You can divide a polynomial by a monomial by dividing each term of the polynomial by the monomial.

Example

1 Find $\dfrac{3x^2y^6 + 18xy^5 - 24x^2y}{3xy^2}$.

$$\frac{3x^2y^6 + 18xy^5 - 24x^2y}{3xy^2} = \frac{3x^2y^6}{3xy^2} + \frac{18xy^5}{3xy^2} - \frac{24x^2y}{3xy^2}$$

$$= xy^4 + 6y^3 - \frac{8x}{y}$$

To divide a polynomial by a polynomial with more than one term, you can use a procedure similar to long division.

Example

2 **Divide $x^2 + 8x + 15$ by $x + 3$.**

Step 1
Divide x^2 by x. The quotient is x.
Multiply $x + 3$ by x.
Subtract $x^2 + 3x$ from $x^2 + 8x$.

$$\begin{array}{r} x \\ x + 3\overline{)x^2 + 8x + 15} \\ \underline{x^2 + 3x} \\ 5x \end{array}$$

Step 2
Divide $5x$ by x. The quotient is 5.
Multiply $x + 3$ by 5.
Subtract $5x + 15$ from $5x + 15$.

$$\begin{array}{r} x + 5 \\ x + 3\overline{)x^2 + 8x + 15} \\ \underline{x^2 + 3x} \\ 5x + 15 \\ \underline{5x + 15} \\ 0 \end{array}$$

Stop dividing when the remainder is 0 or its degree is less than the degree of $x + 3$.

$$\frac{x^2 + 8x + 15}{x + 3} = x + 5$$

Example

3 Divide $y^3 - 8$ by $y + 2$.

$$
\begin{array}{r}
y^2 - \ \ 2y + 4 \\
y + 2 \overline{)y^3 + \ \ 0y^2 + 0y - 8} \qquad \textit{Use } 0y^2 \textit{ and } 0y \textit{ as placeholders.} \\
\underline{y^3 + \ \ 2y^2} \\
-2y^2 + 0y \\
\underline{-2y^2 - 4y} \\
4y - 8 \\
\underline{4y + 8} \\
-16 \qquad \textit{Remainder}
\end{array}
$$

Thus, $\dfrac{y^3 - 8}{y + 2} = y^2 - 2y + 4$ with remainder -16.

This also can be written $y^2 - 2y + 4 + \left(\dfrac{-16}{y + 2} \right)$.

Exercises

Exploratory   Find each quotient.

1. $\dfrac{4y^2 + 6}{2}$

2. $\dfrac{3b^2 + b}{b}$

3. $\dfrac{8a^2b^4 - 14a^2b^3 + 6ab}{ab}$

4. $\dfrac{6xy^2 - 3xy + 2x^2y}{xy}$

5. $\dfrac{a^3b^2 - a^2b + 2a}{-ab}$

6. $\dfrac{6r^2s^2 + 3rs^2 - 9r^2s}{3rs}$

Written   Find each quotient.

1. $\dfrac{x^2 - 10x - 24}{x + 2}$

2. $\dfrac{y^2 + 3y + 2}{y + 1}$

3. $\dfrac{b^2 + 8b - 20}{b - 2}$

4. $\dfrac{2x^2 + 3x - 2}{2x - 1}$

5. $\dfrac{x^3 + 2x^2 - 5x + 12}{x + 4}$

6. $\dfrac{2x^3 - 5x^2 + 22x + 51}{2x + 3}$

7. $\dfrac{10x^2 + 29x + 21}{5x + 7}$

8. $\dfrac{12n^2 + 36n + 15}{6n + 3}$

9. $\dfrac{y^3 - 7y + 6}{y - 2}$

10. $\dfrac{4m^3 + 5m - 21}{2m - 3}$

11. $\dfrac{4t^3 + 17t^2 - 1}{4t + 1}$

12. $\dfrac{2a^3 + 9a^2 + 5a - 12}{a + 3}$

13. $\dfrac{t^3 - 19t + 9}{t - 4}$

14. $\dfrac{3s^3 + 8s^2 + s - 7}{s + 2}$

15. $\dfrac{9d^3 + 5d - 8}{3d - 2}$

9.11   Operations with Rational Expressions

Rational expressions are also called algebraic fractions.

Rational expressions are fractions whose numerator and denominator are polynomials. The expressions $\dfrac{5x + 3}{y}$, $\dfrac{2}{x}$, and $\dfrac{a - 2}{a^2 + 4}$ are examples of rational expressions.

A fraction indicates division. Zero cannot be used as a denominator because division by zero is undefined. Therefore, any value assigned to a variable that results in a denominator of zero must be excluded from the domain of the variable.

For $\dfrac{5}{x}$, exclude $x = 0$.   For $\dfrac{3x + 7}{x + 4}$, exclude $x = -4$.

Example

1 **For each fraction, state the values of the variables that must be excluded.**

a. $\dfrac{5x}{x + 7}$

b. $\dfrac{7}{-4ab^2(c - 1)}$

Exclude the values for which
$x + 7 = 0$.

$$x + 7 = 0$$

$$x = -7$$

Therefore, x cannot equal -7.

Exclude the values for which
$a = 0$, $b^2 = 0$, and $c - 1 = 0$.

$a = 0$ $b^2 = 0$ $c - 1 = 0$

$b = 0$ $c = 1$

Therefore, a cannot equal 0, b cannot equal 0, and c cannot equal 1.

From this point on, it will be assumed that all replacements for variables in rational expressions that result in denominators equal to zero will be excluded.

To multiply fractions, you multiply the numerators and multiply the denominators.

$$\frac{3}{5} \cdot \frac{4}{7} = \frac{3 \cdot 4}{5 \cdot 7}$$
$$= \frac{12}{35}$$

The same method can be used to multiply rational expressions.

Examples

2 Find $\dfrac{5}{a} \cdot \dfrac{b}{7}$ and simplify.

$$\dfrac{5}{a} \cdot \dfrac{b}{7} = \dfrac{5 \cdot b}{a \cdot 7} \qquad \textit{Multiply the numerators.}$$
$$\textit{Multiply the denominators.}$$
$$= \dfrac{5b}{7a}$$

3 Find $\dfrac{2a^2d}{3bc} \cdot \dfrac{9b^2c}{16ad^2}$ and simplify.

$$\dfrac{2a^2d}{3bc} \cdot \dfrac{9b^2c}{16ad^2} = \dfrac{18a^2b^2cd}{48abcd^2}$$
$$= \dfrac{3ab}{8d} \qquad \textit{Change the fraction to simplest form.}$$

Journal

Complete this sentence: "The most challenging thing I learned in this chapter was ...." Tell why.

To add or subtract fractions with *like* denominators, you simply add or subtract the numerators. Then write this sum or difference over the common denominator.

$$\dfrac{3}{7} + \dfrac{2}{7} = \dfrac{5}{7} \qquad\qquad \dfrac{7}{9} - \dfrac{2}{9} = \dfrac{5}{9}$$

These methods can be used to add or subtract rational expressions.

Example

4 Find $\dfrac{3}{2x} + \dfrac{1}{2x}$ and simplify.

$$\dfrac{3}{2x} + \dfrac{1}{2x} = \dfrac{3 + 1}{2x} \qquad \textit{Since the denominators are the same,}$$
$$\textit{add the numerators.}$$
$$= \dfrac{4}{2x}$$
$$= \dfrac{2}{x}$$

When you add or subtract fractions with *unlike* denominators, you must first rename the fractions so the denominators are alike. Any common denominator could be used. However, the computation is easier if the *least common denominator* (LCD) is used.

Suppose you wish to add $\dfrac{5}{36}$ and $\dfrac{7}{24}$. To find the LCD of 36 and 24, use their prime factorization.

$$36 = 2^2 \cdot 3^2$$
$$24 = 2^3 \cdot 3$$

The greatest number of times 3 appears is twice. The greatest number of times 2 appears is three times.

The least common denominator of 36 and 24 contains each prime factor the greatest number of times that it appears.

$$\text{LCD} = 2^3 \cdot 3^2 \quad \text{or} \quad 72$$

Now, rename the fractions using the LCD. Then add.

$$\frac{5}{36} + \frac{7}{24} = \frac{10}{72} + \frac{21}{72} \qquad \frac{5}{36} = \frac{5 \cdot 2}{36 \cdot 2} = \frac{10}{72} \qquad \frac{7}{24} = \frac{7 \cdot 3}{24 \cdot 3} = \frac{21}{72}$$
$$= \frac{31}{72}$$

This method can be used to add or subtract rational expressions with unlike denominators.

Example

5 Find $\dfrac{7x}{15y^2} - \dfrac{y + 1}{18xy}$.

First find the LCD of $15y^2$ and $18xy$.

$$15y^2 = 3 \cdot 5 \cdot y \cdot y \qquad 18xy = 2 \cdot 3 \cdot 3 \cdot x \cdot y$$

The LCD is $2 \cdot 3^2 \cdot 5 \cdot x \cdot y^2$ or $90xy^2$.   *Why?*

$$\frac{7x}{15y^2} - \frac{y + 1}{18xy} = \frac{7x \cdot 6x}{15y^2 \cdot 6x} - \frac{(y + 1) \cdot 5y}{18xy \cdot 5y} \qquad \textit{Why do you multiply } \frac{7x}{15y^2} \textit{ by } \frac{6x}{6x}$$
$$= \frac{42x^2}{90xy^2} - \frac{5y^2 + 5y}{90xy^2} \qquad \textit{and } \frac{y + 1}{18xy} \textit{ by } \frac{5y}{5y}?$$
$$= \frac{42x^2 - (5y^2 + 5y)}{90xy^2}$$
$$= \frac{42x^2 - 5y^2 - 5y}{90xy^2}$$

Exercises

Exploratory For each rational expression, state the excluded values of the variables.

1. $\dfrac{3}{2x}$ 2. $\dfrac{7y}{5a^2}$ 3. $\dfrac{-9}{t+3}$ 4. $\dfrac{m-5}{m-4}$ 5. $\dfrac{xy^2}{2z-11}$ 6. $\dfrac{-2c-1}{3d^2(c+1)}$

Simplify each rational expression.

7. $\dfrac{42y}{18xy}$ 8. $\dfrac{42y^3x}{18y^7}$ 9. $\dfrac{38a^2}{42ab}$ 10. $\dfrac{-3x^2y^5}{18x^5y^2}$

11. $\dfrac{(-2x^2y)^3}{4x^5y}$ 12. $\dfrac{a^3b^2}{(-ab)^3}$ 13. $\dfrac{(2xy)^4}{(x^2y)^2}$ 14. $\dfrac{(-3t^2u)^3}{(6tu^2)^2}$

Written Find each product and simplify.

$i\ 20n^3x^4$

1. $\dfrac{ab}{ac} \cdot \dfrac{c}{d}$ 2. $\dfrac{a^2b}{b^2c} \cdot \dfrac{c}{d}$ 3. $\dfrac{6a^2n}{8n^2} \cdot \dfrac{12n}{9a}$ 4. $\dfrac{10n^3}{6x^3} \cdot \dfrac{12n^2x^4}{25n^2x^2}$

5. $\dfrac{3ab}{4ac} \cdot \dfrac{a-1}{3b^2}$ 6. $\dfrac{2c+3}{21c} \cdot \dfrac{14c^2}{18a^2}$ 7. $\dfrac{6m^3n}{10a^2} \cdot \dfrac{4a^2m}{9n^3}$ 8. $\dfrac{7xy^3}{11z^2} \cdot \dfrac{44z^3}{21x^2y}$

9. $\dfrac{(cd)^3}{a} \cdot \dfrac{ax^2}{xc^2d}$ 10. $\dfrac{(3a)^3}{18b^3} \cdot \dfrac{12a^4b^5}{(3a)^2}$ 11. $\dfrac{8}{m^2}\left(\dfrac{m^2}{2c}\right)^2$ 12. $\left(\dfrac{2a}{b}\right)^2 \dfrac{5c}{6a}$

13. $\dfrac{y-3}{14} \cdot \dfrac{7(y+2)}{y}$ 14. $\dfrac{2r-3}{8p} \cdot \dfrac{r+4}{q}$ 15. $\dfrac{3k+9}{k} \cdot \dfrac{k^2}{3}$

16. $\dfrac{2a-4b}{3a^2} \cdot \dfrac{a+b}{2b^2}$ 17. $\left(\dfrac{r}{3s}\right)^2 \cdot \dfrac{r-s}{2r^3}$ 18. $\dfrac{(x-y)^2}{2x^2} \cdot \dfrac{4ax}{y^2}$

Find each sum or difference and simplify.

19. $\dfrac{y}{2} + \dfrac{y}{2}$ 20. $\dfrac{a}{12} + \dfrac{5a}{12}$ 21. $\dfrac{5x}{24} - \dfrac{3x-8}{24}$ 22. $\dfrac{7t}{p} - \dfrac{8t}{p}$

23. $\dfrac{3}{a} + \dfrac{6}{a}$ 24. $\dfrac{2y}{b} - \dfrac{y+3}{b}$ 25. $\dfrac{3}{x} - \dfrac{7-x}{x}$ 26. $\dfrac{8}{x} + \dfrac{2x-5}{x}$

27. $\dfrac{8}{11} + \dfrac{y}{4}$ 28. $\dfrac{t}{3} + \dfrac{2t}{7}$ 29. $\dfrac{x}{7} - \dfrac{4}{5}$ 30. $\dfrac{2n}{5} - \dfrac{3m}{4}$

31. $\dfrac{5}{2a} + \dfrac{-3}{6a}$ 32. $\dfrac{6}{5} + \dfrac{a}{5b}$ 33. $\dfrac{7}{3a} - \dfrac{3}{6a^2}$ 34. $\dfrac{7}{a} - \dfrac{x+1}{2a}$

35. $\dfrac{5b}{7x} + \dfrac{3a}{21x^2}$ 36. $\dfrac{5}{xy} + \dfrac{6}{yz}$ 37. $\dfrac{3z}{7w^2} - \dfrac{2z+1}{wz}$ 38. $\dfrac{6}{x} - \dfrac{5-y}{x^2y}$

39. $\dfrac{2}{t} + \dfrac{t+3}{st}$ 40. $\dfrac{7}{a} - \dfrac{x-1}{ab^2}$ 41. $\dfrac{b+7}{5b^2} + \dfrac{6}{b}$ 42. $\dfrac{2m^2-3}{2m^3} - \dfrac{1}{m}$

43. $\dfrac{m+1}{5m} + \dfrac{2m+3}{15m}$ 44. $\dfrac{2a-1}{a^2} - \dfrac{a+6}{3a}$ 45. $\dfrac{x+5}{4x^3} - \dfrac{2x+18}{x^2}$

Problem Solving Application: Work Backwards

Most problems are given to you with a set of conditions. Then you must find a solution. However, in some cases, it may be faster to determine how the problem ends and then *work backwards* rather than start from the beginning to find the solution.

Examples

1 **Find the sum of the reciprocals of two numbers whose sum is 2 and whose product is 3.**

Explore   Let x and y represent the two numbers.
We could set up this pair of equations.

$$x + y = 2$$
$$xy = 3$$

Plan   Finding values for x and y which satisfy both equations could be rather complicated. However, we know that the desired outcome is to find the value of $\dfrac{1}{x} + \dfrac{1}{y}$. We can use this information to work backwards and determine the solution.

Solve $\dfrac{1}{x} + \dfrac{1}{y} = \dfrac{y}{xy} + \dfrac{x}{xy}$ *The LCD of x and y is xy.*

$$= \dfrac{x + y}{xy}$$

Looking back to our original equations, we can see that $x + y = 2$ and $xy = 3$. Thus, $\dfrac{x + y}{xy} = \dfrac{2}{3}$.

The sum of the reciprocals is $\dfrac{2}{3}$.

2 **The garden club has a square plot of land for their herb garden. The square is divided into eight congruent rectangular rows for different plants. Each rectangular row has a perimeter of 45 yards. If the club was to place a fence around the entire square, how many yards of fence would they need to buy?**

Explore   Let s = the length of the side of the square.
Then the perimeter of the square = $4s$.

Plan   We must work backwards to find s before we can find the perimeter of the square. We can use the information about the perimeter of the rectangles to find s. The width of each rectangle is the same as the width of the square. Since there are eight congruent rectangles, the length of each rectangle is $\frac{1}{8}$th of the width of the square.

Solve   Thus, the perimeter of one rectangle $= 2s + 2(\frac{1}{8}s)$ or $2s + \frac{1}{4}s$.

$$2s + \tfrac{1}{4}s = 45$$
$$\tfrac{9}{4}s = 45$$
$$s = 20$$

$$\text{perimeter of square} = 4s$$
$$= 4(20)$$
$$= 80$$

The club would need to buy 80 yards of fence to surround the square herb garden.   *Examine this solution.*

Exercises

Written   **Solve each problem.**

1. A pirate found a treasure chest containing silver coins. He buried half of them and gave half of the remaining coins to his mother. If he was left with 4550 coins, how many were in the chest that he found?

2. If the sum of two numbers is 2 and the product of the numbers is 3, find the sum of the squares of the reciprocals of these numbers.

3. Peter, Jeff, and Opal are playing a card game. They have a rule that when a player loses a hand, he must subtract enough points from his score to double each of the other players' scores. First Peter loses a hand, then Jeff, and then Opal. Each player now has 8 points. Who lost the most points?

4. Tom collects model cars. He decides to give them away. First he gives half of them plus a car more to Lisa. Then he gives half of what is left plus one car to Nadine. Then he has one car left which he gives to Aaron. How many cars did Tom start with? (Assume that no car is cut in half.)

5. A firefighter spraying water on a fire stood on the middle rung of a ladder. The smoke lessened, so she moved up 3 rungs. It got too hot, so she backed up 5 rungs. Later, she went up 7 rungs and stayed until the fire was out. Then she climbed the remaining 6 rungs and went in the building. How many rungs does the ladder have?

Portfolio Suggestion

Select one of the assignments from this chapter that you found especially challenging and place it in your portfolio.

Performance Assessment

Simplify $\dfrac{(-3a^3b^2)(4ab)}{2a^7b}$ in at least two ways, verifying each step. Include an explanation of your solutions.

Chapter Summary

1. A **monomial** is a constant, a variable, or a product. (282)
2. The **degree** of a monomial is the sum of the exponents of its variables. (282)
3. If two monomials are the same or differ only by their coefficients, they are like monomials. (282)
4. **Product of Powers:** For all numbers a and all positive integers m and n, $a^m \cdot a^n = a^{m+n}$. (284)
5. **Power of a Power:** For all numbers a and all positive integers m, and n, $(a^m)^n = a^{mn}$. (286)
6. **Power of a Product:** For all numbers a and b and all positive integers m, $(ab)^m = a^m b^m$. (286)
7. A **polynomial** is the sum or difference of two or more monomials. A polynomial with two unlike terms is called a **binomial.** A polynomial with three unlike terms is called a **trinomial.** A polynomial is in simplest form if it does not contain any like terms. (289)
8. The **degree of a polynomial** is the degree of the term of greatest degree in the polynomial. (289)
9. Polynomials whose terms are arranged such that the powers of the variable are in descending order are in **standard form.** (290)
10. Polynomials can be added using the same properties used to simplify expressions. To subtract a polynomial from a polynomial, add its additive inverse. (291)
11. The rules for multiplying monomials and the distributive property can be used to multiply a polynomial by a monomial or a polynomial. (295, 297)

12. The **FOIL** (First, Outer, Inner, Last) method is used to multiply binomials. (298)

13. **Dividing Powers:** For positive integers m and n and nonzero number a, $\dfrac{a^m}{a^n} = a^{m-n}$ if $m > n$, $\dfrac{a^m}{a^n} = 1$ if $m = n$, and $\dfrac{a^m}{a^n} = \dfrac{1}{a^{n-m}}$ if $n > m$. (303)

14. **Definition of a^0:** For every nonzero real number a and every positive integer n, $a^{-n} = \dfrac{1}{a^n}$. (304)

15. **Definition of a^{-n}:** For every nonzero real number a and every positive integer n, $a^{-n} = \dfrac{1}{a^n}$. (304)

 # Chapter Review

9.1 **Simplify each of the following, if possible.**

1. $-25x + 5x$
2. $6ab - 2ab$
3. $4m^2n - 17m^2n$
4. $17x^2 - 5x^2 + 4x^2$
5. $2xy + 3x - 5y$
6. $4a^2b^3 - a^2b^3 + 10a^2b^3$

9.2 **Find and simplify each product.**

7. $a^5 \cdot a^2$
8. $(x^2y)(xy)$
9. $(4x^2y^3)(2xy^4)$
10. $(ax^2y)(5xy^4)$
11. $(2ab)(-5a^2b)$
12. $(2xy^2)(xy^3)(-2x^3y)$

9.3 **Simplify.**

13. $(3^4)^2$
14. $(2x)^3$
15. $(3xz^2)^3$
16. $(4a^2xy^3)^3$
17. $6(ax^2y)^3$
18. $(4a^2b^2)^2 - (2ab)^4$

9.4 **Simplify each polynomial and then state its degree.**

19. $6x + 2x + 3y + 5y$
20. $3x + 4x + 6x^2 - 3x^2$
21. $a^2 + a + 5a^2$
22. $3x + 2y - 5x + 6y$
23. $b^3 + c^2 + 6bc + 3b^3$
24. $3(x^2 + y) - 4(2x^2 - y)$

9.5 **Find each sum.**

25. $(2m + 3n) + (8m - 2n)$
26. $(5a + 3b - 7c) + (8a - b + 4c)$
27. $(2x^2 + 5x - 3) + (8x^2 + 3)$
28. $(-4xy^2 + 2x^2y + 7) + (3xy^2 - 9)$
29. $3x^2 + \quad ax + c^2$
$\underline{2x^2 - 3ax - c^2}$
30. $a \qquad + 3c$
$\underline{3a - 4b - 7c}$

Find each difference.

31. $(7x + 2y) - (3x - 3y)$

32. $(5a - 2b + c) - (8a + 7b - 8c)$

33. $(5n^2 - 4n + 5) - (3n^2 + 7n + 6)$

34. $(3z^2 + 5z - 4) - (3z^2 + 5z + 2)$

35. $\begin{array}{r} 5x^2 - 6x + 7 \\ -2x^2 + 8x - 1 \\ \hline \end{array}$

36. $\begin{array}{r} -4a^2 + 3b - d \\ 7a^2 - 2b + d \\ \hline \end{array}$

9.6 **Find each product.**

37. $3(4x^2 + 3)$

38. $a(-3b^2 + 7b)$

39. $-2y(5x^2 + 2x)$

40. $10x(3x^2y + 5xy^2 - 9)$

41. $7a^2b(-13a^3b^2 - 7ab^4)$

42. $15x^2y^3(10x^3y^2 - 6x^2y^3 + 5x^5)$

9.7 **Find each product.**

43. $(x + 5)(x + 6)$

44. $(x - 7)(x + 4)$

45. $(2x - 9)(x + 8)$

46. $(3a + 5)(a - 7)$

47. $(n - 4)(n - 9)$

48. $(5b - 6)(3b - 7)$

9.8 **Find each product.**

49. $(a + 7)(a^2 - 3a + 2)$

50. $(x - 1)(2x^2 - 4x + 7)$

51. $(3y - 2)(y^2 + 3y - 2)$

52. $(a - 4)(a^3 + 2a^2 + 5a - 6)$

53. $(2m + 1)(m^2 - 2)$

54. $(3y + 2)(2y^2 + 3y - 1)$

9.9 **Find each quotient.**

55. $\dfrac{x^8}{x^2}$

56. $\dfrac{a^2}{a^9}$

57. $\dfrac{9a^2b^2}{3ab^3}$

58. $\dfrac{m^3n^2}{m^3n^2}$

59. $\dfrac{4x^2y^3z^4}{4x^4y^2z^3}$

60. $\dfrac{-3m^2n^2}{27(mn)^3}$

9.10 **Find each quotient.**

61. $\dfrac{4b^2 + 8b}{2b}$

62. $\dfrac{6a^2 - a - 35}{3a + 7}$

63. $\dfrac{27x^2 - 4x + 7}{9x - 2}$

64. $\dfrac{9m^3 + 5m - 5}{3m - 2}$

9.11 **Perform the indicated operations and simplify.**

65. $\dfrac{5}{x} + \dfrac{7}{x}$

66. $\dfrac{3a - 7}{2m^2} - \dfrac{4 - a}{2m^2}$

67. $\dfrac{4}{3y} - \dfrac{5}{9y}$

68. $\dfrac{6t + 5}{8t} + \dfrac{2t - 1}{6t^2}$

69. $\dfrac{b + 1}{3b} \cdot \dfrac{5b}{-2}$

70. $\dfrac{2r + 3}{6p} \cdot \dfrac{r - 4}{q}$

Chapter Test

Simplify.

1. $4y - 9y$ **2.** $-3x^2 + 2x - x^2$ **3.** $7ab + 3ab - ab$

4. x^4x^5 **5.** $(a^3b)(a^5b^2)$ **6.** $(4x^2y)(-2xy^3z)(-2xyz^2)$

7. $(5xyz)^3$ **8.** $(2xy^2z)(3xy)^2$ **9.** $(4x^2y)^2 - 3(xy)^2y^2$

Simplify each polynomial and then state its degree.

10. $3x + 2x - 3y + y$ **11.** $3m - 5n + 2m + 6n$

12. $xy + 14xy - xy^2$ **13.** $2(a^2 + 4) - 3a^2$

Find each sum.

14. $(5m + 2n) + (8m - 3n)$ **15.** $(2x^2 - 5x + 3) + (3x^2 + 8x - 1)$

16. $\quad 10ax^2 - 5ax + 6a$
$\quad\underline{-7ax^2 - 2ax - 8a}$

Find each difference.

17. $(5a - 3b) - (2a - 5b)$ **18.** $4(x + 3y - 7z) - (x + 8y)$

19. $\quad 4a^2 \qquad + 3$
$\quad\underline{-3a^2 + 2a - 1}$

Find each product.

20. $5(7a^2 - 7a)$ **21.** $3y(-2x^2 + 3x - 7)$ **22.** $7x^2y(-x^3y + 8xy^3 - 9x^2y^2)$

23. $(x + 6)(x + 8)$ **24.** $(2x - 7)(x + 4)$ **25.** $(5y + 1)(y - 6)$

26. $(x + 2)(x^2 - 2x + 4)$ **27.** $(a - 2)(3a^2 + 4a - 1)$ **28.** $(2y + 5)(2y^2 + y - 5)$

Find each quotient.

29. $\dfrac{a^8}{a^3}$ **30.** $\dfrac{4x^3y^5}{16x^2y^7}$

31. $\dfrac{y^2 + 3y + 2}{y + 1}$ **32.** $\dfrac{6n^3 + 5n^2 + 12}{2n + 3}$

Perform the indicated operations and simplify.

33. $\dfrac{-3}{2a} + \dfrac{11}{2a}$ **34.** $\dfrac{x^2 - 5}{2x} - \dfrac{x + 2}{5x^2}$ **35.** $\dfrac{3k}{7k} \cdot \dfrac{4k + 9}{2k^2}$

Factoring

Application in Marine Biology

Marine biologists study fish and other marine life. Dolphins are the most elaborate creatures studied by these scientists. Dolphins probably communicate by using various frequencies and amplitudes, and they can swim at a speed of 20 miles per hour almost indefinitely.

In a standard aquatic park pool, a dolphin can reach a speed of 24 feet per second just before breaking the surface of the water for a jump. The **formula** for the height of the dolphin completing such a jump is $h = 24t - 16t^2$ where h is the height of the dolphin in feet and t is the time in seconds. To find the times that the dolphin breaks the surface of the water and returns to the water, solve the equation $0 = 24t - 16t^2$. How long does the dolphin remain in the air?

Group Project: *Physics*

The equation for an object dropped from a still position is $d = t - 4.9t^2$ where d is the distance in meters and t is the time in seconds. With your teacher's permission, drop a ball from a window of the school while a classmate measures the time with a stopwatch. Conduct several trials and record each on a chart. Use the average time to find the height from which the ball is dropped. A negative distance means the direction is downward.

10.1 Factoring Monomials

Suppose it takes 60 square yards of carpet to cover a recreation room. What are the dimensions of the room? Two possibilities are listed below.

6 yards by 10 yards $\Rightarrow$ 60 square yards
5 yards by 12 yards $\Rightarrow$ 60 square yards

You may recall that 6 and 10 are called **factors** of 60 since 6 × 10 = 60. Likewise, 5 and 12 are factors of 60. You can find factors of a number by division. A number is said to be **factored** if it is expressed as the product of two or more integers. But 60 can be factored even further.

$$60 = 6 \cdot 10 \qquad\qquad 60 = 5 \cdot 12$$
$$= 2 \cdot 3 \cdot 2 \cdot 5 \qquad\qquad = 5 \cdot 2 \cdot 6$$
$$\qquad\qquad\qquad\qquad\qquad = 5 \cdot 2 \cdot 2 \cdot 3$$

The only factors of each of the numbers shown in color are the numbers themselves and 1. The numbers 2, 3, and 5 are called **prime numbers.** The prime factors of 60 are 2, 2, 3, and 5.

Definition of Prime Number

> A prime number is an integer, greater than 1, whose only positive factors are 1 and itself.

The product of the prime factors of a number is called the **prime factorization** of the number.

Example

1 Find the prime factorization of 315.

$$315 = 3 \cdot 105 \qquad \textit{2 does not divide 315 evenly. Try 3.}$$
$$= 3 \cdot 3 \cdot 35 \qquad \textit{Try 3 again.}$$
$$= 3 \cdot 3 \cdot 5 \cdot 7$$

Can you continue factoring? *No. 3, 5, and 7 are prime numbers.*
The prime factorization of 315 is 3 • 3 • 5 • 7. *This can also be written $3^2 \cdot 5 \cdot 7$.*

Notice that only integral factors of the coefficient 4 are considered.

Thus, $\frac{1}{8}x \cdot 32x^2$ is <u>not</u> included.

Monomials also can be factored as shown for $4x^3$ below.

$$4x^3 = 2x \cdot 2x^2 \qquad 4x^3 = 4x \cdot x^2 \qquad 4x^3 = 2 \cdot 2 \cdot x \cdot x \cdot x$$

Sometimes two or more numbers have common factors other than 1. Consider 27 and 45.

$$27 = 3 \cdot 9 \qquad\qquad 45 = 3 \cdot 15$$
$$= 3 \cdot 3 \cdot 3 \qquad\qquad = 3 \cdot 3 \cdot 5$$

The integers 27 and 45 have common factors of 3 and $3 \cdot 3$ or 9. The greatest of these factors, 9, is called the **greatest common factor (gcf)** of 27 and 45.

Example

2 **Find the greatest common factor of 72 and 126.**

$$72 = 2 \cdot 2 \cdot 2 \cdot 3 \cdot 3 \qquad \textit{Find the common factors.}$$
$$126 = 2 \qquad\quad \cdot 3 \cdot 3 \cdot 7 \qquad \textit{They are 2, 3, and 3.}$$

The greatest common factor of 72 and 126 is $2 \cdot 3 \cdot 3$, or 18.

The gcf of two or more monomials is the common factor with the greatest numerical factor and the greatest degree.

Examples

3 **Find the gcf of $18x^5$ and $24x^3$.**

$$18x^5 = 2 \cdot 3 \cdot 3 \cdot x \cdot x \cdot x \cdot x \cdot x$$
$$30x^3 = 2 \cdot 3 \cdot 5 \cdot x \cdot x \cdot x$$

The gcf of $18x^5$ and $30x^3$ is $2 \cdot 3 \cdot x \cdot x \cdot x$ or $6x^3$.

4 **Find the gcf of $-60x^3y^5$ and $90x^5y^2$.**

$$-60x^3y^5 = -1 \cdot 2 \cdot 2 \cdot 3 \cdot 5 \cdot x \cdot x \cdot x \cdot y \cdot y \cdot y \cdot y \cdot y$$
$$90x^5y^2 = \quad\; 2 \cdot 3 \cdot 3 \cdot 5 \cdot x \cdot x \cdot x \cdot x \cdot x \cdot y \cdot y$$

The gcf of $-60x^3y^5$ and $90x^5y^2$ is $2 \cdot 3 \cdot 5 \cdot x \cdot x \cdot x \cdot y \cdot y$ or $30x^3y^2$.

Exercises

Exploratory State whether the second integer is a factor of the first integer. If it is a factor, state the other factor associated with it.

Sample:	10, 2
	Yes; 2 is a factor of 10 because $10 \div 2 = 5$. The other factor is 5.

1. 8, 2	**2.** 15, 3	**3.** 29, 7	**4.** 51, 3
5. 98, 7	**6.** 63, 8	**7.** 111, 3	**8.** 210, 7
9. 432, 8	**10.** 142, 3	**11.** 198, 6	**12.** 456, 12

Find the gcf for each pair of integers.

13. 4, 12	**14.** 15, 24	**15.** 8, 20	**16.** -9, 30
17. 10, -50	**18.** 13, 39	**19.** 27, 36	**20.** 12, 18
21. -10, 16	**22.** 5, 50	**23.** 20, 44	**24.** -11, 33

Written Find the prime factorization of each integer. Write each negative integer as the product of -1 and its prime factors.

1. 21	**2.** 40	**3.** 60	**4.** -12
5. 68	**6.** -26	**7.** 80	**8.** -112
9. 304	**10.** 384	**11.** -96	**12.** 280
13. 504	**14.** -900	**15.** 225	**16.** 5005

Complete each of the following.

17. $2x^2y = 2x \cdot \underline{\hspace{2em}}$

18. $2x^2y = xy \cdot \underline{\hspace{2em}}$

19. $18r^3s = -9r^2 \cdot \underline{\hspace{2em}}$

20. $18r^3s = 18rs \cdot \underline{\hspace{2em}}$

21. $-12xy^2z = 6xy \cdot \underline{\hspace{2em}}$

22. $28a^3b^2c = -14ab \cdot \underline{\hspace{2em}}$

23. $-36m^6n^4p^2 = 3m^4n^2p^2 \cdot \underline{\hspace{2em}}$

24. $-42r^6s^3t^2 = 7r^2s^2t^2 \cdot \underline{\hspace{2em}}$

25. $12rst = 6rs \cdot \underline{\hspace{2em}}$

26. $64x^2y^2z^4 = -16xyz^2 \cdot \underline{\hspace{2em}}$

27. $72a^2b^3c = -72b^3a^2c \cdot \underline{\hspace{2em}}$

28. $765x^5y^3 = 15xy \cdot \underline{\hspace{2em}}$

Find the gcf for each of the following.

29. 18, 48

30. 80, -45

31. 216, 384

32. 96, 30, 66

33. $10x^2$, $35x$

34. $22y^2$, $33y^3$

35. $-25ab$, 35

36. $60x^2$, $68a^2$

37. $20x^2y$, $16xy^2$

38. $24ab^2c$, $30a^2b$

39. $42ab$, $54a^2b$

40. $56x^3y$, $49ax^2$

41. $20x^2y^3$, $24x^2z$

42. $24ax^3$, $44ax^2$

43. $12a^2b$, $18ab^3$, $30ab$

44. $8a^2x^3$, a^2x, $13a^2x^2$

45. $14a^2b^2$, $-18a^2b^3$, $2ab^2$

46. $21x^2y^3$, $14x^3y^2$, $35x^4y$

10.2   Factoring Polynomials

You have used the distributive property to multiply a monomial by a binomial.

$$3a(a + 4b) = 3a(a) + 3a(4b)$$
$$= 3a^2 + 12ab$$

Suppose you wish to factor $3a^2 + 12ab$. First, find the greatest common factor of $3a^2$ and $12ab$.

$$3a^2 = \boxed{3} \qquad \cdot \boxed{a} \cdot a$$
$$12ab = \boxed{3} \cdot 2 \cdot 2 \cdot \boxed{a} \cdot b$$

The gcf of $3a^2$ and $12ab$ is $3a$.

Now write each term as a product of the gcf and the remaining factors. Then use the distributive property.

$$3a^2 + 12ab = 3a(a) + 3a(4b)$$
$$= 3a(a + 4b)$$

The binomial $3a^2 + 12ab$ written in factored form is $3a(a + 4b)$.

Journal

Look up the word *factor* in the dictionary. Describe how it relates to the mathematics in this lesson.

Examples

1 **Factor $4x^2 - 3x$.**

First, find the gcf.

$$4x^2 = 2 \cdot 2 \cdot \boxed{x} \cdot x$$
$$3x = \qquad 3 \cdot \boxed{x}$$

The gcf of $4x^2$ and $3x$ is x. Now write the terms as products using the gcf as one factor.

$$4x^2 - 3x = x(4x) - x(3)$$
$$= x(4x - 3)$$

2 **Factor $2a^5 + 4a^3 + 6a^2$.**

$$2a^5 = \boxed{2} \cdot a \cdot \boxed{a \cdot a} \cdot a \cdot a$$
$$4a^3 = \boxed{2} \cdot 2 \cdot \boxed{a \cdot a} \cdot a$$
$$6a^2 = \boxed{2} \cdot 3 \cdot \boxed{a \cdot a} \qquad \textit{The gcf is } 2a^2.$$

$$2a^5 + 4a^3 + 6a^2 = 2a^2(a^3) + 2a^2(2a) + 2a^2(3)$$
$$= 2a^2(a^3 + 2a + 3)$$

Example

3 Circle O is inscribed in square $ABCD$. Express the area of the shaded region in factored form.

$$\frac{\text{area of}}{\text{shaded region}} = \frac{\text{area of}}{\text{square}} - \frac{\text{area of}}{\text{circle}}$$

$$A = 4r^2 - \pi r^2$$
$$= r^2(4 - \pi) \qquad \textit{The gcf of } 4r^2 \textit{ and } \pi r^2 \textit{ is } r^2$$

Exercises

Exploratory Find the gcf for the terms in each polynomial.

1. $7y^2 - 21y$
2. $5x^2 + 15x$
3. $x^2 + y^2$
4. $4xy - 10x$
5. $18mn - 24n$
6. $a^2b - b^2a$
7. $xy^2 - x^2y$
8. $4x^3 + 12x^5$
9. $x^3 + x^2 + x$
10. $14ab + 28b$
11. $16x + 24xy$
12. $14abc^2 + 18c$
13. $4x^2y^2 + 6xz^2$
14. $3x^2y + 12xyz^2$
15. $18xy^2 + 2y^3$
16. $8ab^2 - 16a^2b$
17. $9y^2 - 27y + 36$
18. $2xy^3 + 4x^2y - 9x^3y^2$

Written Factor.

1. $6a + 6b$
2. $10x + 10y$
3. $8c - 8d$
4. $4x - 4y$
5. $8m - 2n$
6. $18c + 27d$
7. $ab + ac$
8. $a^2 + a$
9. $y^3 + y^2$
10. $xy - xz$
11. $25t^2 - 15$
12. $y^2 + 7y$
13. $a^3 + 2a^2$
14. $3m^2 - 15m$
15. $5x^2y - 10xy^2$
16. $18x^2 - 9x$
17. $-15x^2 - 5x$
18. $988x^2 + 2233x$
19. $8m^2 - 16m + 24$
20. $28a^2 - 7a - 14$
21. $9x^5 - 18x^3 + 27$
22. $a^3 + a^2 + a$
23. $x^2y + x^3y$
24. $4y^2 + 2y^3$
25. $4a^2b - 6a^4$
26. $-9c^3 + 27c^2$
27. $15d^2 - 25d + 30$
28. $4ym + 2m$
29. $ab + ac + ab + ac$
30. $rs + ru + tr + ru$

Express the area of the shaded region in each figure in factored form.

31.

32.

33.

34.

10.3   Factoring the Difference of Squares

Suppose you want to find the product of a sum, $x + 4$, and a difference, $x - 4$.

$$\begin{aligned}(x + 4)(x - 4) &= x(x - 4) + 4(x - 4) && \text{\textit{Use the distributive}}\\ &= x^2 - 4x + 4x - 16 && \text{\textit{property.}}\\ &= x^2 - 16\end{aligned}$$

Notice that x^2 is the square of x and 16 is the square of 4. Thus, the product $x^2 - 16$ is called the **difference of two squares.**

Difference of Squares

For numbers a and b,
$$(a + b)(a - b) = a^2 - b^2.$$

Examples

1 Find $(x + 5)(x - 5)$.

$(a + b)(a - b) = a^2 - b^2$

$$\begin{aligned}(x + 5)(x - 5) &= x^2 - 5^2\\ &= x^2 - 25\end{aligned}$$

2 Find $(2m - 3n)(2m + 3n)$.

$(a - b)(a + b) = a^2 - b^2$

$$\begin{aligned}(2m - 3n)(2m + 3n) &= (2m)^2 - (3n)^2\\ &= 4m^2 - 9n^2\end{aligned}$$

Suppose you reverse the sides of the equation $(a + b)(a - b) = a^2 - b^2$.

$$a^2 - b^2 = (a + b)(a - b)$$

You can use this result to factor the difference of two squares.

Examples

3 **Factor $m^2 - 81$.**

$$m^2 - 81 = (m)^2 - (9)^2$$
$$= (m + 9)(m - 9) \qquad a^2 - b^2 = (a + b)(a - b)$$

4 **Factor $9x^2 - 64y^2$.**

$$9x^2 - 64y^2 = (3x)^2 - (8y)^2$$
$$= (3x + 8y)(3x - 8y)$$

Sometimes a common factor can be factored from a polynomial first.

Example

5 **Factor $50a^2 - 98b^2$.**

$$50a^2 - 98b^2 = 2(25a^2 - 49b^2) \qquad \text{\textit{The gcf of $50a^2$ and $98b^2$ is 2.}}$$
$$= 2[(5a)^2 - (7b)^2]$$
$$= 2(5a + 7b)(5a - 7b)$$

Exercises

Exploratory   State whether each binomial can be factored as the difference of squares.

1. $x^2 + y^2$

2. $a^2 + 25$

3. $4a^2 - b^2$

4. $a^2 - 4b^4$

5. $9y^2 + a^2$

6. $9x^2 - 5$

7. $25x^2 - 9$

8. $36a^2 - 49$

9. $16x^2 - 6$

10. $a - 9$

11. $36a^2 + 49$

12. $25a^2 + b^2$

Find each product.

13. $(x - 3)(x + 3)$

14. $(a + 4)(a - 4)$

15. $(d - 4)(d + 4)$

16. $(c - d)(c + d)$

17. $(y + 7)(y - 7)$

18. $(2a + 3)(2a - 3)$

Written   Find each product.

1. $(a - 9)(a + 9)$
2. $(q + 11)(q - 11)$
3. $(s - t)(s + t)$
4. $(x + y)(x - y)$
5. $(2a + 3)(2a - 3)$
6. $(4x + y)(4x - y)$
7. $(3c + 2)(3c - 2)$
8. $(3b - 5)(3b + 5)$
9. $(s - 4t)(s + 4t)$
10. $(3x - 8)(3x + 8)$
11. $(5x - 3)(5x + 3)$
12. $(a - 5b)(a + 5b)$
13. $(2r - 3s)(2r + 3s)$
14. $(3w + 4z)(3w - 4z)$
15. $(3p - 8q)(3p + 8q)$
16. $(m - 6n)(m + 6n)$
17. $(4x + 5)(4x - 5)$
18. $(4a + 7b)(4a - 7b)$
19. $(6f + 5g)(6f - 5g)$
20. $(13v - 10w)(13v + 10w)$
21. $(19w + 24x)(19w - 24x)$
22. $(2a + 9b)(2a - 9b)$
23. $(8 - 3y)(8 + 3y)$
24. $(2 + 7m)(2 - 7m)$

Factor.

25. $x^2 - 49$
26. $y^2 - 81$
27. $m^2 - 100$
28. $a^2 - b^2$
29. $r^2 - 4s^2$
30. $c^2 - 49d^2$
31. $m^2 - 9n^2$
32. $25a^2 - b^2$
33. $36s^2 - 100$
34. $1 - 64r^2$
35. $-9 + 4y^2$
36. $25a^2x^2 - 1$
37. $121b^2y^2 - 1$
38. $4x^2 - 9y^2$
39. $16a^2 - 9b^2$
40. $68g^2 - 17h^2$
41. $2z^2 - 98$
42. $12a^2 - 48$
43. $12c^2 - 12$
44. $7 - 28y^2$
45. $8x^2 - 32y^2$
46. $8x^2 - 18$
47. $144x^2 - 9y^2$
48. $4x^2 - 64y^2$
49. $15x^2 - 60y^2$
50. $45x^2 - 20z^2$
51. $5x^2 - 125y^4$
52. $16x^2 - 196y^4$
53. $(a + b)^2 - m^2$
54. $z^2 - (x - y)^2$

Compute each of the following by expressing each product in the form $(a + b)(a - b)$.

Sample:	$(32)(28) = (30 + 2)(30 - 2)$
	$= 30^2 - 2^2$
	$= 900 - 4$ or 896

55. $(63)(57)$
56. $(56)(44)$
57. $(107)(93)$
58. $(71)(69)$
59. $(199)(201)$
60. $\left(5\dfrac{1}{2}\right)\left(4\dfrac{1}{2}\right)$
61. $(1050)(950)$
62. $(1150)(1250)$

Express the area of the shaded region in each figure in factored form.

63.

64.
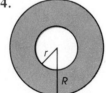

65.
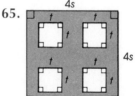

66.
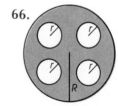

10.4 Factoring Quadratic Polynomials

A polynomial in one variable of degree 2 is called a **quadratic polynomial.** For example, $4n^2$, $5a - 3a^2$, and $x^2 + 5x + 6$ are quadratic polynomials. The term of degree 2 in a quadratic polynomial is called the *quadratic term*. The term of degree 1 is called the *linear term*. The numerical term is called the *constant term*.

Quadratic Polynomial	Quadratic Term	Linear Term	Constant Term
$4n^2$	$4n^2$	$0n$	0
$2a - 3a^2$	$-3a^2$	$2a$	0
$x^2 + 5x + 6$	x^2	$5x$	6

Recall how the FOIL method is used to multiply two binomials.

$$\overset{\textbf{F}\qquad\textbf{O}\qquad\textbf{I}\qquad\textbf{L}}{(x + 2)(x + 3) = x \cdot x + x \cdot 3 + 2 \cdot x + 2 \cdot 3}$$
$$= x^2 + 3x + 2x + 2 \cdot 3$$
$$= x^2 + (3 + 2)x + 2 \cdot 3$$
$$= x^2 + 5x + 6$$

Notice the following pattern in this multiplication.

$$(x + 2)(x + 3) = x^2 + (3 + 2)x + 2 \cdot 3$$
$$(x + m)(x + n) = x^2 + (n + m)x + mn$$
$$= x^2 + (m + n)x + mn$$

This pattern can be used to factor some quadratic polynomials.

To factor $x^2 + 5x + 6$ into the form $(x + m)(x + n)$, you must find the factors of 6 whose sum is 5. That is, find values for m and n such that $m + n = 5$ and $mn = 6$.

pairs of factors of 6	sum of factors
1, 6	$1 + 6 = 7$
$-1, -6$	$-1 + (-6) = -7$
2, 3	$2 + 3 = 5$
$-2, -3$	$-2 + (-3) = -5$

Since the sum must be 5, then 2 and 3 are the factors.

$$(x + m)(x + n) = x^2 + (m + n)x + mn$$
$$(x + 2)(x + 3) = x^2 + 5x + 6 \qquad \textit{Check by using FOIL.}$$

Examples

1 **Factor $y^2 - 8y + 12$.**

First list the possible pairs of factors of 12. Then find the sum for each pair.

pairs of factors of 12	sum of factors
1, 12	$1 + 12 = 13$
$-1, -12$	$-1 + (-12) = -13$
3, 4	$3 + 4 = 7$
$-3, -4$	$-3 + (-4) = -7$
2, 6	$2 + 6 = 8$
$-2, -6$	$-2 + (-6) = -8$

The factors are -2 and -6 since their sum is -8.
Thus, $y^2 - 8y + 12 = (y + (-2))(y + (-6))$ or $(y - 2)(y - 6)$.
Check that the product $(y - 2)(y - 6)$ is $y^2 - 8y + 12$.

2 **Factor $t^2 - 3t - 10$.**

List the possible pairs of factors of -10 and find the sum for each pair.

pairs of factors of -10	sum of factors
1, -10	$1 + (-10) = -9$
$-1, 10$	$-1 + 10 = 9$
2, -5	$2 + (-5) = -3$
$-2, 5$	$-2 + 5 = 3$

The factors are 2 and -5 since their sum is -3.
Thus, $t^2 - 3t - 10 = (t + 2)(t - 5)$.
Check that the product $(t + 2)(t - 5)$ is $t^2 - 3t - 10$.

Consider $2x^2 + 7x + 6$. The coefficient of x^2 is *not* 1. The factors of $2x^2 + 7x + 6$ are of the following form.

$$\overset{\text{factors of 2}}{\overbrace{\quad\quad\quad}}$$
$$(\square x + \square)(\square x + \square)$$
$$\underset{\text{factors of 6}}{\underbrace{\quad\quad\quad}}$$

The possible factors of $2x^2 + 7x + 6$ are as follows.

possible factors	outer + inner terms
$(x + 1)(2x + 6)$	$6x + 2x = 8x$
$(x + 6)(2x + 1)$	$x + 12x = 13x$
$(x - 1)(2x - 6)$	$-6x - 2x = -8x$
$(x - 6)(2x - 1)$	$-x - 12x = -13x$
$(x + 2)(2x + 3)$	$3x + 4x = 7x$
$(x + 3)(2x + 2)$	$2x + 6x = 8x$
$(x - 2)(2x - 3)$	$-3x - 4x = -7x$
$(x - 3)(2x - 2)$	$-2x - 6x = -8x$

Therefore, $2x^2 + 7x + 6 = (x + 2)(2x + 3)$. *Check by using FOIL.*

Example

3 **Factor $3x^2 - 14x - 5$.**

First, find the factors of 3 and -5.
factors of 3: 1, 3   factors of -5: -1, 5
                    -1, -3 1, -5
Now, list and check the possible factors of $3x^2 - 14x - 5$.

possible factors	outer + inner terms
$(x + 1)(3x - 5)$	$-5x + 3x = -2x$
$(x - 1)(3x + 5)$	$5x - 3x = 2x$
$(x + 5)(3x - 1)$	$-x + 15x = 14x$
$(x - 5)(3x + 1)$	$x - 15x = -14x$

Therefore, $3x^2 - 14x - 5 = (x - 5)(3x + 1)$.
Check that the product $(x - 5)(3x + 1)$ is $3x^2 - 14x - 5$.

Exercises

Exploratory   State whether each of the following is a quadratic polynomial. If it is, state the quadratic term, the linear term, and the constant term.

1. $6x^2 + x + 3$

2. $-5t^2 + 4$

3. $6x$

4. $4m^2 - 6m$

5. $-8 + x^2 + 7x$

6. $y^3 + y^2 - 4y + 3$

7. $5x^2$

8. $-8x + 10 - 7x^2$

9. $s^4 - s^3 + 3s + 4$

Find two integers with the given product and sum.

Sample: sum = 72, product = 18
        The integers are 6 and 12 since 6 × 12 = 72 and 6 + 12 = 18.

	product	sum			product	sum
10.	1	2	**11.**	9	10
12.	20	9	**13.**	15	8
14.	11	12	**15.**	28	11
16.	−3	−2	**17.**	−15	−2

Complete the factoring.

18. $x^2 + 6x + 5 = (x + 1)(x + \square)$

19. $y^2 - 8y + 12 = (y - \square)(y - 6)$

20. $n^2 - 2n - 15 = (n + 3)(n - \square)$

21. $p^2 - 2p - 63 = (p \square 7)(p - 9)$

22. $2x^2 + 5x + 3 = (2x + \square)(x + 1)$

23. $3y^2 + 14y + 8 = (3y + 2)(y + \square)$

24. $m^2 + 13m + 30 = (m \square 3)(m + \square)$

25. $x^2 - 19x + 34 = (x \square \square)(x - 2)$

26. $x^2 - 3x - 54 = (x \square \square)(x - 9)$

27. $y^2 - 12y - 28 = (y \square 14)(y + \square)$

28. $8k^2 - 10k + 3 = (\square k - 1)(4k - \square)$

29. $3a^2 + 11a - 20 = (\square a - \square)(a \square 5)$

Written   **Factor each quadratic polynomial. If the polynomial cannot be factored into binomials with integral coefficients, write *cannot be factored*.**

1. $a^2 + 8a + 15$

2. $x^2 + 12x + 27$

3. $c^2 + 12c + 35$

4. $x^2 + 6x - 7$

5. $x^2 + 2x - 3$

6. $y^2 + y - 12$

7. $g^2 - 10g + 21$

8. $m^2 - 15m + 26$

9. $p^2 - 19p + 60$

10. $a^2 - 22a + 21$

11. $s^2 - 7s + 10$

12. $x^2 + 5x + 8$

13. $t^2 + 3t - 54$

14. $a^2 + 6a - 72$

15. $m^2 + 3m - 40$

16. $p^2 + 11p + 28$

17. $t^2 + 14t + 33$

18. $x^2 + 5x - 24$

19. $3a^2 + 8a + 4$

20. $3a^2 + 11a + 6$

21. $7x^2 + 22x + 3$

22. $2x^2 - 3x - 5$

23. $6a^2 - a - 2$

24. $3r^2 - 7r - 6$

25. $3x^2 - 30x + 56$

26. $9g^2 - 12g + 4$

27. $8k^2 - 18k + 9$

28. $36 + 13y + y^2$

29. $m^2 - 4m - 12$

30. $18d^2 - 19d - 12$

31. $x^2 + 6xy + 8y^2$

32. $a^2 + 7ab + 10b^2$

33. $a^2 + 11ab - 42b^2$

34. $s^2 + 16st - 36t^2$

35. $20x^2 + 17xy - 24y^2$

36. $15a^2 - 13ab + 2b^2$

Express the answers in each of the following in terms of x. Give all possible answers.

37. The area of a rectangle is $x^2 + 13x + 40$ square units. Find the length and the width.

38. The area of a square is $x^2 - 16x + 64$ square units. Find the measure of a side of the square.

39. The area of a triangle is $x^2 + 3x$ square units. Find the height and the base.

40. The area of a trapezoid is $x^2 + 2x$ square units. Find the height and the two bases.

10.5   Solving Equations by Factoring

Consider the following products.

$$7 \cdot 0 = 0 \qquad 0(-4) = 0 \qquad 0a = 0 \qquad 0 \cdot 0 = 0$$

What do these products have in common? They are all equal to 0. Notice that in each case *at least one* of the factors is 0. This property may be stated as follows.

Zero Product Property

> For all numbers a and b,
> $ab = 0$ if and only if $a = 0$ or $b = 0$.

Both a and b can be zero.

You can use this property to solve equations expressed in factored form.

Examples

1 Solve $x(x + 3) = 0$.

If $x(x + 3) = 0$, then $x = 0$ or $x + 3 = 0$.

$x = 0 \qquad$ or $\qquad x + 3 = 0$
$\qquad\qquad\qquad\qquad\qquad x = -3$

Check: $x(x + 3) = 0$

$\qquad\qquad 0(0 + 3) \overset{?}{=} 0 \qquad$ or $\qquad -3(-3 + 3) \overset{?}{=} 0$
$\qquad\qquad\qquad 0(3) \overset{?}{=} 0 \qquad\qquad\qquad\qquad -3(0) \overset{?}{=} 0$
$\qquad\qquad\qquad\quad 0 = 0 \qquad\qquad\qquad\qquad\qquad 0 = 0$

0 and -3 are solutions of $x(x + 3) = 0$. The solution set is $\{0, -3\}$.

2 Solve $(3x - 3)(2x + 4) = 0$.

If $(3x - 3)(2x + 4) = 0$, then $3x - 3 = 0$ or $2x + 4 = 0$.

$3x - 3 = 0 \qquad$ or $\qquad 2x + 4 = 0$
$\quad 3x = 3 \qquad\qquad\qquad\qquad 2x = -4$
$\qquad x = 1 \qquad\qquad\qquad\qquad\; x = -2$

Check: $(3x - 3)(2x + 4) = 0$

$\quad (3 \cdot 1 - 3)(2 \cdot 1 + 4) \overset{?}{=} 0 \qquad$ or $\quad (3 \cdot -2 - 3)(2 \cdot -2 + 4) \overset{?}{=} 0$
$\qquad\qquad\qquad\qquad 0(6) \overset{?}{=} 0 \qquad\qquad\qquad\qquad\qquad\quad (-9)(0) \overset{?}{=} 0$
$\qquad\qquad\qquad\qquad\quad 0 = 0 \qquad\qquad\qquad\qquad\qquad\qquad\qquad 0 = 0$

The solution set is $\{1, -2\}$.

In the equation $x^2 - 4x = 0$, the binomial $x^2 - 4x$ is *not* in factored form. To solve this equation, you can factor $x^2 - 4x$.

Examples

3 **Solve $x^2 - 4x = 0$.**

$x^2 - 4x = 0$
$x(x - 4) = 0$ *Factor using the distributive property.*

$\qquad x = 0 \qquad$ or $\qquad x - 4 = 0$
$\qquad\qquad\qquad\qquad\qquad\qquad x = 4$

Check: $x^2 - 4x = 0$

$0^2 - 4 \cdot 0 \stackrel{?}{=} 0 \qquad$ or $\qquad 4^2 - 4 \cdot 4 \stackrel{?}{=} 0$
$\qquad 0 - 0 \stackrel{?}{=} 0 \qquad\qquad\qquad\qquad 16 - 16 \stackrel{?}{=} 0$
$\qquad\qquad 0 = 0 \qquad\qquad\qquad\qquad\qquad\qquad 0 = 0$

The solution set is $\{0, 4\}$.

4 **Solve $x^2 + 5x + 6 = 0$.**

$x^2 + 5x + 6 = 0$
$(x + 3)(x + 2) = 0$ *Factor.*

$\qquad x + 3 = 0 \qquad$ or $\qquad x + 2 = 0$
$\qquad\qquad x = -3 \qquad\qquad\qquad\qquad x = -2$

Check: $x^2 + 5x + 6 = 0$

$(-3)^2 + 5(-3) + 6 \stackrel{?}{=} 0 \qquad$ or $\qquad (-2)^2 + 5(-2) + 6 \stackrel{?}{=} 0$
$\qquad 9 - 15 + 6 \stackrel{?}{=} 0 \qquad\qquad\qquad\qquad 4 - 10 + 6 \stackrel{?}{=} 0$
$\qquad\qquad\qquad 0 = 0 \qquad\qquad\qquad\qquad\qquad\qquad\qquad 0 = 0$

The solution set is $\{-3, -2\}$.

An equation that can be written in the form $ax^2 + bx + c = 0$, where $a \neq 0$, is called a quadratic equation.

When a quadratic equation is written so that one side is zero and the other is a quadratic polynomial in standard form, the quadratic equation is in *standard form*. Quadratic equations must be written in standard form before they can be solved by factoring.

Examples

5 **Solve $t^2 - 8t = 9$.**

$$t^2 - 8t = 9$$
$$t^2 - 8t - 9 = 0 \qquad \textit{Write the equation in standard form.}$$
$$(t - 9)(t + 1) = 0$$

$$t - 9 = 0 \qquad \text{or} \qquad t + 1 = 0$$
$$t = 9 \qquad\qquad\qquad t = -1$$

The solution set is $\{9, -1\}$.

Lab Activity

You can learn how to use a graphing calculator to solve quadratic equations in Lab 10 on page A14.

6 **Solve $x^2 - 50 = 5x$.**

$$x^2 - 50 = 5x$$
$$x^2 - 5x - 50 = 0$$
$$(x - 10)(x + 5) = 0$$

$$x - 10 = 0 \qquad \text{or} \qquad x + 5 = 0$$
$$x = 10 \qquad\qquad\qquad x = -5$$

The solution set is $\{10, -5\}$.

Exercises

Exploratory **State the conditions under which each equation will be true.**

Sample: $x(x + 7) = 0$ will be true if $x = 0$ or $x + 7 = 0$.

1. $x(x - 5) = 0$ **2.** $y(y + 12) = 0$

3. $2n(n - 3) = 0$ **4.** $3m(3m - 24) = 0$

5. $3n(2 - n) = 0$ **6.** $2t(8 - t) = 0$

7. $x(x + 2) = 0$ **8.** $x(x - 7) = 0$

9. $3r(r - 4) = 0$ **10.** $(x - 6)(x + 4) = 0$

11. $(x - 2)(x - 7) = 0$ **12.** $(b - 3)(b - 5) = 0$

13. $d^2 + 12d + 20 = 0$ **14.** $y^2 - 13y + 30 = 0$

15. $m^2 - 9m + 20 = 0$ **16.** $x^2 - x - 2 = 0$

17. $y^2 - y - 42 = 0$ **18.** $2x^2 - 7x + 3 = 0$

Written   Solve each equation and check your answer.

1. $y(y - 7) = 0$
2. $7x(x + 6) = 0$
3. $2x(5x - 10) = 0$
4. $4x(3x + 12) = 0$
5. $(x - 6)(x + 4) = 0$
6. $(b - 3)(b - 5) = 0$
7. $(x + 3)(3x - 12) = 0$
8. $(4x + 4)(2x + 6) = 0$
9. $(5x + 40)(2x - 4) = 0$
10. $(w - 7)(2w - 10) = 0$
11. $(z + 2)(z + 2) = 0$
12. $(x - 3)(x - 3) = 0$
13. $(3x - 9)(5x - 15) = 0$
14. $(y - 9)(y + 9) = 0$
15. $(6x + 36)(7x - 42) = 0$
16. $(x - 12)(x + 12) = 0$
17. $x^2 - 6x = 0$
18. $y^2 = 7y$
19. $x^2 + 8x = 0$
20. $y^2 = -5y$
21. $3x^2 + 15x = 0$
22. $3n^2 + 12n = 0$
23. $2s^2 - 18s = 0$
24. $2x^2 = 50$
25. $4s^2 = -36s$
26. $3x^2 + 21x = 0$
27. $9x^2 - 36x = 0$
28. $x^2 - 25 = 0$
29. $p^2 = 81$
30. $x^2 - 14x + 49 = 0$
31. $y^2 + 8y + 16 = 0$
32. $x^2 + 10x + 9 = 0$
33. $y^2 + 5y - 6 = 0$
34. $m^2 - 7m + 6 = 0$
35. $a^2 - 8a - 9 = 0$
36. $r^2 - 36 = 0$
37. $r^2 - 10r = 0$
38. $x^2 - 8 = 7x$
39. $h^2 + 21 = -10h$
40. $z^2 = 18 + 7z$
41. $p^2 + 5p = 14$
42. $3b^2 - 27 = 0$
43. $m^3 - 64m = 0$
44. $x^3 - 6x^2 + 8x = 0$
45. $m^4 - 8m^2 + 16 = 0$
46. $(x + 3)(x - 2) = 14$
47. $(a - 1)(a - 1) = 36$
48. $h^3 - h^2 - h + 1 = 0$

Mathematical Excursions

Factoring Cubes

You know how to factor the difference of two squares. Can you factor the difference of two cubes? Consider $a^3 - b^3$.

$$a^3 - b^3 = a^3 - a^2b + a^2b - b^3 \qquad \text{Notice that } -a^2b + a^2b = 0.$$
$$= a^2(a - b) + b(a^2 - b^2)$$
$$= a^2(a - b) + b(a - b)(a + b)$$
$$= (a - b)[a^2 + b(a + b)]$$
$$= (a - b)(a^2 + ab + b^2)$$

Example   Factor $y^3 - 27$.

y^3 is the cube of y. 27 is the cube of 3.
Now use the pattern above.
$$y^3 - 27 = (y - 3)(y^2 + 3y + 9)$$

Exercises   Factor using the pattern above.

1. $x^3 - 64$
2. $8x^3 - y^3$
3. $27a^3 - 64b^3$
4. Find a pattern for factoring $a^3 + b^3$ by using a method similar to the one above.

When you solve a quadratic equation by factoring, you use the following principle.

$$\text{If } ab = 0, \text{ then } a = 0 \text{ or } b = 0.$$

Using logic symbolism, you can restate the principle as follows.

$$[ab = 0] \longrightarrow [(a = 0) \lor (b = 0)]$$

Now consider the inequality $ab > 0$. For what values of a and b will this inequality be true? It will be true only when either a and b are both positive or both negative.

$$[ab > 0] \longrightarrow [(a > 0) \land (b > 0)] \lor [(a < 0) \land (b < 0)]$$

Example Solve $(x + 6)(x - 2) > 0$ and graph the solution.

$$[(x + 6 > \quad 0) \land (x - 2 > 0)] \lor [(x + 6 < \quad 0) \land (x - 2 < 0)]$$
$$[(x > -6) \land \qquad (x > 2)] \lor \qquad [(x < -6) \land \qquad (x < 2)]$$
$$(x > 2) \lor \qquad (x < -6) \qquad \textit{Why?}$$

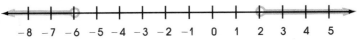

If $ab < 0$, either a must be positive and b negative or b must be positive and a negative.

$$[ab < 0] \longrightarrow [(a > 0) \land (b < 0)] \lor [(a < 0) \land (b > 0)]$$

Example Solve $(x + 7)(x - 8) < 0$ and graph the solution.

$$[(x + 7 > \quad 0) \land (x - 8 < 0)] \lor [(x + 7 < \quad 0) \land (x - 8 > 0)]$$
$$[(x > -7) \land \qquad (x < 8)] \lor \qquad [(x < -7) \land \qquad (x > 8)]$$

This disjunct can never be true.

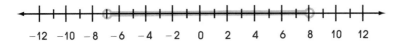

Exercises Solve each inequality and graph the solution.

1. $x(x + 4) > 0$
2. $x(x - 3) > 0$
3. $(x + 2)(x - 3) > 0$
4. $x(x - 5) < 0$
5. $(x + 6)(x - 3) < 0$
6. $x^2 + 5x + 6 > 0$
7. $x^2 - 2x - 15 > 0$
8. $x^2 + 6x + 5 < 0$
9. $y^2 < 9$
10. $y^2 + 2y > 35$
11. $x^2 > 36$
12. $x^2 - 3x \le -2$

10.6   Using Factoring in Problem Solving

Many problems can be solved by using factoring.

Examples

1 The area of the giraffe enclosure at the Sunbrook Zoo is 600 square meters. The length is 25 meters longer than the width. What are the dimensions of the enclosure?

Explore   Let x = the width.
Then $x + 25$ = the length.

Plan
$$\ell \cdot w = A$$
$$(x + 25)x = 600$$

Solve
$$x^2 + 25x = 600$$
$$x^2 + 25x - 600 = 0$$
$$(x + 40)(x - 15) = 0$$

$$x + 40 = 0 \quad \text{or} \quad x - 15 = 0$$
$$x = -40 \qquad\qquad x = 15$$

Since width must be positive, -40 is not a reasonable answer.
The dimensions are 15 meters and $15 + 25$, or 40 meters. *Examine this solution.*

2 Find two consecutive integers whose product is 72.

Explore   Let x = the lesser integer.
Then $x + 1$ = the next integer.

Plan
$$x(x + 1) = 72$$

Solve
$$x^2 + x = 72$$
$$x^2 + x - 72 = 0$$
$$(x + 9)(x - 8) = 0$$

$$x + 9 = 0 \quad \text{or} \quad x - 8 = 0$$
$$x = -9 \qquad\qquad x = 8$$

If $x = -9$, then $x + 1 = -8$. If $x = 8$, then $x + 1 = 9$.
The consecutive integers are -9 and -8 or 8 and 9. *Examine these solutions.*

Examples

3 Chris Mulisano is a professional photographer. She has a photo 8 cm long and 6 cm wide. A customer wants a print of the photo. The print is to be similar to the original but have half the area of the original. Chris plans to reduce the length and width of the photo by the same amount. What are the dimensions of the print?

Explore Let x = the number of centimeters the length and width are reduced.
The dimensions of the print are $(8 - x)$ cm and $(6 - x)$ cm.
The area of the photo is $8 \cdot 6$, or 48 cm^2.
The area of the print is $\dfrac{48}{2}$, or 24 cm^2.

Plan
$$\ell \cdot w = A$$
$$(8 - x)(6 - x) = 24$$

Solve
$$48 - 8x - 6x + x^2 = 24$$
$$x^2 - 14x + 24 = 0$$
$$(x - 12)(x - 2) = 0$$
$$x - 12 = 0 \quad \text{or} \quad x - 2 = 0$$
$$x = 12 \qquad\qquad x = 2$$

Why is 12 cm not a reasonable amount for Chris to reduce the length and width? Chris reduced the length and width by 2 cm each. The dimensions of the print are $8 - 2$, or 6 cm and $6 - 2$, or 4 cm.
Examine this solution.

4 The rectangular pelican pool at the Sunbrook Zoo measures 20 m by 12 m. A walk of uniform width surrounds the pool. The total area of the pool and walk is 560 square meters. How wide is the walk?

Explore Let x = width of the walk.

Plan

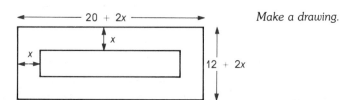

Make a drawing.

The total length is $(20 + 2x)$ meters.
The total width is $(12 + 2x)$ meters.

$$\ell \cdot w = A$$
$$(20 + 2x)(12 + 2x) = 560$$

Solve

$$240 + 40x + 24x + 4x^2 = 560$$
$$4x^2 + 64x - 320 = 0$$
$$4(x^2 + 16x - 80) = 0$$
$$4(x + 20)(x - 4) = 0$$
$$(x + 20)(x - 4) = 0 \qquad \textit{Divide both sides by 4.}$$
$$x + 20 = 0 \quad \text{or} \quad x - 4 = 0$$
$$x = -20 \qquad\qquad x = 4$$

Since -20 is not a reasonable answer, the walk is 4 m wide.   *Why is -20 not a reasonable answer?*

Exercises

Exploratory   **For each problem, define a variable. Then state an equation. Do *not* solve the equation.**

1. When 4 times a number is subtracted from the square of a number, the result is 32. Find the number.

2. When 5 times a number is added to the square of the number, the result is 66. Find the number.

3. Find two numbers whose sum is 17 and whose product is 72.

4. Find a positive integer that is 42 less than its square.

5. The area of a rectangle is 72 m². The rectangle is 6 m longer than it is wide. Find its dimensions.

6. A rectangle is 7 cm longer than it is wide. If its area is 30 cm², find its dimensions.

7. The length of a rectangle is 3 m more than its width. Find its dimensions if its area is 54 m².

8. The length of a rectangular garden is 6 m more than its width. If its area is 247 m², find its dimensions.

Written   **Solve the equation and answer the problem in each of the following Exploratory Exercises.**

1. Exercise 1
2. Exercise 2
3. Exercise 3
4. Exercise 4
5. Exercise 5
6. Exercise 6
7. Exercise 7
8. Exercise 8

For each problem, define a variable. Then write and solve an equation and answer the problem.

9. The height of a parallelogram is 4 in. shorter than its base. If the area of the parallelogram is 96 in.², find the base and height.

10. The height of a triangle is 4 cm less than the base. The area of the triangle is 16 cm². Find the base and height.

11. Find two consecutive integers whose product is 90.

12. Find two consecutive integers whose product is 132.

13. Find two positive consecutive integers whose product is 210.

14. Find two negative consecutive integers whose product is 110.

15. Find two consecutive even integers whose product is 80.

16. Find two consecutive even integers whose product is 224.

17. Find two negative consecutive even integers whose product is 120.

18. Find two positive consecutive even integers whose product is 440.

19. Find two consecutive odd integers whose product is 99.

20. Find two consecutive odd integers whose product is 195.

21. Find two positive consecutive odd integers whose product is 143.

22. Find two negative consecutive odd integers whose product is 399.

23. The sum of the squares of two consecutive positive integers is 145. Find the integers.

24. Find two consecutive integers whose product is 24 more than the square of the first integer.

25. The difference of the squares of two consecutive integers is 19. Find the integers.

26. If the squares of two positive consecutive even integers are added, the result is 100. Find the integers.

27. The sum of the squares of two positive consecutive odd integers is 202. Find the integers.

28. A rectangle is 10 cm by 13 cm. If each dimension is increased by the same amount, find the dimensions of a new rectangle whose area is 50 cm² more than the original rectangle.

29. The Protem Garden Society wants to double the area of its rectangular display of cacti. If it is now 6 m by 4 m, by what equal amount must each dimension be increased?

30. A rectangular garden 25 ft by 50 ft is increased on all sides by the same amount. Its area increases 400 sq ft. By how much is each dimension increased?

31. One leg of a right triangle is 1 cm shorter than the other leg. If the hypotenuse measures 5 cm, find the length of each leg.

32. One leg of a right triangle is 7 feet longer than the other leg. If the hypotenuse measures 13 ft, find the length of each leg.

33. A swimming pool, 20 ft by 40 ft, is surrounded by a walk of uniform width. The area of the walk itself is 1216 sq ft. Find the width of the walk.

34. A rectangular painting is 6 in. by 8 in. It is placed in a frame of uniform width. If the area of the frame itself is 72 sq in., find the width of the frame.

35. Find the width of the frame in exercise 34 if the painting is 3 ft by 4 ft and the total area of the frame and the painting is 20 sq ft.

36. Bob and Ellen's living room has a rug which is 9 ft by 12 ft. A strip of floor of equal width is uncovered on all sides of the room. If the area of the uncovered floor is 270 sq ft, how wide is the strip?

An equation like $x^2 - 4x + 4 = 3$ can be solved in the following way.

$$x^2 - 4x + 4 = 3$$
$$(x - 2)^2 = 3 \qquad \text{Factor } x^2 - 4x + 4.$$
$$\sqrt{(x - 2)^2} = \sqrt{3} \qquad \text{Find the square root of each side.}$$
$$|x - 2| = \sqrt{3} \qquad \sqrt{a^2} = |a|$$
$$x = 2 \pm \sqrt{3} \qquad \text{Why?}$$
$$x = 2 + \sqrt{3} \quad \text{or} \quad x = 2 - \sqrt{3}$$

To solve a quadratic equation using this method, the quadratic expression must be a perfect square. If it is not a perfect square, then a method called **completing the square** may be used.

Consider the pattern for squaring a binominal.

$$(x + 6)^2 = x^2 + x \cdot 6 + 6 \cdot x + 6 \cdot 6 \qquad \text{This can be written } x^2 + 2(6)x + 6^2.$$
$$= x^2 + 12x \qquad\qquad + 36$$

$$\left(\frac{12}{2}\right) \qquad \rightarrow \qquad 6^2 \qquad \text{Notice that 6 is } \frac{1}{2} \text{ of 12 and that 36 is } 6^2.$$

To complete the square for an expression of the form $x^2 + bx$, follow the steps listed below.

Step 1   Find one-half of b, the coefficient of x.
Step 2   Square the result of **Step 1.**
Step 3   Add the result of **Step 2** to $x^2 + bx$.

The method of completing the square can be used to develop a general formula for solving any quadratic equation. Begin with the general form of a quadratic equation, $ax^2 + bx + c$ where $a \neq 0$.

$$ax^2 + bx + c = 0$$
$$x^2 + \frac{b}{a}x + \frac{c}{a} = 0 \qquad \text{Divide by } a \text{ so the coefficient of } x^2 \text{ becomes 1.}$$
$$x^2 + \frac{b}{a}x = -\frac{c}{a} \qquad \text{Subtract } \frac{c}{a} \text{ from each side.}$$

$$x^2 + \frac{b}{a}x + \left(\frac{b}{2a}\right)^2 = -\frac{c}{a} + \left(\frac{b}{2a}\right)^2 \qquad \text{Now complete the square.}$$
$$\left(x + \frac{b}{2a}\right)^2 = -\frac{c}{a} + \frac{b^2}{4a^2} \qquad \text{Factor the left side.}$$
$$\left(x + \frac{b}{2a}\right)^2 = \frac{b^2 - 4ac}{4a^2} \qquad \text{Simplify the right side.}$$

$$x + \frac{b}{2a} = \pm \sqrt{\frac{b^2 - 4ac}{4a^2}}$$ *Find the square root of each side.*

$$x + \frac{b}{2a} = \frac{\pm \sqrt{b^2 - 4ac}}{2a}$$ *Simplify the square root on the right side.*

$$x = \frac{\pm \sqrt{b^2 - 4ac}}{2a} - \frac{b}{2a}$$ *Subtract $\frac{b}{2a}$ from each side.*

$$x = \frac{-b \pm \sqrt{b^2 - 4ac}}{2a}$$

This result is called the **Quadratic Formula** and can be used to solve *any* quadratic equation. ·

Quadratic Formula

> The solutions of a quadratic equation of the form $ax^2 + bx + c = 0$ with $a \neq 0$ are given by:
> $$x = \frac{-b \pm \sqrt{b^2 - 4ac}}{2a}$$

Example   Solve $x^2 - 6x - 3 = 0$ using the Quadratic Formula.

$$x = \frac{-b \pm \sqrt{b^2 - 4ac}}{2a}$$

$$= \frac{-(-6) \pm \sqrt{(-6)^2 - 4(1)(-3)}}{2(1)}$$ $a = 1, b = -6, c = -3$

$$= \frac{6 \pm \sqrt{36 + 12}}{2}$$

$$= \frac{6 \pm \sqrt{48}}{2}$$ $\sqrt{48} = \sqrt{16 \cdot 3}$ or $4\sqrt{3}$

$$= \frac{6 \pm 4\sqrt{3}}{2} \quad \text{or} \quad 3 \pm 2\sqrt{3}$$ *Check this result.*

The solution set is $\{3 + 2\sqrt{3}, 3 - 2\sqrt{3}\}$.

Exercises   Solve each equation using the Quadratic Formula.

1. $x^2 + 7x + 6 = 0$ **2.** $y^2 + 8y + 15 = 0$ **3.** $m^2 + 4m + 2 = 0$

4. $p^2 + 5p + 3 = 0$ **5.** $2r^2 + r - 15 = 0$ **6.** $8t^2 + 10t + 3 = 0$

7. $2t^2 - t = 4$ **8.** $-4x^2 + 8x = -3$ **9.** $2y^2 + 3 = -7y$

10. $3n^2 - 5n = -1$ **11.** $3m^2 + 2 = -8m$ **12.** $3k^2 + 11k = 4$

13. $-x^2 + 5x - 6 = 0$ **14.** $r^2 + 10r + 9 = 0$ **15.** $y^2 - 7y - 8 = 0$

16. $24x^2 - 2x - 15 = 0$ **17.** $21x^2 + 5x - 6 = 0$ **18.** $35x^2 - 11x - 6 = 0$

19. $2x^2 - 0.7x - 0.3 = 0$ **20.** $x^2 - 1.1x - 0.6 = 0$ **21.** $-1.6r^2 - 8.7r = 3.5$

10.7   Additional Factoring Techniques

A factorable polynomial should always be factored completely. That is, continue until you can no longer factor. For example, consider $3x^2 + 33x + 90$.

$$3x^2 + 33x + 90 = 3(x^2 + 11x + 30)$$
$$= 3(x + 6)(x + 5)$$

The polynomial $3x^2 + 33x + 90$ is now factored completely.

Example

1 **Factor $y^4 - 10y^2 + 9$ completely.**

$$y^4 - 10y^2 + 9 = (y^2 - 1)(y^2 - 9)$$
$$= (y - 1)(y + 1)(y - 3)(y + 3)$$

Both $y^2 - 1$ and $y^2 - 9$ are the difference of squares.

Sometimes you can factor a polynomial by grouping the terms and factoring a monomial from the groups. To group the terms, look for common factors.

Example

2 **Factor $x^2 - 3xy + 8x - 24y$.**

$$x^2 - 3xy + 8x - 24y = (x^2 - 3xy) + (8x - 24y) \quad \text{\textit{Group the terms.}}$$
$$= x(x - 3y) + 8(x - 3y) \quad \text{\textit{Factor a monomial from each group.}}$$
$$= (x + 8)(x - 3y) \quad \text{\textit{Use the distributive property.}}$$

Sometimes you can group the terms in more than one way. In the following example, the terms in example 2 are grouped in a different way.

Example

3 Factor $x^2 - 3xy + 8x - 24y$.

$$x^2 - 3xy + 8x - 24y = (x^2 + 8x) + (-3xy - 24y)$$
$$= x(x + 8) - 3y(x + 8)$$
$$= (x - 3y)(x + 8) \qquad \text{The result is the same as in example 2.}$$

You have used many methods to factor polynomials. The following steps can help you decide the method to use when factoring a polynomial which has no like terms.
   1. Check for a common monomial factor.
   2. Check for the difference of two squares.
   3. Check to see if it is a factorable trinomial.
   4. For more than three terms, try factoring by grouping.

Exercises

Exploratory Factor completely.

1. $3x^2 + 9$
2. $3ax + 5ay$
3. $8m^2n - 13m^2p$
4. $4cd^2 + 16bd^2$
5. $4a^2 - 8a^2b$
6. $3m^2n + 6mn + 9mn^2$
7. $2a^2b + 8a^2b^2 + 10a^2b^3$
8. $4x^2 - 81y^2$
9. $m^2 - 4m + 4$
10. $x^2 + 14x + 49$
11. $3x^2 - 7x + 2$
12. $8y^2 - 6y - 9$

Express each of the following in factored form.

13. $a(a + b) - b(a + b)$
14. $m(n + t) - n(n + t)$
15. $5x(x - y) + y(x - y)$
16. $r(r - s) + s(r - s)$
17. $4ab(a - 3b) - 3c^2(a - 3b)$
18. $8xy(x - 3) - 5(x - 3)$
19. $5a(x + y) + (x + y)$
20. $a^2(a - 3b) - (a - 3b)$
21. $3(2a - 3b) + a^2(2a - 3b)$
22. $3ab(2a + 3b) - 5bc(2a + 3b)$
23. $6x(a + b) + 4y(a + b) + 7(a + b)$
24. $7a(x + y - z) + 8b(x + y - z)$

Written Factor completely.

1. $3x^2 - 3y^2$
2. $5a^2 - 5b^2$
3. $7y^2 - 7$
4. $2y^2 - 72$
5. $3b^2 - 147$
6. $ax^2 - ay^2$
7. $y^4 - 16$
8. $x^4 - 81$
9. $x^4 - y^2$

10. $3y^2 + 9$

11. $3a^2 - 27b^2$

12. $m^4 - 1$

13. $4x^2 - 36$

14. $b^3 - 4b$

15. $3x^4 - 48$

16. $5b^2 - 20c^2$

17. $16a^2 - a^2b^4$

18. $x^2y^3 - 25y$

19. $6y^2 + 13y + 6$

20. $3m^2 + 21m - 24$

21. $5x^2 + 15x + 10$

22. $a^2 + 8ab + 16b^2$

23. $3a^2 + 24a + 45$

24. $20y^2 + 34y + 6$

25. $3xy - 6x + 5y - 10$

26. $12ab - 28a - 9b + 21$

27. $5x^2 - 5xy + xy - y^2$

28. $a^2 - ab - ab^2 + b^3$

29. $5y^6 - 5y^2$

30. $y^4 - 10y^2 + 9$

31. $6a^2 + 27a - 15$

32. $4m^2 + 12mn + 9n^2$

33. $6x^2 + 9x - 105$

34. $7mx^2 + 2nx^2 - 7my^2 - 2ny^2$

35. $4ax + 14ay - 10bx - 35by$

36. $r^2 - rt - rt^2 + t^3$

37. $8ax - 6x - 12a + 9$

38. $10x^2 - 14xy - 15x + 21y$

39. $8a^2 + 8ab + 8ac + 3a + 3b + 3c$

40. $2a^2x + 10ax + 4x + a^2y + 5ay + 2y$

41. $2a^2 + 3ab - 4ac - 14a - 21b + 28c$

42. $5a^2x + 4aby + 3acz - 5abx - 4b^2y - 3bcz$

43. $x^3 - 5x^2 - 7x + x^2y - 5xy - 7y$

44. $3x^3 + 2x^2 - 5x + 9x^2y + 6xy - 15y$

═══════ **Mixed Review** ═══════

Perform the indicated operations and simplify.

1. $\dfrac{5}{2x^2} + \dfrac{7}{4x}$

2. $\dfrac{2t - 3}{6t} - \dfrac{5t - 4}{8t}$

3. $\dfrac{y + 12}{2y} \cdot \dfrac{3y}{10}$

Solve each equation.

4. $\dfrac{x + 7}{5} = \dfrac{3}{10}$

5. $\dfrac{y - 1}{6} = \dfrac{y + 3}{4}$

6. $\dfrac{a - 5}{6} = \dfrac{6}{a}$

7. The ratio of the length to the width of a rectangle is 2 : 3. If the area of the rectangle is 216 cm², find its dimensions.

8. The length of a rectangle is 10 m more than twice the width. If the area of the rectangle is 5500 m², find its dimensions.

9. In a certain theater, each row has the same number of seats. The number of rows in the theater is six less than the number of seats in a row. If the theater has 160 seats, find the number of rows.

10. A square and a rectangle have the same area. The length of the rectangle is 8 cm more than length of the square. The width of the rectangle is 4 cm less than the width of the square. Find the dimensions of the square and the rectangle.

Portfolio Suggestion

Review the items in your portfolio. Make a table of contents of the items, noting why each item was chosen. Replace any items that are no longer appropriate.

Performance Assessment

One way to factor a trinomial such as $x^2 - 2x - 3$ is to assign a value such as 10 to x and evaluate the expression. For example, $x^2 - 2x - 3$ becomes $10^2 - 2(10) - 3$ or 77. Since 77 is $(7)(11)$ and $x = 10$, $(7)(11)$ can be expressed as $(x - 3)(x + 1)$. How would you use this method to factor $x^2 - 8x + 15$? Include an explanation of your solution.

Chapter Summary

1. A **prime number** is an integer, greater than 1, whose only positive factors are 1 and itself. (320)
2. The product of the prime factors of a number is called the **prime factorization** of the number. (320)
3. The **greatest common factor** (gcf) of two or more monomials is the common factor with the greatest numerical factor and the greatest degree. (321)
4. The distributive property and the gcf may be used to factor expressions with terms containing common factors. (323)
5. **Difference of Squares:** $(a + b)(a - b) = a^2 - b^2$ (325)
6. A polynomial in one variable of degree 2 is called a **quadratic polynomial.** The term of degree 2 in a quadratic polynomial is the **quadratic term.** The term of degree 1 is the **linear term.** The numerical term is the **constant term.** (328)
7. The pattern $(x + m)(x + n) = x^2 + (m + n)x + mn$ can be used to factor trinomials of the form $x^2 + bx + c$. For trinomials of the form $ax^2 + bx + c$, consider the factors of both a and c. (329)
8. **Zero Product Property:** For all numbers a and b, $ab = 0$ if and only if $a = 0$ or $b = 0$. (332)
9. Sometimes a polynomial can be factored by grouping the terms and factoring a monomial from the group. (343)

Chapter Review

10.1   Find the prime factorization of each integer. Write each negative integer as the product of -1 and its prime factors.

1. 28   **2.** -50 **3.** 124   **4.** 200

Find the gcf for each of the following.

5. 24, 120   **6.** 64, 48   **7.** $-16a^2$, $30a^3$
8. $24xy^3$, $36z^2$ **9.** $12xy^3z$, $72x^2yz$ **10.** $60a^3b^5c^2$, $150a^2b^2c$, $36a^2bc^4$

10.2   Factor.

11. $12x^2 + 18x$ **12.** $3a^2b + 12ab^2$
13. $x^4 + x^3$ **14.** $21x^3y + 15x^2y^2 - 9y^4$

10.3   Find each product.

15. $(a + 2)(a - 2)$ **16.** $(x - 7)(x + 7)$
17. $(f - g)(f + g)$ **18.** $(5a + 3b^2)(5a - 3b^2)$

Factor.

19. $a^2 - 25$ **20.** $4x^2 - 36$
21. $16x^2 - 81y^2$ **22.** $14x^2 - 56y^2$

10.4   Factor.

23. $y^2 + 9y + 14$ **24.** $x^2 - 8x + 15$ **25.** $a^2 + 5a - 6$
26. $b^2 + 2b - 35$ **27.** $2x^2 + x - 15$ **28.** $6y^2 - y - 12$

10.5   Solve each equation and check your answer.

29. $n(n + 12) = 0$ **30.** $(x - 7)(3x + 15) = 0$
31. $x^2 + 13x + 42 = 0$ **32.** $x^2 + 35 = -12x$

10.6   For each problem, define a variable. Then write and solve an equation and answer the problem.

33. A certain number decreased by 3 is multiplied by the same number increased by 3. The product is 27. What is the number?

34. The length of the top of George's bookcase is 30 cm more than the width. The area of the top is 1800 cm². What are its dimensions?

10.7   Factor.

35. $3x^2 - 48$ **36.** $4a^2 + 12a + 8$
37. $2y^3 - 18y^2 + 28y$ **38.** $bx + cx - 64b - 64c$
39. $ay^2 - by^2 - 16a + 16b$ **40.** $m^2p^2 - 4m^2 + n^2p^2 - 4n^2$

Chapter Test

Find the prime factorization of each number.

1. 40

2. 75

Find the gcf for each of the following.

3. 40, 64

4. $28a^2b$, $70a^2b^3c$

5. $12x^2y^3$, $18xy^4z$, $45xy^2z$

Factor.

6. $9a^3b^2 + 12ab^3$

7. $12x^2y^2 - 20x^2y^5 + 8xy^4$

8. $x^2 - 81$

9. $4x^2 - 9y^2$

10. $x^2 + 9x + 8$

11. $m^2 - 4m + 3$

12. $h^2 + 3h - 10$

13. $a^2 - 11a - 12$

14. $3x^2 + x - 14$

15. $10x^2 - 37x + 30$

16. $xk - 8m + xm - 8k$

17. $y^2 - 2y + 3y - 6$

18. $8x^3 + 28x^2 + 12x$

19. $s^4 - 5s^2 + 4$

Solve each equation and check your answer.

20. $7x(2x - 14) = 0$

21. $a^2 + a - 30 = 0$

22. $y^2 + 11y = -24$

For each problem, define a variable. Then write and solve an equation and answer the problem.

23. The sum of the squares of two consecutive integers is 313. Find the integers.

24. The height of a triangle is 8 cm greater than twice its base. If the area of the triangle is 21 square centimeters, find the base and height.

25. A rectangular garden at Overlook Park measures 25 m by 16 m. A walk of uniform width surrounds the garden. The total area of the garden and the walk is 910 square meters. Find the width of the walk.

Introduction to
Probability

Application in Sports

Professional athletes know it takes good strategy as well as skill to win a game. Coaches are responsible for determining the basic strategy of a game. They base their decisions on previous experience and probability.

Roberto Alomar is an outstanding second baseman for the Toronto Blue Jays. If his current batting average is 0.285, what is the **probability** that he will get a hit each of his next two times at bat?

Group Project: *Agriculture*

Use two different packages of seeds for bean plants. You may use different brands or different types of beans. Place the seeds from each package between damp paper towels keeping the different types of seed separated. After several days, check to see how many of each type of seed germinated. Make a chart showing the total number of each type of seed, the number of each that germinated, and the probability that each type of seed will germinate.

11.1   The Probability of an Event

Suppose the weather forecast predicts that the probablity of rain is 95%. Few of us on hearing this would go out without a raincoat or umbrella. But if that probability is near zero, we would not bother with such equipment.

Probability theory can help you to make many decisions.

When a coin is tossed, only 2 **outcomes** or *events* are possible. The coin will land either heads or tails. Each outcome is equally likely to happen. The outcome we want to happen is called a **success.**

Examples

1 **What is the probability that a tossed coin will land *heads?***

The event we want is heads. There is 1 success out of 2 possible outcomes.

$P(heads) = \dfrac{1}{2}$ *P(heads) is read "the probability of heads."*

or $P(H) = \dfrac{1}{2}$ *P(heads) may be named P(H).*

The probability that a tossed coin will land heads is $\dfrac{1}{2}$.

2 **What is the probability that the spinner shown in the drawing will stop on 3?**

$P(3) = \dfrac{1}{5}$ *There is 1 success. There are 5 possible outcomes. The spinner is made so that each outcome is equally likely to happen.*

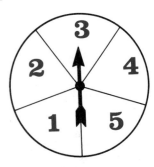

The probability of spinning a 3 is $\dfrac{1}{5}$.

We shall assume that coins, spinners, and so on are **fair,** or **unbiased,** unless we say otherwise. Thus, we assume that a coin will not produce an unlikely outcome. That is, heads will not appear more often than tails. We shall also assume that coins will not land on their rims or dice on their edges.

Example

3 **What is the probability that the spinner in example 2 will stop on an even number?**

Since there are 2 even numbers, the 2 and 4, there are 2 successful events.

$P(even\ number) = \dfrac{2}{5}$ The probability of spinning an even number is $\dfrac{2}{5}$.

Probability of an Event

The probability of an event, $P(E)$, is found by dividing the number of ways that the event can occur, $n(E)$, by the total number of possible outcomes, $n(S)$.

$$P(E) = \dfrac{n(E)}{n(S)}$$

Examples

4 **A card is chosen at random from a deck of 52 cards. What is the probability of choosing a king or queen?**

Use the rule $P(E) = \dfrac{n(E)}{n(S)}$.

$n(E) = 4 + 4 = 8$ *There are four kings and four queens.*
$n(S) = 52$ *There are 52 possible cards to pick.*

$P(K\ or\ Q) = \dfrac{n(E)}{n(S)} = \dfrac{8}{52}$ or $\dfrac{2}{13}$ The probability of a king or queen is $\dfrac{2}{13}$.

5 **What is the probability that the spinner in example 2 will *not* stop on an even number?**

Not stopping on an even number is the same as stopping on an odd. But note that the number of even and odd numbers is the total number of possible outcomes. In general, $n(E) + n(not\ E) = n(S)$.

$P(not\ even) = \dfrac{3}{5}$ *There are 5 possible outcomes and 2 even numbers.*
                            Thus, n(not E) = 3.

Exercises

Exploratory   If a die is fair, find the probability of rolling each of the following.

1. 4
2. an even number
3. an odd number
4. *not* 5

5. 6
6. a prime number
7. an even prime number
8. a 1 or 3

Suppose you choose answers at random on a test. Find the probability of choosing the right answer in each case.

9. a true-false question
10. a multiple-choice question with 5 choices

Express each of the following fractions as a decimal and as a percent.

11. $\dfrac{4}{100}$
12. $\dfrac{1}{2}$
13. $\dfrac{1}{5}$
14. $\dfrac{3}{8}$
15. $\dfrac{65}{100}$

Express each percent as a fraction in simplest form.

16. 75%
17. $66\dfrac{2}{3}$%
18. $12\dfrac{1}{2}$%
19. 32%
20. 7.5%

Written   A spinner and a circle with eight congruent sectors are shown. Find the probability of spinning each of the following.

1. 5
2. 7
3. an even number
4. an odd number
5. *not* 8
6. a prime number
7. a number greater than 5
8. a 5 or 6
9. an even prime number
10. a number greater than or equal to 5

A bag contains 2 red, 1 green, and 3 blue marbles. If you choose a marble at random, find the probability of choosing each of the following.

11. a green marble
12. a red marble
13. a blue marble
14. a red or green marble
15. a red or blue marble
16. a marble which is *not* red

A card is chosen at random from a well-shuffled deck of 52 cards. Find the probability of choosing each of the following.

17. the king of hearts
18. a red ace
19. a queen
20. the jack of spades
21. a club
22. *not* a 7

Challenge   Find the probability of spinning each of the following with a spinner such as the one shown above.

1. at most 3
2. an integer
3. a negative number

11.2   Impossibilities and Certainties

What is the probability of rolling a 10 with one roll of one die? "That's impossible," you say.

What is the probability of the impossible event?

$$P(E) = \frac{n(E)}{n(S)} \qquad \text{Use the rule.}$$

Here, E is "rolling a 10." The numerator, $n(E)$, is the number of ways you can roll a 10 with one die. So $n(E)$ is 0. Since there are six possible outcomes, $n(S)$ is 6.

$$P(10) = \frac{0}{6} \quad \text{or} \quad 0$$

Similar examples of other impossible events lead us to the following conclusion.

Probability of an Impossible Event

> The probability of an impossible event is 0.

Example

1 Each of the letters of the word SMILE is written on a slip of paper. The five slips of paper are placed in a hat. What is the probability of choosing the letter A from the hat?

Of 5 possible outcomes, *none* is A. Therefore, the probability is 0.

If the probability of an event which *cannot* occur is 0, what is the probability of an event which is certain to occur?

What is the probability that a number rolled with a die is less than 15? This event is certain to occur. Every time you roll a die you get a number less than 15. That is, $n(E)$ is equal to $n(S)$. There are six outcomes so $n(S) = 6$. Thus, $n(E) = 6$.

$$P(\text{less than 15}) = \frac{6}{6} \quad \text{or} \quad 1$$

If an event is a certainty, then every outcome is a success.

Since $n(E) = n(S)$, then $\dfrac{n(E)}{n(S)}$ is 1.

**Probability of
a Certainty**

> The probability of an event which is certain to occur is 1.

Example

2 **What is the probability of choosing a consonant from the letters Q, B, X, or Z?**

Since these letters are all consonants, every possible choice is a success. The probability is 1.

Can the probability of any event ever be negative? Since $P(E) = \dfrac{n(E)}{n(S)}$, $P(E)$ will be negative only if either $n(E)$ or $n(S)$ is negative. But this will be true only if there is either a negative number of successes or a negative number of possible outcomes. Both of these are impossible!

Can $P(E)$ be greater than 1? For $P(E)$ to be greater than 1, $\dfrac{n(E)}{n(S)}$ must be greater than 1. This requires that $n(E)$ be greater than $n(S)$. But the number of ways an event can occur cannot be greater than the total number of possible outcomes. These conclusions can be summarized as follows.

**Least and Greatest
Probabilities**

> The probability of an event can never be less than 0 or greater than 1.

*P(not E) is the probability
that event E does not
occur.*

For any event E, there is a corresponding event, *not E,* that consists of all outcomes that *are not* successes for event E. The event *not E* is called the *complement* of event E.

If we know the probability of event E, $P(E)$, then we can determine $P(not\ E)$.

$$P(\text{not } E) = \frac{n(\text{not } E)}{n(S)}$$

$$= \frac{n(S) - n(E)}{n(S)} \qquad n(S) - n(E) = n(\text{not } E). \text{ Why?}$$

$$= \frac{n(S)}{n(S)} - \frac{n(P)}{n(S)} \quad \text{or} \quad 1 - P(E)$$

Rule for *P(not E)* | $P(\text{not } E) = 1 - P(E)$

Examples

3 **What is the probability that the number rolled with a die is not a 5?**

$$P(\text{not } 5) = 1 - P(5) \qquad \textit{There are six possible outcomes and only one 5.}$$

$$= 1 - \frac{1}{6} \qquad \textit{Thus, } P(5) = \frac{1}{6}.$$

$$= \frac{5}{6} \qquad \textit{The probability of not rolling a 5 is } \frac{5}{6}.$$

4 **One letter is chosen at random from the word MANHATTAN. Find the probability that the letter chosen is an A. Then find the probability that the letter chosen is not an N.**

$$P(A) = \frac{3}{9} \qquad \textit{There are 9 possible outcomes and 3 As.}$$

$$= \frac{1}{3}$$

$$P(\text{not } N) = 1 - P(N)$$

$$= 1 - \frac{2}{9} \qquad \textit{There are 9 possible outcomes and 2 Ns.}$$

$$= \frac{7}{9}$$

The probability of choosing an A is $\frac{1}{3}$.

The probability of not choosing an N is $\frac{7}{9}$.

Exercises

Exploratory   A die is rolled. Find the probability of rolling each of the following.

1. 1
2. a number less than 3
3. a multiple of 3
4. 7
5. a number less than 7
6. at least 3

Answer each of the following.

7. What is the probability of an impossible event?

8. What is the probability of an event which is certain to occur?

9. Can the probability of an event be negative?

10. Can the probability of an event be greater than 1?

Written One letter is chosen at random from each of the following words. Find the probability that the letter chosen is an A.

1. HUMAN
2. RADAR
3. BASKETBALL
4. CLAMMY
5. BANANA
6. BASEBALL
7. YAMAHA
8. PANAMANIAN

Each of the 26 letters of the alphabet is written on a slip of paper and placed in a box. Assume that the only vowels are A, E, I, O, and U, and the rest are consonants. Suppose a slip is chosen at random. Find the probability of choosing each of the following.

9. E
10. a vowel
11. B or C
12. a vowel or consonant
13. a consonant
14. neither a vowel nor a consonant
15. a letter from CLAM
16. a letter not in OYSTER

Find the probability of choosing each of the following from a 52-card deck.

17. a red 3
18. a heart, spade, club, or diamond
19. a diamond
20. the 17 of clubs
21. a 4 or 7
22. the 2 or 3 of spades

A box contains 4 black marbles, 3 white marbles, 2 blue marbles, and 1 red marble. Find the probability of choosing each of the following.

23. a black marble
24. a white marble
25. a green marble
26. a red or black marble
27. a red marble
28. a marble that is *not* green

Use the same marbles as in exercises 23-28. Find each of the following.

29. *P(blue)*
30. *P(not blue)*
31. *P(not white)*
32. *P(not red)*
33. *P(not orange)*
34. *P(not black)*
35. *P(yellow)*
36. *P(white or blue)*
37. *P(white or green)*

Challenge A card is chosen at random from a 52-card deck. Let *E* be the event "a 5 is chosen." Let *F* be "a diamond is chosen." Let *G* be "a face card is chosen." Find each of the following   *A face card is a jack, queen, or king.*

1. *P(E)*
2. *P(not E)*
3. *P(F)*
4. *P(G)*
5. *P(not F)*
6. *P(not G)*
7. *P(E or F)*
8. *P(F or G)*
9. *P(E or G)*
10. *P(not E or F)*
11. *P(not F or G)*
12. *P(not E or G)*

11.3   Adding Probabilities

Suppose that on a shelf of 10 books, 3 of the books are red and 1 is green. What is the probability that a book chosen at random is red or green? There are 4 red or green books. Thus, $P(red\ or\ green) = \dfrac{4}{10}$. But $P(red) = \dfrac{3}{10}$ and $P(green) = \dfrac{1}{10}$. Therefore, we see that $P(red\ or\ green) = P(red) + P(green)$.

For *any* events E and F, is it always true that $P(E\ or\ F) = P(E) + P(F)$? Consider the following example.

Example

1 **A card is drawn at random from a deck. What is the probability of drawing a king or a club?**

$P(king) + P(club) = \dfrac{4}{52} + \dfrac{13}{52}$ or $\dfrac{17}{52}$ $P(king) = \dfrac{4}{52}, P(club) = \dfrac{13}{52}$

If $\dfrac{17}{52}$ is $P(king\ or\ club)$, then out of 52 possibilities there are 17 successes. Let us name and count them. First, draw the thirteen clubs one by one.

Now draw the four kings.

Only three kings can be drawn! The king of clubs was drawn with the other clubs. Thus, there are only 16 successes.

The probability is $\dfrac{16}{52}$ or $\dfrac{4}{13}$.   *P(king or club) is not equal to P(king) + P(club).*

In the previous example, difficulty comes in counting one success, the king of clubs, twice. Notice that this card satisfies both conditions. It is both a king and a club.

Example

2 What is the probability that Sara will spin a number greater than 8 or an odd number with the spinner shown at the right?

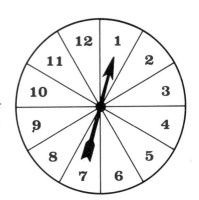

Count the successes to find $n(E)$.

9, 10, 11, 12   *Numbers on the spinner greater than 8.*
1, 3, 5, 7   *Odd numbers not yet counted.*
$n(E) = 8$ *Number of successes.*
$n(S) = 12$ *Number of possible outcomes.*

$P(\text{greater than 8 or odd}) = \dfrac{8}{12}$ or $\dfrac{2}{3}$.

The probability that Sara will win is $\dfrac{2}{3}$.

In example 2, suppose we let $P(E)$ represent $P(\text{greater than 8})$ and $P(F)$ represent $P(\text{odd})$. Is $P(E \text{ or } F) = P(E) + P(F)$?

$$P(\text{greater than 8}) + P(\text{odd}) = \frac{4}{12} + \frac{6}{12}$$
$$= \frac{10}{12} \text{ or } \frac{5}{6}$$

Thus, $P(\text{greater than 8 or odd}) \neq P(\text{greater than 8}) + P(\text{odd})$.

Again some successes were counted twice. They are the 9 and 11. Both are odd and greater than 8. In a correct solution they would be counted only once. Thus, we must subtract from $P(E) + P(F)$ the probability of those successes which were counted twice, that is, the probability of E and F.

Rule for $P(E \text{ or } F)$

$$P(E \text{ or } F) = P(E) + P(F) - P(E \text{ and } F)$$

If events E and F have no outcomes in common, $P(E \text{ and } F)$ is zero. In this case, $P(E \text{ or } F) = P(E) + P(F)$.

In example 2, we could use the rule for $P(E \text{ or } F)$.

$$P(\text{greater than 8 or odd}) = \frac{4}{12} + \frac{6}{12} - \frac{2}{12}$$
$$= \frac{8}{12} \text{ or } \frac{2}{3}$$

Examples

3 If a letter is selected at random from the alphabet, what is the probability it will be a letter from CAT or NAP?

$P(CAT \text{ or } NAP) = P(CAT) + P(NAP) - P(CAT \text{ and } NAP)$

$$= \frac{3}{26} + \frac{3}{26} - \frac{1}{26} \quad \text{or} \quad \frac{5}{26} \qquad \textit{There is one letter in common, A.}$$

The probability is $\frac{5}{26}$.

4 The table on the right lists the name, sex, and political party of members of a certain committee. Suppose one member is chosen at random to chair the committee. What is the probability the person will be a Republican or a woman?

Name	Sex	Party
1. K. Adams	F	Independent
2. M. Barnard	F	Republican
3. S. Cohen	M	Democrat
4. D. Davidson	M	Democrat
5. C. Eastman	M	Conservative
6. S. Falk	M	Republican
7. B. Garber	F	Liberal
8. J. Halloran	M	Independent
9. M. Ianelli	F	Conservative
10. R. Jarmyn	F	Republican

There are 3 Republicans on the committee. There are 5 women. Two of the women are Republicans.

$P(R \text{ or } W) = P(R) + P(W) - P(R \text{ and } W)$

$$= \frac{3}{10} + \frac{5}{10} - \frac{2}{10} \quad \text{or} \quad \frac{6}{10}$$

R stands for Republican.
W stands for Woman.
R and W stands for Republican Woman.

The probability of choosing a Republican or a woman is $\frac{6}{10}$ or $\frac{3}{5}$.

5 The committee in example 4 will select its treasurer by random choice. Find the probability that the person chosen will be a Liberal or a woman.

$P(L \text{ or } W) = P(L) + P(W) - P(L \text{ and } W)$

$$= \frac{1}{10} + \frac{5}{10} - \frac{1}{10}$$

$$= \frac{5}{10} \quad \text{or} \quad \frac{1}{2}$$

One of the events is completely contained in the other.

Exercises

Exploratory **Answer each of the following.**

1. In general, is *P(A or B)* equal to *P(A)* + *P(B)*?

2. Is it ever true that *P(A or B)* equals *P(A)* + *P(B)*?

Suppose the names JACOB, JASON, JIM, MARIA, MARK, RAY, and SALLY are written on slips of paper and placed in a box. A slip of paper is chosen from the box. Find the probability of each of the following choices by counting successes. If it is appropriate, use the general rule *P(E or F)* = *P(E)* + *P(F)* − *P(E and F)*.

3. a boy's name

4. a girl's name

5. a name beginning with J

6. a name ending with Y

7. a 6-letter name

8. a name beginning with J or M

9. a 5-letter name beginning with M

10. a 3-letter name beginning with J

11. a name containing 5 letters or beginning with M

12. a name containing 3 letters or beginning with J

Written **Each letter of the alphabet is written on a slip of paper which is placed in a hat. One is chosen at random. Find the probability of each of the following choices.**

1. B or C

2. a vowel or T

3. a letter from DOG or WALK

4. a letter from CAT or CAR

5. a letter from CAT or SKATE

6. a letter from DOG or DONKEY

7. a letter from BLACK or WHITE

8. a letter from WHITE or WHISK

The committee in example 4 will choose a secretary at random. Find the probability of choosing each of the following.

9. a woman or a Democrat

10. a woman or an Independent

11. a man or a woman

12. a man or a Liberal

13. a man or a Republican

14. a woman who is not a Conservative

Copy these tables for events *A* and *B*. Fill in the blanks.

	P(A)	P(B)	P(A and B)	P(A or B)
15.	0.3	0.4	0.1	____
17.	0.23	0.03	0.17	____
19.	0.4	0.5	____	0.8

	P(A)	P(B)	P(A and B)	P(A or B)
16.	0.5	0.23	0.14	____
18.	0.001	0.125	0.01	____
20.	0.15	0.46	____	0.5

A number *n* is randomly chosen from the set {1, 2, 3, 4, 5, 6, 7, 8, 9, 10}. Find the probability of each of the following.

21. *n* is even

22. *n* is a multiple of 3

23. *n* is a prime number

24. *n* is odd

25. *n* is a multiple of 5

26. $3n = 12$

27. $4n - 5 = 31$

28. $6n \leq 35$

29. $n^2 = 50$

30. $n^2 > 30$

31. $(n - 3)(n - 5) = 0$

32. $n^2 - 6n - 16 = 0$

11.4 Probability and Sets

Important ideas about probability can be expressed in set notation.

A set, or collection of things, is named by a capital letter. The symbol $n(B)$ stands for the number of elements of set B. Suppose $B = \{$Washington, Adams, Jefferson$\}$. Then $n(B) = 3$.

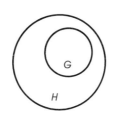

Suppose all the elements of some set G are contained in set H. Then G is a **subset** of H. This is symbolized $G \subset H$.

In the diagram at the left, set H includes the region within the large circle. Set G includes the region within the small circle.

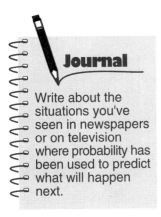

The null set is sometimes called the empty set.

The set that has no members is the *null set* or $\emptyset$. Clearly, $n(\emptyset) = 0$. Also, $\emptyset \subset A$ for any set A, since there can be no element of $\emptyset$ which fails to be in A.

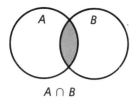

$A \cap B$

Sets can be combined. Suppose A and B are the sets given at the right. The new set formed by taking the elements that A and B have in common is called the **intersection** of A and B, or $A \cap B$.

$A = \{1, 2, 3, 5, 7\}$
$B = \{2, 4, 5, 8, 10\}$
$A \cap B = \{2, 5\}$

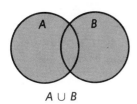

$A \cup B$

The set formed by all the elements found in either set A or set B or both is called the **union** of A and B, or $A \cup B$.

$A \cup B = \{1, 2, 3, 4, 5, 7, 8, 10\}$

An outcome set is also called a sample space.

In a discussion of probability, a set of all the possible outcomes is called an **outcome set**. Outcome sets are generally symbolized by the letter S. The number of elements in an outcome set is $n(S)$. In the case of tossing a coin, $S = \{heads, tails\}$ and $n(S) = 2$. Any subset of the outcome set is called an **event**. Set notation can be used to find probabilities as in the example which follows.

Example

1 **Suppose one die is rolled. Use set notation to find the probability of rolling a prime number.**

Since $E = \{2, 3, 5\}$, $n(E) = 3$.   *There are 3 successes, 2, 3, and 5.*

Since $S = \{1, 2, 3, 4, 5, 6\}$, $n(S) = 6$.   *S is the outcome set.*

$P(prime) = \dfrac{3}{6}$ or $\dfrac{1}{2}$ *Use the rule $P(E) = \dfrac{n(E)}{n(S)}$.*

Suppose in rolling one die you wish to roll a prime or even number. Let $E = \{prime\}$ and $F = \{even\}$. What is "E or F"? It is the event of rolling a prime or even number. That event is the set $\{2, 3, 4, 5, 6\}$. But this set is $E \cup F$. Thus, $E \cup F$ is another way to refer to the *event E or F*.

Consider the event of rolling a prime and an even number. This is the set $\{2\}$ which is also $E \cap F$. Thus, $E \cap F$ is another way to refer to the *event E and F*.

Thus, the rule $P(E \text{ or } F) = P(E) + P(F) - P(E \text{ and } F)$ can be rewritten in set notation.

Rule for P(E ∪ F)

$$P(E \cup F) = P(E) + P(F) - P(E \cap F)$$

Example

2 **Use the notation from example 1 to find the probability of rolling a prime or an even number with the roll of one die. Then, verify the rule for $P(E \cup F)$.**

$P(E \cup F) = \dfrac{n(E \cup F)}{n(S)} = \dfrac{5}{6}$ *Use the rule $P(E) = \dfrac{n(E)}{n(S)}$.*

$P(E \cup F) = \dfrac{n(E)}{n(S)} + \dfrac{n(F)}{n(S)} - \dfrac{n(E \cap F)}{n(S)}$ *Use the rule $P(E \cup F) = P(E) + P(F) - P(E \cap F)$.*

$= \dfrac{3}{6} + \dfrac{3}{6} - \dfrac{1}{6}$ or $\dfrac{5}{6}$

Thus, we see that the rule is verified.

Exercises

Exploratory   Answer each of the following.

1. What does it mean to say that set D is a subset of set E? Give an example.
2. If you know the elements of sets A and B, how do you find $A \cap B$? $A \cup B$?

State an example of a set A where each of the following is true.

3. $n(A) = 3$ 4. $n(A) = 5$ 5. $n(A) = 100$ 6. $n(A) = 0$

Write an outcome set for each of the following situations.

7. tossing a coin
8. choosing a vowel from the alphabet
9. choosing a marble from a box of 3 blue, 2 red, and 1 white marbles

Written   Suppose a letter is chosen at random from the word CHAIR. Let E be the event of choosing a vowel.

1. Write E using set notation.
2. Write $n(E)$.
3. Find $P(E)$.
4. Find $P(not\ E)$.

Suppose we choose a letter at random from the alphabet. Let the following be events: $A = \{A, B, C, D, E\}$, $B = \{A, E, I, O, U\}$, $C = \{M, N, R, S, T, W, X\}$, $D = \{C, D, M, N, P, Q, R\}$.

5. Write $A \cup B$.
6. Write $A \cap B$.
7. Write $B \cap C$.
8. Write $A \cap D$.
9. Write $n(A)$.
10. Write $n(B)$.
11. Write $n(A \cup B)$.
12. Write $n(A \cap D)$.
13. Find $P(A \cup B)$.
14. Find $P(C \cup D)$.
15. Find $P(B \cup C)$.
16. Find $P(A \cup D)$.

Mixed Review

A box contains red, blue and green marbles. There are 5 more red marbles than blue and three times as many green marbles as blue. There are 20 marbles in the box. If you choose a marble at random, find the probability of choosing each of the following.

1. red   2. not blue   3. green or blue   4. red or blue

In each of the following, $P(E)$ is given for a certain event E. Find $P(not\ E)$.

5. 0.2   6. $\frac{2}{3}$ 7. x 8. $2x - y + 1$

9. On an English test, Nancy answered 21 out of 25 questions correctly. What percent of the questions did she answer correctly?

10. One base of a trapezoid is twice the height. This base is also 8 units shorter than the second base of the trapezoid. Write an expression to represent the area of the trapezoid.

11.5 Outcome Sets

Suppose that 2 coins are tossed. What is the probability that both will land heads?

To find *P(2 heads)*, use the rule $P(E) = \dfrac{n(E)}{n(S)}$. What is *n(S)*? You might be tempted to say that *n(S)* = 3. It appears that 3 outcomes are possible. These are 2 heads, 2 tails, and 1 of each. But consider the following situation.

Imagine that a penny and a nickel are tossed. Note that the case of 1 of each includes 2 different outcomes. Thus, there are 4 outcomes.

Further reasoning suggests that the same is true for 2 pennies. Suppose the dates on the pennies are 1988 and 1989. The 4 outcomes are shown below.

These results would not change regardless of when the pennies were minted. Since there are 4 outcomes, *n(S)* = 4.

What is *n(E)*? There is only one way that 2 coins can come up heads. Therefore, *n(E)* = 1.

$$P(2 \ heads) = \frac{1}{4}$$

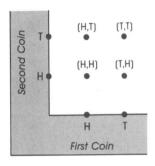

The outcomes of an experiment constitute the outcome set or **outcome space** for the experiment. An outcome set is often depicted as shown at the left.

Note that each *point* in the picture is a particular outcome. It is labeled by an **ordered pair.** For example, (H, T) describes the outcome *head on first coin, tail on second coin.* On the other hand, (T, H) is *tail on first coin, head on second coin.* Ordered pairs are so named because the order of their elements is significant.

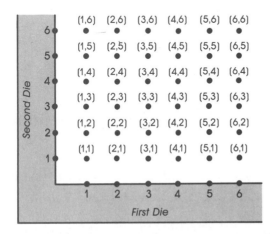

A more complicated situation is rolling two dice. Each die can land in 6 ways. The outcome space for rolling two dice is shown at the left. You can see there are 36 possible outcomes.

The ordered pair (4, 3) describes the outcome 4 on the first die and 3 on the second die. But (3, 4) describes 3 on the first die and 4 on the second die.

Examples

1 **Suppose two dice are rolled. How many ways can the faces show a sum of 4?**

Each of the outcomes (1, 3), (2, 2), and (3, 1) shows a sum of 4 since $1 + 3 = 4, 2 + 2 = 4,$ and $3 + 1 = 4.$

Thus, there are 3 ways that the faces can show a sum of 4.

2 **What is the probability of rolling a sum of 4 with two dice?**

$n(E) = 3$ *The successes are (1, 3), (2, 2), and (3, 1).*
$n(S) = 36$

$P(\text{sum of } 4) = \dfrac{3}{36}$ or $\dfrac{1}{12}$ *Use the rule for P(E).*

The probability of rolling a sum of 4 is $\dfrac{1}{12}$.

Examples

3 **What is the probability of rolling a double?**

$$P(\text{double}) = \frac{6}{36} \quad \text{or} \quad \frac{1}{6}$$ *There are 6 successes, (1, 1), (2, 2), (3, 3), (4, 4), (5, 5), and (6, 6).*

4 **What is the probability of rolling a sum greater than 9 or a double?**

Count the sums greater than 9 in the upper right corner. There are 6. There are 6 doubles. There are 2 sums greater than 9 that are also doubles.

$$P(\text{sum greater than 9 or double}) = \frac{6}{36} + \frac{6}{36} - \frac{2}{36} \quad \textit{Use the rule for P(A or B).}$$

$$= \frac{10}{36} \quad \text{or} \quad \frac{5}{18}$$

Exercises

Exploratory **Solve each problem.**

1. An outcome space for tossing two coins is shown. Copy the diagram and label each point with an ordered pair.
2. What is the difference between the ordered pairs (H, T) and (T, H)?

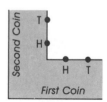

Name the ordered pairs in the diagram that describe each of the following.

3. 2 heads   **4.** 2 tails   **5.** 1 head and 1 tail

Find each of the following probabilities for tossing two coins.

6. *P(2 tails)*   **7.** *P(1 head and 1 tail)*   **8.** *P(no heads)*

Written **Complete each of the following.**

1. Suppose a red die and a green die are rolled. Draw the outcome space.
2. The following table lists sums which can be obtained by rolling two dice. Copy and complete the table. Count the number of ways each sum can be obtained. Then, find the probability of rolling each sum.

Sum	2	3	4	5	6	7	8	9	10	11	12
Number of Successes	1	2									
Probability of Sum	$\frac{1}{36}$	$\frac{2}{36}$									

Find each of the following if a red die and a green die are rolled.

3. P(sum of 7 or 11)

4. P(sum of 3 or 5)

5. P(sum not 7)

6. P(4 on the red die)

7. P(4 on the red and a sum of 10)

8. P(4 on the red or a sum of 10)

Two tetrahedral dice such as the ones in the drawing are rolled. The four congruent triangular faces are numbered from 1 to 4. In the roll of the dice, exactly one of the four faces of each die lands down.

9. Draw a picture of the outcome space. Landing down is an outcome.

10. How many ordered pairs are in the outcome space?

11. Complete the table listing the sums obtained by rolling the two tetrahedral dice. Count the number of ways each sum can be obtained. Find the probability of rolling each sum.

Sum	2	3	4	5	6	7	8
Number of Successes							
Probability of Sum							

Find each of the following probabilities for rolling two tetrahedral dice.

12. P(sum less than 4)

13. P(at least one die lands on 4)

14. P(sum of 2 or 3)

15. P(sum of 5 or 7)

16. P(4 on both dice)

17. P(sum is not 5)

18. P(sum is at least 6)

19. P(sum is at most 3)

20. P(3 on one die or a sum of 5)

21. P(2 on both dice or sum of 4)

Mathematical Excursions

Finding the Odds

In rolling a die, the **odds** of rolling a 3 are 1 to 5. One face of the die has a 3. The other 5 faces do not have 3.

The odds of an event is the ratio of the number of ways the event can occur to the number of ways it cannot occur.

$$\text{Odds} = \frac{n(E)}{n(\text{not } E)}$$

Exercises A card is chosen at random from a 52-card deck. Find the odds of choosing each of the following.

1. the ace of hearts

2. a black jack

3. a 7

4. a heart

5. a face card or 10

6. a 2, 3, or 4

7. not a 2

8. a red card

9. The probability Ted will win is $\frac{2}{3}$. What are the odds he will win?

10. The probability of rain is 60%. What are the odds it will rain?

11.6   The Counting Principle

In the solution of some probability problems, $n(S)$ is more difficult to find than $n(E)$. A **tree diagram** is one method for finding the total number of possible outcomes.

Examples

1 **Gwen has 3 skirts — 1 blue, 1 yellow, and 1 red. She has 4 blouses — 1 yellow, 1 white, 1 tan, and 1 striped. How many skirt-blouse outfits can she choose? What is the probability she will choose the blue skirt and white blouse?**

A tree diagram such as the one shown can be used to find the number of outfits.

The skirt possibilities are shown at the left of the diagram. The four blouse possibilities branch from each skirt possibility.

Count all the branches created at the right to find the total number of outfits. There are 12 possible outfits.

The colored arrows in the diagram show the blue skirt — white blouse outfit.

If we assume each combination is equally likely, the probability that Gwen will select the blue skirt — white blouse outfit is $\frac{1}{12}$.

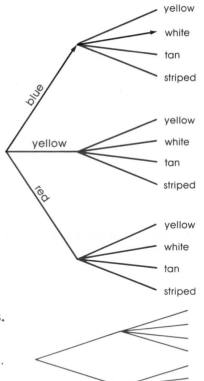

2 **Suppose Gwen has 2 skirts and 4 blouses. How many possible outfits has she?**

Draw a tree diagram and count the branches. There are 2 • 4 branches all together.

The total number of outfits is 8.

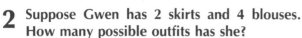

These examples illustrate the **Counting Principle.**

Counting Principle

> Suppose one activity can occur in any of *m* ways. Another activity can occur in any of *n* ways. The total number of ways both activities can occur is given by the product *mn*.

Examples

3 There are 2 roads west from Eastburg to Middleburg and 3 roads west from Middleburg to Westburg. In how many ways can a traveler make the trip from Eastburg to Westburg?

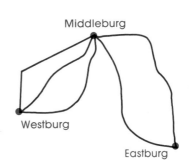

A traveler can choose either of 2 roads to Middleburg and then any of 3 to Westburg.

By the Counting Principle, the total number of choices is 2 · 3 or 6.

4 Suppose in addition to 3 skirts and 4 blouses, Gwen has 2 scarfs. How many different skirt-blouse-scarf outfits can she choose?

By extending the Counting Principle, we find the number is 3 · 4 · 2 or 24.

The Counting Principle can be used to help find probabilities. The probability that Gwen will choose the blue skirt is $\frac{1}{3}$, while the probability of choosing the white blouse is $\frac{1}{4}$. The probability that she will choose both is $\frac{1}{3} \cdot \frac{1}{4}$.

This is an example of a useful principle of probability.

A Probability Principle

> Suppose the probability of event *E* is *r* and the probability of event *F* is *s*. If the occurrence of either event does not affect the occurrence of the other, then the probability of *E* and *F* both occurring is the product *rs*.

Exercises

Exploratory   **A coin is tossed and a die is rolled.**

1. Draw an outcome space.

2. How many possible outcomes are there?

3. Find *P(head on coin).*

4. Find *P(head on coin and 2 on die).*

5. Find *P(3 on die).*

6. Find *P(head on coin and even number on die).*

Solve each problem.

7. Two coins are tossed. Do you have an equal chance of being correct by calling either "both heads" or "not both heads"?

8. Suppose Horatio has 3 suits—black, brown, and blue. He has 2 shirts—white and striped. Draw a tree diagram to show all his possible outfits.

Written   **In exercises 1 and 2 draw a tree diagram to show all possible outcomes.**

1. Two coins are tossed.

2. A die and a coin are tossed.

Solve each problem.

3. A baseball team has 8 pitchers and 3 catchers. How many pitcher-catcher combinations are possible?

4. Horatio has enlarged his wardrobe to include 5 suits, 4 shirts, and 3 pairs of shoes. How many suit-shirt-shoes outfits are possible?

5. A dinner menu offers choices of 7 appetizers, 12 entrees, and 6 desserts. How many ways can you choose one of each course?

6. A car dealer offers a choice of 6 vinyl-top colors, 18 body colors, and 7 upholstery colors. How many color combinations are there?

7. Draw a tree diagram to show the possible combinations for boys and girls in a family with 2 children.

8. Assume that the probability of a child being a boy is $\frac{1}{2}$. In a 2-child family, what is the probability that both children are boys?

9. Draw a tree diagram to show the possible combinations for boys and girls in a family with 3 children.

10. Show the possible combinations in a list of a 3-child family using *ordered triples.*

11. What is the probability that a 3-child family has 3 girls?

12. What is the probability that a 3-child family has at least 1 boy?

Challenge   **Solve each problem.**

1. A certain city has 3456 families with 3 children. About how many of these families would you expect to have 3 girls?

2. Draw a tree diagram to show the possible combinations for boys and girls in a 4-child family. Then, show the possible combinations in a list.

3. What is the probability that a 4-child family has 4 boys? 4 girls?

4. What is the probability for *n* boys in an *n*-child family? For *n* girls?

11.7   Multiplying Probabilities

Probability problems of a special kind are sometimes called "urn problems." Consider the following examples of "urn problems with replacement."

Examples

1 An urn contains 3 red and 5 white marbles. A marble is chosen from the urn at random and its color noted. It is replaced in the urn and another marble is chosen. What is the probability that both marbles are red?

$P(red) = \dfrac{3}{8}.$ *For the first choice, there are 8 marbles and 3 are red.*

$P(red) = \dfrac{3}{8}.$ *For the second choice, the first marble is replaced.*
There are again 8 marbles in the urn.

$P(red, red) = \dfrac{3}{8} \cdot \dfrac{3}{8}$ or $\dfrac{9}{64}.$

The probability that both marbles chosen are red is $\dfrac{9}{64}.$

2 Consider the urn in example 1. A marble is chosen, replaced, and another chosen. Find the probability that both are the same color.

Both marbles the same color means both red or both white. Since these events have no outcomes in common,
$$P(both\ red\ or\ both\ white) = P(both\ red) + P(both\ white).$$

$P(both\ red) = \dfrac{9}{64}.$ *This is the solution in example 1.*

$P(both\ white) = \dfrac{25}{64}.$ *Use the same method to find P(both white).*

$$P(both\ red\ or\ both\ white) = P(both\ red) + P(both\ white)$$
$$= \dfrac{9}{64} + \dfrac{25}{64}$$
$$= \dfrac{34}{64} \quad or \quad \dfrac{17}{32}$$

The probability that both are the same color is $\dfrac{17}{32}.$

Here are examples of "urn problems without replacement."

Examples

3 **Use the same urn containing 3 red and 5 white marbles. This time a marble is chosen and *not* replaced in the urn. Then another marble is chosen. What is the probability that both marbles will be red?**

Draw a tree diagram to show all the outcomes. Each probability, *P,* is written on the appropriate branch.

On the first choice, the probabilities are the same as in example 1. On the second choice, *n(E)* depends on the color of the first marble. If a red marble is chosen the first time, there are only 2 red marbles remaining. But now there are only 7 marbles left in the urn.

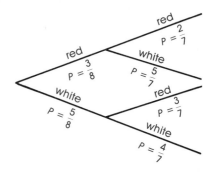

$P(\text{both marbles red}) = P(\text{first red and second red})$

$$= \frac{3}{8} \cdot \frac{2}{7}$$

$$= \frac{6}{56} \quad \text{or} \quad \frac{3}{28} \qquad \text{The probability of both red is } \frac{3}{28}.$$

4 **Clem has 2 navy blue socks and 4 black socks in a drawer. One dark morning he pulls out 2 socks. What is the probability of a matching pair?**

This situation is equivalent to pulling out 1 sock, not replacing it, then pulling out another. The tree diagram shown describes the situation.

A pair of matching socks means 2 of the same color. We want the events (blue, blue) or (black, black).

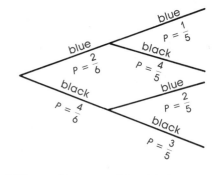

$P(\text{blue, blue}) = \dfrac{2}{6} \cdot \dfrac{1}{5} \quad \text{or} \quad \dfrac{2}{30}$

$P(\text{black, black}) = \dfrac{4}{6} \cdot \dfrac{3}{5} \quad \text{or} \quad \dfrac{12}{30}$

$P(\text{pair}) = \dfrac{2}{30} + \dfrac{12}{30}$ Use P(A or B) = P(A) + P(B) since the events (blue, blue) and (black, black) have no elements in common.

$\qquad = \dfrac{7}{15}$ The probability of a matching pair is $\dfrac{7}{15}$.

Exercises

Exploratory   A bag contains 2 blue and 5 white marbles. A marble is chosen from the bag, its color is noted, and then it is returned to the bag. Another is chosen.

1. Draw a tree diagram for this situation. Write the probabilities on each branch.

2. Find *P(both blue)*.

3. Find *P(both white)*.

4. Find *P(both the same color)*.

5. Find *P(both different colors)*.

Answer each of the following.

6. Give 3 examples of problems that are mathematically equivalent to "urn problems with replacement" and 3 examples of those "without replacement."

7. A tree diagram is given for a situation with the probabilities calculated for every possibility. What is the sum of the probabilities on the last branches?

Written   An urn contains 4 blue and 6 green marbles. Three marbles are chosen, one at a time, with replacement. Find each of the following probabilities.

1. *P(all three blue)*

2. *P(all three green)*

3. *P(first two blue)*

4. *P(exactly 2 blue, 1 green, any order)*

A bag contains 2 white and 3 blue marbles. A marble is chosen, not replaced, and then another is chosen.

5. Draw a tree diagram for this situation. Include probabilities.

6. Find *P(white, white)*.

7. Find *P(blue, blue)*.

8. Find *P(same color)*.

9. Find *P(one white, one blue)*.

Solve exercises 6-9 for a bag with 4 white and 5 blue marbles.

10. Exercise 6   **11.** Exercise 7   **12.** Exercise 8   **13.** Exercise 9

Solve exercises 6-9 for a bag with 2 white and 6 blue marbles.

14. Exercise 6   **15.** Exercise 7   **16.** Exercise 8   **17.** Exercise 9

The great Fumbo claims he can magically choose white rabbits from a hat holding 2 white and 4 brown rabbits. But actually he chooses rabbits at random. Find the following probabilities assuming Fumbo is right with a white rabbit. If he is wrong, the brown rabbit is replaced.

18. *P(wrong the first choice, right the second)*

19. *P(wrong twice)*

20. *P(wrong twice, right the third)*

21. *P(wrong three times)*

Solve exercises 18-21 if the brown rabbit is *not* replaced.

22. Exercise 18   **23.** Exercise 19   **24.** Exercise 20   **25.** Exercise 21

11.8   Permutations

Suppose Mary, Sally, and Clem are passengers in your compact car. They can be seated in one front and two back seats (Seat 1, Seat 2, and Seat 3). How many ways can they be seated?

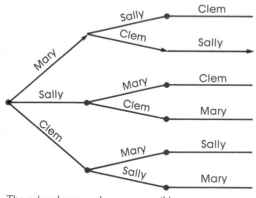

The colored arrows show one possible arrangement.

Look at the top branch of the tree diagram. Suppose Mary is in Seat 1. Then either Sally or Clem must be in Seat 2. Note that once the choice has been made for Seats 1 and 2, only one person remains for Seat 3.

There are 6 possible seating arrangements. There are 3 ways to fill Seat 1. Then, there are 2 ways to fill Seat 2. Then, there is only 1 remaining passenger for Seat 3.

The Counting Principle may be used in this situation. There are 3 • 2 • 1 or 6 different arrangements of 3 passengers in 3 seats.

Use the Counting Principle.

Suppose you can put another passenger in the car. How many ways can you seat 4 passengers in 4 seats? The answer is 4 • 3 • 2 • 1 or 24 ways. How many ways can 5 passengers be seated in 5 seats? The answer is 120.

Now suppose you had to place 100 passengers in a 100-seat airplane. The number of ways has over 150 digits!

$$100 \cdot 99 \cdot 98 \cdot \ldots \cdot 3 \cdot 2 \cdot 1$$

The number of ways r passengers can be seated in r seats is r!

The product of all the numbers from 1 to 100 is called 100 **factorial.** It is written 100! Thus, 100! = 100 • 99 • • • 3 • 2 • 1.

Definition of n Factorial

For any positive integer n,
$$n! = n(n - 1)(n - 2) \cdot \ldots \cdot 3 \cdot 2 \cdot 1$$

Example

1 **Find each value: a. 4!   b. 10!   c. (9 − 4)!   d. 5! − 3**

a. 4! = 4 • 3 • 2 • 1
    = 24

b. 10! = 10 • 9 • 8 • . . . . • 1
     = 3,628,800

c. (9 − 4)! = 5! or 5 • 4 • 3 • 2 • 1
        = 120

d. 5! − 3 = 120 − 3
       = 117

Example

2 Assume that any arrangement of the 5 letter tiles shown below is a *word*. How many different *words* of the 5 letters can be made from the tiles? Repeating the same letter is not allowed.

No English words can be made from the tiles. But suppose that any arrangement, such as QBXRT or QBTRX, is a word.

There are 5 different letters that must be placed in 5 different places in the word.   *This is really equivalent to seating 5 passengers in 5 seats.*

$5! = 5 \cdot 4 \cdot 3 \cdot 2 \cdot 1$ or 120 *Use the Counting Principle.*

The total number of 5-letter *words* is 120.

Calculator Hint

Many calculators have an $\boxed{x!}$ key that calculates the factorial of the last number you entered.

An arrangement of a given number of letters, passengers, or any objects is called a **permutation.** Thus, QBXRT and BTXRQ are different permutations of the letters Q, B, X, R, and T.

The number of ways to arrange 5 things taking them 5 at a time is written $_5P_5$. In general, the symbol $_nP_n$ means that n objects are being arranged, or permuted, n at a time. Thus, $_nP_n = n!$

Example

3 How many ways can John place 4 books on a shelf if a certain book must be placed first?

John has only 1 first choice. Then, he is limited to placing the other 3 books in 3 places.

$$\underline{\quad 1 \quad} \cdot \underline{\quad 3 \quad} \cdot \underline{\quad 2 \quad} \cdot \underline{\quad 1 \quad} = 6$$

1st place   2nd place   3rd place   4th place

John has 6 ways to place 4 books on the shelf if he must place a certain book first.

You have considered permutations of a number of objects taken all at a time. But sometimes not all the objects are used.

Example

4 **How many 3-letter *words* can be made from 5 different letter tiles?**

There are 3 places in which to put 5 different objects. The first place can be filled in any of 5 ways. The second can be filled by any of the 4 remaining letters, and the third in any of 3 ways.

$$\underline{\quad 5 \quad} \cdot \underline{\quad 4 \quad} \cdot \underline{\quad 3 \quad} = 60 \qquad \textit{Use the Counting Principle.}$$
$$\textit{1st place} \quad \textit{2nd place} \quad \textit{3rd place}$$

There are 60 different 3-letter words that can be made from 5 different letters.

The number of permutations of n objects, taken r at a time, is symbolized by $_nP_r$. Thus, $_5P_3$ is the symbol for the number of ways 5 objects may be arranged in 3 places. Example 4 shows that $_5P_3 = 60$. Since 60 is equal to $\dfrac{5 \cdot 4 \cdot 3 \cdot 2 \cdot 1}{2 \cdot 1}$ or $\dfrac{5!}{2!}$, we also know $_5P_3 = \dfrac{5!}{2!}$ or $\dfrac{5!}{(5-3)!}$. This suggests that $_nP_r = \dfrac{n!}{(n-r)!}$.

Exercises

Exploratory Find the value of each of the following.

1. $5!$

2. $6!$

3. $(10 - 3)!$

4. $6! - 4!$

5. $(3!)(2!)$

6. $\dfrac{10!}{8!}$

7. $\dfrac{(8!)(3!)}{5!}$

8. $\dfrac{7!}{(7-3)!}$

How many ways can the letters of each of these words be arranged?

9. MATH

10. CLEAR

11. PRINCE

12. DRAGONS

Find the value of each of the following.

13. $_3P_3$

14. $_6P_6$

15. $_4P_4 - _2P_2$

16. $\dfrac{_6P_6}{_4P_4}$

Written   State whether each of the following statements is *true* or *false*.

1. $5! - 3! = 2!$
2. $6 \cdot 5! = 6!$
3. $(5 - 3)! \cdot 3 = 3!$
4. $3! + 4! = 5 \cdot 3!$
5. $\dfrac{6!}{3!} = 2!$
6. $\dfrac{{}_9P_9}{9!} = 1$

Solve each of the following.

7. How many 5-place numerals can be formed with the digits 2, 3, 4, 5, and 6 if no digit is used more than once in a number?

8. How many arrangements in a row can a photographer make of 14 members of the Westburg field hockey team?

9. How many ways can 9 books be arranged in a row on a shelf?

10. How many ways can 12 people line up to buy tickets for a concert?

11. How many 4-place numerals can be formed using the digits 1, 2, 3, and 4 if no digit is used more than once and each number formed is even?

12. How many different 8-letter combinations can be formed from the letters of HEPTAGON if each word must begin with H and end with N?

13. How many numbers between 2,000 and 5,000 can be written using the digits 2, 3, 7, and 9 if no digit can be used more than once?

14. How many permutations are possible of ABCDefg if each permutation must begin and end with a capital letter?

Suppose you have the letter tiles Q, B, X, R, T, and L and you wish to form 3-letter *words*.

15. How many choices do you have for the first place? Then, how many choices remain for the second letter? How many ways can the third letter be chosen?

16. Find the total number of 3-letter words that can be formed from 6 letters.

Find the value of each of the following.

17. ${}_5P_2$
18. ${}_6P_2$
19. ${}_7P_2$
20. ${}_7P_4$
21. ${}_8P_3$
22. ${}_{10}P_4$
23. ${}_5P_3$
24. ${}_7P_3$
25. ${}_{10}P_2$
26. ${}_7P_5$
27. ${}_{11}P_3$
28. ${}_{13}P_5$
29. ${}_4P_2 + {}_7P_3$
30. ${}_6P_3 - {}_{10}P_3$
31. ${}_9P_9 - {}_8P_2$
32. ${}_5P_3 + {}_{12}P_5$

Challenge   A telephone number has 7 digits. Suppose only the digits 2, 3, 4, 5, 6, 7, and 8 are used. Find the number of different telephone numbers that are possible in each of the following cases.

1. No digit is repeated.
2. Each digit may be repeated.
3. The number begins with 628 and no digit is repeated.
4. The number begins with 555 and a digit may be repeated.

Problem Solving Application: List Possibilities

A useful strategy for solving problems may be to *list possibilities*. When making a list, use a systematic approach so you do not omit any items. This strategy is often helpful when you need to find the number of solutions to a problem or the number of successes of an event.

Examples

1 **How many rectangles have a perimeter of 18 cm, dimensions that are integers, and an area that is greater than 15 cm²?**

Explore   Let ℓ = the length of the rectangle.
Let w = the width of the rectangle.
We need to find all positive integers, ℓ and w, such that $2\ell + 2w = 18$ and $\ell w > 15$.

Plan   First, list all positive integers, ℓ and w, such that $2\ell + 2w = 18$ or $\ell + w = 9$. Then find all the values in the list where $\ell w > 15$.

Solve

ℓ = 1 cm, w = 8 cm, A =   8 cm² ℓ = 8 cm, w = 1 cm, A =   8 cm²
ℓ = 2 cm, w = 7 cm, A = 14 cm² ℓ = 7 cm, w = 2 cm, A = 14 cm²
ℓ = 3 cm, w = 6 cm, A = 18 cm² ℓ = 6 cm, w = 3 cm, A = 18 cm²
ℓ = 4 cm, w = 5 cm, A = 20 cm² ℓ = 5 cm, w = 4 cm, A = 20 cm²

There are four rectangles with a perimeter of 18 cm, dimensions that are integers, and an area greater than 15 cm².

Examine   The rectangles with dimensions 3 cm by 6 cm, 4 cm by 5 cm, 5 cm by 4 cm, and 6 cm by 3 cm each have a perimeter of 18 cm and an area greater than 15 cm². Since no other rectangles with dimensions that are integers satisfy these conditions, the solution is correct.

2 **Find the probability of getting at least 1 dime when you receive change for a quarter.**

Explore   The outcomes are the different ways you can receive change for a quarter.
The event we want is to get at least 1 dime.

Plan   List all the possible ways of receiving change for a quarter. Count all the cases where you get at least 1 dime.

Solve
  1. 25 pennies
  2. 20 pennies, 1 nickel
  3. 15 pennies, 2 nickels
  4. 15 pennies, 1 dime
  5. 10 pennies, 3 nickels
  6. 10 pennies, 1 nickel, 1 dime
  7. 5 pennies, 4 nickels
  8. 5 pennies, 2 nickels, 1 dime
  9. 5 pennies, 2 dimes
  10. 5 nickels
  11. 1 nickel, 2 dimes
  12. 3 nickels, 1 dime

There are 12 possible ways to receive change for a quarter. In 6 cases, you will get at least 1 dime. Thus, the probability of getting at least 1 dime when you receive change for a quarter is $\frac{6}{12}$ or $\frac{1}{2}$.   *Examine this solution.*

Exercises

Written   Solve each problem.

1. How many rectangles have a perimeter of 30 cm, dimensions that are integers, and an area that is greater than 50 cm^2?

2. How many different whole numbers can be written using the digits, 3, 3, 6, and 7 if each number must use all four digits?

3. Find the probability of getting at least 1 dime and 1 quarter when you receive change for a half-dollar.

4. What is the probability of getting at least 2 dimes when you receive change for a dollar if pennies are not used?

5. In how many ways can you write 45 as the sum of positive consecutive integers?

6. How many positive integers less than 50 can be written as the sum of two squares?

7. An ice cream shop makes chocolate, vanilla, and strawberry sundaes. Any sundae can be served with whipped cream, nuts, both, or neither. How many ways can a sundae be served?

8. If you choose a sundae at random from the ice cream shop in Exercise 7, what is the probability that it will not be chocolate and it will not have whipped cream or nuts?

The president, vice president, secretary, and treasurer of a club are to be seated in a row of four chairs. Find the probability of each seating arrangement.

9. The president sits next to the secretary.

10. Someone sits between the vice president and the treasurer.

Portfolio Suggestion

Place your favorite word problem from this chapter in your portfolio with a note explaining why it is your favorite. Be sure to include your solution.

Performance Assessment

Roll a die 50 times and record how many times each number is rolled. How many times would you expect each number on the die to occur? How do your results compare with the expected outcomes for rolling each number? What might you expect if you rolled a die 200 times?

Chapter Summary

1. The **probability** of an event, $P(E)$, is found by dividing the number of ways that the event can occur, $n(E)$, by the total number of possible **outcomes,** $n(S)$. (351)
2. The probability of an impossible event is 0. The probability of an event which is certain to occur is 1. The probability of an event can never be less than 0 or greater than 1. (353, 354)
3. $P(not\ E) = 1 - P(E)$. (355)
4. $P(E\ or\ F) = P(E) + P(F) - P(E\ and\ F)$. (358)
5. In set notation, $P(E\ B\ F) = P(E) + P(F) - P(E \cap F)$. (362)
6. **Counting Principle:** Suppose one activity can occur in any of m ways. Another activity occurs in any of n ways. The total number of ways both activities can occur is given by the product mn. (368)
7. Suppose the probability of event E is r and the probability of event F is s. If the occurrence of either event does not affect the occurrence of the other, then the probability of E and F both occurring is the product rs. (369)
8. For any positive integer n, $n! = n(n - 1)(n - 2) \cdot \ldots \cdot 3 \cdot 2 \cdot 1$. (374)

Chapter Review

11.1 **Each letter from BLEAT is written on a different card and placed in a hat. If one is chosen at random, find the probability of choosing each of the following.**

    1. T   **2.** a vowel   **3.** a consonant   **4.** a letter from BAT

11.2   **Find the probability of drawing each of the following from the letters of BLEAT.**

5. a Z   **6.** a letter from TABLE

11.3   **A letter is chosen from the alphabet. Find the probability that it is contained in each of the following.**

7. MOUSE   **8.** TRAP   **9.** JOUST   **10.** MOUSE or JOUST
11. TRAP or JOUST   **12.** TRAP or PARTY

11.4   **Suppose one die is rolled. Use set notation to find the probability of rolling each of the following.**

13. an odd number   **14.** a prime or odd number

11.5   **Two red marbles and 2 blue marbles are in a vase. Suppose a marble is chosen, put back, and another is drawn. Find each of the following.**

15. *P(both marbles blue)*   **16.** *P(one red, one blue)*
17. *P(one marble yellow)*   **18.** *P(at least one marble blue)*

Suppose 2 dice are rolled. Find the probability of rolling each of the following.
19. sum of 8   **20.** sum of 10   **21.** sum of 8 or 10   **22.** sum of 8 or double

11.6   **Amy has 2 cars, a Slosh and a Vroom, and 3 boyfriends, Tom, Bill, and Otto, who will go out with her whenever she asks.**

23. How many different car-boy com-   **24.** Draw a fully labeled tree diagram
binations can Amy choose?   for this situation.
25. What is the probability Amy picks the Slosh and Bill?

11.7   **An urn contains 2 white and 5 blue marbles. One is chosen at random, not re-placed, and another chosen.**

26. Draw a fully labeled tree diagram   **27.** What is the probability both mar-
for this situation.   bles drawn will be white? Blue?
28. If the marble is replaced after each choice in exercises **26** and **27,** what is the
probability of 3 whites in 3 choices?

11.8   **Find the number of ways the letters in each of the following can be arranged.**
29. BLAH   **30.** WHITE

Find the number of 3-letter *words* that can be formed from each of the following.
31. BAH   **32.** BLAH   **33.** BLACK   **34.** BLANKS

Find the value of each of the following.
35. 3!   **36.** (10 − 6)!   **37.** $_6P_6$ **38.** $_7P_2$

Chapter Test

A letter is chosen at random from PARIS. Find each of the following.

1. *P(R)* **2.** *P(vowel)* **3.** *P(not R)*
4. *P(R or a vowel)* **5.** *P(Q)* **6.** *P(a letter from PAIRS)*

A card is chosen from a 52-card deck. Find the probability of each of the following.

7. an ace **8.** a heart **9.** the ace of hearts **10.** an ace or a heart

Suppose a marble is chosen from a bag of 1 blue, 2 red, and 3 white marbles. Let choosing a red be an event *E*. Use set notation to answer each of the following.

11. Write an outcome set. **12.** Write event *E*.
13. Find *n(E)* and *n(S)*. **14.** Find *P(red)*.

A letter is chosen at random from CAT and a die is rolled.

15. Draw a tree diagram and show all the possible outcomes.
16. Find the probability of choosing a T and rolling a 4.
17. Find the probability of choosing an A and rolling an odd number.

A hat contains 2 green marbles and 6 red marbles.

18. A marble is chosen, replaced, another chosen, replaced, and another chosen. What is the probability of drawing a green marble all three times?
19. Suppose a marble is chosen, not replaced, and another is chosen. Draw a labeled tree diagram. Include all probabilities on the branches.

Find each of the following probabilities in exercise 19.

20. *P(green, green)* **21.** *P(red, red)* **22.** *P(same color)*

Solve each of the following problems.

23. How many different ways can 6 books be arranged on a shelf?
24. How many different ways can 6 books be arranged if the first and last must always be the same two books?

Find the value of each of the following.

25. 5! **26.** (10 − 7)! **27.** $_5P_5$ **28.** $_8P_3$

Find how many 4-letter *words* can be made from the letters in each of the following.

29. CRATE **30.** DECATUR

Introduction to
Statistics

Application in Finance

Stock brokers, financial advisors, and economists use statistics to study trends in the economy. They use this information to advise others on financial matters. They may recommend that a company expand its production of a certain product or help an individual to pick a stock that will likely increase in value.

Lata Mohamed is watching the value of certain stock for a client. The daily closing value of the stock for the past week is $23.25, $23.75, $24.50, $24.25, and $23.75. Find the **mean, median,** and **mode** of the closing values.

Individual Project: *Economy*

Choose a stock and record its daily closing value for two weeks. Make a line graph showing the closing values of the stock. Does your graph show any trend in the value of the stock? Find the mean, median, and mode of the closing values.

12.1   Statistics and Sampling

The methods and applications of **statistics** affect your life in many ways. Perhaps you have taken part in a survey. A statistical survey collects pieces of information called **data.** The data are analyzed to help people make decisions.

Sometimes computers are used to analyze data. Graphs and charts are often used to present statistical information.

Definitions of Statistics and Data

> Statistics is the science of collecting, organizing, analyzing, and presenting numerical data. Data are the pieces of information used in statistics.

Why is it not practical to test every bulb?

Is this a fair process?

Suppose the Stay-Brite Light Bulb Company wishes to know the average life of its light bulbs. Since it cannot test every bulb it manufactures, the company takes a **sample** of its bulbs. The sample bulbs are tested until they burn out. The lifetimes of the sample bulbs are accepted as typical of the lifetimes of all the light bulbs.

This procedure is called sampling. The sample must be a fair one. The bulbs must be chosen at random. No special long-lasting bulbs may be used in the sample. Of course, no defective bulbs would purposely be chosen. The **sample population** must be **representative** of the **general population.**

Example

1 **An agency wants to know how much money is spent by a typical American family to heat its home. Should a sample of families be chosen only from the state of Arizona?**

The climate of Arizona is generally much warmer than that of the United States as a whole. Less fuel is required to heat homes in Arizona.

A sample of families in Arizona is *not* a representative sample of the general population of American families.

▰▰ Example ▰▰▰▰▰

2 **The Crunchy-Wunchy Corporation has a new product, Diet Crunchy-Wunchies. It is to be test marketed in Pine City. After a few months, a sample population of families will be chosen for a survey to see how many use Diet Crunchy-Wunchies. What are some of the questions the Crunchy-Wunchy Corporation people must consider?**

1. Must each family member eat Diet Crunchy-Wunchies in order for the family to be counted?
2. Will only families who regularly eat cereal be counted?
3. Will families with school-age children be given special consideration?
4. How will the families be surveyed?
5. Will families be called back if they are not home when first telephoned?
6. Suppose questionnaires are sent. What will be done about families who do not return them?

7. Suppose people are interviewed personally. Where should this be done?
8. Will it be assumed that everyone who answers is telling the truth?
9. Can the survey be checked to see that it has been fair?

Usually, sampling is used when gathering statistical data. However, there are instances where information is gathered from every member of a group. This procedure is called a census.

Deciding what questions to ask is important in gathering information. Decisions also may include how to record the data. The recording of data is called **tabulation.** Most important of all is how the data are interpreted.

▰▰ Exercises ▰▰▰▰▰

Exploratory **Answer each of the following.**

1. What is meant by a sample population?

2. What is the difference between a sample and a census?

State a reason why or why not each of the following locations would be good for the given survey.

Survey	Location
3. Favorite candidate	Republican headquarters
4. Number of books read	Library
5. Favorite detergent	Laundromat
6. Number of dogs	Apartment building
7. Favorite carpet color	Carpet store
8. Favorite singer	Homes during the day
9. Favorite lunch	School cafeteria
10. Favorite lunch	Pizza parlor

Written A survey is to be taken to find out how many hours the typical ninth grader in your school watches television on a weekday evening. For each of the following procedures, state what problem researchers might have.

1. Giving out a questionnaire in all ninth grade Latin classes.

2. Asking students in a randomly chosen English class to discuss their TV viewing habits in a composition.

3. Sending a questionnaire to parents of 200 students, selected at random, asking them to describe their child's TV viewing habits.

4. Calling randomly chosen students on the evening of December 28 and asking them how many hours of television they watched the night before.

A national survey on a controversial issue must be made. State what problems researchers might have using each of the following surveying methods.

5. Stopping people on their way to work in front of a busy subway station in New York City.

6. Sending a questionnaire to everyone who lives on a randomly chosen street in a randomly chosen town.

7. Calling people at random during a weekday afternoon.

8. Sending questionnaires to randomly chosen people to fill out and return.

9. Sending questionnaires home with randomly chosen students for their parents to fill out.

10. Taking a person-by-person survey of every adult resident of Blakeburg, population 3,416.

11. Placing advertisements in randomly chosen local newspapers asking people to call or write in their opinions.

12. Taking a survey of every person who enters or leaves randomly chosen YWCAs on a given Thursday night.

Challenge Answer each of the following.

1. What qualities should a statistical questionnaire have? Why is each important?

2. What problems might there be in selecting interviewers for person-to-person interviews?

3. What rules should good interviewers observe? Why?

12.2   Averages

Sometimes a number called an **average** is used to represent or typify a whole set of numbers. You may have made computations involving averages such as the ones in the following examples.

Examples

1 **Jonathan has grades of 81, 75, 78, 80, and 82 in five quizzes. What is his average grade on the quizzes?**

$$81 + 75 + 78 + 80 + 82 = 396$$ *Add the grades together.*
$$396 \div 5 = 79.2$$ *Divide by 5, the number of grades.*

Jonathan's quiz average is 79.2.

2 **On three French exams this term, Helen has grades of 83, 75, and 90. She has another exam tomorrow. She wants an average of 85 on the four exams. What grade must she make tomorrow in order to achieve her goal?**

Define a variable. Let x = the minimum grade Helen needs.

Write an equation.
$$\frac{83 + 75 + 90 + x}{4} = 85$$

Solve the equation.
$$248 + x = 340$$
$$x = 92$$

Answer the problem. Helen needs at least a 92. Of course, if she scores higher than 92, her average will be higher than 85.

Mathematicians call this kind of average the **mean.** Other names are **arithmetic mean** and **numerical average.**

Definition of Mean

> The mean of a set of n numbers is the sum of the numbers divided by n.

Other kinds of averages represent some sets of data better than the mean. The **mode** is often used to represent sets of sizes of shoes or clothing. The following example shows that the mode is sometimes a better average to use when one value appears frequently.

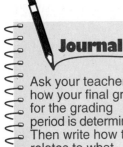

Journal

Ask your teacher how your final grade for the grading period is determined. Then write how this relates to what you've learned in this lesson.

A country store owner finds she must limit her stock of a certain hunting jacket to three jackets. With such a small order, the supplier of these jackets specifies that they must all be of the same size. The store owner asks her assistant to determine the size to order by surveying possible customers.

The assistant finds that there are a dozen such people in the area. Their jacket sizes are as follows.

40 42 42 42 42 42 42 42 44 48 50 52

He takes the mean of these values and obtains 44. He tells his employer to order jacket size 44. But only one of the possible customers wears a 44! *The mode in this situation is 42.*

Definition of Mode

> The mode of a set of numbers is the number which occurs most often.

Examples

3 **What is the mode of 2, 3, 4, 5, 5, 5, 6, 6, and 11?**

Since 5 occurs most frequently, 5 is the mode.

4 **What is the mode of the set of shoe sizes 5, 6, 6, 6, 7, 8, 9, 10, 11, 11, 11, and 12?**

This set has two modes, 6 and 11.

Heights of Seven Students

152 cm
155 cm
157 cm
159 cm
163 cm
187 cm
187 cm

A set of numbers with exactly two modes is said to be **bimodal.** The set in example 4 is bimodal.

The twins, Tim and Bob, are unusually tall with heights of 187 cm. Seven students including Tim and Bob are asked to find a height that represents the heights of students their age. The list at the left shows the heights recorded by the seven students. The mean height is about 166 cm. Is 166 truly representative of the set of measures?

Tim and Bob are the only students who are the same height. Thus, the mode of the set is 187. Is 187 typical of all the measures? Another average is needed.

When data include extreme values, the middle value, or **median,** may represent the data best. The median in this case is 159.

Definition of Median

The median of a set of numbers is the one which lies in the middle when the numbers are arranged in order. If there are two middle values, the median is the mean of these values.

Examples

5 **What is the median of the golf scores, 72, 73, 75, 76, 78, 79, and 98?**

There are 3 scores less than 76 and 3 scores greater than 76.

The median score is 76.   *The mean is about 79. But 76 typifies the scores better because there is one extremely high score.*

6 **Six students paid the following for their lunches: $0.95, $1.00, $1.15, $1.25, $1.50, and $2.00. What is the median amount?**

When there is an even number of values, there are two middle values. The two middle values are 1.15 and 1.25. The mean of 1.15 and 1.25 is 1.20.

Therefore, the median amount is $1.20.

The word *average* may be used to refer to the mean, median, or mode. You should be careful to find out how the word is used and if that is the best average to use.

Exercises

Exploratory   Find the mean of each of the following sets of numbers.

1. 1, 2, 2, 2, 3, 5
2. 2, -2, 3
3. -3, -4, 12, -6, -5
4. 2, 3, 5
5. 0, 0, 0, 500, 0
6. 1, 2, 2, 2, 3, 5000
7. -4, 0, 4
8. -11, 3, 4, -9, 25
9. 1.19, 3.14, 6.02, -2.97
10. 5, 6, 6, -10, 6, -6, -3, 6
11. $7\frac{1}{3}$, $7\frac{1}{3}$, $7\frac{1}{3}$, $7\frac{1}{3}$, $7\frac{1}{3}$
12. $7\frac{1}{8}$, $6\frac{3}{4}$, $2\frac{1}{2}$, $-4\frac{1}{4}$, $-2\frac{1}{8}$

Find the mode for the set of numbers in each of the following.

13. Exercise 1   **14.** Exercise 5   **15.** Exercise 6   **16.** Exercise 10

Find the median for the set of numbers in each of the following.

17. Exercise 1   **18.** Exercise 5   **19.** Exercise 6   **20.** Exercise 10

Written   **For each of the following sets of numbers, find the mean, median, and mode. State if the set is bimodal.**

1. 32, 42, 42, 52, 62   **2.** 2, 2, 2, 2, 2, 2, 3
3. -50, 50, -50, -50 **4.** 4, 2, 4, 0
5. 10, 10, 11, 11, 12, 9, 13   **6.** -12, -11, -4, 1, 2

Find the value of x so that each of the following sets of numbers has the indicated mean. The mean is represented by M.

7. 3, 3, 3, 3, x; $M = 3$ **8.** -7, -6, 0, x, 8; $M = 0$
9. x, $2x$, $3x$; $M = 18$ **10.** 3, 4, 5, 6, x; $M = -10$
11. $2x$, $2x + 5$, $6x - 1$; $M = 8$ **12.** x, $3x + 1$, $2x - 1$, $3x$; $M = 6$
13. x, $x + 1$, $2x + 3$; $M = 6$ **14.** $2x$, $6x$, $9x$, $13x$; $M = 56.25$

Write a set of numbers satisfying the conditions given in each of the following.

15. The mean is greater than the median.
16. The mode is 40 and the median is 20.
17. The mean, median, and mode are all the same number.
18. The group is bimodal and the mean and median are 14.

Solve each of the following.

19. Gwen's grades on French quizzes are $8\frac{1}{2}$, 5, $8\frac{1}{2}$, $7\frac{1}{2}$, 8, 8, 9, 7, 10, and 7. Find the mean.
20. What is Gwen's median grade in exercise 19?

21. Spencer's mean score in Spanish this month is 77. His first two scores are 75 and 79, but he forgets his third and last score. What is it?
22. Seven employees earn $8,000, three earn $10,000, and two earn $40,000. List the salaries of the twelve employees. Find the mean, median, and mode.

Challenge   **Solve each of the following.**

1. If the median of the following set of numbers is 8, what is x? 11, 2, 3, 3.2, 13, 14, 10, 8, x
2. Find the value for y, in terms of x, such that the numbers x, $x + 5$, and $x + y$ have mean $2x + 1$.
3. Find the value for z such that the numbers w, $2w$, $2w + 3$, and $32 + z$ have mean $2w + 8$.
4. What do we mean when we say that the average American family owns a TV? Is this use of the word *average* closest to the mean, median, or mode?

12.3   Statistical Graphs

Graphs are often used to display collected data so that information can be interpreted easily. Different kinds of graphs are chosen for different purposes or types of data.

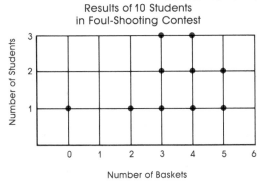

Results of 10 Students
in Foul-Shooting Contest

A **dot-frequency graph** shows how different quantities compare. It is one of the simplest graphs to construct. The graph at the left shows how a group of ten students scored in a foul-shooting contest. Each of ten students took 6 shots at the basket.

One student scored 0 baskets, no students scored exactly 1 basket, one scored exactly 2 baskets, three exactly 3 baskets, and so on.

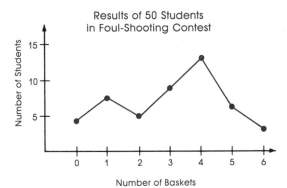

Results of 50 Students
in Foul-Shooting Contest

Similar to the dot-frequency graph is the **line graph.** The graph at the left gives results for fifty students in the same foul-shooting contest described above.

Line graphs are especially useful when the number of dots required for a dot-frequency graph is great. They are also helpful for showing trends or changes.

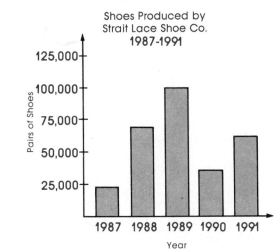

Shoes Produced by
Strait Lace Shoe Co.
1987-1991

One of the most common and important kinds of graphs is the **bar graph.** The lengths of thick bars show how different quantities compare.

The bar graph given at the left shows the number of pairs of shoes produced by the Strait Lace Shoe Company in the years 1987 to 1991. In 1987, the company produced about 23,000 pairs of shoes, and in 1988, it produced about 70,000 pairs.

You can interpret the graph at a glance. You can see that production increased every year except in 1990 when there was a decrease.

A well-drawn graph includes **1.** a title, **2.** scale labels, and **3.** appropriately spaced scales.

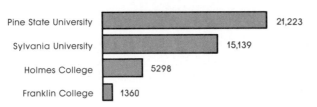

Students Enrolled in Colleges and Universities in Pine County, 1990-1991

A bar graph may have horizontal bars as in the graph at the left.

Sometimes values are written beside each bar.

Example

1 Construct a line graph and a bar graph to display the following data.

Percent of 18-Year-Olds with High School Diplomas

Year	1950	1955	1960	1965	1970	1975	1980	1985
Percent	56	65	72	71	77	74	72	74

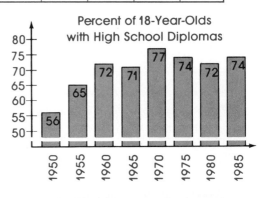

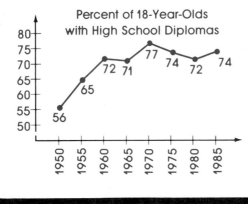

A **picture graph** or **pictograph** is similar to a bar graph. A symbol is used to stand for a given quantity of something. What the symbol is and what it stands for must be stated clearly. These graphs are often used to show information in an appealing way.

Each apple in the graph at the left stands for 50,000 real apples. Parts of pictures can be used. For example, Beck Orchard produced about 75,000 apples.

Students in Various
Foreign Language
Courses

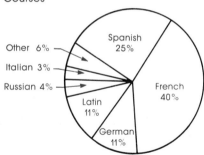

A **circle graph** shows parts of a whole. Percents are often used. Each wedge or sector of the circle represents a percent.

In the circle graph at the left, all the students who took a language at Franklin College are represented by a circle, or 360°. Forty percent of these students took French.

$$0.40 \cdot 360° = 144°$$

Thus, the angle at the center of the sector for the students who took French has a measure of 144°.

Example

2 During a certain year, Pine City spent 65% of its income on services, 25% on public safety, 4% on recreation, and 6% on administration. Construct a circle graph to show how Pine City spent its money.

Find the number of degrees representing each area of spending. Then draw the graph.

Services   65% of 360° = 234°
Safety   25% of 360° = 90°
Recreation   4% of 360° = 14.4°
Administration   6% of 360° = 21.6°

The sum of the degrees in the circle must be 360°.

$234 + 90 + 14.4 + 21.6 = 360$ √

How Pine City
Spent Its Income

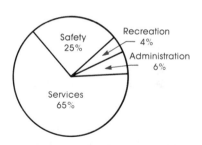

Exercises

Exploratory   Use the bar graph shown below to answer each of the following.

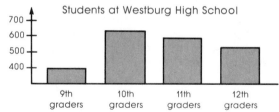

1. About how many tenth graders are there?

2. About how many more eleventh graders are there than twelfth graders?

3. What is the ratio of ninth graders to twelfth graders?

4. What is the ratio of tenth graders to ninth graders?

Use the picture graph at the right to find about how many television sets are in each of the following towns.

5. Appleburg   **6.** Blakeburg
7. Centerburg   **8.** Davisburg

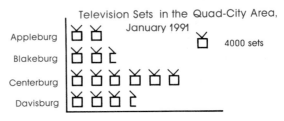

Television Sets in the Quad-City Area, January 1991

4000 sets

Appleburg
Blakeburg
Centerburg
Davisburg

Written   Construct a graph as directed in each problem.

1. Construct a dot-frequency graph to show the number of librarians in the following villages: Baxton, 3; Danton, 5; Centerburg, 3; Southmont, 1; Earlville, 0; and Wilmore, 4.

2. Construct a line graph to show the data in exercise 1, but also include the following data: Westburg, 12; Hudson, 18; and Beechwood, 14.

Construct a bar graph to display the data in each of the following tables.

3. Number of Tons of Peanut Butter Processed

Plant	A	B	C	D	E
Number of tons	32	220	175	90	180

4. Number of Employees of Four Electric Companies

Company	Southern Electric	XYZ Power	Ohm Electric	Iontronics
Number of employees	1120	575	700	1950

5. Amount, in Tons, of Coal Mined by the Acme Coal Company

Year	1950	1955	1960	1965	1970
Amount	1000	10% Increase of 1950	10% Increase of 1955	20% Increase of 1960	15% Decrease of 1965

Construct a line graph for the information in each of the following.

6. Exercise 3   **7.** Exercise 4   **8.** Exercise 5

Use the circle graph at the right to complete exercises 9-22. Find the percent of the Bensons' annual income spent on each of the following.

 9. food   **10.** clothing
11. rent   **12.** items other than rent and food

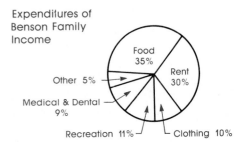

Expenditures of Benson Family Income

Food 35%
Rent 30%
Other 5%
Medical & Dental 9%
Recreation 11%
Clothing 10%

The Bensons' annual income is $20,000. Find the amount spent on each of the following.

13. medical and dental

14. food and clothing

15. recreation

16. other than rent, food, and clothing

Suppose their annual income is $27,500. Find the amount spent on each of the following.

17. clothing

18. other than rent and food

19. rent

20. medical, dental, and recreation

21. List what some of their other expenses might include.

22. Find the number of degrees in the central angle of each sector of the circle graph.

Construct a circle graph to display the data in each of the following tables.

23. Expenditures of the Chong Family

Item	food	rent	clothing	recreation	other
Percent	40%	25%	12%	10%	13%

24. Distribution of Books in the Pine City Library

Classification	reference	fiction	biography	nonfiction
Percent	20%	32%	8%	40%

_____ **Mixed Review** _____

Solve each equation and check your answer.

1. $6x^2 - 15x = 0$

2. $a^2 + 6a + 9 = 0$

3. $r^2 - 64 = 0$

4. $\dfrac{r + 1}{5} = 2r + 13$

5. $\dfrac{b + 11}{10} = \dfrac{6}{b}$

6. $\dfrac{n - 3}{2} = \dfrac{15}{n - 4}$

Each letter of the alphabet is written on a slip of paper which is then placed in a hat. One is chosen at random. Find the probability of each choice.

7. a letter from NUMBER

8. a letter from MATH or ENGLISH

9. a letter from WINTER or SPRING

10. a letter from SPACE or SPICE

For each set of numbers, find the mean, median, and mode. State if the set is bimodal.

11. 12, 13, 14, 14, 18, 19

12. 21, 23, 26, 28, 28, 42

13. 72, 89, 88, 129, 96, 96, 72, 77, 81, 72, 96

14. A rectangular garden, 15 m by 25 m, is surrounded by a walk of uniform width. The area of walk itself is 176 sq m. Find the width of the walk.

15. Larry has 6 brown and 8 black socks in a drawer. If he pulls out 2 socks at random, what is the probability he chooses a matching pair?

16. The base of a triangle is 3 cm longer than the altitude drawn to that base. If the area of the triangle is 20 cm², find the base and the altitude.

12.4 Histograms

The graph given below shows the number of students who scored from 0 to 6 baskets, exactly, in a foul-shooting contest. This graph is a **histogram.** It shows how frequently a certain number of baskets were scored.

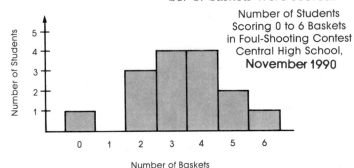

Number of Students Scoring 0 to 6 Baskets in Foul-Shooting Contest Central High School, **November 1990**

A histogram is a bar graph which shows the **frequency** of certain occurrences or scores. In a histogram, the bars are placed next to each other.

Suppose a survey was taken of the number of hours per week 35 ninth graders watched television. Each ninth grader was asked to record the number of hours to the nearest half hour he or she watched television during a certain week.

Calculator Hint

Most graphing calculators have a statistical mode that allows you to enter a set of data, find its mean and median, and graph a histogram or line plot.

The results are shown in the order in which they were recorded.

19	$12\frac{1}{2}$	25	$25\frac{1}{2}$	$17\frac{1}{2}$	16	8
17	27	18	23	13	17	17
$11\frac{1}{2}$	12	$23\frac{1}{2}$	$1\frac{1}{2}$	19	5	$18\frac{1}{2}$
17	$20\frac{1}{2}$	15	$14\frac{1}{2}$	$10\frac{1}{2}$	23	14
3	17	16	23	31	18	14

It is difficult to see any patterns or draw any useful conclusions from this kind of list. A histogram presents the survey data in a more helpful way.

Example

1 **Draw a histogram to display the data from the survey of the television habits of ninth graders.**

First of all, rewrite the data in numerical order.

$$1\frac{1}{2} \quad 3 \quad 5 \quad 8 \quad 10\frac{1}{2} \quad 11\frac{1}{2} \quad 12$$

$$12\frac{1}{2} \quad 13 \quad 14 \quad 14 \quad 14\frac{1}{2} \quad 15 \quad 16$$

$$16 \quad 17 \quad 17 \quad 17 \quad 17 \quad 17 \quad 17\frac{1}{2}$$

$$18 \quad 18 \quad 18\frac{1}{2} \quad 19 \quad 19 \quad 20\frac{1}{2} \quad 23$$

$$23 \quad 23 \quad 23\frac{1}{2} \quad 25 \quad 25\frac{1}{2} \quad 27 \quad 31$$

Patterns can now be seen more readily. The greatest and least can be picked out quickly.

Next, **group** the data in convenient **intervals.**

Each class interval contains scores within a certain range. The number of intervals should not be great. Too many intervals would defeat the purpose of grouping the data. The intervals should be equal width and should not overlap.

No student watched more than 32 hours.

$$32 \div 8 = 4$$

We can separate these scores into eight convenient intervals of $3\frac{1}{2}$ hours each.

Next, **tally** the scores in each interval.

Tally of TV Viewing Scores

Interval	Tally
$0\text{-}3\frac{1}{2}$	ll
$4\text{-}7\frac{1}{2}$	l
$8\text{-}11\frac{1}{2}$	lll
$12\text{-}15\frac{1}{2}$	⊮⊮ ll
$16\text{-}19\frac{1}{2}$	⊮⊮ ⊮⊮ lll
$20\text{-}23\frac{1}{2}$	⊮⊮
$24\text{-}27\frac{1}{2}$	lll
$28\text{-}31\frac{1}{2}$	l

Finally, construct a histogram from the information in the table.

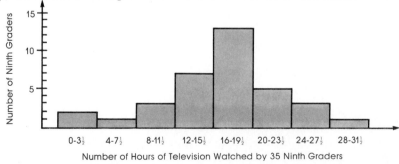

Number of Hours of Television Watched by 35 Ninth Graders

Exercises

Exploratory   Answer each of the following.

1. How is a histogram different from a bar graph?

2. When intervals for histograms are being chosen, what must be considered?

Each of the following is a title for a graph. State whether you would use a histogram or some other kind of graph to present the data for each. Explain your answer.

3. the number of questions answered correctly by 12 students on a five-question quiz

4. the number of lawn mowers produced by four lawn mower manufacturers in 1980

5. the effective life of 1000 flashlight batteries

6. the heights of 220 incoming freshmen at Pine City State College

7. the uses to which one dollar is put by the XYZ Corporation

8. the scores of 10,000 students on a standardized test

Use the histogram at the right to answer each of the following.

9. What is the width of each interval? How do you think the intervals were chosen?

10. About how many bulbs had lives of between 1001 and 1050 hours?

11. How many bulbs had lives of over 1250 hours?

12. About how many more bulbs burned for between 1051 and 1100 hours than between 951 and 1000 hours?

13. What percent of the bulbs, approximately, burned out between 1101 and 1150 hours?

14. About how many bulbs lasted more than 1100 hours?

15. About how many bulbs lasted fewer than 1151 hours?

16. What percent of the bulbs lasted fewer than 1101 hours?

17. What percent of the bulbs lasted fewer than 1201 hours?

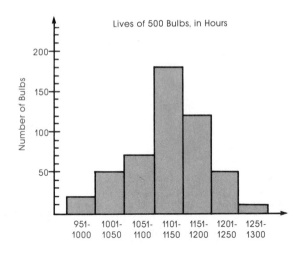

Written   For each of the following sets of data, complete each of the following processes.

a. Rearrange the data in order from least to greatest.

b. Group the data in convenient class intervals.

c. Tally the data according to your intervals.

d. Construct a histogram to display the data.

1. Test scores in a mathematics examination.

73	88	68	91	79	88	70	90	75	72	66	63
65	85	81	81	78	96	80	59	86	87	74	73
81	78	87	83	87	80	77	75	86	84	82	78

2. The weights, to the nearest ounce, of laboratory animals.

171	161	159	153	159	166	173	168	201	177
161	119	180	140	155	114	163	166	149	210
169	164	190	181	153	157	156	129	155	169
162	150	101	175	166	156	131	177	180	161
162	171	158	160	188	177	170	169	156	128

3. The lives of 50 bulbs from the assembly line, to the nearest hour.

1222	1052	1081	931	944	1289	1070	1009	1116	912
1133	1074	955	1172	957	1148	1033	1030	1306	1272
1023	1120	1115	1066	1211	998	996	1156	887	784
1104	1108	1294	1318	1221	1053	1172	1182	1380	1079
1057	1199	1014	1033	1232	863	1230	1063	959	1129

4. The number of questions answered correctly by 90 students on an examination.

74	76	91	80	86	78	77	54	64	90	68	77	76	70	73
84	86	74	74	75	58	97	66	78	76	72	88	85	76	75
88	80	79	69	67	77	74	73	87	90	61	65	82	81	78
90	90	76	77	75	89	63	78	79	76	94	63	72	74	60
76	89	87	85	76	95	67	84	87	78	78	70	71	72	77
74	76	76	85	62	58	78	70	86	85	81	55	79	92	82

Mathematical Excursions

Misleading Graph Scales

The two graphs below were made to show the results of a survey in which consumers compared two brands of frozen pizza.

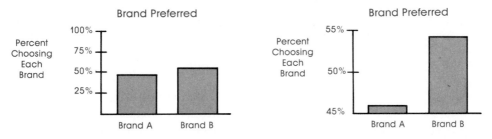

Different scales change the appearance of the graphs, which show the same results. Sometimes different scales help to make graphs easier to read. But other times the scales are chosen purposely to mislead.

12.5   Cumulative Frequency Histograms

Interval	Tally
41-50	I
51-60	III
61-70	ℍℍℍ ℍℍℍ ℍℍℍ ℍℍℍ
71-80	ℍℍℍ ℍℍℍ ℍℍℍ ℍℍℍ ℍℍℍ ℍℍℍ ℍℍℍ
81-90	ℍℍℍ ℍℍℍ ℍℍℍ I
91-100	ℍℍℍ

Suppose that a quiz is given to 80 students. A tally of the grades, grouped by intervals, is shown at the left.

Here is a histogram based on this information.

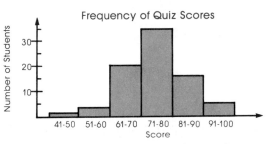

Frequency of Quiz Scores

The histogram shows that 20 students scored in the interval 61-70. Sixteen students scored in the 81-90 interval. The histogram shows the frequency of the scores in an interval.

Suppose you want to know the number of students who scored 80 or less. You can add the numbers of students in all the intervals of the graph up to and including the 71-80 interval. There is a kind of histogram, however, that gives this kind of information quickly. It is called a **cumulative frequency histogram.**

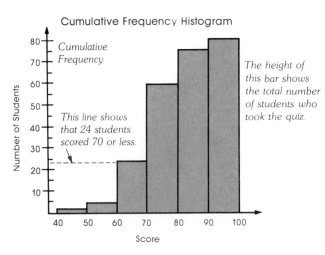

Cumulative Frequency Histogram

The height of this bar shows the total number of students who took the quiz.

This line shows that 24 students scored 70 or less.

This cumulative frequency histogram describes the same data as the first histogram. The difference is that in preparing this graph, the number of scores in every interval is added to the number of scores in the interval before it. Each bar is built by starting from the top of the one before it. Each bar, therefore, will show the number of scores which fall below the score indicated by the right vertical boundary of that bar.

You can read other information from cumulative frequency histograms. The column at the far left of the diagram on the next page gives the **cumulative relative frequency** of the students' scores. The cumulative relative frequency is the cumulative frequency divided by the total number of test takers. It is given as a percent.

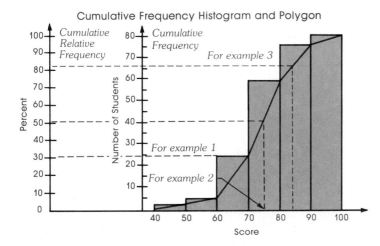

Cumulative Frequency Histogram and Polygon

The colored line drawn in the histogram from the lower left to the upper right corner is called the **cumulative frequency polygon.** This device aids in making important approximations from the graph.

Examples

1 **What is the cumulative relative frequency for the students who scored 70 or less on the quiz?**

Look at the cumulative frequency column in the diagram. The cumulative frequency for those who scored 70 or less is 24.

$$\text{The cumulative relative frequency for "70 or less"} = \frac{\text{cumulative frequency}}{\text{total number of quiz takers}}$$

$$= \frac{24}{80} \text{ or } \frac{3}{10} \text{ or } 30\%$$

This value can also be read directly from the far-left column.

Thus, 30% of all the students who took the test scored 70 or less.

2 **Find the median score of the students who took the quiz.**

The median is the score that 50% of the quiz takers scored at or below. You need to use the cumulative frequency polygon drawn in the figure. To find the median, locate 50% in the cumulative relative frequency column. Follow the horizontal line to where it meets the cumulative frequency polygon. From this point, go straight down to the approximate test score that produced this cumulative relative frequency.

The median appears to be about 75.

These methods are only approximations. They work best when there are many data.

The second quartile is the median since the median is the measure at or below which 50% of all the students scored.

There are two other measures that can be found from the diagram. The **first quartile** is the measure at or below which 25% of all the students scored. The **third quartile** is the score at or below which 75% of all the students scored. In the example, the first quartile is about 67 and the third, about 82.

Statisticians often use **percentiles** to describe certain data. For example, the 82nd percentile is the measure at or below which 82% of the measures of a given set fall.

Definition of nth percentile

The *n*th percentile of a set of data is the score at or below which *n*% of the scores lie.

Example

3 **Find an approximate value for the 82nd percentile.**

Locate 82% in the cumulative relative frequency column. Follow a horizontal line to where it meets the cumulative frequency polygon.

A vertical line shows that 82% of the students scored about 85 or less.

Exercises

Exploratory Answer each of the following.

1. Explain how to construct a cumulative frequency histogram.

2. What is a cumulative frequency polygon?

3. What information does the cumulative frequency polygon help you to determine? How?

4. In terms of your class standing at school, would you rather be closer to the first or the third quartile?

Written Answer each of the following.

1. Give the percentile ranks for the first quartile, third quartile, and median.

2. On various nationwide tests, students are often given their percentile rating as well as their regular, or "raw" scores. Why is this done?

Refer to the figure at the right to answer exercises 3-17. State how many parts lasted the following times.

3. less than 2100 hours
4. 2300 hours or less
5. between 2300 and 2399 hours

Answer each of the following.

6. Between which two lengths of time did most of the parts tested last?

7. How many more parts lasted 2400 to 2499 hours than 2500 to 2599 hours?

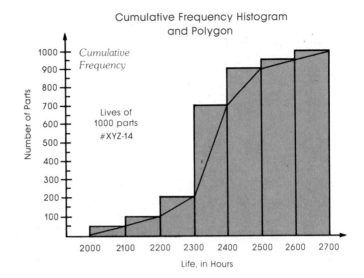

Cumulative Frequency Histogram and Polygon

Cumulative Frequency

Lives of 1000 parts #XYZ-14

Number of Parts

Life, in Hours

Find approximate values for each of the following measures.

8. the median

9. the first quartile

10. the third quartile

Answer each of the following.

11. The first quartile and the median seem to be very close together. Explain this.

12. The top of the bar for 2600 to 2699 hours is at the maximum height of the histogram. Why is this the case?

13. Can you use the polygon to find the number of hours at or below which 60% of the parts wore out? How?

14. Generalize your result in exercise 13.

Use a straightedge to find an approximate value for each of the following.

15. 58th percentile

16. 21st percentile

17. 85th percentile

Construct a cumulative frequency histogram and polygon for each of the following sets of measures. Mark the first and third quartiles and the median.

18. Test scores in a mathematics examination.

73	88	68	91	79	88	70	90	75	72	66	63
65	85	81	81	78	96	80	59	86	87	74	73
81	78	87	83	87	80	77	75	86	84	82	78

19. The lives of 50 bulbs from the assembly line, to the nearest hour.

1222	1052	1081	931	944	1289	1070	1009	1116	912
1133	1074	955	1172	957	1148	1033	1030	1306	1272
1023	1120	1115	1066	1211	998	996	1156	887	784
1104	1108	1294	1318	1221	1053	1172	1182	1380	1079
1057	1199	1014	1033	1232	863	1230	1063	959	1129

12.6   Empirical Probability

Without rolling a die, you know that the probability of rolling a 5 is $\frac{1}{6}$. $P(rolling\ a\ 5) = \frac{1}{6}$ is an example of a **theoretical probability.**

Suppose an ordinary thumbtack falls to the floor. Can you determine the theoretical probability, $P(point\ up)$?

You probably know that such a probability cannot be found. Furthermore, the probability may change depending on the thumbtack. For example, do you think both tacks in the figure at the left have the same probability of landing point up?

You can find an approximation of $P(point\ up)$ by tossing the tack 100 times and counting the number of points up. Suppose the probability could be found theoretically by some means to be $\frac{1}{4}$. Then the tack should land point up about $\frac{1}{4} \cdot 100$ or 25 times.

Suppose, however, you toss the tack 100 times and the tack lands with its point up 62 times. You conclude that the theoretical probability of point up is about 0.62.

Suppose you toss the tack 500 times. The tack lands point up 314 times. You find the probability is about 0.63 as follows.

$$P(point\ up) \approx \frac{314}{500} \text{ or } 0.628$$

If you toss the tack 1000 times and get 623 points up, you say the probability is about 0.62. This number is called an **empirical probability.** The more times the tack is tossed, the closer you expect to get to the theoretical probability.

Definitions of Empirical Probability and Empirical Evidence

> A probability determined by observation or experimentation is said to be an empirical probability. The evidence, or data, obtained in such an investigation is called empirical evidence.

You can determine that a coin is fair by empirical methods. If you toss the coin many times, you assume it is fair if it lands heads about half the time. Suppose you toss a coin 1000 times. It lands heads 750 times. You then conclude the coin may have been tampered with. In this case, $P(heads) \approx \frac{3}{4}$.

Exercises

Exploratory   For each of the following experiments, find the theoretical probability as indicated.

1. Toss 2 fair coins. *P(2 heads)*
2. Toss 3 fair coins. *P(3 heads)*
3. Toss 4 fair coins. *P(4 heads)*
4. Toss *n* fair coins. *P(n heads)*
5. Choose a card from a deck. *P(ace)*
6. Choose a card from a deck. *P(club)*.
7. Choose a card, return it, choose another. *P(2 aces)*
8. Choose a card, do not return it, choose another. *P(2 aces)*
9. Choose a letter at random from MATH-EMATICS. *P(vowel)*
10. Roll 3 fair dice. *P(sum is 3)*
11. Choose a state at random from the 50 United States. *P(state has 2 senators)*
12. Choose a state at random from the 50 United States. *P(state's name begins with Q)*.
13. Roll 2 fair dice. *P(sum is even)*
14. Roll 3 fair dice. *P(3 sixes)*
15. Choose a day at random from the days in a year. *P(day in September)*
16. Roll 3 fair dice. *P(sum is 7)*

Written   State if each of the following probabilities can be determined theoretically or empirically.

1. The probability of picking 2 queens when 2 cards are picked.
2. The probability that Player A wins when Player A and Player B play a game of chess.
3. The probability that it will rain on November 4, 1992.
4. The probability that a family with 3 children has all girls.
5. The probability that a cone-shaped cup lands point up instead of on its side when it is thrown into the air.
6. The probability that a person will win the state lottery.
7. Clem tosses a penny 3 times and gets 2 heads. He says, "For this penny, $P(heads) = \frac{2}{3}$." How would you respond to Clem?
8. You have reason to believe that a die is unfair. Describe a process you would use to test your theory.
9. Two fair dice are rolled together 60 times. About how many times would you expect to get a sum of 7?
10. If you got forty 7's in exercise 9, what would be your temporary guess about *P(sum of 7)* for those dice? How would you confirm your guess?
11. Use empirical methods to explain how to justify that $P(tails) = \frac{1}{2}$ for a fair coin.
12. For the coin tossed in exercise 11, would you expect to get exactly 1000 tails for 2000 tosses? Why or why not?

12.7 Organizing Experimental Data

Data from experiments can be organized to help you find empirical probabilities. Suppose a class must find an approximation for the probability that a thumbtack will land point up. Ten students are each asked to toss the tack 20 times. Each toss is a **trial.**

Each student tallies his or her results. The tally at the left shows that student 1 tossed 11 points up in 20 trials.

A table like the one below is constructed to organize the trial results. You will be asked to complete the table as an exercise. Explanations of the column headings follow the table.

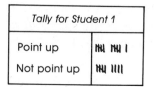

Tally for Student 1	
Point up	ⵑⵑ ⵑⵑ I
Not point up	ⵑⵑ IIII

For the first student, the relative frequency is $\frac{11}{20}$ since out of 20 trials, 11 tossed points up.
For the second student who tossed 13 points up, the cumulative frequency is 11 + 13 or 24. The total number of trials is 40, so the cumulative relative frequency is $\frac{24}{40}$.

	Table for Thumbtack Point Up			
Student	Frequency	Relative Frequency	Cumulative Frequency	Cumulative Relative Frequency
1	11	$\frac{11}{20}$	11	$\frac{11}{20}$
2	13	$\frac{13}{20}$	24	$\frac{24}{40}$
3	9		33	
4	13		46	
5	12		58	$\frac{58}{100}$
6	13		71	
7	12		83	
8	10		93	
9	11		104	
10	10		114	$\frac{114}{200}$

Explanations of Column Headings

Frequency is the number of times *point up* occurs for each student.

Relative frequency is the number of occurrences of *point up* divided by the number of trials.

Cumulative frequency is the total of all the frequencies recorded up to that time.   *For example, after student 5 records her tally, the cumulative frequency is 58.*

Cumulative relative frequency is the cumulative frequency divided by the total number of trials.   *For example, after student 5, the cumulative relative frequency is $\frac{58}{100}$.*

In the preceding experiment, the students made a total of 200 trials. The cumulative relative frequency for 200 trials is $\frac{114}{200}$ or 0.57. Based on these results, we conclude that the probability for point up for this tack is about 0.57.

Suppose twenty students toss the tack 50 times each, for a total of 1000 tosses. The cumulative relative frequency obtained is a closer estimate of the real probability than after 200 trials. Suppose there are a million trials. The cumulative relative frequency is even closer to the actual probability than after 1000 tosses.

Cumulative Relative Frequency and Theoretical Probability

As the number of trials increases, the cumulative relative frequency of an event gets closer to the theoretical probability.

Example

1 Describe how tossing a fair coin shows that the cumulative relative frequency gets closer to the theoretical probability as the number of trials increases.

Suppose a coin is tossed 10 times. The relative frequency for 10 tosses could be $\frac{4}{10}$, or 0.4.

A result of 4 heads and 6 tails would not be unexpected.

Suppose the coin is tossed 100 times. To obtain a relative frequency of $\frac{2}{5}$, only 40 heads could be tossed. But we would expect the number of heads to be closer to 50, say 45. The cumulative relative frequency would then be $\frac{45}{100}$, or 0.45. By similar reasoning, after 1000 trials, the number of heads would be closer to 500 than 450. Perhaps it would be 470 heads. The cumulative relative frequency would then be $\frac{470}{1000}$ or 0.47.

For a great number of trials, the cumulative relative frequency should begin to get close to about $\frac{1}{2}$. The theoretical probability, *P(heads)*, is $\frac{1}{2}$.

▬ Exercises ▬

Exploratory   Solve each of the following.

1. Copy and complete the table for thumb-tack point up.

2. Is it possible to toss a fair coin 10 times and get 10 tails?

Written   Solve each of the following.

1. Construct a table for the number of sixes recorded when six people each roll a die 40 times. Give values for relative frequency, cumulative frequency, and cumulative relative frequency. Use the following data. Use the table for thumbtack point up as a guide.

Person	1	2	3	4	5	6
Frequency	23	19	20	21	21	18

2. Why does the cumulative relative frequency appear to jump back and forth in exercise 1?

3. What do you suspect about the die based on the cumulative relative frequencies you found in the table?

A coin has been tampered with so that $P(heads)$ **no longer equals** $\frac{1}{2}$. **To find an approximation of** $P(heads)$, **ten students tossed the coin 30 times each. Their tallies are shown at the right.**

4. Use the tallies to construct a table showing frequency, relative frequency, cumulative frequency, and cumulative relative frequency of heads.

5. Give an estimate of $P(head)$ for this coin.

6. How could you get a better estimate?

Tally for Student 1	Tally for Student 2
H ₥ ₥ ₥ ₥ I T ₥ IIII	H ₥ ₥ ₥ ₥ II T ₥ III
Tally for Student 3	**Tally for Student 4**
H ₥ ₥ ₥ III T ₥ ₥ II	H ₥ ₥ ₥ ₥ T ₥ ₥
Tally for Student 5	**Tally for Student 6**
H ₥ ₥ ₥ ₥ I T ₥ IIII	H ₥ ₥ ₥ ₥ I T ₥ IIII
Tally for Student 7	**Tally for Student 8**
H ₥ ₥ ₥ IIII T ₥ ₥ I	H ₥ ₥ ₥ ₥ II T ₥ III
Tally for Student 9	**Tally for Student 10**
H ₥ ₥ ₥ III T ₥ ₥ II	H ₥ ₥ ₥ ₥ T ₥ ₥

For each experiment in the following table, tell, if you can, about how many times you would expect the specific outcome to result. If you cannot do this, say why not.

Experiment	Number of Trials	Specific Outcome
7. A fair die is rolled.	6000	A 3 comes up.
8. A "loaded" die is rolled.	1000	A 3 comes up.
9. Sally takes 5 foul shots.	100	She makes all 5.
10. Three fair coins are tossed.	1000	Exactly 2 heads come up.

Toss two dice 100 times. Record the sums of the faces. Use your results to find each probability. Compare your probabilities to the corresponding theoretical probabilities.

11. *P(sum of 2)*   **12.** *P(sum of 7)*   **13.** *P(sum of 10)*

14. *P(sum of 8)*   **15.** *P(sum less than 5)* **16.** *P(sum greater than 7)*

17. What sum occurred most often? Does this sum have the highest theoretical probability?

18. What sum occurred least often? Does this sum have the lowest theoretical probability?

Toss three coins 100 times. Record the results and use them to find each probability. Compare your probabilities to the corresponding theoretical probabilities.

19. *P(three heads)*   **20.** *P(two heads)*   **21.** *P(one head)*

22. *P(zero heads)*   **23.** *P(at least two heads)* **24.** *P(at most two heads)*

25. What result occurred most often? Does this result have the highest theoretical probability?

26. What result occurred least often? Does this result have the lowest theoretical probability?

Mixed Review

Perform the indicated operations and simplify.

1. $\dfrac{x-5}{2x} + \dfrac{3x+10}{4x}$ **2.** $\dfrac{n-3}{5n} - \dfrac{n+3}{2n^2}$ **3.** $\dfrac{4a-1}{3a^2} \cdot \dfrac{2a}{a-6}$

A bag contains 3 red, 2 white, and 5 blue marbles. If you choose a marble at random, find the probability of choosing each of the following.

4. a red marble   **5.** a blue marble   **6.** a white marble

7. a red or a blue marble   **8.** a marble that is not red

A marble is chosen from the bag in exercises 4-8, not replaced, and then another marble is chosen from the bag. Find each probability.

9. *P(blue, blue)*   **10.** *P(one white, one red)*

11. *P(one red, one blue)*   **12.** *P(same color)*

13. How many 5-digit zip codes can be formed if no digit is used more than once in a zip code?

14. How many permutations are possible of ABCDEF if each permutation must begin and end with a consonant?

15. Five ferry boats make the crossing from Bremerton to Seattle. How many ways can travelers make a round trip if they must make the trip on two different ferry boats?

16. How many different 8-letter combinations can be formed from the letters of TRIANGLE if each word must begin with T and end with LE?

Problem Solving Application: Using Tables and Charts

Many problems can be solved by using a table to organize relevant information. Tables can make it easier to read and interpret the information provided in the problem. In some cases, the solution to a problem can be taken directly from the information in the table.

Examples

1 **Michael is competing in a weekly sales contest at the car dealership where he works. He receives 5 points for each deluxe model he sells and 3 points for each standard model. If Michael received 38 points last week, how many of each model did he sell?**

Explore   Each deluxe model is worth 5 points. Each standard model is worth 3 points. We want to find all combinations of models sold that yield 38 points.

Plan   List the possible combinations in a table and examine the results. We can narrow the list since we know he cannot have sold more than 7 deluxe or 12 standard models.   *Why?*

Solve

No. of Dlx. Models	7	6	6	5	5	4	3	3	2	2	1
No. of Std. Models	1	2	3	4	5	6	7	8	9	10	11
Total Points	38	36	39	37	40	38	36	39	37	40	38

Michael could have sold 7 deluxe models and 1 standard model, 4 deluxe models and 6 standard models, or 1 deluxe model and 11 standard models.

Examine   Since $7 \cdot 5 + 1 \cdot 3 = 38$, $4 \cdot 5 + 6 \cdot 3 = 38$, and $1 \cdot 5 + 11 \cdot 3 = 38$, the solutions are correct.

2 **The cost to send a small parcel includes a certain amount for the first 3 ounces and $0.17 for each additional ounce. It costs $1.23 to send a 5-ounce parcel. What would it cost to send 3-ounce and 8-ounce parcels?**

Explore   We know that it costs $1.23 to send a 5-ounce parcel and it costs $0.17 for each additional ounce over 3 ounces.

Plan   Write the specific case (5-ounce parcel costs $1.23) in a table. Then generate more cases in the table using the information about the price per additional ounce.

Solve   It costs $1.23 − $0.17 or $1.06   It costs $1.23 + $0.17 or $1.40
to send a 4-ounce parcel.   to send a 6-ounce parcel.

Number of Ounces	3	4	5	6	7	8
Cost (in dollars)	0.89	1.06	1.23	1.40	1.57	1.74

It would cost $0.89 to send a 3-ounce parcel and $1.74 to send an 8-ounce parcel.   *Examine this solution.*

Exercises

Written   Solve each problem.

1. Eggs can be purchased in cartons of 12 or 18. If Sondra has 78 eggs, what combinations of cartons could she have purchased?

2. It costs 25¢ for a stamp and 18¢ for a postcard. What combinations of each can you buy if you spend between $1.80 and $2.10 and you must buy at least five stamps?

3. Salads Unlimited charges $1.69 for an 8-ounce salad and $2.84 for a 13-ounce salad. At that rate, how much should an 11-ounce salad cost?

4. Robert Hart was charged $3.12 for a 20-minute call to Buffalo. He was also charged $3.34 for a 22-minute call. How much should Mr. Hart be charged for an hour call?

5. Mei tore a sheet of paper in half. Then she tore each of the resulting pieces in half. If she continued this process 15 more times, how many pieces of paper would she have at the end?

6. Find the sum of the first three consecutive odd integers and of the first four consecutive odd integers. Then find the sum of the first fifty consecutive odd integers.

7. To determine a grade point average, four points are given for an A, three for a B, two for a C, one for a D, and zero for an F. If Greg has a total of 13 points for 5 classes, what combinations of grades could he have received?

8. To determine total bases in softball, 4 bases are given for a home run, 3 for a triple, 2 for a double, and 1 for a single. If Cathy has 15 total bases on 8 hits, what combinations of hits could she have?

9. Water freezes at 32°F or 0°C. Water boils at 212°F or 100°C. Based on this information, at what temperature will the readings on the Fahrenheit and Celsius scales be the same?

10. The measure of the perimeter of a rectangle is the same as the measure of its area. What are the possible dimensions of the rectangle if the measures of the length and width are natural numbers?

Portfolio Suggestion

Select an item from your work that shows your creativity and place it in your portfolio.

Performance Assessment

The opening day scores in a local golf tournament are 78, 83, 70, 84, 89, 67, 84, 92, 78, 91, 85, 77, 68, 80, 71, 78, 99, 81, 75, 88, 90, 71, and 73.

a. Find the mean, median, and mode for this set of scores.

b. Suppose a score of 110 were added to this data. Would the mean or the median be more affected? Explain.

c. Separate the data into intervals and make a cumulative frequency histogram. Explain your procedure.

Chapter Summary

1. **Statistics** is the science of collecting, organizing, analyzing, and presenting numerical **data.** (384)
2. **Data** are the pieces of information used in statistics. (384)
3. The **mean** of a set of n numbers is the sum of the numbers divided by n. (387)
4. The **mode** of a set of numbers is the number which occurs most often. (387)
5. The **median** of a set of numbers is the one which lies in the middle when the numbers are arranged in order. If there are two middle values, the median is the mean of these values. (389)
6. The nth **percentile** of a set of data is the score at or below which n% of the scores lie. (402)
7. A probability determined by observation or experimentation is said to be an **empirical probability.** The evidence, or data, obtained in an investigation of empirical probability is called **empirical evidence.** (404)
8. As the number of trials increases, the **cumulative relative frequency** of an event gets closer to the **theoretical probability.** (407)

Chapter Review

12.1 **A cat food company is surveying cat owners about a new product. State what problems each surveying method might produce.**

   1. Interviewing people in front of a bank at noon on June 1.
   2. Advertising in randomly chosen newspapers for written opinions.
   3. Sending a questionnaire to everyone in a randomly chosen apartment complex in a randomly chosen city.

12.2 **Find the mean, median, and mode of each set of numbers.**

   4. 5, 4, 6, 10, 3, 6, 6
   5. 8, 8, 8, 2, 0, 4

12.3 **Answer each of the following.**

   6. What 2 types of graphs are shown at the right?
   7. Construct a bar graph using the information in figure 1.
   8. What percent of Paoli's medical dollar is spent for costs other than doctor or dentist fees?

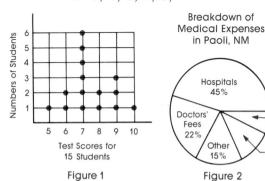

Figure 1

Figure 2

12.4 **For each problem, use the following set of numbers of days absent for**
and **44 employees at XYZ Company during the current year.**
12.5

0 10 8 5 8 9 3 3 2 9 7 1 0 0 4 2 4 6 9 2 3 11
1 5 5 7 2 6 5 3 4 7 1 1 5 3 3 3 1 4 5 5 1 2

   9a. Rearrange the data in numerical order from least to greatest.
   b. Tally the numbers of employees absent for 2-day intervals beginning with 0-1 and ending with 10-11.
   c. Draw a histogram showing the information in the tallies.
   10a. Draw a cumulative frequency histogram for the data in Exercise 9.
   b. Draw a cumulative frequency polygon in the graph in Exercise 10a.
   c. Find the values of the median and third quartile.

12.6 **State whether probabilities can be found theoretically or only empirically for each of the following experiments.**

   11. tossing a suspicious coin
   12. rolling 3 fair dice
   13. choosing a card from a deck
   14. making a field goal

12.7 **Answer each of the following.**

   15. What is meant by cumulative relative frequency?
   16. When does cumulative relative frequency get close to theoretical probability?

Chapter Test

State a problem people conducting surveys might have with each method.

1. Interviewing people at a bus stop on their way home from work.

2. Sending questionnaires to people chosen at random.

Solve each of the following.

3. Find the mean, median, and mode of 7, 5, 5, 7, 4, 5, 3, 5, and 9.

4. Find the value of x if 12, $2x - 1$, and 33 have a mean of 20.

Use the figure at the right to answer each of the following.

5. What percent of the juniors did not choose art or music?

6. If there were 800 juniors at Pine City High, how many chose a foreign language?

7. What is the measurement of the central angle for the leisure skills sector?

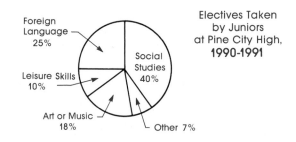

Electives Taken by Juniors at Pine City High, **1990-1991**

Foreign Language 25%

Social Studies 40%

Leisure Skills 10%

Art or Music 18%

Other 7%

Use the histogram given below to solve each of the following.

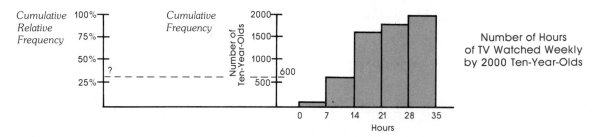

Number of Hours of TV Watched Weekly by 2000 Ten-Year-Olds

8. In making the graph, state how the cumulative frequency and cumulative relative frequency were found.

9. What entry belongs where the question mark is in the cumulative relative frequency column?

10. In which interval did the least number of ten-year-olds watching TV occur?

11. About how many ten-year-olds watched TV 0 to 6 hours?

12. About how many watched TV 14 to 20 hours?

13. About how many watched TV 13 hours or less?

14. In which interval is the median?

15. In which interval is the first quartile?

Answer each of the following.

16. What is empirical evidence?

17. Describe a method for finding an approximate value of *P(heads)* of a suspicious coin.

Introduction to
Coordinate Geometry

Application in Aviation

The most critical times of flying an aircraft are the takeoffs and landings. At these times, the pilot must be careful to use the correct slope of ascent or descent.

A hot-air balloon is 15 meters above ground and is rising at a rate of 18 meters per minute. What is the **slope** of the ascent of the balloon? Write an equation that describes the height *(h)* of the balloon after *t* minutes.

Group Project

Pick a city outside of the United States. Call three long-distance telephone companies to find out their standard evening rate to that city. For each company, write an equation that describes the cost *(c)* for a call that takes *t* minutes. Graph the three equations on the same coordinate plane. Is one company's cost always the lowest? Which company has the best rates to the city you picked?

13.1   Graphing

Two perpendicular number lines separate a plane into four regions called **quadrants.** The horizontal number line is called the **x-axis.** The vertical number line is called the **y-axis.** Their intersection is called the **origin** and is labeled *O*.

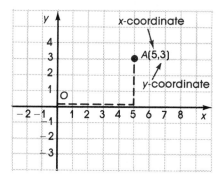

Locating points in a plane can be done by using **ordered pairs.** The figure at the left shows the graph of the ordered pair (5, 3). The first number, 5, is called the **x-coordinate** of point *A*. It tells the number of units to the left or right of the origin. The second number, 3, is called the **y-coordinate** of point *A*. It tells the number of units above or below the origin. A plane which has been set up in this way is referred to as a **coordinate plane.**

When graphing points on the coordinate plane, use the following guidelines.

In an ordered pair, find the horizontal or *x*-coordinate first, then the vertical or *y*-coordinate.

	positive	negative
x-axis	right	left
y-axis	up	down

Note that these directions are from the origin.

Examples

1 **Graph point *B* at (2, −4).**

Start at the origin. Move 2 units to the right. Then, move 4 units down.

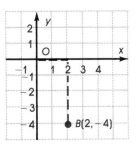

2 **Graph point *C* at (−2, 3).**

Start at the origin. Move 2 units to the left. Then, move 3 units up.

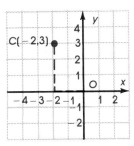

3 **Graph point *D* at (−3, −4).**

Start at the origin. Move 3 units to the left. Then, move 4 units down.

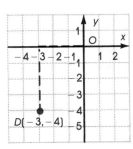

Exercises

Exploratory **Tell what directions are given by each of the following ordered pairs.**

1. (2, 5)   **2.** (2, −5)   **3.** (−2, 5)   **4.** (−2, −5)

5. (5, 2)   **6.** (5, −2)   **7.** (−5, 2)   **8.** (−5, −2)

Name a set of three ordered pairs in each of the following quadrants.

 9. I   **10.** II   **11.** III   **12.** IV

13. What are the coordinates of the origin?

Name the ordered pair for each point on the graph below.

14. *S*   **15.** *T*
16. *U*   **17.** *V*
18. *W*   **19.** *X*
20. *Y*   **21.** *Z*
22. *A*   **23.** *B*

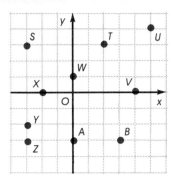

Written   **On a coordinate plane, locate each of the following points. Name them using the letter given.**

1. $A(0, 2)$ **2.** $B(2, -1)$ **3.** $C(4, 0)$ **4.** $D(-4, 1)$
5. $E(3, 3)$ **6.** $F(3, -3)$ **7.** $G(\frac{1}{2}, 5)$ **8.** $H(-\frac{1}{2}, \frac{1}{2})$
9. $J(-2, -5)$ **10.** $K(6, -1.5)$ **11.** $L(4, \sqrt{2})$ **12.** $M(-\sqrt{17}, 0.2)$

Graph each of the following sets of ordered pairs. Then, state whether each set of points are the vertices of a rectangle.

13. $(0, 0), (0, 3), (7, 0), (7, 3)$ **14.** $(-4, 0), (-2, 5), (1, 5), (-1, 0)$
15. $(2, 5), (7, 4), (8, 3), (1, 2)$ **16.** $(0, 1), (7, 1), (0, -3), (7, -3)$
17. $(-1, -1), (-1, -5), (3, -1), (3, -5)$ **18.** $(-4, 6), (-2, 6), (-3, 7), (-1, 7)$
19. $(-3, 5), (4, -2), (-3, -2), (4, 5)$ **20.** $(2, 2), (2, 3), (2, 4), (2, 5)$

Graph each of the following sets of ordered pairs, and draw the triangle which they determine. Then, find the area of each triangle.

21. $(0, 0), (4, 0), (1, 3)$ **22.** $(0, 0), (-3, 0), (0, -4)$
23. $(-1, 0), (4, 0), (0, 6)$ **24.** $(-1, 0), (4, 0), (3, 6)$
25. $(1, -1), (1, 5), (-2, 1)$ **26.** $(3, 1), (4, 1), (4, 2)$
27. $(-3, 1), (-2, 2), (0, 2)$ **28.** $(0, 0), (5, 2), (4, 0)$

On a coordinate plane, graph each of the following ordered pairs. Then, connect the points in order.

29. $A(-3, \frac{1}{2})$ **30.** $B(1, 0)$ **31.** $C(4, 0)$
32. $D(7, -4)$ **33.** $E(8, -4)$ **34.** $F(7, 0)$
35. $G(10, 0)$ **36.** $H(11, -2)$ **37.** $J(12, -2)$
38. $K(11\frac{1}{2}, 0)$ **39.** $L(13, \frac{1}{2})$ **40.** $M(11\frac{1}{2}, 1)$
41. $N(12, 3)$ **42.** $O(11, 3)$ **43.** $P(10, 1)$
44. $Q(7, 1)$ **45.** $R(8, 5)$ **46.** $S(7, 5)$
47. $T(4, 1)$ **48.** $U(1, 1)$ **49.** $V(-3, \frac{1}{2})$
50. What does the graph look like?

13.2 Linear Equations

The solutions for equations such as $y = 3x - 2$ are ordered pairs. By substituting different values for x you can find the corresponding values for y. Some of the solutions for $y = 3x - 2$ are shown in the table below.

x	$3x - 2$	y	(x, y)
-1	$3(-1) - 2$	-5	$(-1, -5)$
0	$3(0) - 2$	-2	$(0, -2)$
1	$3(1) - 2$	1	$(1, 1)$
2	$3(2) - 2$	4	$(2, 4)$

Figure I below shows the graph of some of the solutions of $y = 3x - 2$. Figure II shows a line containing the graphs of all ordered pairs which are solutions of $y = 3x - 2$.

Calculator Hint

Graphing calculators can graph linear equations as well as other types of equations. You determine the range and domain of the display and enter the equation in the form, $y = \underline{?}$. The calculator uses the equation to calculate the points automatically and connects them to form a line.

I 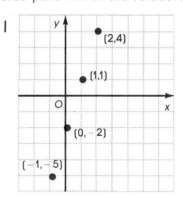 II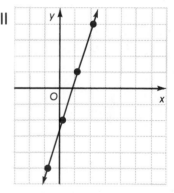

An equation whose graph is a straight line is called a **linear equation.**

Definition of Linear Equation

An equation is linear if and only if it can be written in the form $Ax + By = C$, where A, B, and C are any numbers and A and B are not both 0.

The equations $5x + 6y = 8$, $3x = 4y + 9$, $5m - n = \frac{1}{2}$, and $x = 4$ are linear equations. Each can be written in the form $Ax + By = C$. The equations $4x + 5y^2 = 7$ and $\frac{1}{y} + x = 3$ are *not* linear equations. *Why?*

Examples

1 **Graph $y = 2x + 1$.**

Make a table of values for x and y. Graph the ordered pairs and connect the points with a line.

x	$2x + 1$	y	(x, y)
-1	$2(-1) + 1$	-1	$(-1, -1)$
0	$2(0) + 1$	1	$(0, 1)$
1	$2(1) + 1$	3	$(1, 3)$
2	$2(2) + 1$	5	$(2, 5)$

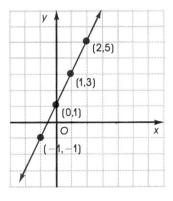

2 **Show that $y = 2x$ is a linear equation. Then draw its graph.**

Rearrange the terms into the form $Ax + By = C$.

$$y = 2x$$
$$0 = 2x - y$$
$$2x - y = 0$$

The values of A, B, and C are 2, -1, and 0 respectively. Thus, the equation $y = 2x$ is a linear equation.

Make a table of values for x and y. Graph the ordered pairs and connect the points with a line.

x	$2x$	y	(x, y)
-2	$2(-2)$	-4	$(-2, -4)$
-1	$2(-1)$	-2	$(-1, -2)$
0	$2(0)$	0	$(0, 0)$
1	$2(1)$	2	$(1, 2)$
2	$2(2)$	4	$(2, 4)$

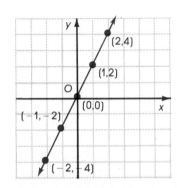

Example

3 Graph $x + y = 5$.

Rearrange the equation so that y is on one side by itself.

$$x + y = 5$$
$$y = 5 - x$$

Make a table of at least two values for x and y. Then, graph the ordered pairs and connect the points with a line.

x	$5 - x$	y	(x, y)
0	$5 - 0$	5	$(0, 5)$
2	$5 - 2$	3	$(2, 3)$

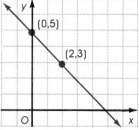

Since only two points are needed to determine a line, only two values for x and y are necessary. Finding more values for x and y will serve as a check.

Exercises

Exploratory State whether each of the following is a linear equation.

1. $y = 2x + 5$ **2.** $x = 3$ **3.** $x^2 + y^2 = 8$

4. $y = -2x$ **5.** $3x + 4y = 5$ **6.** $y = 8$

7. $\dfrac{1}{x} + \dfrac{1}{y} = 1$ **8.** $\dfrac{x + y}{2} = 13$ **9.** $y = -\dfrac{1}{3}x + 2$

10. $2.3x + 0.5y = 4$ **11.** $-y + 4x = 1$ **12.** $x - 0.3y - \sqrt{2} = 0$

In each problem, state which of the three given points, if any, lie on the graph of each linear equation.

13. $y = x + 1$ **a.** $(1, 1)$ **b.** $(1, 0)$ **c.** $(0, 1)$

14. $x + y = 5$ **a.** $(-2, -3)$ **b.** $(0, -5)$ **c.** $(1, 6)$

15. $2x - y = 5$ **a.** $(0, -4)$ **b.** $(1, -3)$ **c.** $(3, -2)$

16. $y = \dfrac{1}{2}x + 2$ **a.** $(6, 9)$ **b.** $(-4, 0)$ **c.** $\left(7, 3\dfrac{1}{2}\right)$

Written Graph each of the following.

1. $x + y = 4$ **2.** $x - y = 3$ **3.** $x + 2y = 4$

4. $2x - y = 4$ **5.** $x + 3y = 3$ **6.** $y = 3x$

7. $y = 0.5x$ **8.** $y = -4$ **9.** $x = \sqrt{2}$

10. $y = \frac{1}{2}x - 1$ **11.** $y = -\frac{1}{3}$ **12.** $x - 2y = 4$

13. $2x + 3y = 1$ **14.** $y = -\frac{1}{6}x - \frac{5}{6}$ **15.** $x + y = -11$

16. $0.3x + 0.2y = 0.1$ **17.** $x - 3y = -2$ **18.** $0.1x + 1.3 = 5y$

In each problem, state which of the three points named, if any, lie on the graph of each linear equation.

19. $x - y = 5$ **a.** $(2, 3)$ **b.** $(14, 9)$ **c.** $(9, 14)$

20. $3x + 2y = 12$ **a.** $(0, 6)$ **b.** $(4, 0)$ **c.** $(\frac{1}{3}, \frac{1}{2})$

21. $x - 5y = -1$ **a.** $(21, 4)$ **b.** $(1, 0)$ **c.** $(24, 5)$

22. $x = y + 1$ **a.** $(1, 1)$ **b.** $(0, 1)$ **c.** $(0, -1)$

23. $8x - y = 3$ **a.** $(\frac{1}{2}, 11)$ **b.** $(-0.5, -5)$ **c.** $(\frac{1}{8}, 2)$

24. $x + y = \sqrt{2}$ **a.** $(\sqrt{2}, \sqrt{2})$ **b.** $(\sqrt{2}, -\sqrt{2})$ **c.** $(2\sqrt{8}, -3\sqrt{2})$

25. $x - 3y = 10.5$ **a.** $(3, -2.5)$ **b.** $(-3, 2.5)$ **c.** $(0.5, -\frac{10}{3})$

Mathematical Excursions

Midpoint Formula

In a coordinate plane, to find the coordinates of the midpoint, (x, y), of a line segment *average* the x-coordinates of the endpoints. Then average the y-coordinates of the endpoints.

$$(x, y) = \left(\frac{x_1 + x_2}{2}, \frac{y_1 + y_2}{2}\right)$$

For example, $\overline{CD}$ has endpoints $C(-2, 2)$ and $D(6, 4)$.

$$(x, y) = \left(\frac{-2 + 6}{2}, \frac{2 + 4}{2}\right) \text{ or } (2, 3)$$

The coordinates of the midpoint are $(2, 3)$.

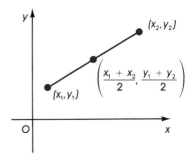

Exercises Find the coordinates of the midpoint of each line segment with the given endpoints.

1. $A(8, 4), B(12, 2)$ **2.** $C(9, 5), D(17, 3)$ **3.** $E(11, 4), F(9, 2)$

4. $S(17, 9), T(11, -3)$ **5.** $M(14, 4), N(2, 0)$ **6.** $P(19, -3), Q(11, 5)$

7. $R(4, 2), S(8, -6)$ **8.** $J(-6, 5), K(8, -11)$ **9.** $V(-6, -5), W(8, 11)$

13.3   Slope

A ramp installed to give handicapped people access to a certain building has a base 36 meters long and an elevation of 3 meters. What is the slope of the ramp?

The measure of the **slope** or steepness of a line is found by using the following ratio.

$$\text{slope} = \frac{\text{change in vertical units}}{\text{change in horizontal units}}$$

The ramp has a slope of $\frac{3}{36}$ or $\frac{1}{12}$.

In the graph of $y = 3x + 1$ shown below, the vertical change of 3 units occurs with a horizontal change of 1 unit.

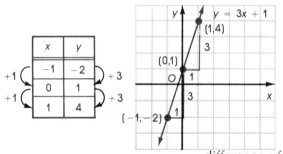

The slope of the graph of $y = 3x + 1$ is $\frac{3}{1}$ or 3. Notice that the vertical change is the difference of the y-coordinates. The horizontal change is the difference of the corresponding x-coordinates.

$$\text{slope} = \frac{\text{difference of the } y\text{-coordinates}}{\text{difference of the corresponding } x\text{-coordinates}}$$

Example

1 **Determine the slope of the line shown below.**

Choose any two points on the line. Find the following ratio.

$$\text{slope} = \frac{\text{difference of the } y\text{-coordinates}}{\text{difference of the corresponding } x\text{-coordinates}}$$

$$= \frac{5 - (-1)}{2 - (-2)} \text{ or } \frac{6}{4}$$

The ordered pairs (2, 5) and (−2, −1) were chosen.

The slope is $\frac{6}{4}$ or $\frac{3}{2}$.

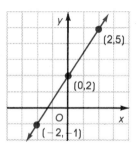

Example 1 suggests that the slope of a nonvertical line can be determined by using *any* points on that line.

Determining Slope Given Two Points

Given two points on a line, (x_1, y_1) and (x_2, y_2), the slope of the line can be found as follows.

$$\text{slope} = \frac{y_2 - y_1}{x_2 - x_1} \text{ where } x_2 \neq x_1$$

y_2 is read "y sub 2." The 2 is a subscript.

Examples

2 **Determine the slope of the line passing through (3, −9) and (4, −12).**

$$\text{slope} = \frac{y_2 - y_1}{x_2 - x_1}$$

$$= \frac{-12 - (-9)}{4 - 3}$$

$$= \frac{-3}{1}$$

$$= -3 \qquad \text{\textit{Notice that} } \frac{-9 - (-12)}{3 - 4} \text{ \textit{is also equal to} } -3.$$

The slope of the line is −3.

3 **Determine the value of *r* so the line through (*r*, 4) and (9, −2) has a slope of $-\frac{3}{2}$.**

$$\text{slope} = \frac{y_2 - y_1}{x_2 - x_1}$$

$$-\frac{3}{2} = \frac{-2 - 4}{9 - r} \qquad \text{\textit{Replace each variable with its appropriate value.}}$$

$$-\frac{3}{2} = \frac{-6}{9 - r}$$

$$-3(9 - r) = 2(-6) \qquad \text{\textit{Cross multiply.}}$$

$$-27 + 3r = -12 \qquad \text{\textit{Solve for r.}}$$

$$3r = 15$$

$$r = 5$$

Example

4 **Determine the slope of the following lines.**

a.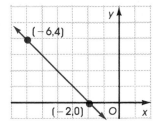
(−6,4)
(−2,0) O x

b.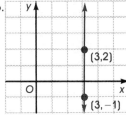
(3,2)
O x
(3,−1)

c.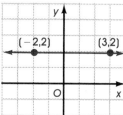
(−2,2)   (3,2)
O x

$$\text{slope} = \frac{y_2 - y_1}{x_2 - x_1}$$

a. slope $= \dfrac{4 - 0}{-6 - (-2)}$

$= \dfrac{4}{-4}$ or -1

The slope is -1.

b. slope $= \dfrac{2 - (-1)}{3 - 3}$

$= \dfrac{3}{0}$

The slope is undefined.

c. slope $= \dfrac{2 - 2}{-2 - 3}$

$= \dfrac{0}{-5}$ or 0

The line has a slope of 0.

The following conclusions can be made about slope.

Lines with positive or negative slope are often called oblique lines.

If the line rises to the right, then the slope is positive.

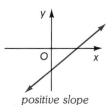

positive slope

If the line rises to the left, then the slope is negative.

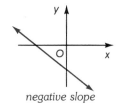

negative slope

The graph of the linear equation $Ax + By = C$ is an oblique line if $A \neq 0$ and $B \neq 0$, a horizontal line if $A = 0$, and a vertical line if $B = 0$.

If the line is vertical, then the slope is undefined.

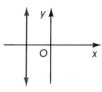

undefined slope

If the line is horizontal, then the slope is zero.

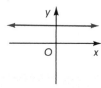

zero slope

Exercises

Exploratory Describe the slope of each of the following.

1. vertical lines
2. horizontal lines
3. lines which rise to the right
4. lines which rise to the left

For each table, state the change in y and the change in x. Then, determine the slope of the line passing through the graphs of the ordered pairs listed.

5.
x	y
4	4
3	3
2	2
1	1
0	0

6.
x	y
-2	-9
-1	-6
0	-3
1	0
2	3

7.
x	y
-2	-5
-1	-3
0	-1
1	1
2	3

8.
x	y
0	0
1	2
2	4
3	6
4	8

Written Find the slope, if it exists, of the line passing through each pair of points named by the ordered pairs below.

1. $(5, 5), (11, 11)$
2. $(5, 3), (4, 0)$
3. $(6, -1), (4, -1)$
4. $(11, 3), (1, 1)$
5. $(2, -3), (0, -5)$
6. $(-1, 4), (0, 3)$
7. $(0, 3), (4, 4)$
8. $(3, 8), (3, 1)$
9. $(1, -9), (-3, 1)$
10. $(4, -4), (3, 5)$
11. $(9, 8), (7, 1)$
12. $(9, -8), (10, -1)$
13. $(2, 3), (11, 14)$
14. $(-6, 1), (7, 11)$
15. $(18, 17), (17, 7)$
16. $(11\frac{3}{4}, \sqrt{5}), (11\frac{3}{4}, 3\sqrt{19})$
17. $(6\frac{2}{3}, 5\sqrt{7}), (3\frac{1}{3}, 5\sqrt{7})$

Find the slope, if it exists, of the graphs of the following linear equations.

18. $3x - 4 = y$
19. $y = 0.3$
20. $x - 1.5 = 4$
21. $x + y = 2$
22. $x - y = 4$
23. $4x + y = 3$
24. $2y - 2x = 5$
25. $x - 6 = 0$
26. $5y - x = 45$
27. $2x + 4y = 8$
28. $3x - y = 24$
29. $x = y$

Determine the value of r so that a line through the points named by the ordered pairs has the given slope.

30. $(r, -5), (5, 3)$; slope $= \frac{2}{3}$
31. $(9, r), (6, 3)$; slope $= \frac{1}{3}$
32. $(5, 1), (r, -3)$; slope $= -4$
33. $(r, 3), (7, r)$; slope $= 1$
34. $(4, -7), (-2, r)$; slope $= \frac{8}{3}$
35. $(r, 7), (11, r)$; slope $= -\frac{1}{5}$
36. $(r, 3), (5, 9)$; slope $= 2$
37. $(8, r), (12, 6)$; slope $= \frac{1}{2}$
38. $(r, 9), (4, r)$; slope $= \frac{3}{2}$
39. $(r, 6), (3, r)$; slope $= 2$

13.4   Equation of a Line

The graph of $y = 2x - 3$ is shown below.

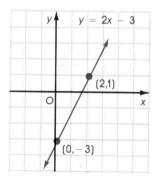

The ordered pairs $(0, -3)$ and $(2, 1)$ can be used to find the slope of this line.

$$\text{slope} = \frac{1 - (-3)}{2 - 0}$$

$$= \frac{4}{2} \text{ or } 2$$

Lab Activity

You can learn how to use a graphing calculator to find an intercept in Lab 11 on page A15.

Compare the slope of the line to the coefficient of x in the equation $y = 2x - 3$. They are the same. The graph crosses the y-axis when y is -3. This value, -3, is called the **y-intercept** of the line.

When a linear equation is written in the form $y = mx + b$, the slope of its graph is m and the y-intercept is b. A linear equation in this form is said to be in the **slope-intercept form.**

Slope-Intercept Form of a Linear Equation

The equation of a line having slope m and y-intercept b is $y = mx + b$.

Example

1 **Find the slope and y-intercept of $5x - 3y = 7$.**

Rewrite the equation in the slope-intercept form, $y = mx + b$.

$$5x - 3y = 7$$

$$-3y = -5x + 7 \qquad \text{\textit{Subtract 5x.}}$$

$$y = \frac{-5x}{-3} + \frac{7}{-3} \qquad \text{\textit{Divide by }} -3.$$

$$y = \frac{5}{3}x - \frac{7}{3} \qquad \text{\textit{Simplify}}$$

The slope is $\dfrac{5}{3}$. The y-intercept is $-\dfrac{7}{3}$.

Examples

2 **Write an equation of a line having slope -1 and y-intercept 4.**

Substitute the values for slope and y-intercept into the slope-intercept form.

$$y = mx + b$$
$$= -1x + 4 \qquad \textit{Substitute } -1 \textit{ for m and 4 for b.}$$

$$= -x + 4 \qquad \textit{Simplify.}$$

An equation is $y = -x + 4$.

3 **The following lines are parallel. Find the slope of each line.**

slope of $s = \dfrac{2 - 0}{-1 - (-4)}$

$\qquad = \dfrac{2}{3}$

slope of $t = \dfrac{0 - (-2)}{2 - (-1)}$

$\qquad = \dfrac{2}{3}$

The slopes are the same.

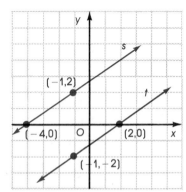

4 **Determine the equation of each line in example 3.**

Both lines have a slope of $\frac{2}{3}$. Substituting this value of m into the equation $y = mx + b$ gives the equation $y = \frac{2}{3}x + b$. To find the y-intercept, b, substitute the coordinates of one point from each line into $y = \frac{2}{3}x + b$.

$y = \frac{2}{3}x + b$ *Use the ordered pair $(-4, 0)$ from line s.*

$0 = \frac{2}{3}(-4) + b$

$b = \frac{8}{3}$ *Simplify.*

The equation of line s is $y = \frac{2}{3}x + \frac{8}{3}$.

$y = \frac{2}{3}x + b$ *Use the ordered pair $(-1, -2)$ from line t.*

$-2 = \frac{2}{3}(-1) + b$

$b = -\frac{4}{3}$ *Simplify.*

The equation of line t is $y = \frac{2}{3}x - \frac{4}{3}$.

Nonvertical lines which are parallel have the same slope, but their y-intercepts are different.

Exercises

Exploratory   State the slope and y-intercept of the graph of each of the following.

1. $y = -2x + 5$
2. $y = -4x + 6$
3. $-2x + y = 15$
4. $y = -\frac{3}{4}x - 3$
5. $y = \frac{1}{3}x + 5$
6. $0.34x - 0.34y = 0.34$
7. $0.2x = y + 5$
8. $y = 7x - 1$
9. $5x - 3y = 0.6$
10. $y = 8$
11. $x = -2$
12. $y = mx + b$

State an equation of the line given the following information.

13. $m = 5$, y-intercept $= -3$
14. $m = -1$, y-intercept $= 3$
15. $m = \frac{1}{3}$, through a point at $(1, 3)$
16. $m = -\frac{2}{3}$, through a point at $(2, 4)$
17. passes through points at $(4, 5)$ and $(1, -1)$

Written   Rewrite each of the following equations in the form $y = mx + b$.

1. $x + y = 3$
2. $x + y = -15$
3. $-2x + y = 1$
4. $y + 8x = 1$
5. $\frac{1}{4}x = 3 + y$
6. $y - \frac{1}{2}x = 1$
7. $y - \dfrac{x}{3} = -1$
8. $-18x - 18y = 18$
9. $3 - y = 2x$
10. $2y = 6x + 1$
11. $2y = 8 - x$
12. $3y - 2x = 9$

Find the slope and y-intercept of the graph of each of the following.

13. $y = x + 2$
14. $y = x - 5$
15. $y = 2x + 1$
16. $2y = 4x - 1$
17. $y = \frac{1}{5}x$
18. $y = -2x$
19. $3x - 4y = 1$
20. $8x = 1$
21. $8x - y = -1$
22. $x - \frac{1}{2}y = 2$
23. $2x - \frac{1}{3}y = -5$
24. $\frac{1}{5}x - \frac{1}{3}y = \frac{1}{7}$

Find the equation of the line which has slope m and passes through the point named.

25. $(2, 3)$; $m = 1$
26. $(4, 6)$; $m = 2$
27. $(-1, 4)$; $m = -2$
28. $(2, 5)$; $m = -3$
29. $(4, 3)$; $m = 8$
30. $(-3, 5)$; $m = -\frac{2}{3}$

In each of the following groups of equations there is one equation of a line which is not parallel to the other two. Find that equation.

31. $x + y = 4$
$x + y = 5$
$x - y = 4$

32. $x - y = 1$
$x + y = 1$
$x + y = 100$

33. $y = \frac{1}{5}x$
$y = 4 - \frac{1}{5}x$
$y = 4 + \frac{1}{5}x$

34. $8 - y = 2x$
$y = -2x + 1$
$y = 2x - 8$

35. $8x - 13y = 1$
$8x + 13y = -1$
$y = \frac{8}{13}x - \frac{1}{13}$

36. $y = -kx + m$
$y = -kx - m$
$y = kx + 2m$

Write an equation of the line which passes through the points named.

37. $(2, 3)$, $(1, 5)$
38. $(2, 5)$, $(3, 6)$
39. $(1, 2)$, $(3, 7)$
40. $(1, 0)$, $(2, 5)$
41. $(-1, 5)$, $(2, 3)$
42. $(2, 13)$, $(8, 13)$
43. $(1, 2)$, $(3, \frac{1}{2})$
44. $(0, 0)$, $(3, 1)$
45. $(3, \sqrt{2})$, $(1, \sqrt{2})$

13.5   Systems of Equations

Tom bought two apples and one pear. The total cost was 11¢. A friend bought one apple and three pears. His total cost was 18¢. What is the cost of one apple and one pear?

If *x* represents the cost of one apple and *y* represents the cost of one pear, then the following equations describe the situation.

$$2x \quad + \quad y \quad = \quad 11 \qquad\qquad x \quad + \quad 3y \quad = \quad 18$$

| cost of 2 apples at x¢ each | cost of 1 pear at y¢ | total cost | cost of 1 apple at x¢ | cost of 3 pears at y¢ each | total cost |

The graphs of $2x + y = 11$ and $x + 3y = 18$ are shown below.

Lab Activity

You can learn how to use a graphing calculator to solve a system of equations in Lab 12 on page A16.

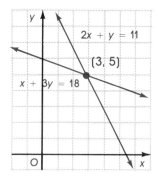

Notice that the graphs intersect at (3, 5). Since this point lies on the graph of each equation, its coordinates are a solution to both $2x + y = 11$ and $x + 3y = 18$. Therefore, one apple costs 3¢ and one pear costs 5¢. You can check this in the following way.

$$\begin{array}{ll} 2x + y = 11 & x + 3y = 18 \\ 2(3) + 5 \stackrel{?}{=} 11 & 3 + 3(5) \stackrel{?}{=} 18 \\ 11 = 11 & 18 = 18 \end{array}$$

The equations $2x + y = 11$ and $x + 3y = 18$ together are called a **system of equations.** The solution of this system of equations is (3, 5).

Example

1 **Graph the equations $3x - 2y = 5$ and $2x + 5y = 16$. Then, find the solution of the system of equations.**

The graphs intersect at $(3, 2)$. Therefore, $(3, 2)$ is the solution of the system of equations $3x - 2y = 5$ and $2x + 5y = 16$.

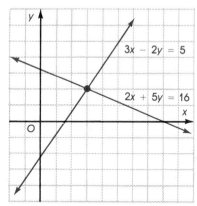

Not all systems of equations have exactly one solution.

Example

2 **Graph the equations $x - 2y = -10$ and $y = \frac{1}{2}x - 2$. Then, find the solution of the system of equations.**

The graphs do not intersect since the lines are parallel. There is *no* solution to this system of equations.

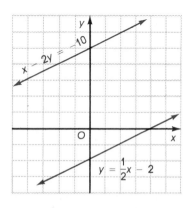

Example

3 Graph the equations $2x - y = 9$ and $4x - 2y = 18$. Then, find the solution of the system of equations.

The graphs are the same line. Such lines are said to be *coincident*. The solutions of the system of equations $2x - y = 9$ and $4x - 2y = 18$ are ordered pairs of the form $(x, 2x - 9)$.

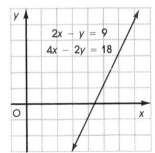

In general, when a system of two equations is graphed on a coordinate plane, the solution can be described in one of the following ways.

The lines intersect at exactly one point. There is exactly one solution.

The lines are distinct and parallel. There are no points of intersection and no solutions.

The equations name the same line. Every point on the line represents a solution to the system of equations.

Exercises

Exploratory Describe the solution set for each of the following systems of equations whose graph is described below.

1. parallel lines   **2.** coincident lines

3. lines which are neither parallel nor coincident

State the letter(s) of the ordered pair that is a solution for each equation.

4. $3x + y = 6$ **a.** $(6, 0)$ **b.** $(2, 6)$ **c.** $(2, 0)$ **d.** $(3, 0)$

5. $y = 8$ **a.** $(8, 3)$ **b.** $(1, 8)$ **c.** $(2, 2)$ **d.** $(2, 1)$

6. $x - 7y = 0$ **a.** $(7, 1)$ **b.** $(7, 0)$ **c.** $(-7, 1)$ **d.** $(1, 7)$
7. $y = 4x$ **a.** $(1, -4)$ **b.** $(-1, 4)$ **c.** $(4, 1)$ **d.** $(0, 0)$
8. $x = -2y + 4$ **a.** $(1, 2)$ **b.** $(2, 1)$ **c.** $(-2, 1)$ **d.** $(-1, 2)$
9. $x + 2y = 7$ **a.** $(3, 2)$ **b.** $(2, 3)$ **c.** $(0, 7)$ **d.** $(3, 4)$
10. $3x = 15$ **a.** $(5, 1)$ **b.** $(5, 0)$ **c.** $(0, 5)$ **d.** $(5, 8)$

Written   Solve each of the following systems of equations by graphing each on the same coordinate plane.

1. $x + y = 6$
  $x - y = 2$

2. $y = x - 1$
  $x + y = 11$

3. $3x - 2y = 10$
  $x + y = 0$

4. $x + 2y = 7$
  $y = 2x + 1$

5. $y = x + 3$
  $3y + x = 5$

6. $y = 4x$
  $x + y = 5$

7. $x + y = 3$
  $x + y = 1$

8. $x + y = 5$
  $x - y = -1$

9. $x + y = 3$
  $2x + y = 4$

10. $x = 3$
  $y = 4$

11. $x + y = 2$
  $x = 0$

12. $x - y = 2$
  $2x - 2y = 4$

13. $y = 9x$
  $x + 8 = y$

14. $y = \frac{1}{8}x$
  $x = 8y$

15. $2x + y = 7$
  $x - y = 5$

16. $x - 2y = -5$
  $y = x + 3$

17. $x - y = 2$
  $x + y = -6$

18. $x + y = 6$
  $y = 2x$

19. $y = 2x + 1$
  $y = -2x + 5$

20. $2x - y = 5$
  $3y = 6x - 15$

21. $y = x - 3$
  $3x + 5y = 9$

Mixed Review

The points $A(2, 3)$, $B(5, 7)$, $C(9, 4)$, and $D(6, 0)$ are the vertices of a rectangle.

1. Find the slope of the line that contains side $\overline{AB}$.

2. Find the y-intercept of the line that contains side $\overline{BC}$.

3. Find an equation of the line that contains diagonal $\overline{AC}$.

4. Find an equation of the line that contains diagonal $\overline{BD}$.

5. Find the point of intersection of diagonals $\overline{AC}$ and $\overline{BD}$.

The number of points scored by players on the Knickerbockers in a recent game was 24, 15, 10, 2, 4, 14, 6, 12, 2, 9, and 23.

6. Find the mean for this data.

7. Find the mode for this data.

8. Find the median for this data.

9. Find the first quartile for this data.

10. Find the third quartile for this data.

A bag contains 2 white and 5 blue marbles. A marble is chosen, not replaced, and then another is chosen. Find each probability.

11. P(same color)

12. P(exactly 1 blue, 1 white, any order)

13.6   Algebraic Solutions

A system of equations can be solved by using algebraic methods as well as by graphing. Two such methods are the **substitution method,** shown in example 1, and the **elimination method.**

Examples

1 Mr. Hyde's will specifies that the Pine City Library is to receive three times as much money as the Pine City Museum. If the total amount given to both is $146,000, how much does each receive?

Let x be the amount given to the Library and y be the amount given to the Museum. Therefore the system of equations $x + y = 146,000$ and $x = 3y$ describe the situation.

The second equation gives a value for x in terms of y. Substitute this value, $3y$, for x in the first equation.

$$x + y = 146,000$$
$$3y + y = 146,000 \qquad \textit{Substitute 3y for x.}$$

The resulting equation has only one variable, y. Solve the equation.

$$3y + y = 146,000$$
$$4y = 146,000$$
$$y = 36,500$$

Now, find x by substituting 36,500 for y in the second equation, $x = 3y$.

$$x = 3y$$
$$x = 3(36,500)$$
$$x = 109,500$$

Therefore, the Library receives $109,500 and the Museum, $36,500.

2 Use the elimination method to solve the system of equations $x + 5y = 20.5$ and $-x + 3y = 13.5$.

Add the second equation to the first equation.

$$\begin{aligned} x + 5y &= 20.5 \\ -x + 3y &= 13.5 \\ \hline 8y &= 34 \\ y &= 4.25 \end{aligned}$$

The variable x is eliminated.

Now substitute 4.25 for y in $x + 5y = 20.5$ to find the value of x.

$$x + 5y = 20.5$$
$$x + 5(4.25) = 20.5$$
$$x = -0.75$$

4.25 could have been substituted in either equation.

Since $x = -0.75$ and $y = 4.25$, the solution is $(-0.75, 4.25)$.

Example

3 Use the elimination method to solve the system of equations
$3x - 2y = -1$ and $2x + 5y = 12$.

In this case, adding or subtracting the two equations will not eliminate a variable. However, suppose both sides of the first equation are multiplied by 2, and both sides of the second equation are multiplied by -3. Then, the system can be solved by adding the equations.

$3x - 2y = -1$ | Multiply by 2 > $6x - 4y = -2$ *Note that the coefficients*
                                                         of x are additive
$2x + 5y = 12$ | Multiply by -3 > $-6x - 15y = -36$ *inverses.*

Now, add to eliminate x. Then solve for y.

$$\begin{array}{r} 6x - 4y = -2 \\ -6x - 15y = -36 \\ \hline -19y = -38 \\ y = 2 \end{array}$$ *The variable x is eliminated.*

Finally, substitute 2 for y in the second equation. Then, solve for x.

$$2x + 5y = 12$$
$$2x + 5(2) = 12$$
$$2x + 10 = 12$$
$$2x = 2$$
$$x = 1$$

Since $x = 1$ and $y = 2$, the solution is $(1, 2)$.

Exercises

Exploratory   Solve each equation for y.

1. $y + 3 = x$

2. $y - 6 = 3x$

3. $-2x + y = \frac{3}{4}$

4. $2y + 3x = -3$

5. $-y = 4x + 2$

6. $3x - 2y = 8$

State whether adding or subtracting the two equations will eliminate a variable.

7. $2x + y = 8$
$x - y = 2$

8. $2x + y = 7$
$3x - 2y = 7$

9. $\frac{1}{2}x + \frac{1}{4}y = 1$
$\frac{1}{3}x - \frac{1}{4}y = 6$

10. $x + 3y = 4$
$-x + 2y = 1$

11. $3x + 4y = 8$
$3x + y = 5$

12. $x + 2y = 1$
$2x + y = 1$

Written   Solve each system of equations by the substitution method. Show your work.

1. $y = 8$
 $7x = 1 - y$

2. $y = x - 1$
 $4x - y = 19$

3. $y = x - 4$
 $2x + y = 5$

4. $x - y = -5$
 $x + y = 25$

5. $x = y + 10$
 $2y = x - 6$

6. $3x + 4y = -7$
 $2x + y = -3$

7. $9x + y = 20$
 $3x + 3y = 12$

8. $x + 2y = 5$
 $2x + y = 7$

9. $y = -2x + 4$
 $x = -2y + 4$

10. $2x - y = 7$
 $\frac{3}{4}x - \frac{1}{2}y = 3$

11. $3x + 5y = 2x$
 $x + 3y = y$

12. $x + \frac{1}{2}y = 4$
 $\frac{1}{3}x - y = -1$

Solve each system of equations by the elimination method. Show your work.

13. $x + y = 7$
 $x - y = 9$

14. $3x + y = 13$
 $2x - y = 2$

15. $2x - 3y = -9$
 $-2x - 2y = -6$

16. $-5 = 4x + 3y$
 $3 = -4x - 2y$

17. $2x + 3y = 6$
 $2x - 5y = 22$

18. $3x + 4y = -7$
 $2x + y = -3$

19. $x - y = 6$
 $2x + 3y = 7$

20. $2x - 5y = 1$
 $3x - 4y = -2$

21. $2x + 2y = 8$
 $5x - 3y = 4$

22. $2x + y = 7$
 $3x - 2y = 7$

23. $-4x + 3y = -1$
 $8x + 6y = 10$

24. $3x + 4y = -25$
 $2x - 3y = 6$

Use either the elimination or the substitution method to solve each system of equations. Show your work.

25. $4x + 3y = 6$
 $-2x + 6y = 7$

26. $x + 2y = 400$
 $x - 100 = y$

27. $x + 5y = 20.25$
 $x + 3y = 13.05$

28. $3x - 2y = 7$
 $2x + 5y = 9$

29. $4x + 2y = 10.5$
 $2x + 3y = 10.75$

30. $x + y = 40$
 $0.2x + 0.45y = 10.5$

Mathematical Excursions

Puzzle

The diagram shows 12 straws, each with the same length. The arrangement can be changed in several ways.

Exercises   Show how to complete each of the following rearrangements.

1. Remove 4 straws so that 1 square remains.

2. Remove 4 straws so that 2 squares remain.

3. Remove 2 straws so that 2 squares remain.

4. Move 3 straws and have 3 squares.

Mathematical Excursions

Aerial photographers often use coordinates when determining the area of a region being studied. A photograph is first placed on a coordinate grid. Then a polygon that encloses the region in the photograph is drawn. The area of the polygon can then be determined using the coordinate method described below.

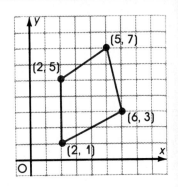

Example Find the area of the polygon shown at the right.

Step 1 List the ordered pairs for the vertices in counterclockwise order, repeating the first ordered pair at the bottom of the list.

Step 2 Find D, the sum of the downward diagonal products (from left to right).

$$D = (5 \cdot 5) + (2 \cdot 1) + (2 \cdot 3) + (6 \cdot 7)$$
$$= \quad 25 \quad + \quad 2 \quad + \quad 6 \quad + \quad 42 \quad \text{or} \quad 75$$

(5, 7)

(2, 5)

(2, 1)

(6, 3)

(5, 7)

Step 3 Find U, the sum of the upward diagonal products (from left to right).

$$U = (2 \cdot 7) + (2 \cdot 5) + (6 \cdot 1) + (5 \cdot 3)$$
$$= \quad 14 \quad + \quad 10 \quad + \quad 6 \quad + \quad 15 \quad \text{or} \quad 45$$

Step 4 Use the formula $A = \frac{1}{2}(D - U)$ to find the area.

$$A = \tfrac{1}{2}(D - U)$$
$$= \tfrac{1}{2}(75 - 45)$$
$$= \tfrac{1}{2}(30) \quad \text{or} \quad 15$$

The area is 15 square units. Count the number of square units enclosed by the polygon. Does this result seem reasonable?

Exercises Graph each polygon. Use the coordinate method to find its area.

1. triangle ABC with $A(1, 1)$, $B(3, 7)$ and $C(8, 2)$
2. triangle DEF with $D(2, 2)$, $E(6, 3)$, and $F(-3, -2)$
3. quadrilateral $GHJK$ with $G(1, 1)$, $H(3, 7)$, $J(4, 4)$, and $K(8, 1)$
4. quadrilateral $LMNP$ with $L(1, 2)$, $M(5, 6)$, $N(8, 6)$, and $P(7, 0)$
5. pentagon $RSTUW$ with $R(-4, -1)$, $S(2, 4)$, $T(6, 2)$, $U(5, -1)$, and $W(1, -2)$

13.7 Using Systems of Equations to Solve Problems

Many types of problems can be solved by using systems of equations.

Example

1 Flora has invested part of $6000 in Pine City Municipal Bonds which pay 6% annual interest. She has invested the rest of the $6000 in Oak City Municipal Bonds which pay 8% annual interest. If her total income from these two investments is $460, how much did she invest in each?

Money Market Checking			Money Market Savings	
4.55	4.65	up to $1,000	5.10	5.22
4.85	4.96	$1,000-$5,000	5.95	6.11
5.10	5.22	$10,000-$24,000	6.20	6.38
5.25	5.38	$25,000 and over		
		$25,000-$49,999	6.20	6.38
		$50,000-$99,999	6.40	6.59
		$100,000 and over	6.70	6.91
True Blue			8.26	8.58
Select-Money Market Reserve			7.95	8.25
Regular Savings			5.10	5.22

Let x be the amount invested in the 6% bond. The amount of interest the bond earns is 6% of x or $0.06x$.

Let y be the amount invested in the 8% bond. The amount of interest the bond earns is 8% of y or $0.08y$.

$$x + y = 6000 \qquad \text{the total amount invested}$$
$$0.06x + 0.08y = 460 \qquad \text{the total income in interest}$$

To produce an equation which does not contain decimals multiply both sides of the second equation by 100.

$$6x + 8y = 46{,}000$$

To solve the system of equations $x + y = 6000$ and $6x + 8y = 46{,}000$, multiply the first equation by -6 and solve for y.

$$
\begin{aligned}
-6x + -6y &= -36{,}000 \\
6x + 8y &= 46{,}000 \qquad \text{Add.} \\
\hline
2y &= 10{,}000 \\
y &= 5000
\end{aligned}
$$

Substitute 5000 for y in the first equation and solve for x.

$$
\begin{aligned}
x + y &= 6000 \\
x + 5000 &= 6000 \\
x &= 1000
\end{aligned}
$$

Flora has invested $1000 at 6% and $5000 at 8%.

Problems relating to numbers can also be solved by using systems of equations.

Example

2 **Find two positive numbers whose sum is 11 and whose difference is 3.**

Let x = one number.
Let y = the other number.

$x + y = 11$ *Their sum is 11.*
$x - y = 3$ *Their difference is 3.*

Add the system of equations and solve for x.

$$\begin{aligned} x + y &= 11 \\ x - y &= 3 \\ \hline 2x &= 14 \\ x &= 7 \end{aligned}$$

Solve for y.

$$\begin{aligned} x + y &= 11 \\ 7 + y &= 11 \\ y &= 4 \end{aligned}$$ *Substitute 7 for x.*

The solution for the system is (7, 4).
The numbers are 7 and 4.

Systems of equations can be used to solve problems involving mixtures.

Example

3 **In order to complete his science experiment, Marty needs 20 liters of a solution which is 25% alcohol. The only two alcohol solutions available in the school laboratory are a solution which is 40% alcohol and another which is 15% alcohol. How many liters of each of these two solutions must he mix together to produce the required solution?**

Let x = the number of liters of the 40% alcohol solution.
Let y = the number of liters of the 15% alcohol solution.

The total number of liters of the two solutions is 20.

$$x + y = 20$$

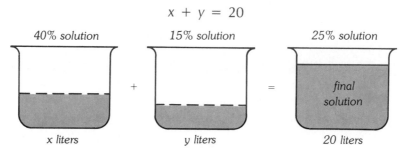

40% solution   15% solution   25% solution

+   final solution   =

x liters   y liters   20 liters

Now, consider the number of liters of alcohol contained in each mixture.

$$0.40x + 0.15y = 0.25(20) \text{ or } 40x + 15y = 500$$

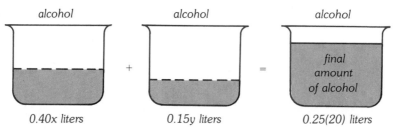

alcohol   alcohol   alcohol

+   final amount of alcohol   =

0.40x liters   0.15y liters   0.25(20) liters

Solve the system of equations.

$$x + y = 20$$
$$40x + 15y = 500$$

Multiply by −40

$$-40x - 40y = -800$$
$$\underline{40x + 15y = 500}$$ *Add.*
$$-25y = -300$$
$$y = 12$$

Use the first equation to find x.

$$x + y = 20$$
$$x + 12 = 20$$
$$x = 8$$

Marty must mix 8 liters of the 40% solution with 12 liters of the 15% solution.

Exercises

Exploratory **For each problem, define the variables. Then state a system of equations.**

1. The sum of two numbers is 33. The greater number is 3 more than the lesser number. Find the numbers.

2. The sum of two numbers is 74. The greater number is 2 more than twice the lesser number. Find the numbers.

3. A rectangle has a perimeter of 22 feet. One dimension is 1 foot less than twice the other. Find its dimensions.

4. What are the dimensions of a rectangle which is twice as long as it is wide and whose perimeter is 208 kilometers?

5. Mr. Sanchez buys 5 shirts and 3 ties for $34. At the same store, Mr. Anderson buys 3 shirts and 6 ties for $33. Find the price of a shirt and the price of a tie.

6. Susan receives a total of $45.80 each year on her investment of $800. Part of the $800 is invested at 5% and the rest at 6%. How much is invested at each rate?

7. Two angles are supplementary. Find their measures if one is 20° less than 3 times the other.

8. Pat buys 38 stamps for $7.50. If the stamps are either 15¢ stamps or 25¢ stamps, how many of each kind did he buy?

9. A two-digit number is 5 times its units digit. The sum of the digits is 7. Find the number. (Hint: Any two-digit number can be represented by 10t + u.)

10. The tens digit of a two-digit number is 6 more than the units digit. The number is 2 more than 8 times the sum of the digits. Find the number.

Written Solve each system of equations in Exploratory Exercises 1-10.

1. Exercise 1
2. Exercise 2
3. Exercise 3
4. Exercise 4
5. Exercise 5
6. Exercise 6
7. Exercise 7
8. Exercise 8
9. Exercise 9
10. Exercise 10

Solve each problem.

11. The perimeter of a rectangle is 94 centimeters. The length is 7 centimeters longer than the width. Find the dimensions of the rectangle.

12. Two numbers differ by 71. If twice the lesser number is added to the other, the result is 500. Find the numbers.

13. The sum of two numbers is 14. If the greater is subtracted from 3 times the lesser, the result is 6. Find the numbers.

14. Find a two-digit number whose tens digit is 3 more than the units digit and whose value is 3 less than 7 times the sum of the digits.

15. At Selita's Boutique, 3 bracelets cost $35.85. What is the cost of 1 bracelet and 1 pair of earrings if 1 bracelet and 4 pairs of earrings cost $43.75?

16. The Hayes family has invested $10,000. Part of the money is invested at 5% and part at 7%. If the total income from these investments is $540, how much is invested at each rate?

17. A dealer wishes to prepare 50 pounds of blended coffee to sell at $2.50 a pound. How many pounds of coffee selling at $4 a pound and how many pounds of coffee selling at $2 a pound should she use?

18. Acid solutions are made up of a certain percentage of acid and the rest water. How many ounces of a solution which is 30% acid and how many ounces of a solution which is 60% acid should be mixed to produce 120 ounces of a 50% solution?

13.8　Graphing Linear Inequalities

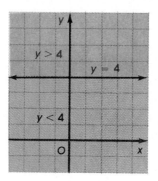

The graph of the equation $y = 4$ is a line which separates the coordinate plane into two regions. Each region is called a **half-plane.** The line for $y = 4$ is called the *boundary line* for each half-plane. The inequality $y > 4$ describes the set of points in the half-plane above the line. The y-coordinate of every point in that region is greater than 4. The graph of $y < 4$ is the half-plane below the line for $y = 4$.

Examples

1 **Graph $y > -2x - 3$.**

First graph $y = -2x - 3$. Draw it as a broken line since this boundary line is *not* part of the graph. Then, test a point on either side of the boundary line. This will tell which half-plane is the graph. The origin is an easy point to check. The coordinates of the ordered pair (0, 0) satisfy the inequality $y > -2x - 3$. Therefore, the coordinates of all points on that side of the line satisfy the inequality.

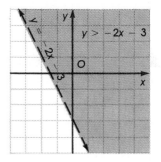

2 **Graph $x - 2y \geq 2$.**

First solve for y.

$$x - 2y \geq 2$$
$$-2y \geq -x + 2$$
$$y \leq \tfrac{1}{2}x - 1$$

Then graph $y = \tfrac{1}{2}x - 1$. Draw it as a solid line since this boundary line is part of the graph. Then test a point to determine which half-plane is the graph. The origin is not part of the graph. Therefore, the graph is the right half-plane including the boundary line.

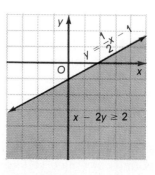

Consider the following system of inequalities.

$$y \leq 2x - 3$$
$$y \leq -2x + 2$$

To solve this system, you must find the ordered pairs that satisfy both inequalities. One way is to graph each inequality and find the overlap of the two graphs.

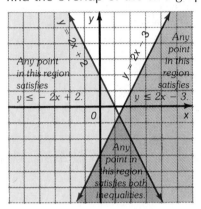

The graphs of $y = 2x - 3$ and $y = -2x + 2$ are the boundaries of the regions. The solution of this system of inequalities is the darkest shaded region.

Examples

3 **Solve the system $y \leq -2$ and $y < x - 3$ by graphing.**

The graphs of $y = -2$ and $y = x - 3$ are the boundaries of the regions. The graph of the equation $y = x - 3$ is a broken line. It is not included in the solution of the system.

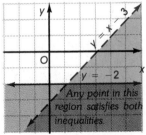

4 **Solve the system $y > x + 3$ and $y < x - 2$ by graphing.**

The graphs of $y = x + 3$ and $y = x - 2$ are the boundaries of the regions. Their graphs are broken lines. Notice that the regions which satisfy the inequalities do not overlap. Therefore, no ordered pair will satisfy the system of inequalities.

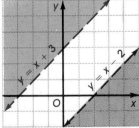

▰▰▰ Exercises ▰▰▰

Exploratory The three lines shown in the figure at the right separate the plane into seven regions. The regions are labeled *A* through *G*. Name the region, or regions, which satisfy the inequalities in each of the following exercises.

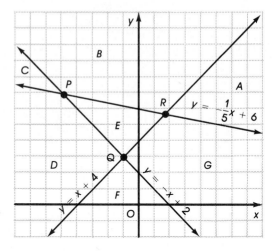

1. $x + y > 2$ **2.** $y > x + 4$
3. $x + y > 2$ and $y < x + 4$
4. $y > x + 4$ and $y < -x + 2$
5. $x + 5y > 30$ and $y > x + 4$
6. $x + 5y < 30$ and $x + y < 2$
7. $y < -x + 2$
8. $y < -x + 2$ and $x + 5y > 30$
9. $x + 5y < 30$, $y > x + 4$, and $x + y > 2$
10. $y < x + 4$, $x + y > 2$, and $x + 5y < 30$

Written   Graph each of the following linear inequalities.

1. $x > 1$	**2.** $x \geq -2$	**3.** $y < -3$
4. $y > 3x$	**5.** $y \geq x + 1$	**6.** $y < x - 2$
7. $2x + y > 3$	**8.** $y \leq -x + 2$	**9.** $x < y - 1$
10. $3x + y \leq 4$	**11.** $y < -\frac{1}{5}x$	**12.** $x - y < -1$
13. $x - 5y > 5$	**14.** $3x + y < 1$	**15.** $-x < y - 1$

Solve each system of inequalities by graphing.

16. $y \geq x$ $\quad y \geq -x$	**17.** $x \geq 1$ $\quad x + y \geq 2$	**18.** $y > 3x$ $\quad x + y > 5$
19. $y \leq x - 1$ $\quad x > 5$	**20.** $y < \frac{1}{2}x$ $\quad y \geq 0$	**21.** $x < 1$ $\quad y > 3x$
22. $2x + 3y < 6$ $\quad y > -x + 4$	**23.** $y \leq x - 2$ $\quad y < 5x$	**24.** $x < -2$ $\quad x > 1$
25. $y > x$ $\quad y < x - 3$	**26.** $3x + y \leq 8$ $\quad y < x + 4$	**27.** $x + y > 5$ $\quad x + y \leq 5$
28. $5x > y + 2$ $\quad x - y \leq 4$	**29.** $2y - x \geq 3$ $\quad y < -\frac{1}{6}x$	**30.** $y \geq 3$ $\quad x \leq 5$
31. $y \leq 2$ $\quad y \geq 2x$ $\quad y \geq x$	**32.** $y < 2x + 1$ $\quad y > 2x - 2$ $\quad 3x + y > 8$	**33.** $x \geq -1$ $\quad x \leq -1$ $\quad y > 2$

Challenge   Graph each of the following nonlinear inequalities.

1. $y \leq |x|$ **2.** $|x| + |y| < 5$ **3.** $y > |x^2 - 4|$

Problem Solving Application: Using Direct Variation

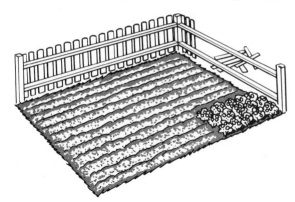

Mrs. Kellar wants to enclose her square garden with a picket fence. The amount of fence she will need depends upon the length of each side of the square. The following table relates the length, in feet, of each side (s) and the total amount of fence, in feet, needed (P) to enclose the garden.

s	2	5	10	15	20
P	8	20	40	60	80

It is also true that as the length of each side decreases, the total amount of fence needed decreases.

Notice that as the length of each side (s) increases, the total amount of fence needed (P) also increases.

The total amount of fence needed by Mrs. Kellar depends *directly on* the length of each side of her garden. However, the ratio of P to s (or s to P) does not vary for corresponding values of P and s. The value of $\frac{P}{s}$ from the table above is 4.

The relationship between the length of each side and the total amount of fence needed is shown by the equation $P = 4s$. Such an equation is called a **direct variation.** We say that P varies directly as s or s and P vary directly.

Definition of Direct Variation

A direct variation is described by an equation of the form $y = kx$, where $k \neq 0$. The variable k is called the *constant of variation*.

Thus, $k = \frac{y}{x}$.

The constant of variation, k, is equal to the ratio $\frac{y}{x}$ for corresponding values of y and x.

Example

1 **The weight of an object on the moon varies directly as its weight on Earth. A certain astronaut weighs 174 pounds on Earth and 29 pounds on the moon. If Kristin weighs 108 pounds on Earth, what would she weigh on the moon?**

Explore   Let x = weight on Earth.
              Let y = weight on the moon.

Plan   First find the constant of variation, k, using the weights for the astronaut on Earth and the moon. Then use this constant and Kristin's weight on Earth to find her weight on the moon.

Solve $k = \dfrac{y}{x}$

$k = \dfrac{29}{174}$ *Substitute the astronaut's weights for x and y.*

$k = \dfrac{1}{6}$ *The constant of variation is $\dfrac{1}{6}$.*

$y = kx$

$y = \dfrac{1}{6}(108)$ *Substitute $\dfrac{1}{6}$ for k and 108 for x. Why?*

$y = 18$

Kristin would weigh 18 pounds on the moon.   *Examine this result.*

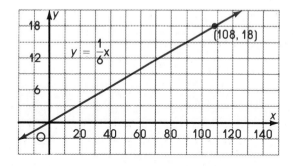

Direct variation is a special case of the linear equation $y = mx + b$ where $m \neq 0$ and $b = 0$. The graph of the linear equation that represents a direct variation will always be an oblique line passing through the origin. The slope of this line is the constant of variation, k.

The graph of the direct variation from Example 1 is shown at the left.

Exercises

Exploratory **State whether each equation represents a direct variation. Write *yes* or *no*. For each direct variation, state the constant of variation.**

1. $P = 3s$ **2.** $y = -5x$ **3.** $a^2 = b$ **4.** $6t = r$

5. $y = 0$ **6.** $xy = 11$ **7.** $y = 2x + 3$ **8.** $x + 2y = 0$

9. $C = \pi d$ **10.** $n = \dfrac{1}{3}m$ **11.** $\dfrac{2}{5}y = -\dfrac{3}{4}x$ **12.** $v = \dfrac{7}{u}$

Write an equation of the form $y = kx$ to represent the relationship shown in each of the following tables.

13.

t	1	2	5	7.2
d	55	110	275	396

14.

r	$3.85	$4.00	$5.00	$6.75
A	$30.80	$32.00	$40.00	$54.00

Written Solve each problem. Assume that y varies directly as x.

1. If $y = 12$ when $x = 3$,
   find y when $x = 7$.

2. If $y = 3$ when $x = 15$,
   find y when $x = -25$.

3. If $y = -6$ when $x = -14$,
   find x when $y = -4$.

4. If $y = 12$ when $x = 15$,
   find x when $y = 21$.

5. If $y = 2\frac{2}{3}$ when $x = \frac{1}{4}$,

   find y when $x = 1\frac{1}{8}$.

6. If $y = \frac{5}{6}$ when $x = -3$,

   find x when $y = \frac{4}{27}$.

7-12. For Exercises 1-6, write an equation of the form $y = kx$ for each variation. Then graph each equation.

Suppose that interest (I) varies directly as the annual interest rate (r). Find how the interest changes if the rate changes as follows.

13. doubles   14. triples   15. halves   16. multiplies by h

Solve each problem.

17. IF 2.5 m of copper wire weighs 325 g, how much do 75 m weigh?

18. If 6 pounds of sugar costs $2.25, how much do 40 pounds cost?

19. Mia's wages vary directly as the time she works. If her wages for 4 days are $110, what are her wages for 17 days?

20. Kelvin Clark hit 20 home runs in 72 games. At that rate, how many home runs will he hit in 162 games?

21. A car uses 6 gallons of gasoline to travel 171 miles. How much gasoline will the car use to travel 304 miles?

22. The interest on $525 is $35.25. At that rate, what is the interest on $875?

23. Buck drives his car at an average speed of 50 miles per hour. How long will it take him to travel 720 miles?

24. Paul and Paula drove 194 miles in 4 hours. At that rate, how far will they travel in 6.5 hours?

25. A 25-acre field yields 2985 bushels of corn. At that rate, how much corn would a 40-acre field yield?

26. Last season, Cheryl had a free throw shooting percentage of 0.785. How many free throws would she make if she had 1400 free throw attempts?

27. In an electrical transformer, voltage varies directly as the number of turns of the coil. If 110 volts comes from a coil with 55 turns, what would be the voltage produced by a coil with 66 turns?

Portfolio Suggestion

Select some of your work from this chapter that shows how you used a calculator or computer. Place it in your portfolio.

Performance Assessment

The graphs of each group of equations have at least one characteristic in common. Name the characteristic(s) and graph each group of equations on the same axes to verify your answer.

a. $y = x, y = 3x, y = 5x$
b. $y = 2x, y = 2x + 1, y = 2x - 3$
c. $2y = 6x, 4y - 12x = 0, 3x - y = 0$

Chapter Summary

1. Points in a plane can be located by **ordered pairs.** In an ordered pair, the first number is called the **x-coordinate.** The second number is called the **y-coordinate.** (416)
2. The **x-axis** and **y-axis** separate the plane into four **quadrants.** (416)
3. **Definition of Linear Equation:** An equation is linear if and only if it can be written in the form $Ax + By = C$, where A, B, and C are any numbers and A and B are not both 0. (419)
4. Given two points on a line, (x_1, y_2) and (x_2, y_2), the **slope** of the line can be found by using the formula, $m = \dfrac{y_2 - y_1}{x_2 - x_1}$, where $x_2 \neq x_1$. (423)
5. **Slope-Intercept Form of a Linear Equation:** The equation of a line having a slope m and a y-intercept b is $y = mx + b$. (427)
6. Two or more equations together with the same variables are called a **system of equations.** (430)
7. Two algebraic methods of solving a system of equations are the **substitution** method and the elimination method. (431)
8. An **inequality** describes the set of points above or below the boundary line. (442)

Chapter Review

13.1 Graph each of the following on a coordinate plane.

1. $(5, 2)$ **2.** $(-1, 4)$ **3.** $(0, 3)$ **4.** $(-2, -1)$

13.2 Which of the point(s) named lie on the graph of the given equation?

5. $y = 3x + 4$ **a.** $(0, 0)$ **b.** $(-1, 1)$ **c.** $(-2, -2)$
6. $5x - y = -1$ **a.** $(1, -4)$ **b.** $(-1, 4)$ **c.** $(2, 9)$

7. Write $2y = 3 - 5x$ in the form $Ax + By = C$.

8. Graph $x - 3y = 6$.

13.3 **9.** Find the slope of the line determined by $(5, 1)$ and $(3, 7)$.

10. Find the slope of the line given by the equation $2x - 3y = 6$.

13.4 **11.** Write $x + 5y = 10$ in the form $y = mx + b$.

12. Find the slope and y-intercept of $2x - 3y = 12$.

13. Write an equation of the line through $(1, 2)$ and $(8, -3)$.

14. Write the letter of the equation whose graph is parallel to the graph of $x - 2y = 5$.

a. $x + 2y = 5$ **b.** $y = \frac{1}{2}x - 3$ **c.** $y = -\frac{1}{2}x - 3$ **d.** $y = 2x - 3$

13.5 Solve each system of equations by graphing.

15. $x + 3y = 7$
$x - y = -1$

16. $x + y = 5$
$x + 3y = 3$

13.6 Solve each system of equations by the substitution method.

17. $y = x - 4$
$2x + y = 5$

18. $y = \frac{1}{2}x + 5$
$3x + 2y = 2$

Solve each system of equations by the elimination method.

19. $2x + 3y = 13$
$2x - 3y = -17$

20. $2x + y = 4$
$6x + 2y = 9$

13.7 **21.** Find two numbers whose sum is 20 and whose difference is 6.

22. A certain book comes in hardcover and softcover. Three softcovers and one hardcover cost $12 while one softcover and two hardcovers cost $14. What is the cost of each type of book?

23. The sum of the digits of a two-digit number is 9. If 45 is subtracted from the number, the digits of the number are reversed. Find the number.

13.8 **24.** Graph $y \le 3$.

25. Solve the system $y \ge x + 1$ and $x + y \ge 3$ by graphing.

Chapter Test

1. In which quadrant does every point have a negative first coordinate and a positive second coordinate?

State whether each of the following is a linear equation.

2. $x + y = 12$ **3.** $xy = 12$ **4.** $\dfrac{x}{2} + \dfrac{y}{2} = 12$ **5.** $y = 12x$

Find the slope of the graph of each of the following.

6. $x = 2$ **7.** $y = 2$ **8.** $x + y = 0$ **9.** $x + y = -5$

State whether the graph of each equation is parallel to the graph of $y = \dfrac{1}{3}x + 2$.

10. $y = 3x + 2$ **11.** $3y + x = 5$ **12.** $-x + 3y - 4 = 0$

13. Graph $y \le -2$ on a coordinate plane.

14. Find the solution set for the system $x + y = 9$ and $x - y = 7$.

15. Write the equation $x = \frac{1}{2}y - 1$ in the form $Ax + By = C$.

16. Write the equation $2x - y = 4$ in the form $y = mx + b$.

17. Find the slope and y-intercept of the equation $x + y = 5$.

18. Find the value of k if $(3, k)$ lies on the graph of $x + 2y = -1$.

19. Solve the system of equations $4x + y = 8$ and $y = x + 3$ by graphing.

20. Solve the system of equations $3x + 2y = 11$ and $5x + 3y = 16$ algebraically.

21. The measure of an acute angle of a right triangle is 30 more than twice the measure of the other acute angle. Find the measures of the angles of the triangle.

22. How much of a 20% solution of alcohol and how much of a 50% solution should be mixed to give 12 liters of a 30% solution?

23. Write an equation of the line through $(-1, -6)$ and $(7, -3)$.

24. Determine the value of r so that a line through $(2, r)$ and $(5, -4)$ has slope -2.

25. On a fully labeled coordinate plane, find and label the region satisfying the system $y \le x$ and $2x + y \ge 3$.

Application in Art

Artists often use line symmetry, line reflections, translations, rotations, and dilations to create their works of art.

Maurits C. Escher (1898–1972) was a renowned Dutch artist who frequently used **geometric transformations** in his work. Many of his works involve repeated images that cover an entire surface. While his artworks appear to be drawings, they were actually done as woodcarvings or lithographs. Color was often added later in reproduction of the art. One example of his art, called *Liberation*, is shown at the right. What transformations did he use in this drawing?

Individual Project: *Design*

Find at least one example of each of the following: line symmetry, line reflection, translation, rotation, and dilation. Use pictures of art work or pieces of wallpaper or fabric. Make a poster of your findings, explaining what each represents. Display your posters around the room.

14.1 Mappings and Functions

The chart at the right lists students' names and test grades. Notice that each name is paired with exactly one grade. This is an example of a mapping from the set of names to the set of grades.

Student	Grade
Harry	73
Maria	85
Bernie	65
Adrienne	73
Sheila	92

Definition of Mapping

> A **function,** or **mapping,** pairs each member of one set with exactly one member of the same, or another set. The first set is called the **domain.** The second set is called the **range.**

In the mapping described by the chart above, the domain is {Harry, Maria, Bernie, Adrienne, Sheila}. The range is {73, 85, 65, 92}.

Another way to represent a mapping is by making an arrow diagram.

"Harry ⟶ 73" is read
"Harry maps to 73"

Domain	**Range**
Harry	73
Maria	85
Bernie	65
Adrienne	
Sheila	92

Consider the following mapping.

Domain	**Range**
−3	9
2	4
3	
4	16

Notice that each number in the domain is assigned its square in the range.
$$x \longrightarrow x^2$$

Lowercase letters such as f, g, or h are used to name mappings. For example, the mapping shown above in the arrow diagram can be written as follows.

$$f: x \longrightarrow x^2 \quad \text{or} \quad x \xrightarrow{\ f\ } x^2$$

It is read "f is the function or mapping that maps x to x^2". In this mapping, 2 maps to 4. We write $f: 2 \longrightarrow 4$ or $2 \xrightarrow{\ f\ } 4$. The **image** of 2 is 4 and 2 is the **preimage** of 4. Preimages are always in the domain of a mapping. Images are always in the range.

Example

1 **Find the images of −1, 0, and 1 if *h*: *x* ⟶ *x²* + *x* + 1.**

h: $-1 \longrightarrow (-1)^2 + (-1) + 1$ or h: $-1 \longrightarrow 1$
h: $0 \longrightarrow 0^2 + 0 + 1$ or h: $0 \longrightarrow 1$
h: $1 \longrightarrow 1^2 + 1 + 1$ or h: $1 \longrightarrow 3$

The image of −1 and 0 is 1 and the image of 1 is 3.

Another way to represent a mapping is by an equation that expresses the relationship between each member of the domain and its image. Usually, the variable *x* represents members of the domain and *y* represents members of the range. The mapping in the previous example may be represented by $y = x^2 + x + 1$.

Example

2 **Draw an arrow diagram for the function defined by $y = x^2 - 2$ if the domain is {−1, 0, 1, 2}.**

Replace *x* by each member of the domain and find the corresponding image.

		Domain	**Range**
Let $x = -1$.	$y = (-1)^2 - 2$ or -1	-1	-1
Let $x = 0$.	$y = 0^2 - 2$ or -2	0	-2
Let $x = 1$.	$y = 1^2 - 2$ or -1	1	
Let $x = 2$.	$y = 2^2 - 2$ or 2	2	2

Some equations do *not* represent mappings. For example, the equation $y^2 = x$ does *not* represent a mapping because members of the domain correspond to more than one member of the range.

Domain	**Range**
4	−2
	2
9	−3
	3

Assume the domain is $\mathcal{W}$

Domain		Range
1	$\xrightarrow{\;\;f\;\;}$	5
2	$\xrightarrow{\;\;f\;\;}$	3
3	$\xrightarrow{\;\;f\;\;}$	18

one-to-one

Domain		Range
4	$\xrightarrow{\;\;g\;\;}$	a
5	$\xrightarrow{\;\;g\;\;}$	b
6	$\xrightarrow{\;\;g\;\;}$	

<u>not</u> one-to-one

Consider the mappings f and g shown at the left. In mapping f, each member of the range has exactly one preimage. In mapping g, b has two preimages, 5 and 6.

Definition of One-to-One Mapping

A mapping is a **one-to-one mapping** if and only if each member of the range has exactly one preimage.

Example

3 The domain is $\mathbb{Z}$. Is the mapping represented by $y = |x|$ a one-to-one mapping?

Consider two members of the domain, -1 and 1.

$$-1 \longrightarrow 1 \quad \text{and} \quad 1 \longrightarrow 1$$

The image 1 has two preimages. Thus, $y = |x|$ does *not* represent a one-to-one mapping if the domain is $\mathbb{Z}$. *What if the domain is $\mathbb{W}$?*

Exercises

Exploratory State whether each of the following represents a mapping. If so, state the domain and range of the mapping.

1.

Degrees Celsius	Degrees Fahrenheit
100	212
35	95
20	68
0	32

2.

Cost of Telegram	Number of Words
$5.00	1, 2, 3, or 4
$10.00	5, 6, 7, or 8

3.

Table	Students
1	John, Ted, Rosemary
2	Denny, Laura

4.

City	Population
Kenton	22,358
McGuffey	4,225
Alger	4,225

5.
5 ⟶ 8
9 ⟶ 6
17 ⟶

6.
8 ⟶ x
13 ⟶ y
14 ⟶ z

State whether each of the following represents a one-to-one mapping.

7. $2 \longrightarrow 3$
$4 \longrightarrow 5$
$6 \longrightarrow 7$

8. $a \longrightarrow I$
$b \nearrow$
$c \longrightarrow II$

9.

x	-3	4	5	8	9
y	0	1	3	0	1

10. Suppose h is a one-to-one mapping. If its domain has 21 members, how many members are in its range?

Written   Use the arrow diagram below to answer each of the following.

1. Write the domain of f.
2. Write the range of f.

$$-3 \xrightarrow{\ f\ } -10$$
$$-2 \longrightarrow -5$$
$$-1 \nearrow 0$$
$$0 \longrightarrow -1$$
$$1$$
$$2$$

Find the image of each of the following.

3. 0   **4.** 1   **5.** -2 **6.** -1

Find all preimages of each of the following.

7. -10 **8.** -5 **9.** 0   **10.** -1

If $g: x \longrightarrow 3x - 5$, find the image of each of the following.

11. 0   **12.** 1   **13.** 2   **14.** $\frac{1}{3}$ **15.** -2 **16.** -3

The domain is $\{-3, -2, -1, 0, 1, 2, 3\}$. Draw an arrow diagram for each of the following mappings and then state the range.

17. $g: x \longrightarrow 2x$ **18.** $x \xrightarrow{\ f\ } x - 5$ **19.** $m: x \longrightarrow x^2$

20. $t: x \longrightarrow x^2 - x + 1$ **21.** $x \xrightarrow{\ s\ } |x - 2|$ **22.** $f: x \longrightarrow |x| - 2$

The domain is $\{-2, -1, 1, 2\}$. Draw an arrow diagram for the mapping represented by each of the following and then state the range.

23. $y = 4x$ **24.** $y = x + 5$ **25.** $y = x^2$

26. $y = x^3$ **27.** $y = \frac{1}{2}x$ **28.** $y = \frac{4}{x}$

The domain is $\mathbb{Z}$. State whether each of the following represents a one-to-one mapping.

29. $x \longrightarrow x$ **30.** $x \longrightarrow 4x$ **31.** $x \longrightarrow x - 7$
32. $x \longrightarrow x^2$ **33.** $y = 5 - x$ **34.** $y = x^2 - 9$
35. $y = |x| - 1$ **36.** $y = |x - 1|$ **37.** $y = 2x - 1$
38. $y = x^2 + 4$ **39.** $y = x^2 - 6x + 1$ **40.** $y = |x| - 6$

The domain is $\mathbb{Z}$. State whether each of the following defines a mapping.

41. $y = x - 4$ **42.** $y = x$ **43.** $y = |x|$ **44.** $x^2 + y^2 = 4$

Challenge   Write an equation to represent each of the following.

1.

x	0	2	4	5	6
y	0	6	12	15	18

2.

x	-3	-2	-1	0
y	-7	-5	-3	-1

3.

x	1	2	3	4
y	3	12	27	48

14.2 Line Reflections and Line Symmetry

The domain and range of some mappings is the set of all points in a plane. These mappings are used in geometry.

Definition of Transformation

A **transformation** is a one-to-one mapping whose domain and range are the set of all points in the plane.

When water in a lake is perfectly calm you often can see mirror-like reflections of things on the shore. This situation suggests the idea of a **line reflection.**

In the figure below, point A is flipped, or reflected over line ℓ onto point A′. We read A′ as A *prime*. We say that A′ is the **image** of A and B′ is the image of B. A is the **preimage** of A′ and B is the preimage of B′. Notice that point C, which is on line ℓ, is its own image.

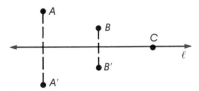

Line ℓ is called the **line of reflection.** Notice that line ℓ is perpendicular to $\overline{AA'}$ and $\overline{BB'}$ and also bisects each segment. In other words, ℓ is the perpendicular bisector of the segment joining a point and its image.

A line reflection is a one-to-one mapping since each point maps to a unique image. Therefore, a line reflection is a transformation.

Definition of Line Reflection

A reflection in line ℓ (or reflection over line ℓ) is a transformation that maps each point P onto a point P′ in one of the following ways.
1. If P is on line ℓ, then the image of P is P.
2. If P is not on line ℓ, then ℓ is the perpendicular bisector of $\overline{PP'}$.

The symbol $A \xrightarrow{R_\ell} A'$ means that A maps to A' for a reflection over line ℓ.

Example

1 **Find the reflection of $\triangle ABC$ over line m.**

Construct perpendiculars from A, B, and C through m. Locate A', B', and C' so that line m is the perpendicular bisector of $\overline{AA'}$, $\overline{BB'}$, and $\overline{CC'}$. A', B', and C' are the corresponding images of A, B, and C. The reflection of $\triangle ABC$ is found by connecting the points A', B', and C'.

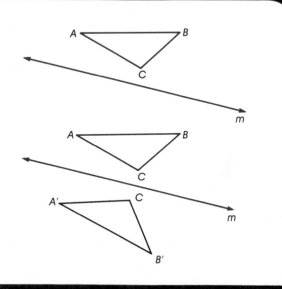

$$\triangle ABC \xrightarrow{R_m} \triangle A'B'C'$$

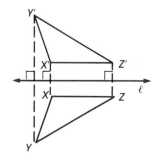

Consider the line reflection at the left. $\triangle X'Y'Z'$ is the reflection of $\triangle XYZ$. The reflections, or images, of $\overline{X'Y'}$, $\overline{Y'Z'}$, and $\overline{X'Z'}$ are the same length as $\overline{XY}$, $\overline{YZ}$, and $\overline{XZ}$, respectively. Also, the measures of the corresponding angles are equal. Thus, *line reflections preserve angle measure and distance measure.*

A line can be drawn through many plane figures so that the figure on one side is a reflection of the figure on the opposite side. In such a case, the line of reflection is called a **line of symmetry.** The figure is said to be *symmetrical with respect to the line of reflection* or have line symmetry. The figures below have line symmetry.

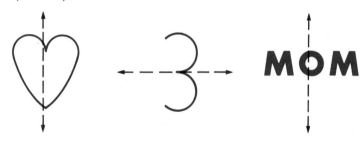

Example

2 **Are ℓ and m lines of symmetry for rectangle ABCD?**

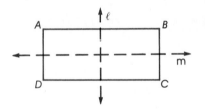

By measuring it can be seen that ℓ is the perpendicular bisector of both $\overline{AB}$ and $\overline{CD}$. Likewise, m is the perpendicular bisector of $\overline{AD}$ and $\overline{BC}$. Any point to the left of ℓ on the rectangle has its image to the right of ℓ on the rectangle. Also, any point above m on the rectangle has its image below m on the rectangle. Therefore, ℓ and m are lines of symmetry for rectangle ABCD.

Exercises

Exploratory For each of the following, name its image for a reflection over line ℓ.

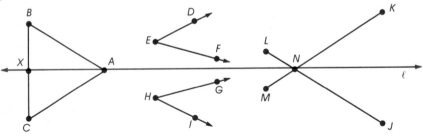

1. A	**2.** B	**3.** $\overline{AB}$	**4.** F	**5.** M	**6.** K
7. $\overline{KM}$	**8.** ∠DEF	**9.** $\overline{EF}$	**10.** △BXA	**11.** N	**12.** $\overline{DE}$

Written Copy each figure below. Then draw the image of each figure for R$_m$. Label all image points using prime notation.

1.

2.

3.

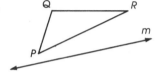

4.

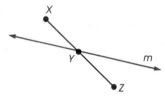

5.

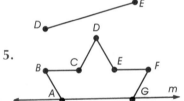

6.

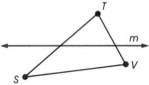

Copy each figure below. Then use only compass and straightedge to construct the image of each figure in a reflection over line ℓ.

7.

8.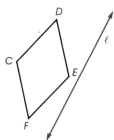

9.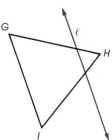

Copy each figure below. Draw all possible lines of symmetry. If none exist, write *none*.

10.

11.

12. **DAD**

13.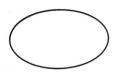

14.

15.

16.

17.

18. **BOOK**

Use the figure at the right to answer each of the following.

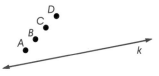

19. Draw the images of points A, B, C, and D for a reflection over line k.

20. Explain how the images of points A, B, C, and D suggest that line reflections preserve collinearity.

21. Explain how the images of points A, B, C, and D suggest that line reflections preserve betweenness of points.

Answer each of the following.

22. Can a transformation map two different points to the same point? Explain.

23. Can a transformation map each point to itself? Explain.

24. Find a property of geometric figures *not* preserved by line reflections. (Hint: Think of a line reflection of a clock.)

14.3 Line Reflections in the Coordinate Plane

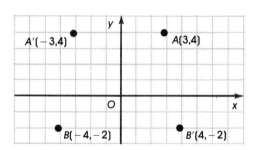

The axes in the coordinate plane also can be used for reflection. What are the images of $A(3, 4)$ and $B(-4, -2)$ when reflected over the y-axis? The images are $A'(-3, 4)$ and $B'(4, -2)$. In each case the y-coordinate of the original point and its image are the same. The x-coordinate of the image is the opposite of the original x-coordinate.

Reflection over the y-axis

If point $P(x, y)$ is reflected over the y-axis, its image is $P'(-x, y)$.

$$(x, y) \xrightarrow{R_{y\text{-axis}}} (-x, y)$$

Example

1 Find and graph the image of a triangle whose vertices are $A(-1, 2)$, $B(-4, 1)$, and $C(-3, -2)$ when reflected over the y-axis.

$$A(-1, 2) \xrightarrow{R_{y\text{-axis}}} A'(1, 2)$$

$$B(-4, 1) \xrightarrow{R_{y\text{-axis}}} B'(4, 1)$$

$$C(-3, -2) \xrightarrow{R_{y\text{-axis}}} C'(3, -2)$$

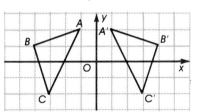

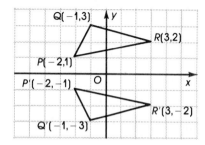

Consider a reflection over the x-axis.

$$P(-2, 1) \xrightarrow{R_{x\text{-axis}}} P'(-2, -1)$$

$$Q(-1, 3) \xrightarrow{R_{x\text{-axis}}} Q'(-1, -3)$$

$$R(3, 2) \xrightarrow{R_{x\text{-axis}}} R'(3, -2)$$

Reflection over the x-axis

If point $P(x, y)$ is reflected over the x-axis, its image is $P'(x, -y)$.

$$(x, y) \xrightarrow{R_{x\text{-axis}}} (x, -y)$$

Example

2 Find and graph the image of a triangle whose vertices are $A(-2, 1)$, $B(1, 4)$, and $C(3, 2)$ when reflected over the line for $x = 3$.

Count the number of horizontal units each point is from the line for $x = 3$. The x-coordinate of each image is the same number of horizontal units from the line for $x = 3$ on the other side. The y-coordinate of each image is the same as the y-coordinate of its preimage.

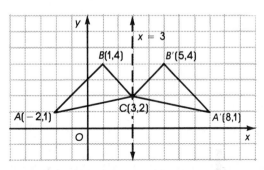

Exercises

Exploratory State whether each of the following figures is symmetric with respect to the y-axis, the x-axis, both or neither.

1.

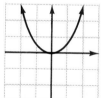

2.

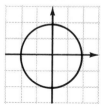

3.

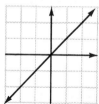

4.

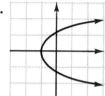

5.

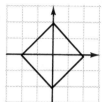

6.

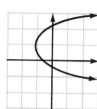

7.

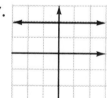

8.

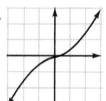

9.

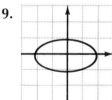

Written   Graph each of the following points on graph paper. Then graph the image of each point when reflected over the y-axis. Write the coordinates of each image.

1. $A(2, 3)$ **2.** $B(-4, 6)$ **3.** $C(0, 6)$ **4.** $D(-4, -4)$

5. $E(3, -5)$ **6.** $F(-6, 0)$ **7.** $G(-2, 4)$ **8.** $H(1, 3)$

9. $W(-1, -2)$ **10.** $J(7, -4)$ **11.** $K\left(2\frac{1}{2}, 7\right)$ **12.** $L\left(5\frac{1}{2}, -6\right)$

Graph each of the following points on graph paper. Then graph the image of each point when reflected over the x-axis. Write the coordinates of each image.

13. $M(1, 3)$ **14.** $N(0, 5)$ **15.** $P(-3, 4)$ **16.** $Q(5, -3)$

17. $R(5, 0)$ **18.** $S(-4, 6)$ **19.** $T(-1, -4)$ **20.** $U(7, -2)$

21. $V(-5, -7)$ **22.** $X\left(3\frac{1}{2}, 3\frac{1}{2}\right)$ **23.** $Y(-8, 3)$ **24.** $Z\left(-2\frac{1}{2}, -6\right)$

Graph each of the following points on graph paper. Then graph the image of each point when reflected over the line for $x = 4$. Write the coordinates of each image.

25. $A(1, 4)$ **26.** $B(-2, 3)$ **27.** $C(-2, -4)$ **28.** $D(5, -3)$

Graph each of the following points on graph paper. Then graph the image of each point when reflected over the line for $y = -2$. Write the coordinates of each image.

29. $E(4, 2)$ **30.** $F(3, -1)$ **31.** $G(-3, 4)$ **32.** $H(-2, -3)$

Graph each of the following points on graph paper. Then graph the image of each point when reflected over the line for $y = x$. Write the coordinates of each image.

33. $M(2, 1)$ **34.** $J(4, 0)$ **35.** $K(-5, -1)$ **36.** $L(-2, 7)$

37. Use the results of exercises 33-36 to complete the following. $(x, y) \xrightarrow{R_{y=x}}$ _____

Graph the quadrilateral with vertices $A(2, 1)$, $B(5, -2)$, $C(9, 4)$ and $D(6, 7)$. Find the image of $ABCD$ for each of the following reflections.

38. $R_{y\text{-axis}}$ **39.** $R_{x\text{-axis}}$ **40.** reflection over the line for $x = 1$

41. reflection over the line for $x = -2$ **42.** reflection over the line for $y = -3$

Draw a figure on graph paper that satisfies each of the following conditions.

43. symmetric about the y-axis, but not the x-axis

44. symmetric about the y-axis and the x-axis

45. symmetric about the line for $y = x$, but not either of the two coordinate axes

46. symmetric about the line for $x = 4$ and the line for $y = 3$

Find a line of symmetry for the graph of each equation. If there is no line of symmetry, write *none*.

47. $y = x^2 + 3$ **48.** $y = |x|$ **49.** $y = x^2 + 4x - 1$ **50.** $x = y^2$

14.4   Translations

A race car speeds along a track to the finish line. The result of a movement in one direction is a transformation called a **translation.**

Consider moving △*ABC* two centimeters in the direction given.

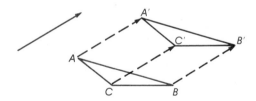

In this case, *A* maps to *A'*, *B* maps to *B'*, and *C* maps to *C'*. Note that the segments connecting each point and its image are parallel and congruent.

Definition of Translation

A translation is a transformation that maps every point in the plane to its image by moving each point the same distance in the same direction.

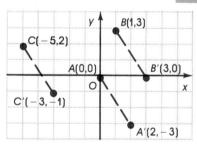

Consider a translation in the coordinate plane. In the diagram, each point is moved two units to the right and three units down.

$$A(0, 0) \longrightarrow A'(2, -3)$$
$$B(1, 3) \longrightarrow B'(3, 0)$$
$$C(-5, 2) \longrightarrow C'(-3, -1)$$

In general, this translation maps $P(x, y)$ as follows.

$$P(x, y) \longrightarrow P(x + 2, y - 3)$$

The symbol $A \xrightarrow{T_{a,b}} A'$ means that $A(x, y)$ maps to $A'(x + a, y + b)$ for translation $T_{a,b}$. Thus, in the previous example, $T_{2,-3}$ may be used to symbolize the translation.

Examples

1 Find and graph the image of a triangle whose vertices are $A(1, -2)$, $B(4, -1)$, and $C(-3, 2)$ and for $T_{-2,5}$.

$$P(x, y) \xrightarrow{T_{-2,5}} P(x - 2, y + 5)$$

$A(1, -2) \longrightarrow A'(-1, 3)$
$B(4, -1) \longrightarrow B'(2, 4)$
$C(-3, 2) \longrightarrow C'(-5, 7)$

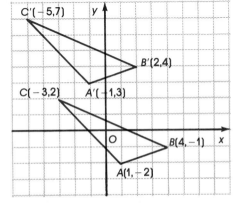

2 Find the distance between the origin and its image for $T_{-4,-6}$.

$$O(0, 0) \xrightarrow{T_{-4,-6}} O'(-4, -6)$$

Note that figure OXO' is a right triangle.

$OX = 4$

Let d = the number of units O is moved. $XO' = 6$

$d^2 = 4^2 + 6^2$ *Use the Pythagorean Theorem.*

$d^2 = 16 + 36$

$d = \sqrt{52}$ or $2\sqrt{13}$ *Why is $-\sqrt{52}$ not a possible answer?*

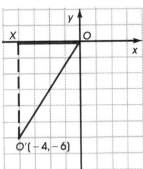

The distance between the origin and its image is $2\sqrt{13}$.

The translation $T_{0,0}$ maps every point in the plane to itself. Thus, it is called an **identity transformation**.

Exercises

Exploratory In the figure at the right, suppose *D* is the image of *C* for a translation. Name the images of each of the following.

1. *J*
2. *B*
3. $\overline{FG}$
4. $\angle GFJ$
5. $\angle EIJ$
6. $\overrightarrow{BF}$

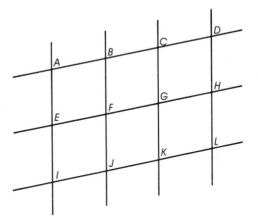

In the figure at the right, suppose *E* is the image of *J* for a translation. Name the images of each of the following.

7. *H*
8. *K*
9. $\overline{GH}$
10. $\overline{JF}$
11. $\angle KLH$
12. $\overrightarrow{KJ}$

Written Copy the figure at the right. For a certain translation, *A'* is the image of *A*. Draw the images of each of the following.

1. *B*
2. *C*
3. *D*
4. *E*

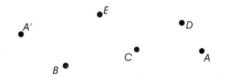

Copy the triangle at the right and draw its image for each of the following.

5. a translation 3 cm to the right
6. a translation 4 cm in the direction of $\overrightarrow{PR}$
7. a translation 2 cm to the left followed by a translation 3 cm down
8. a translation 3 cm down followed by a translation 2 cm to the left

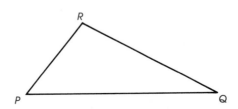

Find the image of *P*(− 2, 9) under each of the following transformations.

9. $T_{2,3}$
10. $T_{-5,6}$
11. $T_{4,-8}$
12. $T_{-2,-3}$

13. $T_{-7,1}$
14. $T_{0,-1}$
15. $T_{-3,-10}$
16. $T_{4,7}$

17. $R_{y\text{-axis}}$
18. $R_{x\text{-axis}}$
19. $R_{y=x}$
20. $R_{x=3}$

Find a rule for a translation which maps *A* to *B* in each of the following.

21. *A*(1, 1); *B*(2, 5)
22. *A*(1, − 3); *B*(− 3, 1)
23. *A*(− 5, 12); *B*(0, 0)
24. *A*(− 2, − 7); *B*(4, − 3)
25. *A*(14, − 2); *B*(3, − 6)
26. *A*(4√2, 1); *B*(√2, − 1)

Find the distance between the origin and its image for each of the following translations.

27. $T_{2,4}$ **28.** $T_{3,-6}$ **29.** $T_{-4,3}$ **30.** $T_{-5,-2}$

31. a. Graph the triangle with vertices $A(-2, 5)$, $B(1, -1)$, and $C(3, 3)$.

    b. Graph the image of $\triangle ABC$ for $T_{-2,3}$. Label the image $\triangle A'B'C'$.

    c. Graph the image of $\triangle A'B'C'$ for $T_{-3,4}$. Label the image $\triangle A''B''C''$.

    d. To what point is (x, y) finally mapped for $T_{-3,4}$ following $T_{-2,3}$?

32. a. Graph the triangle with vertices $A(-5, -2)$, $B(-4, 1)$ and $C(-1, -3)$.

    b. Graph the image of $\triangle ABC$ for a reflection over the y-axis. Label the image $\triangle A'B'C'$.

    c. Graph the image of $\triangle A'B'C'$ for a reflection over the line $x = 7$. Label the image $\triangle A''B''C''$.

    d. Is there a single transformation which maps $\triangle ABC$ to $\triangle A''B''C''$? Explain.

Tell whether each of the following is preserved by translations. Write *yes* or *no*.

33. collinearity   **34.** betweenness of points

35. angle measure   **36.** distance measure

Mixed Review

Choose the best answer.

1. Which expression is the factored form of $x^2 - 5x - 6$?

    a. $(x - 2)(x - 3)$ **b.** $(x - 5)(x + 1)$ **c.** $(x - 6)(x + 1)$ **d.** $(x - 6)(x - 1)$

2. What is the probability of rolling a double or a sum less than 6 with two dice?

    a. $\dfrac{4}{9}$ **b.** $\dfrac{7}{18}$ **c.** $\dfrac{1}{2}$ **d.** $\dfrac{2}{9}$

3. What is the value of x if 4, $3x + 7$, $5x - 3$, and 32 have a mean of 22?

    a. 6   **b.** 5   **c.** 2.25   **d.** 10.5

4. What is an equation of the line through $(-2, 8)$ and $(1, 3)$?

    a. $y = -\dfrac{3}{5}x + \dfrac{18}{5}$ **b.** $y = -\dfrac{5}{3}x + 6$ **c.** $y = -\dfrac{5}{3}x + \dfrac{14}{3}$ **d.** $y = \dfrac{5}{3}x + \dfrac{4}{3}$

5. What is the image of $(-1, 2)$ for $R_{y\text{-axis}}$?

    a. $(1, -2)$ **b.** $(2, -1)$ **c.** $(1, 2)$ **d.** $(-1, -2)$

6. Which of the following expressions is undefined when $x = 5$?

    a. $\dfrac{1}{x}$ **b.** x^{-5} **c.** $\dfrac{1}{x - 5}$ **d.** $\dfrac{1}{x + 5}$

14.5 Rotations

A ferris wheel suggests another type of transformation called a **rotation.**

Definition of Rotation

A rotation is a transformation that maps every point in the plane to its image by rotating the plane around a fixed point. The fixed point is called the **center** of the rotation and is its own image.

In the diagram at the right, $\overline{AB}$ is rotated about A through an angle of $65°$ counterclockwise. $\overline{AB'}$ is the image of $\overline{AB}$.

The symbol $\text{Rot}_{A,65°}$ stands for a rotation with center A through an angle of $65°$ *counterclockwise.*

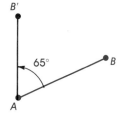

A maps to A since it is the center of the rotation.
B maps to B'.

$$\overline{AB} \xrightarrow{\text{Rot}_{A,65°}} \overline{AB'}$$

For a rotation with center A through an angle of $65°$ *clockwise,* the symbol $\text{Rot}_{A,-65°}$ is used.

Examples

1 **Draw the image of △ABC for Rot$_{O,-100°}$.**

First find the image of A.
1. Draw $\overrightarrow{OA}$.
2. Draw a 100° angle with vertex O and sides $\overrightarrow{OA}$ and $\overrightarrow{OX}$.
3. Locate A' on $\overrightarrow{OX}$ such that $OA' = OA$. A' is the image of A.

Then find the images of B and C in the same manner.

Connect A', B', and C' to

form $\triangle A'B'C'$.

$$\triangle ABC \xrightarrow{\text{Rot}_{O,-110°}} \triangle A'B'C'$$

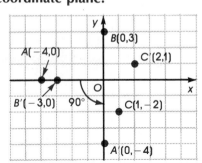

2 **Find the images of $A(-4, 0)$, $B(0, 3)$, and $C(1, -2)$ for a counterclockwise rotation of 90° about the origin in the coordinate plane.**

Imagine turning the coordinate plane 90° counterclockwise.

$$A(-4, 0) \xrightarrow{\text{Rot}_{O,90°}} A'(0, -4)$$

$$B(0, 3) \xrightarrow{\text{Rot}_{O,90°}} B'(-3, 0)$$

$$C(1, -2) \xrightarrow{\text{Rot}_{O,90°}} C'(2, 1)$$

The measurement of the angle must be greater than 0° and less than 360°.

Some plane figures can be rotated through a certain angle such that they are the images of themselves. A figure which has this property is said to have **rotational symmetry**. Each of the following figures has rotational symmetry.

In each case, the center of rotation is C.

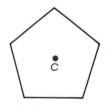
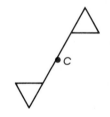

The figure on the left has rotational symmetry since it is its own image for a rotation of either 120° or 240° clockwise or counter-clockwise. The figure in the center has rotational symmetry since it is its own image for a rotation, in either direction, of $n \cdot 72°$ for $n = 1, 2, 3, 4,$ or 5. The figure on the right has rotational symmetry. It is its own image for a rotation of 180°.

Exercises

Exploratory In the diagram at the right, $\triangle ABC \longrightarrow \triangle DEF$ for a rotation. Name each of the following.

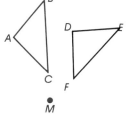

1. center of rotation   A15-25—ID
2. image of A
3. image of B
4. image of C
5. Use a protractor to measure the angle of rotation. What is the angle of rotation?

Written Copy the figure at the right. Draw the image of $\overline{XY}$ for each of the following.   A15-26—ID

1. $\text{Rot}_{P,90°}$ 2. $\text{Rot}_{P,-90°}$

3. $\text{Rot}_{P,-240°}$ 4. $\text{Rot}_{P,60°}$

5. $\text{Rot}_{X,35°}$ 6. $\text{Rot}_{Y,-120°}$

Copy the figure at the right. Draw the image of $\triangle ABC$ for each of the following.

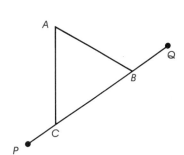

7. $\text{Rot}_{P,90°}$ 8. $\text{Rot}_{P,-75°}$

9. $\text{Rot}_{Q,-100°}$ 10. $\text{Rot}_{Q,65°}$

11. $\text{Rot}_{P,180°}$ 12. $\text{Rot}_{Q,-270°}$

Use graph paper. Find the image of each point for $\text{Rot}_{O,-90°}$.

13. $(0, -4)$ 14. $(5, 0)$ 15. $(-2, 3)$ 16. $(-4, -6)$

Use graph paper. Find the image of each point for $\text{Rot}_{O,180°}$.

17. $(4, 0)$ 18. $(0, 5)$ 19. $(3, -1)$ 20. $(-2, -4)$

Complete each of the following.

21. $(x, y) \xrightarrow{Rot_{O,90°}}$ ___?___
(Hint: see example 2)

22. $(x, y) \xrightarrow{Rot_{O,-90°}}$ ___?___
(Hint: see exercises 13-16)

23. $(x, y) \xrightarrow{Rot_{O,180°}}$ ___?___
(Hint: see exercises 17-20)

State whether each of the following figures has rotational symmetry. If so, name a rotation which shows it.

24.   **25.**   **26.**   **27.**   **28.**

29.   **30.**   **31.**   **32.**   **33.**

A figure that is its own image for some 180° rotation with center *P* is said to have *point symmetry* with respect to *P*. State whether the figure in each exercise has point symmetry. Write *yes* or *no*.

34. Exercise 24 **35.** Exercise 25 **36.** Exercise 26 **37.** Exercise 27
38. Exercise 28 **39.** Exercise 29 **40.** Exercise 30
41. Exercise 31 **42.** Exercise 32 **43.** Exercise 33

Tell whether each of the following is preserved by rotations. Write *yes* or *no*.

44. collinearity **45.** betweenness of points
46. angle measure **47.** distance measure

48. Copy the figure at the right.

    a. Draw the image of $\triangle ABC$ for $Rot_{P,-60°}$. Label the image $\triangle A'B'C'$.

    b. Draw the image of $\triangle A'B'C'$ for $Rot_{P,-40°}$. Label the image $\triangle A''B''C''$.

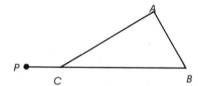

    c. Is there a single transformation which maps $\triangle ABC$ to $\triangle A''B''C''$? Explain.

49. Sketch a figure that has rotational symmetry but does not have line symmetry.
50. Sketch a figure that has line symmetry but does not have rotational symmetry.

Mathematical Excursions

A rotation of 180° about a point P is called a **half-turn** with center P. It is a special case of a rotation.

If a figure is its own image under a half-turn with center P, then it has **point symmetry with respect to P.**

Example Find the image of AB for a half-turn about center P.

First find the image of A for a 180° rotation about P. Then find the image of B.

$$\overline{AB} \xrightarrow{\text{Rot}_{P,180°}} \overline{A'B'}$$

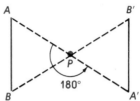

Notice that P is the midpoint of $\overline{AA'}$ and $\overline{BB'}$. Thus, a half-turn with center P is a transformation that maps every point A to a point A' such that P is the midpoint of $\overline{AA'}$.

Example Find the image of $A(2, 3)$ and $B(-3, 5)$ for a half-turn about the origin.

$$A(2,\ 3) \xrightarrow{\text{Rot}_{O,180°}} A'(-2,\ -3)$$

$$B(-3,\ 1) \xrightarrow{\text{Rot}_{O,180°}} B'(3,\ -1)$$

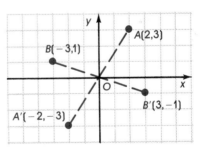

In general, the following rule can be used to find the image of any point with a half-turn about the origin.

$$(x,\ y) \xrightarrow{\text{Rot}_{O,180°}} (-x,\ -y)$$

Exercises Copy the figure at the right. Find the image of $\triangle PQR$ for a half-turn about each of the following centers.

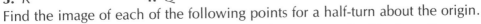

1. P **2.** C
3. R **4.** Q

Find the image of each of the following points for a half-turn about the origin.

5. $A(3, 2)$ **6.** $B(-5, 4)$ **7.** $C(0, -3)$ **8.** $D(-1, -6)$

14.6 Dilations

A transformation which preserves distance is called an **isometry.** Thus, line reflections, translations, and rotations are examples of isometries.

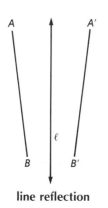

line reflection

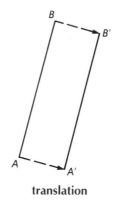

translation

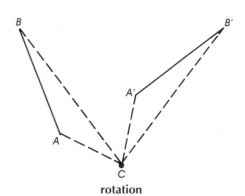

rotation

Suppose a geometric figure is enlarged or reduced. The figure does *not* change its shape. However, it is altered in size. This type of transformation is called a **dilation.**   *Is a dilation an isometry?*

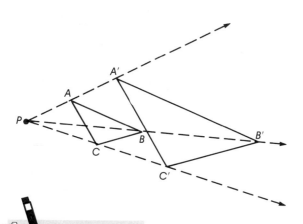

In the figure at the left, $\triangle A'B'C'$ is the image of $\triangle ABC$. The measure of the distance from P to a point on $\triangle A'B'C'$ is twice the measure of the distance from P to the corresponding point of $\triangle ABC$.

$$PA' = 2 \cdot PA$$
$$PB' = 2 \cdot PB$$
$$PC' = 2 \cdot PC$$

This transformation is an example of a dilation with **center** P and a **scale factor** of 2. The symbol $D_{P,2}$ is used to represent the dilation.

$$\triangle ABC \xrightarrow{D_{P,2}} \triangle A'B'C'$$

Although distance is *not* preserved by dilations, the ratios of the measures of corresponding parts are equivalent.

$$\frac{AB}{A'B'} = \frac{AC}{A'C'} = \frac{BC}{B'C'}$$

This means that $\triangle ABC \sim \triangle A'B'C'$.

Journal

Look back over what you have learned this year in mathematics class. What things did you enjoy learning about most this year?

Example

1 Find the image of $\triangle PQR$ for a dilation with center C and a scale factor of $\frac{3}{4}$.

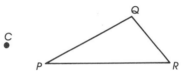

Draw $\overrightarrow{CP}$, $\overrightarrow{CQ}$, and $\overrightarrow{CR}$. Find P', Q', and R' so that $CP' = \frac{3}{4}(CP)$, $CQ' = \frac{3}{4}(CQ)$, and $CR' = \frac{3}{4}(CR)$.

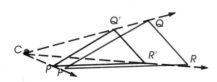

$$\triangle PQR \xrightarrow{D_{C,\frac{3}{4}}} \triangle P'Q'R'$$

It is possible for the scale factor of a dilation to be negative. When this is the case, the image of a point is on the ray in the opposite direction. For example, $\triangle D'E'F'$ is the image of $\triangle DEF$ for a dilation with center O and a scale factor of -2.

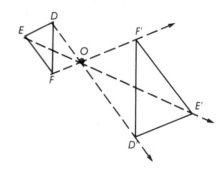

The origin of the coordinate plane can be used as the center of a dilation.

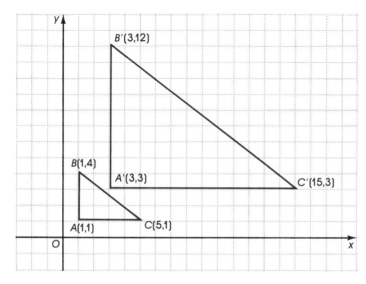

In the figure at the left,

$$\triangle ABC \xrightarrow{D_{O,3}} \triangle A'B'C'.$$

$$A(1, 1) \xrightarrow{D_{O,3}} A'(3, 3)$$

$$B(1, 4) \xrightarrow{D_{O,3}} B'(3, 12)$$

$$C(5, 1) \xrightarrow{D_{O,3}} C'(15, 3)$$

In general, the following rule can be used to find the image of any point for a dilation with the origin as its center and a scale factor of k.

$$(x, y) \xrightarrow{D_{O,k}} (kx, ky)$$

Exercises

Exploratory Name the image of *S* for a dilation with center *M* for each of the following scale factors.

1. 1
2. 2
3. $\frac{2}{3}$
4. $1\frac{1}{3}$
5. −2
6. −1
7. $1\frac{1}{6}$
8. $-\frac{5}{6}$
9. $-1\frac{1}{2}$
10. $1\frac{2}{3}$
11. $-\frac{1}{2}$
12. $-1\frac{1}{3}$

Written In a dilation with center *C* and a scale factor of *k*, *A′* is the image of *A* and *B′* is the image of *B*. Find *k* for each of the following conditions.

1. $CA' = 10, CA = 5$
2. $CB' = 18, CB = 9$
3. $CA' = 6, CA = 4$
4. $CA = 2, CA' = 10$
5. $AB = 3, A'B' = 1$
6. $AB = 3, A'B' = 4$

Copy the figure at the right. Then draw the image of △*PQR* for the given scale factor and center.

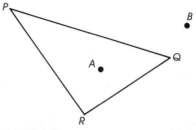

7. center *B*, scale factor of 2
8. center *B*, scale factor of −1
9. center *A*, scale factor of $-\frac{1}{2}$
10. center *A*, scale factor of 1
11. center *Q*, scale factor of $\frac{1}{2}$

Graph each of the following points. Connect the points in order. Draw the image of each figure for a dilation with the origin as the center and a scale factor of 2.

12. (0, 2), (4, 0)
13. (3, −3), (−2, −2)
14. (3, 4), (6, 10), (−3, 5)
15. (6, 5), (4, 5), (3, 7)
16. (−1, 4), (0, 1), (2, 3)
17. (1, 2), (3, 3), (3, 5), (1, 4)

Draw the image of each geometric figure in each of the following exercises for a dilation with the origin as the center and a scale factor of $\frac{1}{2}$.

18. Exercise 12
19. Exercise 14
20. Exercise 16
21. Exercise 17

Find the image of *A*(3, −4) for each of the following dilations.

22. $D_{O,2}$
23. $D_{O,-3}$
24. $D_{O,4}$
25. $D_{O,-14}$
26. $D_{O,\frac{1}{2}}$

Tell whether each of the following is preserved by dilations. Write *yes* or *no*.

27. collinearity
28. betweenness of points
29. angle measure
30. distance measure

Let *p*, *q*, and *r* represent the following statements.
  p: "Transformation *t* is an isometry."
  q: "Transformation *t* is a dilation with a scale factor of 1."
  r: "Transformation *t* does not preserve distance."
State the truth value of each of the following.

31. $p \longrightarrow r$
32. $q \longrightarrow r$
33. $\sim r \longrightarrow q$
34. $\sim q \longrightarrow \sim r$
35. $\sim p \longrightarrow r$

Problem Solving Application: Networks

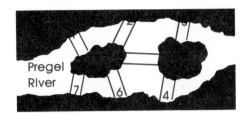

Pregel
River

The study of *graph theory* began with a now famous problem about bridges. The city of Königsberg in Europe had seven bridges connecting both sides of the Pregel River to two islands in the river. The problem was to find a path that would take you over all seven bridges without crossing the same bridge twice.

In the eighteenth century, the problem caught the attention of the Swiss mathematician Leonhard Euler (pronounced OY-lur). He analyzed the problem by modeling it as a graph. In this graph, called a **network,** the *edges* represent the seven bridges, and the *vertices* represent the sides of the river and the two islands. Copy the graph. Can you trace this network with your pencil without lifting your pencil or retracing an edge?

A network can be defined as a figure consisting of vertices and edges. A network may have more than one edge connecting the same pair of vertices. One way of examining a network is to count the number of edges that meet at each vertex. This is called the **degree of the vertex.**

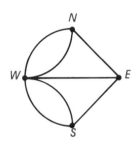

W and E represent the islands. N and S are the sides of the river.

Example

1 Find the degree of each node in the network of the Königsberg Bridges.

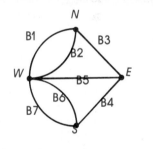

Vertex	Edges at the Vertex	Degree of Vertex
N	B1, B2, B3	3
E	B3, B5, B4	3
S	B4, B6, B7	3
W	B1, B2, B5, B6, B7	5

While studying the Königsberg Bridges, Euler discovered two interesting properties about the traceability of a network.

**Traceability
Test of a
Network**

A network can be traced in one continuous path without retracing any edge if one of the following is true.
1. The graph only has vertices with even degrees.
2. The graph has exactly two vertices with odd degrees.

Look again at Example 1. Neither of the two conditions of traceability is met in the bridge problem. Therefore, there is no path that you could walk to cross all seven bridges without repeating at least one of the bridges on your journey. The network representing the bridges is not traceable.

Example

2 Determine if each graph passes the traceability test for networks.

a.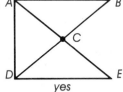
yes
Exactly two vertices, A and D, have odd degrees.

b.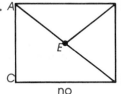
no
Four vertices, A, B, D, and E, have odd degrees.

c.
no
All vertices have odd degrees.

Only the graph in a. passes the traceability test.

Exercises

Written Determine whether each network is traceable. If so, copy it on your paper and show the direction of the paths taken to trace the graph.

1.

2.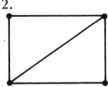

3.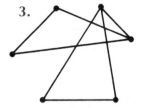

4.

5. Could the people of Königsberg have walked over all the bridges if one bridge had been added? If so, where?

6. Could the people of Königsberg have walked over all the bridges if one of the bridges had been removed? If so, where?

7. If the graph at the right represents the streets of a small town, find how many different routes there are from *A* to *B*. You may move only east and north.

Portfolio Suggestion

Review the items in your portfolio. Make a table of contents of the items, noting why each item was chosen. Replace any items that are no longer appropriate.

Performance Assessment

Draw a △GHI and two lines *m* and *n* that intersect at an angle of 30°. Use a composite of reflections with respect to *m* and *n* to produce a rotation image of △GHI. Explain your solution.

Chapter Summary

1. A **function,** or **mapping,** pairs each member of one set with exactly one member of the same, or another set. The first set is called the **domain.** The second set is called the **range.** (452)
2. The members in the domain of a mapping are called **preimages.** The members in the range are called **images.** (452)
3. A mapping is a **one-to-one mapping,** if and only if each member of the range has exactly one preimage. (454)
4. A **reflection** in line ℓ (or reflection over line ℓ) is a transformation that maps each point P onto a point P' in the following ways.
   (1) If P is on line ℓ, then the image of P is P.
   (2) If P is not on line ℓ, then ℓ is the perpendicular bisector of PP'.
   (456)
5. A **line of symmetry** is a line of reflection. (457)
6. If point $P(x, y)$ is reflected over the y-axis, its image is $P'(-x, y)$.
   That is, $(x, y) \xrightarrow{\;R_{y\text{-axis}}\;} (-x, y)$. (460)
7. If point $P(x, y)$ is reflected over the x-axis, its image is $P'(x, -y)$.
   That is, $(x, y) \xrightarrow{\;R_{x\text{-axis}}\;} (x, y)$. (460)

8. A **translation** is a transformation that maps every point in the plane the same distance in the same direction. (463)
9. A transformation that maps every point in the plane to itself is called an **identity transformation.** (464)
10. A **rotation** is a transformation that maps every point in the plane to its image by rotating the plane around a fixed point. The fixed point is called the **center of the rotation** and is its own image. (467)
11. A figure that can be rotated through a certain angle such that it is the image of itself has **rotational symmetry.** (488)
12. A **dilation** is a transformation that changes the size of a figure but not its shape. (472)

Chapter Review

14.1 **State whether each of the following represents a mapping. If so, state the domain and range of the mapping.**

1. $4 \longrightarrow a$
   $5 \nearrow$
   $6 \longrightarrow b$

2. $3 \longrightarrow a$

3. $A \rightleftarrows 1$
   $B \searrow 2$
   $C \longrightarrow 3$

Use the mapping represented in the table below to answer each of the following.

4. Write the domain of the mapping.
5. Write the range of the mapping.
6. Find the image of 4.
7. Find the preimage of 4.

x	-2	4	0	7
y	4	3	-1	4

The domain is $\{-4, -2, 0, 2, 4\}$. Draw an arrow diagram for the mapping represented by each of the following. Then state whether it is a one-to-one mapping.

8. $x \longrightarrow 2x - 1$
9. $y = x^2$
10. $y = -3x$

14.2 **Find all the lines of symmetry in each of the following. If none, write *none*.**

11.

12.

13.

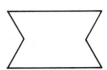

Copy each of the following. Then draw the image of each figure for R_ℓ. Label all image points using prime notation.

14.
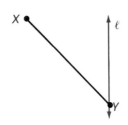

15.
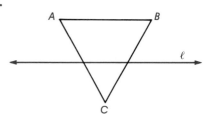

14.3   Complete each of the following.

16. $(2, 3) \xrightarrow{R_{x\text{-axis}}}$ __?__

17. $(-1, 5) \xrightarrow{R_{y\text{-axis}}}$ __?__

18. $(-3, -4) \xrightarrow{R_{x=y}}$ __?__

19. __?__ $\xrightarrow{R_{x\text{-axis}}} (5, 7)$

20. Graph the triangle with vertices $A(1, -1)$, $B(2, 4)$, and $C(4, 0)$. Then find the image of $\triangle ABC$ for a reflection over the line for $y = -2$.

14.4   Complete each of the following.

21. $(1, 7) \xrightarrow{T_{4,1}}$ __?__

22. $(-2, 3) \xrightarrow{T_{-4,-8}}$ __?__

23. __?__ $\xrightarrow{T_{4,3}} (-6, 8)$

24. $(-2, 5) \underline{\quad?\quad} (1, 3)$

25. Graph the quadrilateral with vertices $A(-1, 3)$, $B(-2, 5)$, $C(2, 7)$, and $D(4, 2)$. Then graph its image for $T_{-6,5}$.

14.5   Find the image A' of $A(1, -3)$ for each of the following.

26. $\text{Rot}_{O,90°}$

27. $\text{Rot}_{O,-90°}$

28. $\text{Rot}_{O,180°}$

29. $\text{Rot}_{O,-270°}$

30. Copy the figure at the right. Find the image of $\triangle PQR$ for $\text{Rot}_{C,60°}$.

State whether each of the following figures has rotational symmetry.

31.

32.

33.

14.6   Complete each of the following.

34. $(-1, 5) \xrightarrow{D_{O,3}}$ __?__

35. $(2, -6) \xrightarrow{D_{O,-4}}$ __?__

36. __?__ $\xrightarrow{D_{O,2}} (8, 2)$

37. $(2, 4) \xrightarrow{\quad?\quad} (1, 2)$

38. Graph the triangle with vertices $A(-2, 1)$, $B(0, 3)$, and $C(1, -4)$. Find the image of $\triangle ABC$ for $D_{O,2}$.

Chapter Test

State whether each of the following is *always*, *sometimes*, or *never* true.

1. A transformation preserves distance measure.

2. A line reflection maps a point onto itself.

3. A scalene triangle has rotational symmetry.

4. A line reflection is an isometry.

5. A translation is an isometry.

6. A dilation is an isometry.

Label three noncollinear points *A*, *B*, and *C*. Draw each of the following.

7. image of *A* for a reflection over $\overleftrightarrow{BC}$

8. image of *A* for a translation that maps *B* onto *C*

9. image of $\overline{AB}$ for $\text{Rot}_{C,-60°}$

State whether each figure has line symmetry, rotational symmetry, neither, or both.

10.   **11.** **B** **12.** **8** **13.** **HIH** **14.**

Find the image of $(-2, 4)$ for each of the following transformations.

15. $R_{x\text{-axis}}$ **16.** $R_{x=y}$ **17.** $T_{3,-5}$ **18.** $T_{-4,3}$ **19.** $\text{Rot}_{O,180°}$ **20.** $D_{O,3}$

Graph the triangle with vertices $A(2, 1)$, $B(5, -2)$, and $C(9, 4)$. Then graph the image of $\triangle ABC$ for each of the following transformations.

21. $R_{y\text{-axis}}$ **22.** $R_{x=4}$ **23.** $T_{-2,3}$ **24.** $D_{O,2}$ **25.** $\text{Rot}_{O,90°}$ **26.** $D_{O,-2}$

Name the transformation for which $\triangle A'B'C'$ is the image of $\triangle ABC$.

27. 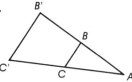 **28.** 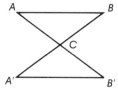 **29.**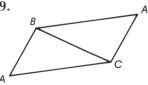

30. 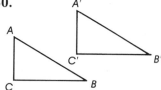 **31.** 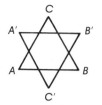 **32.**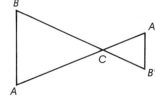

Symbols

$=$	is equal to		π	pi
$\neq$	is not equal to		$|a|$	absolute value of a
$>$	is greater than		$\sqrt{}$	principal square root
$<$	is less than		$a:b$	ratio of a to b
$\geq$	is greater than or equal to		$\{\ \}$	set
$\leq$	is less than or equal to		$\in$	is a member of
$\approx$	is approximately equal to		$\cup$	union
$\cdot$	times		$\cap$	intersection
$-$	negative		$\wedge$	conjunction
$+$	positive		$\vee$	disjunction
$\pm$	positive or negative		$\rightarrow$	conditional; is mapped onto
$\overleftrightarrow{AB}$	line containing points A and B		$\leftrightarrow$	biconditional
$\overrightarrow{AB}$	ray with endpoint A passing through B		$n!$	n factorial
$\overline{AB}$	line segment with endpoints A and B		$x \xrightarrow{f} x^2$	f is the function that maps x to x^2
AB	measure of AB		R_{ℓ}	reflection over line ℓ
$\overset{\frown}{AB}$	arc with endpoints A and B		$T_{2,3}$	translation 2 units to the right and 3 units up
$m\overset{\frown}{AB}$	measure of arc AB		$\text{Rot}_{A,65°}$	rotation with center A through an angle of 65° counterclockwise
$\angle$	angle		$D_{P,2}$	dilation with center P and a scale factor of 2
$m\angle A$	measure of angle A		$\mathcal{N}$	set of natural numbers
$°$	degree		$\mathcal{W}$	set of whole numbers
$\triangle$	triangle		$\mathcal{Z}$	set of integers
$\square$	parallelogram		$\mathcal{Q}$	set of rational numbers
$\odot P$	circle with center P		$\mathcal{R}$	set of real numbers
$\cong$	is congruent to			
$\sim$	is similar to; negation			
$\parallel$	is parallel to			
$\perp$	is perpendicular to			

Squares and Approximate Square Roots

n	n²	√n	n	n²	√n
1	1	1.000	51	2601	7.141
2	4	1.414	52	2704	7.211
3	9	1.732	53	2809	7.280
4	16	2.000	54	2916	7.348
5	25	2.236	55	3025	7.416
6	36	2.449	56	3136	7.483
7	49	2.646	57	3249	7.550
8	64	2.828	58	3364	7.616
9	81	3.000	59	3481	7.681
10	100	3.162	60	3600	7.746
11	121	3.317	61	3721	7.810
12	144	3.464	62	3844	7.874
13	169	3.606	63	3969	7.937
14	196	3.742	64	4096	8.000
15	225	3.873	65	4225	8.062
16	256	4.000	66	4356	8.124
17	289	4.123	67	4489	8.185
18	324	4.243	68	4624	8.246
19	361	4.359	69	4761	8.307
20	400	4.472	70	4900	8.367
21	441	4.583	71	5041	8.426
22	484	4.690	72	5184	8.485
23	529	4.796	73	5329	8.544
24	576	4.899	74	5476	8.602
25	625	5.000	75	5625	8.660
26	676	5.099	76	5776	8.718
27	729	5.196	77	5929	8.775
28	784	5.292	78	6084	8.832
29	841	5.385	79	6241	8.888
30	900	5.477	80	6400	8.944
31	961	5.568	81	6561	9.000
32	1024	5.657	82	6724	9.055
33	1089	5.745	83	6889	9.110
34	1156	5.831	84	7056	9.165
35	1225	5.916	85	7225	9.220
36	1296	6.000	36	7396	9.274
37	1369	6.083	87	7569	9.327
38	1444	6.164	88	7744	9.381
39	1521	6.245	89	7921	9.434
40	1600	6.325	90	8100	9.487
41	1681	6.403	91	8281	9.539
42	1764	6.481	92	8464	9.592
43	1849	6.557	93	8649	9.644
44	1936	6.633	94	8836	9.695
45	2025	6.708	95	9025	9.747
46	2116	6.782	96	9216	9.798
47	2209	6.856	97	9409	9.849
48	2304	6.928	98	9604	9.899
49	2401	7.000	99	9801	9.950
50	2500	7.071	100	10000	10.000

Glossary

absolute value (48)   The absolute value of a number is the number of units it is from zero on the number line.

acute angle (146)   An angle whose degree measure is less than 90.

acute triangle (159)   A triangle with three acute angles.

additive inverse (62)   For any rational number a, the additive inverse is $-a$ if $a + (-a) = 0$.

adjacent angles (149)   Two angles which have a common side, the same vertex, and no interior points in common.

algebraic expression (38)   A mathematical phrase containing one or more variables and operations.

altitude (247)   A perpendicular segment to the base from a vertex.

angle (144)   The union of two rays that have the same endpoint.

antecedent (16)   The statement which follows the word *if* in a conditional.

area of a polygon (246)   The measurement of the region formed by the polygon and its interior.

arrow diagram (452)   A diagram used to represent a mapping.

associative property for addition (61)   For any rational numbers a and b, $a + (b + c) = (a + b) + c$.

associative property for multiplication (61)   For any rational numbers a and b, $a(bc) = (ab)c$.

axiom (154)   A statement which is accepted as true without proof.

bar graph (391)   A statistical graph using bars to show how different quantities compare.

biconditional (28)   A statement formed by the conjunction of the conditionals $p \rightarrow q$ and $q \rightarrow p$.

binomial (289)   A polynomial with two unlike terms.

bisector of an angle (149)   A ray which is the common side of two adjacent angles having the same measure.

center of a circle (257)   See circle.

chord of a circle (257)   A segment whose endpoints are points of the circle.

circle (257)   The set of all points in a plane a given distance from a point in the plane called the center.

circle graph (393)   A statistical graph used to show how parts are related to the whole.

circumference (257)   The distance around a circle.

coefficient (75)   The numerical factor in a term, or monomial.

collinear (144)   Two or more points are collinear if and only if they lie on the same line.

commutative property for addition (61)   For any rational numbers a and b, $a + b = b + a$.

commutative property for multiplication (61)   For any rational numbers a and b, $ab = ba$.

complementary (149)   Two angles are complementary if and only if the sum of their degree measures is 90.

compound inequality (125)   A sentence which is the conjunction of two inequalities.

conditional (16)   A compound statement formed by joining two statements with the words *if . . ., then.*

congruent angles (191)   Angles having the same measure.

congruent circles (257)   Two circles that have congruent radii.

congruent figures (191)   Figures which have the same shape and the same size.

congruent polygons (191)   Two polygons with the sides and angles of one congruent to the corresponding sides and angles of the other.

congruent segments (191)   Line segments having the same measure.

conjunct (8)   Each of the individual statements contained in a conjunction.

conjunction (8)   A compound statement formed by joining two statements with the word *and*.

consecutive integers (96)   Integers in counting order.

consequent (16)   The statement which follows the word *then* in a conditional.

construction (198)   The process of producing a figure that will satisfy certain given conditions, using only a straightedge and compass.

contrapositive (25)   A new conditional formed by negating both the antecedent and consequent of a conditional and switching their order.

converse (24)   A new conditional formed by switching the antecedent and consequent in a conditional.

coordinate (43)   The number that corresponds to a point on the number line.

coordinate plane (416)   A plane which has been separated into four quadrants by two perpendicular number lines.

cumulative frequency (406)   The total of all frequencies up to a certain point.

cumulative frequency histogram (400)   A histogram whose bars show the cumulative frequency at certain intervals.

cumulative frequency polygon (401)   The line segments connecting the bars in a cumulative frequency histogram.

cumulative relative frequency (400)   The cumulative frequency divided by the total number of values.

data (384)   Pieces of information used in statistics.

degree (145)   A unit of angle measure. A degree is $\frac{1}{360}$ of a complete rotation of a ray.

degree of a monomial (282)   The sum of the exponents of the variables of the monomial.

degree of a polynomial (289)   The same as the term of the polynomial which has the greatest degree.

density property (206)   An infinite number of rational numbers exist between any two rational numbers.

diameter of a circle (257)   A chord that contains the center of the circle.

dilation (472)   A transformation in which the size of a plane figure is altered based on a center and a scale factor.

disjunct (12)   Each of the individual statements contained in a disjunction.

disjunction (12)   A compound statement formed by joining two statements with the word *or*.

domain (2)   The set of all possible replacements for the placeholder or variable in an open sentence.

dot-frequency graph (391)   A statistical graph which shows how different quantities compare.

drawing (198)   The process of producing a figure using measuring devices as well as straightedge and compass.

edge of a polyhedron (264)   See polyhedron.

empirical probability (404)   A probability determined by observation or experimentation.

empty set (3)   The set having no members.

equation (78)   A statement of equality between two mathematical expressions.

equilateral triangle (159)   A triangle whose sides have the same measure.

equivalence (29)   A set of statements which are always either both true or both false.

equivalent equations (80)   Two or more equations which have the same solution set.

equivalent ratios (176) Two or more ratios which name the same number.

event (361) The set containing all the successes of a probability experiment.

exponent (39) A numeral used to tell how many times a number or variable is used as a factor.

extremes (179) In the proportion $\frac{a}{b} = \frac{c}{d}$, the terms a and d are the extremes.

face of a polyhedron (264) See polyhedron.

factor (320) The numbers or variables multiplied in a multiplication expression.

factorial (374) The expression $n!$ (n factorial) is the product of all the numbers from 1 to n for any positive integer n.

FOIL rule for multiplying binomials (298) The product of two binomials is the sum of the products of
F the first terms,
O the outer terms,
I the inner terms, and
L the last terms.

formula (106) An equation that states a rule for the relationship between certain quantities.

frequency (396) The number of values in a certain interval.

function (452) A relation in which each member of the domain is paired with exactly one member of the range.

geometric solid (264) An enclosed region of space.

gram (272) See metric system.

graph of set (43) To graph a set of numbers means to locate the points named by those numbers on the number line.

greatest common factor (321) The gcf of two or more monomials is the common factor with the greatest numerical factor and the greatest degree.

grouping symbol (41) Parentheses, brackets, and fraction bars may be used to state the order of operations.

half-plane (442) Part of a plane on one side of a straight line drawn in the plane.

histogram (396) A vertical bar graph whose bars are next to each other.

hypotenuse (230) The side opposite the right angle in a right triangle.

identity (98) An equation which is true for every replacement of the variable.

identity transformation (464) A transformation which maps every point in the plane to itself.

image (452) If A is mapped to A', then A' is called the image of A. A is the preimage of A'.

inequality (113) A mathematical sentence that contains a symbol such as $<$, $>$, $\leq$, $\geq$, or $\neq$.

integer (43) The set of numbers $\{\ldots, -3, -2, -1, 0, 1, 2, 3, \ldots\}$.

intersection (141) The intersection of two sets is the set containing all the elements which both sets have in common.

inverse (24) A new conditional formed by negating both the antecedent and consequent of a conditional.

irrational number (213) A number which cannot be represented by a repeating decimal.

isometry (472) A transformation for which a geometric figure and its image are congruent.

isosceles triangle (159) A triangle with at least two sides that have the same measure.

least common denominator (LCD) (56) The least positive expression that is a multiple of the denominator of two or more fractions.

leg of a right triangle (230)   Any side of a right triangle *not* opposite the right angle.

line (141)   One of the basic undefined terms of geometry. A line extends indefinitely and has no thickness or width. It is represented by a double arrow and named either by a lower case letter or two points of the line.

linear equation (419)   An equation which can be written in the form $Ax + By = C$, where A, B, and C are any numbers and A and B are not both 0.

line graph (391)   A statistical graph which shows trends or changes.

line reflection (456)   A reflection in line ℓ is a transformation that maps each point P onto a point P' in one of the following ways.
1. If P is on line ℓ, then the image of P is P.
2. If P is not on line ℓ, then ℓ is the perpendicular bisector of $\overline{PP'}$.

line segment (141)   Part of a line that consists of two endpoints and all the points between them.

line symmetry (457)   A plane figure is said to have line symmetry if a line can be drawn through it so that the figure on one side is a reflection of the figure on the opposite side.

liter (272)   See metric system.

mapping (452)   A relation in which each member of the domain is paired with exactly one member of the range.

mean (387)   The mean of a set of n numbers is the sum of the numbers divided by n.

means (179)   In the proportion $\dfrac{a}{b} = \dfrac{c}{d}$, the terms b and c are the means.

median (388)   The number which lies in the middle when a set of numbers is arranged in order. If there are two middle values, the median is the mean of these values.

meter (242)   See metric system.

metric system (242)   A decimal system of weights and measures based on three basic units; the meter for length, liter for volume, and the gram for mass (weight).

mode (387)   The number(s) which occurs most often in a set of numbers.

monomial (292)   An expression that is a constant, a variable, or a product.

multiplicative inverse (62)   For every rational number $\dfrac{a}{b}$, if a and b are not 0, the multiplicative inverse is $\dfrac{b}{a}$ if $\dfrac{a}{b} \cdot \dfrac{b}{a} = 1$.

negation (5)   If a statement is represented by p, then *not p* is the negation of that statement.

obtuse angle (146)   An angle whose degree measure is between 90 and 180.

obtuse triangle (159)   A triangle with one obtuse angle.

one-to-one mapping (454)   A mapping in which each member of the range has exactly one preimage.

open sentence (2)   A sentence containing a placeholder(s) to be replaced in order to determine if the sentence is true or false.

opposite rays (144)   Two rays that head in opposite directions from a common endpoint.

ordered pair (416)   A pair of numbers in which the order is specified. An ordered pair is used to locate points in a plane.

origin (416)   The point of intersection of the two axes of the coordinate plane.

outcome (350)   A possible result of a probability experiment.

outcome set (361)   A set of all the possible outcomes in a probability experiment.

outcome space (365)   The outcomes of a probability experiment.

parallel lines (154)   Two lines that lie in the same plane and do *not* intersect.

parallelogram (160)   A quadrilateral with two pairs of parallel sides.

percent (184)   The ratio of a number to 100.

percentage (184)   A number which is compared to another number (base) in the percent proportion.

percentile (402)   The *n*th percentile of a set of data is the score at or below which *n*% of the scores lie.

perfect square (217)   A rational number whose square root is a rational number.

permutation (375)   An arrangement of a given number of objects.

perpendicular lines (146)   Two lines which intersect to form right angles are perpendicular to each other.

picture graph (392)   A statistical graph used to show information in an appealing way.

plane (141)   One of the basic undefined terms of geometry. A plane extends indefinitely in all directions and has no thickness. It is represented by a four-sided figure and by a capital script letter.

point (140)   A basic undefined term of geometry. A point has no size, is represented by a dot, and is named by a capital letter.

polygon (158)   A simple closed curve composed entirely of line segments.

polyhedron (264)   A geometric solid formed by parts of planes which are polygons. The polygons and their interiors are called faces. The intersection of pairs of faces are line segments called edges. Three or more edges intersect at a point called the vertex.

polynomial (289)   Expressions containing the sum or difference of two or more monomials.

power (39)   A number written with exponents.

preimage (452)   See image.

prime factorization (320)   The product of the prime factors of a number.

prime factors (320)   Factors that are prime numbers.

prime number (320)   An integer, greater than 1, whose only positive factors are 1 and itself.

prism (265)   A polyhedron that has two parallel faces called bases, and three or more other faces called lateral faces that are parallelograms.

proportion (179)   An equation of the form $\frac{a}{b} = \frac{c}{d}$ which states that two ratios are equivalent.

pyramid (266)   A polyhedron formed by joining the vertices of a polygon to a point not in the plane of the polygon.

quadrant (416)   One of the four regions into which two perpendicular number lines separate a plane.

quadratic polynomial (328)   A polynomial in one variable of degree two.

radical sign (216)   The symbol $\sqrt{}$ used to denote the positive square root of a number.

radicand (216)   The expression under the radical sign.

radius of a circle (257)   Any segment whose endpoints are the center and a point of the circle.

ratio (176)   A comparison of two numbers by division.

rational number (56)   Any number that can be written in the form $\frac{a}{b}$ where *a* is any integer and *b* is any integer except 0.

ray (144)   Part of a line consisting of one endpoint and all the points on one side of the endpoint.

real numbers (213)   The set consisting of all rational numbers and all irrational numbers.

rectangle (160)   A parallelogram with four right angles.

rectangular solid (264)   A polyhedron whose faces are rectangles.

reflex angle (146)   An angle whose degree measure is greater than 180.

regular polygon (158) A polygon in which all sides have the same measure and all angles have the same measure.

relative frequency (407) The number of times a success has occurred divided by the total number of trials.

repeating decimal (207) A decimal in which a digit or a group of digits repeats.

repetend (207) The digit or group of repeating digits in a repeating decimal.

replacement set (2) See domain.

rhombus (160) A parallelogram with all sides having the same measure.

right angle (146) An angle whose degree measure is 90.

right circular cone (268) A solid figure with one circular base whose axis from its vertex is perpendicular to the base.

right circular cylinder (268) A solid figure with two circular bases whose axis is also the altitude of the figure.

right prism (265) A prism with lateral faces that are rectangles.

right triangle (159) A triangle with one right angle.

root (80) A replacement for the variable which makes an open sentence true.

rotation (467) A transformation that maps every point in the plane to its image by rotating the plane around a fixed point.

rotational symmetry (468) A plane figure is said to have rotational symmetry if it can be rotated through a certain angle and be the image of itself.

sample (384) Part of a population which is studied to determine properties of the whole population.

scalene triangle (159) A triangle with no sides that have the same measure.

similar figures (187) Figures that have the same shape but not necessarily the same size.

similar polygons (187) Two polygons whose corresponding angles have the same measure and whose corresponding sides have measures in proportion.

simple closed curve (158) A figure which can be traced without lifting the pencil from the paper and not tracing any point other than the starting point more than once.

skew lines (154) Two lines that do *not* lie in the same plane and do *not* intersect.

slope (423) The slope of a line is the ratio of the change in y to the corresponding change in x.

slope-intercept form (427) The slope-intercept form of the equation of a line is $y = mx + b$. The slope of the line is m and the y-intercept is b.

solution (38) A replacement from the domain which makes an open sentence true.

solution set (2) The set of all replacements from the domain which make an open sentence true.

sphere (268) The set of all points that are a given distance from a given point in space.

square (160) A rectangle with all sides having the same measure.

square root (216) If $x^2 = y$, then x is the square root of y.

standard form of a polynomial (290) A polynomial whose terms are arranged so that the powers of the variable are in descending order.

statement (2) Any sentence which is true or false, but not both.

straight angle (146) An angle whose degree measure is 180.

subset (361) A set whose elements are all contained in another set.

success (350) A desired outcome of a probability experiment.

supplementary (149) Two angles are supplementary if and only if the sum of their degree measures is 180.

tautology (20)   A compound statement which is true regardless of the truth values of the statements of which it is composed.

terminating decimal (208)   A specific type of repeating decimal in which only a zero repeats.

theorem (150)   A mathematical statement which can be proved.

theoretical probability (404)   A probability which is based on theory.

transformation (456)   A one-to-one mapping whose domain and range are the set of all points in the plane.

translation (463)   A transformation that maps every point in the plane to its image by moving each point the same distance in the same direction.

transversal (154)   A line that intersects two other lines in the same plane in two different points.

trapezoid (160)   A quadrilateral with exactly one pair of parallel sides.

tree diagram (368)   A diagram used to show the total number of possible outcomes in a probability experiment.

triangle (158)   A polygon with three sides.

trinomial (289)   A polynomial with three unlike terms.

truth value (2)   The truth or falsity of a statement.

undefined term (140)   A term whose meaning is accepted without formal definition. *Point, line,* and *plane* are undefined terms of geometry.

union (142)   The union of two sets is the set containing all the elements which are in either (or both) of the original sets.

variable (2)   The symbol or placeholder to be replaced in an open mathematical sentence.

vertex of a polyhedron (264)   See polyhedron.

vertical angles (150)   Two nonadjacent angles formed by two intersecting lines.

x-axis (416)   The horizontal line of the two perpendicular lines in a coordinate plane.

x-coordinate (416)   The first number in an ordered pair.

y-axis (416)   The vertical line of the two perpendicular lines in a coordinate plane.

y-coordinate (416)   The second number in an ordered pair.

y-intercept (427)   The y-intercept of a line is the y-coordinate of the point at which the graph of an equation crosses the y-axis.

▬ Selected Answers ▬

CHAPTER 1   INTRODUCTION TO LOGIC

Page 4   Lesson 1.1

Exploratory **1.** no; command **3.** no; question
5. yes; true **7.** no; question **9.** yes; false
11. yes; false **13.** yes; true **15.** yes; false
Written **1.** {Pennsylvania} **3.** {Massachusetts}
5. ∅ **7.** {12} **9.** {15} **11.** {6} **13.** {5}
15. {2} **17.** $\mathcal{W}$ **19.** {3} **21.** {4} **23.** {7, 8,
9, . . .} **25.** {7} **27.** ∅ **29.** {2}

Pages 6-7   Lesson 1.2

Exploratory **1.** $7 \neq 4 + 2$ **3.** $9 = 3 + 2$
5. The product of 8 and 11 is not an odd
number. **7.** Maine is not the smallest state in
area in the United States. **9.** Dolley Madison
was the wife of President Madison. **11.** false,
true **13.** true, false **15.** false, true **17.** false,
true **19.** false, true **Written** **1.** p **3.** $\sim q$
5. 8 is a prime number. **7.** Lincoln was not a
U.S. president. **9.** Lincoln was a U.S.
president. **13.** opposite **15.** same **17.** p
19. {2} **21.** $\mathcal{W}$ **23.** $x + 5 \neq 7$; {0, 1, 3, 4,
5, . . .} **Mixed Review** **1.** {4, 5, 6, . . .}
2. {0, 1, 2, 3, 4} **3.** {0} **4.** {9, 10, 11, . . .}
5. {22, 23, 24, . . .} **6.** {0, 1, . . . , 16,
17} **7.** ∅ **8.** $\mathcal{W}$ **9.** {3, 4, 5, . . .} **10.** {0,
1, 2, 3} **11.** {0, 1, 2, 3, 4, 5} **12.** {2, 3,
4, . . .} **13.** {6, 7, 8, . . .} **14.** {0, 1, 2, 3,
4} **15.** {0}

Pages 10-11   Lesson 1.3

Exploratory **1.** true **3.** false **5.** false
Written **1.** $p \wedge q$, true **3.** $q \wedge \sim r$, true
5. $r \wedge \sim p$, false **7.** Longfellow was a poet
and Shakespeare wrote Hamlet. true
9. Longfellow was a poet and Mozart was not a
painter. true **11.** Longfellow was not a poet
and Shakespeare did not write Hamlet. false
13. Longfellow was not a poet and Mozart was
not a painter. false **15.** Longfellow was a poet
and Shakespeare wrote Hamlet, and Mozart
was a painter. false **17.** It is false that
Shakespeare wrote Hamlet and Mozart was a
painter. true **23.** {4} **25.** {0, 1} **27.** ∅
29. {6, 7, 8, . . .} **31.** {3, 4, 5} **33.** false
35. false **37.** false **39.** true **41.** false
43. false **45.** false

Pages 14-15   Lesson 1.4

Exploratory **1.** true **3.** true **5.** false **7.** true
Written **1.** $p \vee q$, true **3.** $q \vee \sim r$, true
5. $r \vee \sim p$, false **7.** The Thames flows through
London or the Seine flows through Paris. true
9. The Seine flows through Paris or the Nile
flows through Rome. true **11.** The Thames
does not flow through London or the Seine
does not flow through Paris. false **13.** The
Thames does not flow through London or the
Nile does not flow through Rome. true
15. Either the Thames flows through London or
the Seine flows through Paris, or the Nile flows
through Rome. true **17.** It is false that the
Thames flows through London or that the Seine
flows through Paris. false **27.** {0, 1, 2, 3, 4, 5,
6} **29.** {7, 8, 9. . .} **31.** $\mathcal{W}$ **33.** $\mathcal{W}$ **35.** {0,
1, 2, 4, 5, 6, . . .} **37.** true **39.** true
41. true **43.** false **45.** true **47.** true

Pages 18-19   Lesson 1.5

Exploratory **1.** It rains today; I will stay home.
3. Joe earns an A in math; he will be on the
honor roll. **5.** Mark is out of town; Garnet will
date Paul. **7.** false **9.** false **11.** false if today
is Wednesday, true if today is not Wednesday
Written **1.** If June has 30 days, then 1983 is a
leap year. false **3.** If June has 30 days, then
Labor Day is in September. true **5.** If June has
30 days, then Labor Day is not in September.
false **7.** If June does not have 30 days, then
1983 is a leap year. true **9.** If June has 30
days and 1983 is a leap year, then Labor Day
is in September. true **11.** If June has 30 days
and 1983 is a leap year, then Labor Day is not
in September. true **13.** If June has 30 days,
then 1983 is a leap year and Labor Day is not
in September. false **15.** If 1983 is a leap year,
then June does not have 30 days or Labor Day
is in September. true
17. $M \rightarrow \sim H$ **19.** $\sim J \rightarrow F$
21. $(C \vee H) \rightarrow S$ **23.** $\sim S \rightarrow (F \rightarrow C)$
29. no conclusion **31.** no conclusion
33. true **35.** false **37.** true

Exploratory **1.** true **3.** false **5.** true **7.** true

9. true   **11.** false   **13.** false   **15.** true
Written   **1.** yes   **3.** yes   **5.** yes   **7.** yes
9. yes   **11.** yes   **13.** yes   **15.** yes   **17.** no
19. $4 + 3 = 5$ **21.** true   **23.** yes

Pages 26-27   Lesson 1.7
Exploratory   **1.** $q \rightarrow p$ **3.** $\sim t \rightarrow \sim m$
5. $\sim p \rightarrow \sim q$ **7.** $m \rightarrow p$ **9.** $\sim q \rightarrow \sim p$
11. $t \rightarrow m$ **Written**   **1.** If $3^2 = 9$, then
$2^2 = 4$.   **3.** If Mark Twain wrote *Tom Sawyer*,
then $2 + 2 = 4$.   **5.** If you collect $200, then
you pass GO.   **7.** If Louisa Alcott wrote *Little
Women*, then 2 is not a prime number.   **9.** If
$2^2 \neq 4$, then $3^2 \neq 9$.   **11.** If $2 + 2 \neq 4$, then
Mark Twain did not write *Tom Sawyer*.   **13.** If
you do not pass GO, then you do not collect
$200.   **15.** If 2 is a prime number, then Louisa
Alcott did not write *Little Women*.   **17.** If
$3^2 \neq 9$, then $2^2 \neq 4$.   **19.** If Mark Twain did
not write *Tom Sawyer*, then $2 + 2 \neq 4$.
21. If you do not collect $200, then you do
not pass GO.   **23.** If Louisa Alcott did not
write *Little Women*, then 2 is a prime number.
27. false, true, true, false   **29.** true, true,
true, true   **31.** true, true, true, true
33. no conclusion   **35.** yes   **37.** no
39. no conclusion   **41.** yes

Pages 29-30   Lesson 1.8
Exploratory   **1.** true   **3.** true   **5.** false   **7.** true
Written   **1.** New Mexico is in Central America
if and only if $3 + 4 < 8$. false   **3.** $3 + 4 < 8$
if and only if pink is a primary color. false
5. New Mexico is not in Central America if and
only if pink is not a primary color. true
7. New Mexico is in Central America or pink is
a primary color if and only if $3 + 4 < 8$. false
9. New Mexico is in Central America or
$3 + 4 < 8$ if and only if New Mexico is in
Central America or pink is a primary color.
false   **11.** $R \leftrightarrow (\sim P \wedge \sim H)$ **13.** $T \leftrightarrow B$
15. $(F \leftrightarrow W) \wedge (W \leftrightarrow A)$ **17.** no   **19.** yes
21. yes   **23.** equivalent   **25.** John does
not live in Cleveland or he lives in Ohio.
27. $p \vee [q \vee (\sim r \wedge s)]$

Page 32   Problem Solving Application
Exploratory   **1.** p is true; q is true.   **3.** r is
true; s is false.   **5.** e is true.   **7.** a is false.
9. p is true.   **Written**   **1.** no conclusion

3. no conclusion   **5.** no conclusion   **7.** b is
true.   **9.** p is true.   **11.** s is false.   **13.** Alan:
soup; Bill: salad; Cathy: sandwich
15. perpendicular

Pages 34-35   Chapter Review
1. false   **3.** true   **5.** false   **7.** $\{11\}$ **9.** $\emptyset$
11. Lincoln was a U.S. president.
13. $8 - 6 = 5$ **15.** false, true   **17.** true, false
19. Water boils at 100°C and France is not in
Europe. false   **21.** Water does not boil at
100°C and France is not in Europe. false
23. France is not in Europe or Mozart wrote
Hamlet. false   **25.** France is in Europe or
Mozart did not write Hamlet. true   **33.** $\{6, 7,$
$8, \ldots\}$ **35.** If water boils at 100°C, then
France is not in Europe. false   **37.** If water
boils at 100°C or France is not in Europe, then
Mozart wrote Hamlet. false   **43.** $R \rightarrow (W \vee H)$
45. $(P \wedge C) \rightarrow S$ **47.** yes   **49.** yes   **51.** $t \rightarrow s$,
true; $\sim s \rightarrow \sim t$, true; $\sim t \rightarrow \sim s$, false
53. $\sim t \rightarrow \sim s$, false; $s \rightarrow t$, false; $t \rightarrow s$, true
55. If I drive less, then gas is expensive. If gas
is not expensive, then I do not drive less. If I
do not drive less, then gas is not expensive.
57. Water boils at 100°C if and only if France is
not in Europe. false   **59.** Mozart did not write
Hamlet if and only if water boils at 100°C and
France is not in Europe. false

CHAPTER 2   OPERATIONS AND NUMBERS

Page 40   Lesson 2.1
Exploratory   **1.** 3^3 **3.** 5^2 **5.** y^5 **7.** $4n^2$
9. $18z^4$ **11.** $5^2 b^3$ **13.** 64   **15.** 625   **17.** 1024
19. 6   **Written**   **1.** 17   **3.** 4   **5.** 2   **7.** 14
9. 15   **11.** 12   **13.** 18   **15.** 44   **17.** 13   **19.** 0
21. 11   **23.** 23   **25.** 6   **27.** 2   **29.** 96   **31.** 80
33. 25   **35.** 121   **37.** 100   **39.** 245   **41.** 900
43. 88   **45.** 294,912   **47.** 2,560,000   **49.** $\{4\}$
51. $\{0, 1, 2, 3, 4\}$ **53.** $\{8\}$ **55.** $\{2\}$

Page 42   Lesson 2.2
Exploratory   **1.** Add 7 and 6, subtract 4, add
3.   **3.** Multiply 6 and 4, multiply 2 and 2, add
the results.   **5.** Multiply 3 and 6, add 5.
7. Divide 32 by 4, multiply the result by 2.
9. Divide 56 by 4, divide the result by 2.

11. Subtract 2 from 9, multiply by 7.
13. Multiply 4 and 4, subtract 2.   **15.** Add 4
and 2, add 8 and 3, multiply the results.
Written **1.** 13 **3.** 21 **5.** 7 **7.** 34 **9.** 1
11. 13 **13.** 160 **15.** 576 **17.** 1 **19.** 4
21. 8 **23.** 12 **25.** 36 **27.** 144 **29.** 20,736
31. 2 **33.** 21 **35.** 3 **37.** 33 **39.** 3 **41.** 4
43. 5 **45.** 61

Pages 44-45   Lesson 2.3
Exploratory **1.** 5 **3.** $^-6$ **5.** $^-10$ **7.** 9 **9.** 7
11. 3 **13.** $\{^-2, 1, 3\}$ **15.** $\{^-4, ^-3, ^-2, ^-1, 0\}$
17. $\{0, 1, 2, 3, 4\}$ **19.** $\{^-2, ^-1, 0, 1, 2\}$
Written **1.** $^-10$ **3.** $+500$ **5.** $+8$ **7.** $+5$
9. $+8$ **11.** $^-280$ **25.** true **27.** false
29. true **31.** true **33.** false **35.** false
37. true **39.** $^-1$ **41.** does not exist **Mixed**
Review **1.** false **2.** false **3.** true **4.** false
5. true **6.** 70 **7.** 142 **8.** 220 **9.** 175
10. 10 **11.** 5 **12.** 58 **13.** 178 **14.** 65
15. 5

Page 47   Lesson 2.4
Exploratory **1.** $^-2 + 5$ **3.** $4 + ^-2$
5. $3 + ^-6$ **7.** $^-5 + 2$ **9.** $5 + ^-6$
Written **1.** 13 **3.** $^-22$ **5.** $^-6$ **7.** $^-5$
9. 0 **11.** 4 **13.** $^-7$ **15.** 11 **17.** 9 **19.** 5
21. 0 **23.** 7 **25.** $^-17$ **27.** $^-25$
29. $4 + ^-7 = ^-3$ **31.** $^-8 + 14 = 6$
33. $^-11 + 7 = ^-4$ **35.** $6 + ^-4 = 2$

Page 49   Lesson 2.5
Exploratory **1.** 4 **3.** 12 **5.** 8 **7.** 14 **9.** 9
11. $+$ **13.** $-$ **15.** $+$ **17.** $+$
Written **1.** 11 **3.** 10 **5.** 8 **7.** 0 **9.** 15
11. $^-11$ **13.** 11 **15.** 2 **17.** $^-5$ **19.** 15
21. 612 **23.** 121 **25.** 2 **27.** $^-8$ **29.** $^-3$
31. 3 **33.** gain $72

Page 51   Lesson 2.6
Exploratory **1.** -7 **3.** -8 **5.** 13 **7.** b
9. 0 **11.** 4 **13.** 15 **15.** 10
Written **1.** $10 + (-2)$ **3.** $12 + (-7)$
5. $-2 + (-11)$ **7.** $-8 + 6$ **9.** 4 **11.** 16
13. -3 **15.** 13 **17.** 164 **19.** 6 **21.** 0
23. -9 **25.** 22 **27.** -10 **29.** sometimes
31. sometimes

Pages 54-55   Lesson 2.7
Exploratory **1.** $-$ **3.** $+$ **5.** $-$ **7.** $-$ **9.** $+$
11. $-$ **13.** $-$ **15.** $+$ **17.** $3 \times 3 = 9$

19. $-4 \times -3 = 12$ **Written** **1.** -72
3. -48 **5.** -96 **7.** 85 **9.** -154 **11.** 66
13. -210 **15.** -720 **17.** 4 **19.** -6
21. -8 **23.** -13 **25.** 15 **27.** 13 **29.** 2
31. 4 **33.** -12 **35.** 24 **37.** -21 **39.** -3
41. 2 **43.** 8 **45.** 8 **47.** -14 **49.** -7
51. -9 **53.** 24 **55.** 4 **57.** sometimes
59. always **61.** sometimes **63.** sometimes
65. always **67.** always **69.** sometimes
71. sometimes **73.** always **75.** always

Pages 58-60   Lesson 2.8
Exploratory **1.** $\frac{3}{5}$ **3.** $\frac{9}{10}$ **5.** $\frac{-2}{3}, \frac{2}{-3}$
7. $-\frac{2}{11}, \frac{2}{-11}$ **9.** $<$ **11.** $>$ **Written** **1.** $\frac{3}{4}$
3. $\frac{1}{2}$ **5.** $\frac{2}{3}$ **7.** $-\frac{9}{5}$ **9.** $-\frac{3}{5}$ **11.** $1\frac{1}{6}$ **13.** $\frac{2}{9}$
15. $-\frac{7}{24}$ **17.** $1\frac{1}{6}$ **19.** $\frac{5}{12}$ **21.** $-\frac{5}{24}$ **23.** 3
25. $-2\frac{11}{12}$ **27.** -6.5 **29.** 1.3 **31.** -4.1
33. -4.85 **35.** 0.4 **37.** 91 **39.** $>$ **41.** $>$
43. $>$ **45.** $<$ **47.** $-50, 0, 8$ **49.** $-\frac{2}{3}, -\frac{1}{2},$
$-\frac{1}{8}$ **51.** $\frac{15}{1}$ **53.** $\frac{2}{10}$ **55.** $\frac{5}{1}$ **57.** sometimes
59. always **61.** always **63.** always **65.** $-7\frac{2}{3}$
67. 2.3 **69.** 11.9 **71.** $-\frac{1}{2}$ **73.** $-2\frac{1}{2}$ **75.** $\frac{23}{36}$
77. 1.973 **79.** 3 **Mixed Review** **1.** true
2. false **3.** false **4.** true **5.** true **6.** false
7. true **8.** true **9.** true **10.** -20 **11.** -192
12. 1152 **13.** -36 **14.** 16 **15.** 400
16. -2 **17.** 9 **18.** 5 **19.** $556 **20.** 15°C

Pages 63-64   Lesson 2.9
Exploratory **1.** 7 **3.** 2.36 **5.** $1\frac{1}{4}$ **7.** $-\frac{1}{9}$
9. $\frac{5}{3}$ **11.** $\frac{9}{1}$ **13.** $-\frac{34}{10}$ **15.** $-\frac{2}{1}$
17. associative property for addition
19. commutative property for multiplication
21. commutative property for addition
23. additive identity property
25. multiplicative identity property **27.** no
29. no **Written** **1.** 4, commutative property
for multiplication **3.** $\frac{1}{3}$, associative property for
addition **5.** 4, commutative property for
multiplication **7.** 3, commutative property for

multiplication   **9.** 3, associative property for multiplication   **11.** 0, additive identity property
13. 1, multiplicative identity property   **15.** 0, additive identity property   **17.** 11, additive inverse property   **19.** 0, multiplicative property of zero   **21.** 1, multiplicative identity property
23. 24   **25.** 126   **27.** 20   **29.** -5 **31.** 0
33. $-\frac{3}{4}$ **35.** 12   **37.** -4 **39.** -2

Page 66   Lesson 2.10
Exploratory   1. $4a + 4b$ **3.** $3r + 3s$
5. $7c + 7d$ **7.** $-45 + 9x$ **9.** $-16 + 8d$
11. $28 + 7y$ **13.** $8x$ **15.** $9y$ **17.** q **19.** 0
21. $18mn$ **23.** $22x^2$ **Written   1.** 152
3. -42 **5.** 400   **7.** 140   **9.** 90   **11.** $-4y$
13. $0.5a$ **15.** $32b$ **17.** $-\frac{1}{5}t$ **19.** $22y^2 + 3$
21. $15a + 2b$ **23.** $14c + 4$ **25.** $5 + 11y$
27. $2x + 6y$ **29.** $18a + 18b$

Page 68   Problem Solving Application
Exploratory   1. $4 \cdot (5 - 2) + 7 = 19$
3. $10 - (4 \cdot 2 - 1) = 3$
5. $(3 + 6 \cdot 4) \cdot 2 = 54$ **Written   1.** bottle:
$1.03; cork: $0.07   **3.** 5 stamps, 4 postcards
5. 8 ducks, 3 cows   **7.** 18   **9.** Don
11. $A = 2, B = 1, C = 9, D = 7, E = 8$

Pages 70-71   Chapter Review
1. 13   **3.** 3   **5.** 16   **7.** 2   **9.** 12   **11.** 11
13. 49   **15.** 21   **17.** 3   **21.** -5 **23.** 2   **25.** 3
27. -4 **29.** -46 **31.** -17 **33.** -9
35. -2 **37.** -44 **39.** -12 **41.** 13   **43.** $\frac{13}{14}$
45. $-\frac{1}{36}$ **47.** 20   **49.** -7.5
51. commutative property for addition
53. associative property for multiplication
55. multiplicative identity property   **57.** $36x$
59. $5x + 8$

CHAPTER 3   INTRODUCTION TO ALGEBRA

Pages 76-77   Lesson 3.1
Exploratory   1. $y + 7$ **3.** $4 + x$ **5.** $8r$
7. $n - 3$ **9.** $r + 6$ **11.** $\frac{1}{2}r$ **13.** xy **15.** $\frac{x}{6}$
17. 9   **19.** 6   **21.** 3   **23.** -1 **Written   1.** $8x$

3. $-8a$ **5.** a **7.** $-28ab$ **9.** $4c$ **11.** $19x$
13. $8a$ **15.** $-7b$ **17.** $-6y$ **19.** $-17d$
21. $6y + 4$ **23.** $6y + 5$ **25.** $-3x + 6$
27. $-4y + 5$ **29.** $\frac{3}{2}x - 5$ **31.** $-2x + 4$
33. $3y + 4$ **35.** $-y - 6$ **37.** $-a + 2$
39. $-6 + 7x$ **41.** $3x + 3$ **43.** $5b + 10c$
45. $2x + 8$ **47.** $7(5 + y)$ **49.** $6c - 4$
51. $2(n + 5)$ **53.** $x =$ Clint's age; $x + 3$
55. $x =$ Suzi's age; $x - 7$ **57.** $x =$ books Lee sold; $x + 14$ **59.** $x =$ miles Ann walked; $x + 2$ **61.** $x =$ length; $x - 5$
63. $10d + 25(d + 5)$ **65.** $100t$ **67.** $s \div 4$
69. $x + 6$ **71.** $2d + 4$ **73.** $3 + 1.5b$
75. $12 + 0.10c$ **77.** $600 + 100y$

Page 79   Lesson 3.2
Exploratory   1. expression   **3.** equation
5. neither   **7.** expression   **9.** neither
11. neither   **Written   1.** $x + 5 = 15$
3. $5n = 35$ **5.** $2y - 6 = -14$
7. $4y - 3 = 5$ **9.** $x - 7 = 123$
11. $n - 4 = 70$ **13.** $8n = 88$
15. $n + 15 = 36$ **17.** $\frac{n}{6} = 18$
19. $2n - 10 = 26$ **21.** $n + 26 = 3n$
23. $2w + 4w = 60$
25. $25q + 10(q - 2) = 225$
27. $23 + 0.18x = 55$

Page 82   Lesson 3.3
Exploratory   1. -9 **3.** 7   **5.** 12.   **7.** -7
9. -18 **11.** 18   **13.** yes   **15.** yes   **17.** yes
Written   1. 10   **3.** 20   **5.** 12   **7.** -5 **9.** -4
11. 12   **13.** 4   **15.** 0   **17.** 11   **19.** 6
21. -12 **23.** 3   **25.** -6 **27.** -14 **29.** 13
31. -3 **33.** -28 **35.** -1808 **37.** 0   **39.** $\frac{2}{3}$
41. $15\frac{1}{3}$ **43.** -0.8 **45.** -3.09 **47.** yes
49. no   **51.** no

Page 86   Lesson 3.4
Exploratory   1. $\frac{1}{4}$ **3.** $\frac{1}{2}$ **5.** 3   **7.** -9
Written   1. 3   **3.** 16   **5.** -15 **7.** 0   **9.** -13
11. -12 **13.** -14 **15.** $-1\frac{1}{3}$ **17.** $\frac{1}{8}$ **19.** $\frac{1}{4}$
21. $\frac{1}{3}$ **23.** -7 **25.** 24   **27.** -8 **29.** -72
31. $1\frac{1}{4}$ **33.** -10 **35.** $\frac{2}{3}$ **37.** 0.3   **39.** 40

Pages 88-89   Lesson 3.5
Exploratory   1. $n + 53 = 96$ **3.** $8n = -112$
5. $x + 100 = 300$ **7.** $8n = 96$
9. $12x = 0.96$ **11.** $x + 161 = 329$
Written   1. 43   **3.** -14 **5.** 200 km   **7.** 12
9. $0.08   **11.** 168   **13.** 172   **15.** 32
17. 56.2 seconds   **19.** 8 m   **21.** 24 lb
23. 74 m   **25.** 87, 29   **27.** 82 cm, 41 cm

Pages 91-92   Lesson 3.6
Exploratory   1. Subtract 3 from both sides.
Divide both sides by 4.   **3.** Add 8 to both
sides. Divide both sides by 2.   **5.** Add 5 to
both sides. Divide both sides by -3.
7. Subtract 5 from both sides. Multiply both
sides by 3.   **9.** Add 5 to both sides. Multiply
both sides by 4.   **11.** Multiply both sides by 4.
Add 3 to both sides.   **Written   1.** 5   **3.** 11
5. 2   **7.** 4   **9.** -6 **11.** 2   **13.** 8   **15.** 10
17. 9   **21.** -45 **23.** 12   **25.** 20   **27.** -4
29. 1   **31.** 8   **33.** -4.1 **35.** $\frac{3}{4}$ **37.** -2
39. 2   **41.** 18   **43.** -13 **45.** 23   **47.** -35
49. $x = 20$; 20 m by 43 m   **51.** $x = 5$; 5 cm,
8 cm, 10 cm, 10 cm, 8 cm

Pages 94-95   Lesson 3.7
Exploratory   1. $2x + 3 = 29$
3. $5 + 3x = -13$ **5.** $3x + 24 = 129$
7. $3w + 5, 8w + 10$ **Written   1.** 13   **3.** -6
5. $35   **7.** $84   **9.** 50   **11.** 120 yd, 65 yd
13. 135   **15.** 22 cm, 33 cm, 47 cm
17. 57, 114, 121   **19.** 36   **21.** 24

Page 97   Lesson 3.8
Exploratory   1. $x + (x + 1) = 17$
3. $x + (x + 2) = -34$
5. $x + (x + 1) + (x + 2) = 39$
7. $x + (x + 2) = 36$ **Written   1.** 8, 9
3. $-18, -16$ **5.** 12, 13, 14   **7.** 17, 19
9. 21, 22   **11.** 23, 24, 25   **13.** 22, 24
15. $-14, -12, -10$ **17.** 19, 21   **19.** 41, 43, 45

Pages 100-101   Lesson 3.9
Exploratory   1. Subtract $3x$ from both sides.
3. Subtract $5x$ from both sides.   **5.** Subtract $8m$
from both sides.   **Written   1.** 4   **3.** 15   **5.** 2
7. -360 **9.** 0   **11.** -2 **13.** -5 **15.** $-1\frac{1}{2}$
17. 2   **19.** -5 **21.** -1 **23.** -6 **25.** 7

27 5   **29.** -1 **31.** $1\frac{1}{15}$ **33.** 48   **35.** no
solution   **37.** $-10, -2$ **39.** $-8, -3$
41. -6 **43.** 9   **45.** 11, 22   **47.** 14, 16
49. 25   **51.** 5, 19   **53.** 44   **55.** $164, $123
Mixed Review   1. c   **2.** b   **3.** d   **4.** b

Page 103   Chapter Review
1. $n - 5$ **3.** $\frac{18}{x}$ **5.** $20m$ **7.** $-5y + 2$
9. $8x - 28$ **11.** $2n - 3 = 7$ **13.** 13   **15.** $\frac{13}{20}$
17. 72   **19.** 22 kg   **21.** $-\frac{13}{3}$ **23.** $280
25. $-27, -26$ **27.** 1   **29.** 8

CHAPTER 4   USING FORMULAS AND INEQUALITIES

Pages 108-109   Lesson 4.1
Exploratory   1. 77°F   **3.** 32°F   **5.** 5°C   **7.** × 3
9. × 8   **11.** × 0.5   **13.** × 4   **15.** × 25
17. × 0.25   **Written   1.** 110   **3.** 225   **5.** 32
7. 7   **9.** -5 **11.** 212°F   **13.** $3.70   **15.** × 2
17. × 15   **19.** $A = 4.50t$ **21.** $C = 0.15x$
23. $C = 20x + 17y$ **25.** $L = 3h + 6$
27. $C = x + 500y$ **29.** $C = rx + sy$

Pages 111-112   Lesson 4.2
Exploratory   1. $\frac{A}{w}$ **3.** $\frac{C}{\pi}$ **5.** $P - b - c$ **7.** 0
9. 5   **11.** -4 **13.** $-\frac{3}{2}$ **Written   1.** 10°
3. $-5°$ **5.** 100°   **7.** 20 m/s   **9.** 90 km/h
11. 1.67 m/s   **13.** 240 mi   **15.** 980 km/h
17. $5\frac{1}{3}$ h   **19.** $\frac{i}{12}$ **21.** $\frac{l}{pr}$ **23.** $2b - 4c$
25. $4c - a$ **27.** $5 - 4\ell$ **29.** $\frac{3}{1 + r}$
31. $\frac{7 - 2x}{3}$ **33.** $\frac{x - m}{T}$ **35.** $\frac{L + tb}{t}$ **37.** $\frac{xy - 5}{M}$
39. $\frac{3H + 10}{2}$ **41.** $\frac{a\ell - 7}{a}$ **43.** 0   **45.** -1
47. 0   **49.** 0   **51.** $-\frac{1}{4}$

Pages 114-115   Lesson 4.3
Exploratory   1. true   **3.** true   **5.** false   **7.** true
9. true   **11.** false   **13.** true   **15.** true
Written   1. $x < 2$ **3.** $x < 4$ **5.** $x \geq 4$
7. $x \neq -1.5$ **21.** $x + 2 > 8$
23. $2x + 3 > 21$ **25.** $2x + 2 \geq 26$

27. $\{-5, -4, -3, -2, -1, 0, 1, 2\}$
29. $\{-5, -4, -3\}$ **31.** $\{2, 3\}$
33. $\{-3, -2, -1, 0, 1, 2, 3\}$ **35.** $\{1, 2, 3\}$
37. $\{-5, -4, -3, -2, -1, 0, 1, 2, 3\}$
39. $\{-5, -4, -3, -2, -1, 0, 1\}$ **41.** $\emptyset$

Page 117   Lesson 4.4
Exploratory   1. -4 **3.** 4 **5.** 3 **7.** 2.3
9. -4 **11.** $2\frac{1}{2}$ **Written   1.** $y < 1$ **3.** $x > 5$
5. $x < 3$ **7.** $x > 3.7$ **9.** $m > -8$ **11.** $m \geq 3$
13. $y > 4$ **15.** $x \geq -2$ **17.** $x \geq 0$ **19.** $x > 4$
21. $x < -8$ **23.** $x \leq 3$

Page 120   Lesson 4.5
Exploratory   1. $-56 < -16$ **3.** $2 < 3$
5. $-20x > -60$ **7.** $x < -27$ **9.** $y < \frac{1}{6}$
Written   1. $x > -6$ **3.** $x > 10$ **5.** $x < -7$
7. $y \leq -20$ **9.** $t \leq -11$ **11.** $t \geq 15$
13. $t \leq 4$ **15.** $y \leq 3$ **17.** $x < 5$ **19.** $x < \frac{10}{3}$
21. $x \leq -5$ **23.** $x < 7$ **25.** $m < 15$
27. $y > 3$ **29.** $a \geq -\frac{3}{2}$ **31.** $y < 2$
33. $n \geq -5$ **35.** $y > 2$ **37.** $m > \frac{4}{9}$
39. $y < -1$ **41.** always   **43.** sometimes
45. sometimes   **47.** sometimes   **49.** always
51. always

Pages 123–124   Lesson 4.6
Exploratory   1. $5x + 4 \geq 19$
3. $x + 3x \leq 260$ **5.** $2x + 2(5x - 6) \geq 84$
7. $x + (x + 2) \leq 20$ **Written   1.** $x \geq 3$
3. $x \leq 65$ **5.** $x \geq 8$ **7.** $x \leq 9$ **9.** 7
11. 33 **13.** less than 32   **15.** 9 oz

Page 128   Lesson 4.7
Exploratory   1. $\{5, 6, 7, 8\}$ **3.** $\{-4, -3, -2, -1\}$ **5.** $\{-4, -3, -2, 2, 3\}$ **7.** $\{-1, 0, 1, 2, 3\}$ **9.** $-3 \leq x \leq 2$
11. $(x < -1) \vee (x \geq 2)$ **13.** $1 \leq x < 4$
Written   13. $-2 < y < 7$ **15.** $-1 < x < 5$
17. $x \leq -5$ **19.** $(y < 1) \vee (y > 6)$
21. $3 < x < 6$ **23.** $2 < y \leq 4$
25. $-3 < x < 3$

Pages 130-131   Lesson 4.8
Exploratory   1. $\frac{5}{16}$ **3.** $\frac{5}{2}$ **5.** $10x$ **7.** $4x + 2$
9. $12x + 25$ **11.** 6 **13.** 8 **15.** 100 **17.** 225

19. 100 **Written   1.** 15 **3.** $x > 20$ **5.** 32
7. 3 **9.** -5 **11.** 25 **13.** 10 **15.** $-\frac{5}{11}$
17. 6 **19.** $x < 3$ **21.** $x < -2$ **23.** $2\frac{1}{2}$
25. -22 **27.** $\frac{2}{9}$ **29.** $x > 5\frac{5}{7}$ **31.** 6
33. -7.2 **35.** -1.25 **37.** 0.375 **39.** 5
Mixed Review   1. -1.7 **2.** 26 **3.** -1.56
4. -11.6 **5.** 0.75 or $\frac{3}{4}$ **6.** -3 **7.** false
8. false   **9.** true   **10.** false

Pages 133-134   Lesson 4.9
Exploratory   1. 240 km, 80x km, $80(x + 1)$
km   **3.** 50 km/h, 60 km/h, $\frac{600}{x}$ km/h   **5.** $6\frac{2}{3}$ h,
$5\frac{1}{3}$ h   **7.** $40(3.25) + 25(3.80)$ **Written   1.** 3
3. $11\frac{1}{2}$ **5.** 40 lb   **7.** 12 **9.** when less than 38
checks/month are used   **11.** 271 or less
13. more than 325

Pages 136-137   Chapter Review
1. 13 **3.** $\times$ 4   **5.** $a - 2b$ **7.** $\frac{c - b}{a}$
13. $y \geq 1$ **15.** $m > -7$ **17.** $x < 9$
19. $y \leq -4$ **21.** $x < 7$ **23.** 8
31. $2 < y < 6$ **33.** -30 **35.** $x > 0.5$
37. 8 P.M.   **39.** 21 or less boxes

CHAPTER 5   ASPECTS OF GEOMETRY

Pages 142-143   Lesson 5.1
Exploratory   1. point   **3.** line   **5.** line   **7.** line
9. plane   **11.** line   **Written   1.** $\overleftrightarrow{AF}$, $\overleftrightarrow{FB}$, or n
3. A **5.** A, E, or D **7.** $\overleftrightarrow{AB}$, $\overleftrightarrow{AF}$, $\overleftrightarrow{FB}$ **9.** true
11. true   **13.** false   **19.** $\overline{EP}$ **21.** $\overrightarrow{BK}$ **23.** $\overleftrightarrow{BK}$
25. E **27.** E **29.** $\emptyset$ **31.** $\{1, 2, 3, 4, 5, 6, 8\}$
33. $\{1, 2, 3, 4, 5, 6, 8, \text{Akron}\}$ **35.** S or R
37. R **39.** P and Q

Pages 147-148   Lesson 5.2
Exploratory   1. yes   **3.** yes   **5.** no   **7.** yes
9. no   **11.** yes   **13.** 170 **15.** 120 **17.** 75
19. 25 **21.** 60 **23.** 105 **Written   1.** acute
3. straight   **5.** right   **7.** obtuse   **9.** $30°$
11. $90°$ **15.** yes   **17.** $\overrightarrow{QS}$ and $\overrightarrow{QT}$

19. $\angle PQV$, $\angle SQV$, $\angle 3$, $\angle 4$ **21.** V **23.** 18°
25. 160°   **27.** 60°   **29.** $\frac{31}{40}$ **31.** $\frac{1}{9}$ **37.** false
39. false

Pages 151-153   Lesson 5.3
Exploratory   1. 60   **3.** 3   **5.** $85\frac{1}{2}$ **7.** 83.95

9. 10   **11.** 90   **13.** $152\frac{1}{3}$ **15.** 18.7

Written   5. 60, 120   **7.** 18, 72   **9.** 130, 50
11. 60, 120   **13.** 60, 30   **15.** 62, 118
17. 60, 30   **19.** 75, 15   **21.** 25, 65   **23.** 45,
45   **25.** 65, 25   **27.** 120, 60   **29.** 150, 30
31. 70, 110   **33.** 90, 90   **35.** 132.5, 47.5
37. 40°   **39.** 40°   **41.** 136°   **43.** 40°   **45.** 30°
47. 17°   **49.** 39, 51   **51.** 36, 144
53. $73 - x$, $163 - x$ **Mixed Review   1.** 7
2. -1 **3.** $a \le 3$ **4.** $b < 7$ **5.** 8   **6.** $\frac{38}{17}$

7. $-2 < p \le 4$ **8.** -2 **9.** $c < 4$ or $c > 6$
10. $d < -3$ **11.** ray   **12.** line   **13.** ray
14. line segment   **15.** ray   **16.** line segment
17. 20°   **18.** 19 m × 41 m

Pages 156-157   Lesson 5.4
Exploratory   1. parallel   **3.** skew or parallel
5. intersecting or parallel   **7.** parallel
9. intersecting   **11.** $\angle 1$ and $\angle 5$, $\angle 2$ and $\angle 6$,
$\angle 3$ and $\angle 7$, $\angle 4$ and $\angle 8$, $\angle 9$ and $\angle 13$, $\angle 10$
and $\angle 14$, $\angle 11$ and $\angle 15$, $\angle 12$ and $\angle 16$
13. $\angle 1$ and $\angle 9$, $\angle 2$ and $\angle 10$, $\angle 3$ and $\angle 11$,
$\angle 4$ and $\angle 12$, $\angle 5$ and $\angle 13$, $\angle 6$ and $\angle 14$, $\angle 7$
and $\angle 15$, $\angle 8$ and $\angle 16$ **Written   5.** $m\angle 1 =$
$m\angle 3 = m\angle 4 = m\angle 7 = 127$; $m\angle 2 = m\angle 5$
$= m\angle 6 = 53$ **7.** $m\angle 1 = m\angle 3 = m\angle 4 =$
$m\angle 7 = 105$; $m\angle 2 = m\angle 5 = m\angle 6 = 75$
9. $x = 20$; 70 or 110   **11.** $x = 10.5$; 108 or
72   **13.** $x = 40$; 60 or 120   **15.** $x = 110$; 125
or 55   **17.** sometimes   **19.** sometimes
21. always   **23.** never

Pages 160-161   Lesson 5.5
Exploratory   1. yes   **3.** no   **5.** no   **7.** yes
9. yes   **11.** no   **13.** no   **15.** yes
Written   1. pentagon, not regular   **3.** triangle,
regular   **5.** triangle, not regular   **7.** octagon,
not regular   **9.** equilateral, acute
11. isosceles, acute   **13.** scalene, obtuse
15. scalene, right   **17.** sometimes   **19.** never
21. never   **23.** always   **25.** sometimes

27. always   **29.** parallelogram, rectangle,
rhombus, square   **31.** rhombus, square

Pages 163-164   Lesson 5.6
Exploratory   1. 85   **3.** 25   **5.** 85   **7.** 53
Written   1. 90, 50   **3.** 90, 10   **5.** 60, 40, 80
7. 30, 35   **9.** 76, 38, 66   **11.** 60, 50, 70
13. 30   **15.** 26, 29, 125   **17.** never   **Mixed**
Review   1. $q = \frac{p - b}{m}$ **2.** $X = \frac{Y - y}{s} + x$

3. $g = \frac{-ta}{(1 - t)}$ **4.** 5   **5.** $-1 \le y < 15$

6. $m\angle 1 = m\angle 2 = 71$
7. $m\angle 1 = m\angle 2 = 81$ **8.** $m\angle 1 = m\angle 2 = 53$
9. $m\angle 1 = 61$, $m\angle 2 = 29$ **10.** $m\angle 1 = 16$,
$m\angle 2 = 164$ **11.** $14 < x < 32$

Pages 166-167   Lesson 5.7
Exploratory   1. 50   **3.** 87   **5.** 75, 75
Written   1. 60   **3.** 100, 100   **5.** $x = 90$; 110,
100, 70, 80   **7.** $x = 100$; 110, 135, 35
9. $x = 45$; 135, 45, 135, 45   **11.** $x = 80$;
80, 130, 90, 60   **13.** 75, 105, 75, 105

Pages 172-173   Chapter Review
1. $\overleftrightarrow{BC}$, $\overleftrightarrow{AC}$, m **3.** E **5.** E **7.** $\overline{AC}$ **9.** obtuse
11. acute   **13.** reflex   **15.** acute   **17.** $\overrightarrow{CA}$, $\overrightarrow{CE}$
19. D **21.** 50, 40   **23.** 65, 115   **25.** 15, 15
27. 35° or 145°   **29.** 68° or 112°   **31.** $x = 15$;
45, 135   **33.** quadrilateral, not regular
35. quadrilateral, regular
37. $m\angle D = 40$, $m\angle E = 50$, $m\angle F = 90$
39. 85   **41.** $x = 50$; 120, 40, 120, 80

CHAPTER 6   GEOMETRIC RELATIONSHIPS

Pages 177-178   Lesson 6.1
Exploratory   1. $\frac{3}{5}$, 3 : 5, $3 \div 5$, or 0.6   **3.** $\frac{8}{5}$,

8 : 5, $8 \div 5$, or 1.6   **5.** $\frac{4}{1}$, 4 : 1, $4 \div 1$, or 4

7. $\frac{13}{100}$, $13 \div 100$, or 0.13   **9.** $\frac{40}{25}$, 40 : 25,

$40 \div 25$, or 1.6   **11.** $\frac{100}{100}$, 100 : 100, 100 ÷

100, or 1   **13.** $\frac{100}{100}$ **15.** $\frac{20}{20}$ **Written   1.** 2

3. 600   **5.** 11.1   **7.** $\frac{5}{3}$ **9.** $\frac{3}{2}$ **11.** $\frac{2}{3}$ **13.** $\frac{3}{8}$

15. $\frac{7}{9}$ **17.** $\frac{1}{50}$ **19.** $\frac{10}{7}$ **21.** $\frac{1}{5}$ **23.** $\frac{2}{3}$ **25.** $\frac{2}{3}$

27. $\frac{3}{5}$ **29.** $\frac{3}{100}$ **31.** $\frac{1}{100}$ **Mixed Review**
1. $x \le \frac{-13}{2}$ **2.** $y > 6$ **3.** $1 < x < 4$ **4.** 53,
53 **5.** 41, 49 **6.** 108.5, 71.5 **7.** 49, 34, 97
8. 65, 7, 103, 85

Page 181   Lesson 6.2
Exploratory 1. 12, 3x **3.** 12, 8y **5.** 8k, 60
7. 75, 25x **Written 1.** 6 **3.** 14 **5.** 60
7. $10\frac{1}{2}$ **9.** $3\frac{1}{4}$ **11.** 5 **13.** 4 **15.** 0.01 **17.** 8
19. 6, 8 **21.** 70 for, 5 against **23.** 30, 150

Pages 185-186   Lesson 6.3
Exploratory 1. 0.075 **3.** 36.5% **5.** 0.1625
7. 173% **9.** 1.25 **Written 1.** 9 **3.** $33\frac{1}{3}$
5. 80 **7.** 6 **9.** 3 **11.** 200 **13.** 40 **15.** 305
17. $60 **19.** 540 seats **21.** $33\frac{1}{3}$% **23.** $26
25. 1872 **Mixed Review 1.** 3 **2.** −4 **3.** 2
4. $45\frac{5}{6}$ **5.** 26 **6.** 19 **7.** true **8.** true
9. false **10.** 132°, 48°

Page 189   Lesson 6.4
Exploratory 3. proportional **Written 1.** ∠A
and ∠K, ∠B and ∠M, ∠C and ∠L, $\overline{AB}$ and
$\overline{KM}$, $\overline{AC}$ and $\overline{KL}$, $\overline{BC}$ and $\overline{ML}$ **3.** ∠R and ∠A,
∠S and ∠B, ∠T and ∠C, $\overline{RS}$ and $\overline{AB}$, $\overline{ST}$ and
$\overline{BC}$, $\overline{RT}$ and $\overline{AC}$ **5.** ∠M and ∠Q, ∠A and ∠S,
∠T and ∠D, ∠H and ∠F, $\overline{MA}$ and $\overline{QS}$, $\overline{AT}$ and
$\overline{SD}$, $\overline{TH}$ and $\overline{DF}$, $\overline{HM}$ and $\overline{FQ}$ **7.** ∠M and ∠B,
∠T and ∠K, ∠A and ∠C, $\overline{MT}$ and $\overline{BK}$, $\overline{TA}$ and
$\overline{KC}$, $\overline{AM}$ and $\overline{CB}$ **9.** 7.5 **11.** x = 3, y = 4.8
13. 0.2 **15.** $b = c = 6\frac{2}{3}$

Page 193   Lesson 6.5
Exploratory 1. same measure
3. corresponding sides and angles congruent
Written 1. ∠A and ∠K, ∠B and ∠L, ∠C and
∠M, $\overline{AB}$ and $\overline{KL}$, $\overline{BC}$ and $\overline{LM}$, $\overline{AC}$ and $\overline{KM}$
3. ∠A and ∠R, ∠B and ∠T, ∠C and ∠Q, $\overline{AB}$
and $\overline{RT}$, $\overline{BC}$ and $\overline{TQ}$, $\overline{AC}$ and $\overline{RQ}$ **5.** $\overline{AC} \cong \overline{KM}$
7. $\overline{RT} \cong \overline{VZ}$, $\overline{RS} \cong \overline{VW}$, ∠R ≅ ∠V; $\overline{TS} \cong \overline{ZW}$,
$\overline{RT} \cong \overline{VZ}$, ∠T ≅ ∠Z; or $\overline{TS} \cong \overline{ZW}$, $\overline{RS} \cong \overline{VW}$,
∠S ≅ ∠W **9.** m∠R = m∠S = 70
11. m∠Q = 40, m∠R = 70 **13.** m∠Q = 36,
m∠R = m∠S = 72 **15.** m∠Q = 70,

m∠R = m∠S = 55 **17.** m∠Q = 140,
m∠R = m∠S = 20

Pages 196-197   Lesson 6.6
Exploratory 1. ∠A ≅ ∠K by ASA or $\overline{BC} \cong \overline{LM}$
by SAS **3.** ∠C ≅ ∠M by ASA or $\overline{AB} \cong \overline{KL}$ by
SAS **Written 1.** sometimes **3.** always
5. always **7.** sometimes **9.** ∠A and ∠K, ∠B
and ∠L, ∠C and ∠M, $\overline{AB}$ and $\overline{KL}$, $\overline{BC}$ and $\overline{LM}$,
$\overline{AC}$ and $\overline{KM}$ **11.** 13 **13. 1.** Given
2. Definition of congruent segments
3. Definition of vertical angles **4.** Theorem 5-
1 **5.** Definition of congruent angles **6.** ASA

Page 201   Problem Solving Application
Written 1. 21 feet 3 inches **3.** 16 pieces
5. 15 games **7.** 3 circuits, 4 circuits

Page 203   Chapter Review
1. $\frac{7}{10}$, 0.7 **3.** $\frac{3}{16}$, 0.1875 **5.** 77 **7.** 17
9. 35, 14 **11.** 0.405 **13.** ∠K **15.** $\overline{DF}$
17. proportional **19.** sometimes
21. sometimes **23.** sometimes

CHAPTER 7   THE REAL NUMBERS

Pages 208-209   Lesson 7.1
Exploratory 1. $\frac{3}{100}$ **3.** $-\frac{6666}{10,000}$ **5.** $-\frac{2}{3}$
7. $\frac{15}{16}$ **9.** $\frac{1}{4}, \frac{1}{3}, \frac{1}{2}$ **11.** 0.0013, 0.01, 0.10
13. $-\frac{1}{5}, -\frac{1}{6}, -\frac{2}{15}$ **15.** 0.5 **17.** 0.75
19. 1.2 **Written 1.** 60.5 **3.** 4.25 **5.** $\frac{9}{40}$
7. $-\frac{9}{40}$ **9.** 140.015 **11.** $0.\overline{6}$ **13.** $0.\overline{4}$
15. $0.\overline{5}$ **17.** $0.41\overline{6}$ **19.** $0.\overline{63}$ **21.** 0.32
23. −2.2 **25.** 0.205 **27.** 3.2 **Mixed**
Review 1. c **2.** b **3.** a **4.** b **5.** d

Page 212   Lesson 7.2
Exploratory 1. $\frac{4}{5}$ **3.** $\frac{63}{100}$ **5.** $\frac{88}{25}$ **7.** $\frac{111}{200}$ **9.** $\frac{4}{9}$
11. $\frac{37}{99}$ **13.** $\frac{2}{3}$ **15.** $\frac{7}{3}$ **Written 1.** $\frac{12}{5}$ **3.** $\frac{11}{4}$
5. $\frac{93}{37}$ **7.** $\frac{5}{33}$ **9.** $\frac{-22}{45}$ **11.** $\frac{-287}{15}$ **13.** $\frac{256413}{49995}$
15. $\frac{40123}{4950}$ **17.** $\frac{12}{5}$ **19.** $\frac{21}{5}$ **21.** $\frac{527}{100}$ **23.** $\frac{11}{4}$

25. $6.\overline{9}$ **27.** $3.3\overline{9}$ **29.** $10.74\overline{9}$

Page 215   Lesson 7.3
Exploratory   1. repeating **3.** repeating **5.** no
7. yes **Written   1.** rational **3.** irrational
5. rational **7.** irrational **9.** true **11.** true
13. false **15.** true **17.** $0.\overline{6}$ **19.** 0 **21.** $1.\overline{89}$
23. 3.565665666 . . . **25.** $0.\overline{1}$
27. 1.7357355 . . . **29.** $7.\overline{5}$ **31.** $x < 2$
33. $x \geq 7$ **35.** $x > 6$ **37.** -3 **39.** $y < -6$
41. $-\dfrac{28}{9}$ **43.** $t \leq \dfrac{22}{19}$ **45.** $p \geq \dfrac{5}{4}$

Pages 217-218   Lesson 7.4
Exploratory   1. 6 **3.** $\dfrac{2}{3}$ **5.** 0.1 **7.** $\dfrac{10}{11}$ **9.** 5
11. 12 **13.** -8 **15.** -7 **Written   1.** 20
3. 0 **5.** $\dfrac{2}{5}$ **7.** $\dfrac{7}{15}$ **9.** $-\dfrac{28}{5}$ **11.** $\dfrac{2}{3}$ **13.** 4, 9,
16, 25, 36, 49, 64, 81 **15.** $x \geq 0$ **17.** yes;
rational **19.** no **21.** no **23.** yes; irrational
25. yes; rational **27.** yes; rational **29.** $\{-6,$
$6\}$ **31.** $\{-6, 6\}$ **33.** $\{-15, 15\}$ **35.** 3 **37.** 0
39. 17 **41.** 38 **43.** -3 **45.** -15.6 **47.** $-\dfrac{4}{5}$
49. $\pm\dfrac{15}{16}$ **51.** ± 3.1 **53.** $-|x|$

Pages 222-223   Lesson 7.5
Exploratory   1. 1, 2 **3.** 3, 4 **5.** 6, 7 **7.** 8, 9
9. 11, 12 **11.** 1, 2 **13.** 4, 5 **15.** 0, 1
Written   1. 2, 3 **3.** $-5, -4$ **5.** 12, 13
7. $-10, -9$ **9.** 6.9 **11.** 4.4 **13.** 9.3
15. 10.1 **17.** 0.5 **19.** 24.2 **21.** 9.1
23. 15.8 **25.** 7.5 **27.** 5 **29.** 4.6 **31.** 4.6
33. 5.5 **35.** 5 **37.** 7.9 cm **39.** 28 cm
41. 3.9 cm² **43.** 29.2 cm² **Mixed**
Review   1. $-6\dfrac{1}{3}$ **2.** $\dfrac{7}{9}$ **3.** ± 12 **4.** ± 7
5. $a < -\dfrac{3}{8}$ **6.** $b \leq \dfrac{5}{7}$ **7.** 24.5 **8.** 12.8
9. 33.5 **10.** -2

Pages 225–226   Lesson 7.6
Exploratory   1. $2\sqrt{2}$ **3.** $3\sqrt{2}$ **5.** $2\sqrt{10}$
7. $3\sqrt{5}$ **Written   1.** $2\sqrt{7}$ **3.** $3\sqrt{6}$ **5.** $3\sqrt{7}$
7. $3\sqrt{11}$ **9.** $11\sqrt{2}$ **11.** $15\sqrt{2}$ **13.** $40\sqrt{2}$
15. $2\sqrt{5}$ **17.** $\sqrt{2}$ **19.** $\sqrt{3}$ **21.** $x\sqrt{x}$
23. $3y^2$ **25.** $2\sqrt{3}$ **27.** $18\sqrt{2}$ **29.** 60 **31.** 12
33. $30\sqrt{6}$ **35.** 120 cm² **37.** 24 cm²
39. $72\sqrt{3}$ cm² **41.** $\{-6\sqrt{2}, 6\sqrt{2}\}$

43. $\{-4\sqrt{3}, 4\sqrt{3}\}$ **45.** $\{4\sqrt{3}, -4\sqrt{3}\}$
47. $\{-6\sqrt{3}, 6\sqrt{3}\}$ **49.** ± 2.8 **51.** ± 4.9
53. ± 5.5

Pages 228-229   Lesson 7.7
Exploratory   1. $6\sqrt{3}$ **3.** $7\sqrt{6}$ **5.** $10\sqrt{7}$
7. 0 **9.** $9\sqrt{3}$ **Written   1.** $5\sqrt{5}$ **3.** $3\sqrt{3}$
5. $14\sqrt{6}$ **7.** $6\sqrt{10}$ **9.** $9\sqrt{x}$ **11.** $-2\sqrt{m}$
13. $8\sqrt{3}$ **15.** $4\sqrt{3}$ **17.** $\sqrt{2}$ **19.** 0
21. $14\sqrt{5}$ **23.** $6\sqrt{3} + 15\sqrt{7}$ **25.** $7\sqrt{2}$
27. $12\sqrt{3}$ **29.** $-5\sqrt{6x}$ **31.** $6 - 3\sqrt{2}$
33. $5 + 5\sqrt{2}$ **35.** $2\sqrt{10}$ **37.** $96\sqrt{2} + 624$
39. $10\sqrt{5} - 10\sqrt{2}$ **41.** $9\sqrt{21} - 18\sqrt{3}$
43. 9.7 cm **45.** 13.7 cm **47.** 6.3 cm²

Page 232   Lesson 7.8
Exploratory   1. 5 **3.** 17 **5.** 16 **7.** 8 **9.** 12
Written   1. 5 **3.** 8 **5.** 8 **7.** $\sqrt{3}$ **9.** 13
11. 4.9 **13.** 3.5 **15.** 8.6 **17.** 8.6 **19.** 11.2
21. 17' **23.** 25 yd **25.** no **27.** yes **29.** no
31. 127.3 ft **33.** 30.0 ft

Pages 234-235   Lesson 7.9
Exploratory   1. $\sqrt{6}$ **3.** 3 **5.** 3 **7.** $\dfrac{\sqrt{5}}{5}$
Written   1. 2 **3.** $\sqrt{7}$ **5.** $\dfrac{\sqrt{2}}{2}$ **7.** $\dfrac{\sqrt{2}}{3}$
9. $\dfrac{3\sqrt{5}}{5}$ **11.** $\sqrt{5}$ **13.** $\dfrac{2\sqrt{5}}{5}$ **15.** $\dfrac{3\sqrt{10}}{10}$
17. $\sqrt{3}$ **19.** $\dfrac{3}{2}$ **21.** $\dfrac{\sqrt{3}}{2}$ **23.** $\dfrac{2\sqrt{14}}{3}$
25. $\dfrac{b\sqrt{3}}{2}$ **27.** $\dfrac{2\sqrt{3} + 3}{3}$ **29.** $2 + \sqrt{3}$
31. -1 **33.** $\dfrac{m^2\sqrt{7}}{7}$ **35.** $\dfrac{3\sqrt{6r}}{r}$ **37.** $\sqrt{105}$
39. $\dfrac{\sqrt{6b}}{6}$ **41.** $\dfrac{\sqrt{10}}{5}$ **43.** $\dfrac{\sqrt{6}}{2}$ **45.** $2\sqrt{15}$
47. $\dfrac{4\sqrt{14}}{7}$ **49.** $\dfrac{3\sqrt{10} + 6\sqrt{2}}{4}$
51. $\dfrac{2\sqrt{14} - 5\sqrt{6}}{2}$ **53.** $5 - 2\sqrt{2}$
55. $83 - 8\sqrt{5}$ **57.** $3\dfrac{2}{3} - \dfrac{2\sqrt{6}}{3}$

Page 237   Problem Solving Application
1. 11, 13, 15 **3.** 10,100 **5.** 45 handshakes
7. 15 rectangles **9.** 55

Page 239   Chapter Review
1. $0.58\overline{3}$ **3.** 0.335 **5.** 3.4 **7.** $\dfrac{109}{50}$ **9.** $\dfrac{581}{100}$

11. $x \le \dfrac{22}{19}$ **13.** 9 **15.** 1.386 **17.** 8.1

19. 9.4 **21.** $2\sqrt{6}$ **23.** 32 **25.** $10\sqrt{7}$

27. $6y\sqrt{2}$ **29.** 2.6 **31.** $\dfrac{5\sqrt{7}}{7}$ **33.** $\dfrac{7\sqrt{15}}{9}$

35. $-\dfrac{5}{12}\sqrt{6} + 6\sqrt{2}$

CHAPTER 8   MEASUREMENT AND GEOMETRY

Pages 244-245   Lesson 8.1
Exploratory 1. m **3.** m **5.** km **7.** mm
9. cm **11.** mm **13.** cm **15.** mm **17.** km
19. m **Written 1.** 200 **3.** 8000 **5.** 100,000
7. 300 **9.** 120 **11.** 0.3 **13.** 273,500
15. 1738.1 **17.** 2000 **19.** 7 **21.** 33.5
23. 3.47 **25.** 0.042 **27.** 8300 **29.** 0.0001
31. 1,200,000 **33.** 5 cm, 50 mm, 0.05 m
35. 4 cm, 40 mm, 0.04 m **37.** 2.5 cm, 25
mm, 0.025 m **39.** 5.7 cm, 57 mm, 0.057 m
41. 0.9 cm, 9 mm, 0.009 m **43.** 30 **45.** 200
47. 8 **49.** 0.032 **51.** 8300 **53.** 30 **55.** 120
57. 30.01 **59.** 0.803 **61.** 24 **63.** 0.000001
65. 0.0000083

Pages 249-250   Lesson 8.2
Exploratory 1. 6 cm^2 **3.** 3.6 m^2 **5.** 300 ft^2
7. 5.39 m^2 **9.** 10 cm **11.** 8.4 m **13.** 70 ft
15. 12 m **17.** 10,000 **19.** 1,000,000
Written 1. 3.1408 cm^2 **3.** 1500 mm^2 **5.** 8
ft^2 **7.** $\sqrt{14} \text{ cm}^2$ **9.** 30 cm^2 **11.** $\sqrt{14} \text{ km}^2$
13. 248 cm^2 **15.** 615.4 cm^2 **17.** 4392
19. 25.45 **21.** 356.38 **23.** 1.3 **25.** 18
27. $3\sqrt{35}$ **29.** insufficient information given
31. 204 cm^2 **33.** 28.8 cm^2 **35.** 24 cm^2
37. 345 cm^2 **39.** 50 cm^2

Pages 252-253   Lesson 8.3
Exploratory 1. 6 **3.** 48 **5.** $\dfrac{1}{16}$ **7.** $2\sqrt{3}$

Written 1. 30 cm^2 **3.** 12 cm^2 **5.** 0.3212
cm^2 **7.** 28.395 cm^2 **9.** 90 cm^2 **11.** 14 cm^2
13. 137.5 cm^2 **15.** 234 cm^2 **17.** sometimes
19. always **21.** sometimes

Page 255   Lesson 8.4
Exploratory 1. true **3.** false **5.** If a figure is
a polygon, then it is a trapezoid. false **7.** If a
figure is a parallelogram, then it is a polygon.
true **9.** If a figure is not a trapezoid, then it is
not a polygon. false **11.** If a figure is not a
polygon, then it is not a parallelogram. true
Written 1. 150 **3.** 1.04 **5.** 292.5 **7.** 0.55
9. 15 **11.** 4.12 **13.** 40 m^2 **15.** 22.5 cm^2
17. 20 ft^2

Pages 259-260   Lesson 8.5
Exploratory 1. H **3.** $\overline{HD}, \overline{HE}, \overline{HG}$ **5.** $\overline{GD}$
7. C **9.** 10π km **11.** π km **13.** 12.5π cm
15. 14π in. **17.** 6π in. **19.** 1.4π cm
Written 1. 31.4 km **3.** 3.1 km **5.** 39.3 cm
7. 44.0 in. **9.** 18.8 in. **11.** 4.4 cm **13.** 6
cm, 12 cm **15.** 5 ft, 10 ft **17.** 0.5 cm, 1 cm
19. 1.6 cm, 3.2 cm **21.** always
23. sometimes **25.** always **27.** 5π cm
29. π^2 units **31.** $\times 2$ **33.** $\times k$

Pages 262-263   Lesson 8.6
Exploratory 1. $\pi \text{ units}^2$ **3.** $49\pi \text{ cm}^2$
5. $1.96\pi \text{ km}^2$ **7.** $\dfrac{1}{4}\pi \text{ in}^2$ **9.** $84\pi \text{ mm}^2$

Written 1. 314 cm^2 **3.** 201 m^2 **5.** 1962.5
in^2 **7.** 113 ft^2 **9.** $61,544 \text{ m}^2$ **11.** 3.8 mm^2
13. 764.2 km^2 **15.** $28.3r^2 \text{ cm}^2$ **17.** 6 cm
19. 1.1 m **21.** s m **23.** $\dfrac{\sqrt{15\pi}}{\pi}$ ft **25.** $\dfrac{\sqrt{B\pi}}{\pi}$
cm **27.** 125.6 in. **29.** 119.3 yd **31.** 56.5 m
33. 58.9 in. **35.** 98.3 km **37.** $\times 16$ **39.** $\times$
100 **41.** $\times 2.25$ **43.** $2\pi \text{ in}^2$ **45.** $428\pi \text{ in}^2$

Page 267   Lesson 8.7
Exploratory 1. $\triangle ABC, \triangle BCD, \triangle BDF, \triangle BFA,$
$\triangle ACE, \triangle CDE, \triangle DFE, \triangle FAE; \overline{AB}, \overline{AC}, \overline{BC}, \overline{CD},$
$\overline{BD}, \overline{FD}, \overline{BF}, \overline{FA}, \overline{AE}, \overline{CE}, \overline{DE}, \overline{FE}; A, B, C, D, E,$
F **3.** $\triangle QVU, \triangle QUT, \triangle QTS, \triangle QSR, \triangle QRV,$
pentagon $VRSTU; \overline{QV}, \overline{QU}, \overline{QT}, \overline{QS}, \overline{QR}, \overline{VU},$
$\overline{UT}, \overline{TS}, \overline{SR}, \overline{RV}; Q, R, S, T, U, V$ **5.** 27 cm^3
7. 1 m^3 **9.** 2.197 cm^3 **Written 1.** 12
3. $17\dfrac{1}{2}$ **5.** 3 **7.** 300 **9.** 450 **11.** 1,200,000
13. 200,000 **15.** 1280 cm^3

Page 270   Lesson 8.8
Exploratory 1. $3\pi \text{ cm}^3$ **3.** $27\pi \text{ cm}^3$ **5.** 500π
cm^3 **7.** $53.9\pi \text{ cm}^3$ **9.** $100\pi \text{ m}^3$ **11.** $36\pi \text{ m}^3$

13. $\frac{50\pi}{3}$ m^3 **15.** 2000π m^3 **Written 1.** 250π cm^3 **3.** 50π cm^3 **5.** 14π cm^3 **7.** 2304π cm^3 **9.** 18.8 cm^3 **11.** 942 cm^3 **13.** 460.5 cm^3 **15.** 10.5 m^3 **17.** 4521.6 mm^3 **19.** $\frac{4000\pi}{3}$ cm^3 **21.** 36π m^3 **23.** $57\frac{1}{6}\pi$ m^3 **25.** $4\pi\sqrt{3}$ m^3 **27.** $\frac{\pi}{48}$ in^3 **29.** $\frac{32r^3\pi}{3}$ cm^3 **31.** $\times$ 4 **33.** 36π cm^3

Page 274 Lesson 8.9

Exploratory 1. kg **3.** mL **5.** mL **7.** kg **9.** L **11.** g **13.** tons **15.** g **17.** mL **19.** L **21.** L **Written 1.** 5 **3.** 0.142 **5.** 3.5 **7.** 3 **9.** 4600 **11.** 4.2 **13.** 3000 **15.** 3080 **17.** 8050 **19.** 0.23 **21.** 2300 **23.** 23,000 **25.** 0.948 **27.** 28 **29.** 2 **31.** 52.65 L

Pages 275-276 Problem Solving Application

1. 34th **3.** 102 **5.** 300 **7.** $\frac{n(n-1)}{2}$ **Mixed**

Review 1. $\pm 2\sqrt{10}$ **2.** $\pm 2\sqrt{2}$ **3.** $\pm 12\sqrt{3}$ **4.** $\pm 6\sqrt{3}$ **5.** 120 cm^2

Pages 278-279 Chapter Review

1. 3 cm, 30 mm **3.** 50 **5.** 800 **7.** 4.5 **9.** 0.025 **11.** 4250 **13.** 6000 m^2 **15.** $21\sqrt{2}$ cm^2 **17.** $12\sqrt{14}$ m^2 **19.** 5 **21.** 4.5 **23.** $\frac{43}{3}\sqrt{3}$ **25.** 24 m^2 **27.** 30 cm^2 **29.** 88 cm^2 **31.** K **33.** $\overline{LM}, \overline{JM}, \overline{JL}$ **35.** 75.4 cm **37.** 40 cm **39.** 960 cm^3 **41.** 840 cm^3 **43.** 108π cm^3 **45.** 270 m^3 **47.** 2.4 **49.** 1800 **51.** 4000 **53.** 2800 **55.** 8

CHAPTER 9 POLYNOMIALS

Page 283 Lesson 9.1

Exploratory 1. yes; 7 **3.** yes; -8 **5.** no **7.** yes; $\frac{11}{7}$ **9.** yes; $-\frac{1}{11}$ **11.** 1 **13.** 0 **15.** 3 **17.** 0 **19.** 1 **Written 1.** $-7y$ **3.** $16x^2$ **5.** $-2a^3$ **7.** x^2 **9.** $-21xy$ **11.** $2mn$ **13.** $9xy^2$ **15.** $13r^3s$ **17.** $9ab$ **19.** $st^2 + 6 - a$ **21.** $-6y^2 - 3$ **23.** $17.4x^2y$

Page 285 Lesson 9.2

Exploratory 1. true **3.** false **5.** true **7.** false **9.** true **11.** false **Written 1.** x^5 **3.** b^8 **5.** r^9

7. d^4 **9.** y^5 **11.** $2x^3$ **13.** $-6xy$ **15.** $-8abc$ **17.** $-8a^2b$ **19.** $12x^3y$ **21.** $10m^2n^2$ **23.** a^5b^3 **25.** $-10m^3n^3$ **27.** $10x^6y^8$ **29.** $32x^2y^2z^3$ **31.** mn^2p^2r **33.** $-\frac{3}{5}a^3b^2$ **35.** $2\sqrt{6}\,x^3y^5$ **37.** $24x^3y^2$ **39.** $18a^4b$

Pages 287-288 Lesson 9.3

Exploratory 1. true **3.** false **5.** false **7.** true **9.** false **11.** true **Written 1.** 729 **3.** m^9 **5.** $27a^3$ **7.** a^2b^2 **9.** $32c^5$ **11.** $-27y^3$ **13.** $\frac{1}{8}a^3b^3$ **15.** $16b^{12}$ **17.** $16x^2y^4$ **19.** $36x^6y^2$ **21.** $-8x^6y^3z^3$ **23.** $-108x^4$ **25.** $-48x^3y^2$ **27.** $-24x^7y$ **29.** a^8b^4 **31.** $-18x^3y^2$ **33.** $4a^{17}b^{22}$ **35.** $4x^{10}y^8z^2$ **37.** x^6y^4 **39.** $18a^4b^3$ **41.** $40x^7y^{11}$ **43.** $8x^3$ **45.** $\frac{8}{27}x^3y^6$ **Mixed**

Review 1. 11 **2.** $-64x$ **3.** $3\sqrt{6}$ **4.** $10y\sqrt{5y}$ **5.** $60\sqrt{10}$ **6.** $-10\sqrt{6}$ **7.** $30 + 60\sqrt{3}$ **8.** 21 cm^2 **9.** 9 cm **10.** 18 m^2 **11.** 15π cm **12.** $4\sqrt{5}$ mm **13.** $457\frac{1}{3}\pi$ in^3 **14.** yes **15.** $48\sqrt{133}$ cm^2

Page 290 Lesson 9.4

Exploratory 1. monomial **3.** binomial **5.** trinomial **7.** trinomial **9.** no **11.** binomial **13.** binomial **15.** no **17.** trinomial **Written 1.** 2 **3.** 1 **5.** 4 **7.** 5 **9.** $5x^2 + 2x$; 2 **11.** $y - 3$; 1 **13.** $-2y^4 - 8y^2 - 6y + 4$; 4 **15.** $3x^3 + 2x^2 - x + 1$; 3 **17.** $29x^3 + 17x^2 + 5$; 3 **19.** $9x - 2$; 1 **21.** $3m + 19n$; 1 **23.** $3y^2 + 7y$; 2 **25.** $8a + b$; 1 **27.** $6y^2 + y$; 2 **29.** $8b + 8c$; 1 **31.** $-2b^3 + 7b^2$; 3 **33.** $4y$; 1 **35.** $5x^2 - 7y$; 2 **37.** 2 **39.** 2 **41.** -147

Pages 292-294 Lesson 9.5

Exploratory 1. $-a$ **3.** $-7a - 6b$ **5.** $8m - 7n$ **7.** $-x^2 + 2xy - y^2$ **9.** $-ab + 5a^2 - b$ **11.** $6x^2 + 2xy - 12$ **Written 1.** $11x + y$ **3.** $16x + 2y$ **5.** $20xy$ **7.** 0 **9.** $8a^2 - 10a - 19$ **11.** $-4a + 9b - 19c$ **13.** $-9x - y - 2z$ **15.** $3a - b$ **17.** $2x - 5$ **19.** $2a - 43c + 5x - 37$ **21.** $11x - 3y$ **23.** $-4b^2 + 13b - 15$ **25.** $-12ax^3 - 11ax^2 - ax$ **27.** $4m - 2n$ **29.** $-9x - 11x^2$ **31.** $-2r + 4s$ **33.** $-x - 4z$ **35.** $4a^2 - 2a + 10b$ **37.** $-4t^2 + 8t - 7$ **39.** 0 **41.** $7n^2 - 8nt +$

$4t^2$ **43.** $-3x^3y - xy^3 + 6$ **45.** $3y^2 + 4y + 3$
47. $-a + 2b - 8c$ **49.** $6m^2n^2 + 10mn - 23$
51. $7y + 3$ **53.** $7y + 2$ **55.** $(6y + 3)$ m
Page 296   Lesson 9.6

Exploratory   1. $12x$ **3.** $40r^8$ **5.** $-200am^5$
7. $2y - 6$ **9.** $-12a - 18b$ **11.** $-3x + 3y$
Written   1. $12a^3b$ **3.** $x^3 - 5x^2$ **5.** $8x^3 -$
$20x^3y$ **7.** $-7a^5 + 7a^3b$ **9.** $15x - 5x^2$
11. $7x^4 + 21x^2 - 13x$ **13.** $2x^3 - 6x^2 - 8x$
15. $8x - 6$ **17.** $15m^3n^3 - 30m^4n^3 + 15m^5n^6$
19. $30c^4d + 20c^3d^3 + 40c^2d^3$
21. $-68b^5d^4 - 187b^6d^5 - 85b^4d^6$
23. $\dfrac{x^3 + 11x^2}{2}$ units

25. $\left(3x^5 + 15x^4 + \dfrac{3x^3}{2} + \dfrac{15x^2}{2} \right)$ units^2

27. $6x^2 + 3x; 3x^2 - 3x; 8x^2 + 4x; 4x^2 - 4x$
29. -23 **31.** 6 **33.** $\dfrac{1}{3}$ **35.** $z \le -\dfrac{12}{11}$
Page 299   Lesson 9.7

Exploratory   1. $x; 7$ **3.** $a; 6$ **5.** $y + 7; y +$
7 **7.** $n - 3m; n - 3m$ **9.** $y^2 + 8y + 15$
Written   1. $x^2 + 9x + 14$ **3.** $m^2 + 15m +$
54 **5.** $m^2 - 2m - 35$ **7.** $x^2 - 11x + 30$
9. $2a^2 + 7a + 3$ **11.** $6y^2 + 7y - 3$
13. $a^2 - 18a + 81$ **15.** $4a^2 - 4ab + b^2$
17. $10x^2 + 11x + 3$ **19.** $6t^2 - 31t + 35$
21. $2r^2 + 7rs + 6s^2$ **23.** $x^2 - 9$ **25.** $4x^2 -$
49 **27.** $a^2 + 20a + 100$ **29.** $4x^2 + 4xy +$
y^2 **31.** $7 + 3\sqrt{3}$ **33.** $-3x^2 + x\sqrt{5} + 10$
35. $5b^2\sqrt{3} - 7b - 2\sqrt{3}$ **37.** 1 **39.** $-\dfrac{4}{3}$

41. -8 **43.** -5 **45.** $y < -\dfrac{25}{4}$ **47.** $x > -3$

Pages 301-302   Lesson 9.8

Exploratory   1. $10a^2 - 35$ **3.** $3a^3 + 2a$
5. $24k^3 + 42k$ **7.** $7a^2 - 14a + 21$
9. $12d^5 - 16d^4 + 12d^3$ **11.** $3a^5 + 6a^4 -$
$18a^3$ **Written   1.** $x^3 + 5x^2 + 7x + 3$
3. $2x^3 + 3x^2 + 3x + 1$ **5.** $y^4 - 3y^3 + y^2 -$
$5y + 6$ **7.** $y^3 + 3y^2 + 5y + 3$ **9.** $x^4 +$
$x^2 - 30$ **11.** $6y^3 - 26y^2 + 38y - 20$
13. $a^3 - 3a^2b + 3ab^2 - b^3$ **15.** $x^4 + 4x^3 +$
$6x^2 + 4x + 1$ **17.** $x^5 - x^2 + x - 1$
19. $-15y^4 - 8y^3 + 12x^2y^2 + 3xy^2 - 2x^2y$
21. $-3y^2 - 6y + 9$ **23.** $8y^3 - 11y^2 + 8y$

25. -2 **27.** 2 **29.** -2 **31.** $\dfrac{1}{2}$ **33.** $8, 9$

35. 4 m, 12 m   **37.** any two consecutive
integers   **39.** 30 ft $\times$ 56 ft   **41.** 9 cm $\times$ 4 cm
43. any two consecutive integers
Mixed Review   1. c   **2.** a   **3.** b   **4.** c   **5.** b
Page 305   Lesson 9.9

Exploratory   1. x^4 **3.** a^2 **5.** $\dfrac{1}{x}$ **7.** k^5 **9.** 1

Written   1. x^5 **3.** an **5.** b^3c^3 **7.** $-x^3y^2$
9. $\dfrac{1}{m^4}$ **11.** $\dfrac{a}{n^2}$ **13.** $\dfrac{1}{y^3}$ **15.** $4a^7$ **17.** $\dfrac{x^2}{7}$ **19.** $\dfrac{y^3}{2}$
21. $4b^2c^3$ **23.** $-3s^6$ **25.** $-\dfrac{ab^4}{12}$ **27.** $\dfrac{1}{y}$
29. $-\dfrac{1}{4y^4}$ **31.** $\dfrac{5}{8b^5c}$ **33.** $\dfrac{1}{4^6}$ **35.** r^3 **37.** $\left(\dfrac{x}{y} \right)^3$
39. $\dfrac{1}{m^8n^{10}}$ **41.** $\dfrac{1}{(3.4)^2}$ **43.** $\dfrac{1}{4^6}$
Page 308   Lesson 9.10

Exploratory   1. $2y^2 + 3$ **3.** $8ab^3 - 14b^2 + 6$
5. $-a^2b + a - \dfrac{2}{b}$ **Written   1.** $x - 12$
3. $b + 10$ **5.** $x^2 - 2x + 3$ **7.** $2x + 3$
9. $y^2 + 2y - 3$ **11.** $t^2 + 4t - 1$ **13.** $t^2 +$
$4t - 3 - \dfrac{3}{t - 4}$ **15.** $3d^2 + 2d + 3 - \dfrac{2}{3d - 2}$
Page 312   Lesson 9.11

Exploratory   1. 0 **3.** -3 **5.** $5\dfrac{1}{2}$ **7.** $\dfrac{7}{3x}$
9. $\dfrac{19a}{21b}$ **11.** $-2xy^2$ **13.** $16y^2$ **Written   1.** $\dfrac{b}{d}$
3. a **5.** $\dfrac{a - 1}{4bc}$ **7.** $\dfrac{4m^4}{15n^2}$ **9.** cd^2x **11.** $\dfrac{2m^2}{c^2}$
13. $\dfrac{(y - 3)(y + 2)}{2y}$ **15.** $k(k + 3)$ **17.** $\dfrac{r - s}{18rs^2}$
19. y **21.** $\dfrac{x + 4}{12}$ **23.** $\dfrac{9}{a}$ **25.** $\dfrac{x - 4}{x}$
27. $\dfrac{32 + 11y}{44}$ **29.** $\dfrac{5x - 28}{35}$ **31.** $\dfrac{2}{a}$
33. $\dfrac{14a - 3}{6a^2}$ **35.** $\dfrac{5bx + a}{7x^2}$
37. $\dfrac{3z^2 - 14wz - 7w}{7w^2z}$ **39.** $\dfrac{2s + t + 3}{st}$
41. $\dfrac{31b + 7}{5b^2}$ **43.** $\dfrac{5m + 6}{15m}$ **45.** $\dfrac{-8x^2 - 71x + 5}{4x^3}$
Page 314   Problem Solving Application

Written   1. 18,200 coins   **3.** Peter   **5.** 23
rungs

Pages 316-317 Chapter Review

1. $-20x$ **3.** $-13m^2n$ **5.** $2xy + 3x - 5y$
7. a^7 **9.** $8a^3y^7$ **11.** -10^3b^2 **13.** 6561
15. $27x^3z^6$ **17.** $6a^3x^6y^3$ **19.** $8x + 8y$; 1
21. $6a^2 + a$; 2 **23.** $4b^3 + 6bc + c^2$; 3
25. $10m + n$ **27.** $10x^2 + 5x$ **29.** $5x^2 - 2ax$
31. $4x + 5y$ **33.** $2n^2 - 11n - 1$ **35.** $7x^2 -$
$14x + 8$ **37.** $12x^2 + 9$ **39.** $-10x^2y - 4xy$
41. $-91a^5b^3 - 49a^3b^5$ **43.** $x^2 + 11x + 30$
45. $2x^2 + 7x - 72$ **47.** $n^2 - 13n + 36$
49. $a^3 + 4a^2 - 19a + 14$ **51.** $3y^3 + 7y^2 -$
$12y + 4$ **53.** $2m^3 + m^2 - 4m - 2$ **55.** x^6
57. $\dfrac{3a}{b}$ **59.** $\dfrac{xy}{x^2}$ **61.** $2b + 4$ **63.** $3x + \dfrac{2x + 7}{9x - 2}$
65. $\dfrac{12}{x}$ **67.** $\dfrac{7}{9y}$ **69.** $-\dfrac{5b + 5}{6}$

CHAPTER 10 FACTORING

Page 322 Lesson 10.1

Exploratory **1.** yes; 4 **3.** no **5.** yes; 14
7. yes; 37 **9.** yes; 54 **11.** yes; 33 **13.** 4
15. 4 **17.** 10 **19.** 9 **21.** 2 **23.** 4
Written **1.** $3 \cdot 7$ **3.** $2^2 \cdot 3 \cdot 5$ **5.** $2^2 \cdot 17$
7. $2^4 \cdot 5$ **9.** $2^4 \cdot 19$ **11.** $-1 \cdot 2^5 \cdot 3$ **13.** 2^3
$3^2 \cdot 7$ **15.** $3^2 \cdot 5^2$ **17.** xy **19.** $-2rs$
21. $-2yz$ **23.** $-12m^2n^2$ **25.** $2t$ **27.** -1
29. 6 **31.** 24 **33.** $5x$ **35.** 5 **37.** $4xy$
39. $6ab$ **41.** $4x^2$ **43.** $6ab$ **45.** $2ab^2$

Page 324 Lesson 10.2

Exploratory **1.** $7y$ **3.** 1 **5.** $6n$ **7.** xy **9.** x
11. $8x$ **13.** $2x$ **15.** $2y^2$ **17.** 9
Written **1.** $6(a + b)$ **3.** $8(c - d)$
5. $2(4m - n)$ **7.** $a(b + c)$ **9.** $y^2(y + 1)$
11. $5(5t^2 - 3)$ **13.** $a^2(a + 2)$ **15.** $5xy(x - 2y)$
17. $-5x(3x + 1)$ **19.** $8(m^2 - 2m + 3)$
21. $9(x^5 - 2x^3 + 3)$ **23.** $x^2y(1 + x)$
25. $2a^2(2b - 3a^2)$ **27.** $5(3d^2 - 5d + 6)$
29. $2a(b + c)$ **31.** $r^2(25 - \pi)$ **33.** $r^2(\pi - 2)$

Pages 326-327 Lesson 10.3

Exploratory **1.** no **3.** yes **5.** no **7.** yes
9. no **11.** no **13.** $x^2 - 9$ **15.** $d^2 - 16$
17. $y^2 - 49$ **Written** **1.** $a^2 - 81$ **3.** $s^2 - t^2$

5. $4a^2 - 9$ **7.** $9c^2 - 4$ **9.** $s^2 - 16t^2$
11. $25x^2 - 9$ **13.** $4r^2 - 9s^2$ **15.** $9p^2 - 64q^2$
17. $16x^2 - 25$ **19.** $36f^2 - 25g^2$
21. $361w^2 - 576x^2$ **23.** $64 - 9y^2$
25. $(x - 7)(x + 7)$ **27.** $(m - 10)(m + 10)$
29. $(r - 2s)(r + 2s)$ **31.** $(m - 3n)(m + 3n)$
33. $4(3s - 5)(3s + 5)$ **35.** $(-3 + 2y) \cdot$
$(-3 + 2y)$ **37.** $(11by - 1)(11by + 1)$
39. $(4a - 3b)(4a + 3b)$ **41.** $2(z - 7)(z + 7)$
43. $12(c - 1)(c + 1)$ **45.** $8(x - 2y)(x + 2y)$
47. $(12x - 3y)(12x + 3y)$ **49.** $15(x - 2y) \cdot$
$(x + 2y)$ **51.** $5(x - 5y^2)(x + 5y^2)$
53. $(a + b - m)(a + b + m)$ **55.** 3591
57. 9951 **59.** 39,999 **61.** 997,500
63. $(s - t)(s + t)$ **65.** $4(2s - t)(2s + t)$

Pages 330-331 Lesson 10.4

Exploratory **1.** yes; $6x^2$, x, 3 **3.** no **5.** yes;
x^2, $7x$, -8 **7.** yes; $5x^2$, 0, 0 **9.** no **11.** 9, 1
13. 5, 3 **15.** 7, 4 **17.** -5, 3 **19.** 2 **21.** $+$
23. 4 **25.** -17 **27.** $-$, 2 **29.** 3, 4, $+$
Written **1.** $(a + 5)(a + 3)$ **3.** $(c + 7)(c + 5)$
5. $(x + 3)(x - 1)$ **7.** $(g - 7)(g - 3)$
9. $(p - 4)(p - 15)$ **11.** $(s - 5)(s - 2)$
13. $(t + 9)(t - 6)$ **15.** $(m + 8)(m - 5)$
17. $(t + 11)(t + 3)$ **19.** $(3a + 2)(a + 2)$
21. $(7x + 1)(x + 3)$ **23.** $(2a + 1)(3a - 2)$
25. cannot be factored **27.** $(2k - 3)(4k - 3)$
29. $(m - 6)(m + 2)$ **31.** $(x + 4y)(x + 2y)$
33. $(a + 14b)(a - 3b)$ **35.** $(4x - 3y)(5x + 8y)$
37. width = $x + 5$, length = $x + 8$
39. base = $2x$, height = $x + 3$; base = $x + 3$,
height = $2x$; base = 2, height = $x^2 + 3x$;
base = $x^2 + 3x$, height = 2; base = x,
height = $2x + 6$; base = $2x + 6$, height = x

Pages 334-335 Lesson 10.5

Exploratory **1.** $x = 0$ or $x - 5 = 0$ **3.** $2n =$
0 or $n - 3 = 0$ **5.** $3n = 0$ or $2 - n = 0$
7. $x = 0$ or $x + 2 = 0$ **9.** $3r = 0$ or $r - 4 =$
0 **11.** $x - 2 = 0$ or $x - 7 = 0$
13. $d + 10 = 0$ or $d + 2 = 0$ **15.** $m - 5 =$
0 or $m - 4 = 0$ **17.** $y - 7 = 0$ or $y + 6 =$
0 **Written** **1.** $\{0, 7\}$ **3.** $\{0, 2\}$ **5.** $\{-4, 6\}$
7. $\{-3, 4\}$ **9.** $\{-8, 2\}$ **11.** $\{-2\}$ **13.** $\{3\}$
15. $\{-6, 6\}$ **17.** $\{0, 6\}$ **19.** $\{-8, 0\}$
21. $\{-5, 0\}$ **23.** $\{0, 9\}$ **25.** $\{-9, 0\}$ **27.** $\{0,$
$4\}$ **29.** $\{-9, 9\}$ **31.** $\{-4\}$ **33.** $\{-6, 1\}$
35. $\{-1, 9\}$ **37.** $\{0, 10\}$ **39.** $\{-7, -3\}$

41. $\{-7, 2\}$ **43.** $\{-8, 0, 8\}$ **45.** $\{-2, 2\}$
47. $\{-5, 7\}$

Pages 339-340   Lesson 10.6
Exploratory   1. $x^2 - 4x = 32$
3. $x(17 - x) = 72$ **5.** $x(x + 6) = 72$
7. $x(x + 3) = 54$ **Written   1.** -4 or 8 **3.** 8,
9 **5.** 6 m $\times$ 12 m **7.** 6 m $\times$ 9 m **9.** base =
12 in., height = 8 in. **11.** -10, -9 or 9, 10
13. 14, 15 **15.** -10, -8 or 8, 10 **17.** -12,
-10 **19.** -9, -11 or 9, 11 **21.** 11, 13
23. 8, 9 **25.** -10, -9 or 9, 10 **27.** 9, 11
29. 2 m **31.** 4 cm, 3 cm **33.** 8 ft **35.** 6 in.

Pages 344-345   Lesson 10.7
Exploratory   1. $3(x^2 + 3)$ **3.** $m^2(8n - 13p)$
5. $4a^2(1 - 2b)$ **7.** $2a^2b(1 + 4b + 5b^2)$
9. $(m - 2)(m - 2)$ **11.** $(3x - 1)(x - 2)$
13. $(a - b)(a + b)$ **15.** $(5x + y)(x - y)$
17. $(4ab - 3c^2)(a - 3b)$ **19.** $(5a + 1)(x + y)$
21. $(3 + a^2)(2a - 3b)$ **23.** $(6x + 4y + 7) \cdot$
$(a + b)$ **Written   1.** $3(x - y)(x + y)$
3. $7(y - 1)(y + 1)$ **5.** $3(b - 7)(b + 7)$
7. $(y - 2)(y + 2)(y^2 + 4)$ **9.** $(x^2 - y)(x^2 + y)$
11. $3(a - 3b)(a + 3b)$ **13.** $4(x - 3)(x + 3)$
15. $3(x - 2)(x + 2)(x^2 + 4)$ **17.** $a^2(4 + b^2)$
$(2 - b)(2 + b)$ **19.** $(2y + 3)(3y + 2)$
21. $5(x + 2)(x + 1)$ **23.** $3(a + 3)(a + 5)$
25. $(3x + 5)(y - 2)$ **27.** $(5x + y)(x - y)$
29. $5y^2(y - 1)(y + 1)(y^2 + 1)$
31. $3(2a - 1)(a + 5)$ **33.** $3(2x - 7)(x + 5)$
35. $(2a - 5b)(2x + 7y)$ **37.** $(2x - 3)(4a - 3)$
39. $(8a + 3)(a + b + c)$ **41.** $(a - 7) \cdot$
$(2a + 3b - 4c)$ **43.** $(x + y)(x^2 - 5x - 7)$
Mixed Review   1. $\dfrac{10 + 7x}{4x^2}$ **2.** $-\dfrac{7}{24}$

3. $\dfrac{3y + 36}{20}$ **4.** -5.5 **5.** -11 **6.** 9 or -4
7. 12 cm $\times$ 18 cm **8.** 110 m $\times$ 50 m **9.** 10
rows **10.** 8 cm $\times$ 8 cm; 16 cm $\times$ 4 cm

Page 347   Chapter Review
1. $2^2 \cdot 7$ **3.** $2^2 \cdot 31$ **5.** 24 **7.** $2a^2$ **9.** $12xyz$
11. $6x(2x + 3)$ **13.** $x^3(x + 1)$ **15.** $a^2 - 4$
17. $f^2 - g^2$ **19.** $(a - 5)(a + 5)$
21. $(4x - 9y)(4x + 9y)$ **23.** $(y + 7)(y + 2)$
25. $(a + 6)(a - 1)$ **27.** $(2x - 5)(x + 3)$
29. $\{-12, 0\}$ **31.** $\{-7, -6\}$ **33.** 6 or -6
35. $3(x - 4)(x + 4)$ **37.** $2y(y - 7)(y - 2)$
39. $(y - 4)(y + 4)(a - b)$

CHAPTER 11   INTRODUCTION TO
PROBABILITY

Page 352   Lesson 11.1
Exploratory   1. $\dfrac{1}{6}$ **3.** $\dfrac{1}{2}$ **5.** $\dfrac{1}{6}$ **7.** $\dfrac{1}{6}$ **9.** $\dfrac{1}{2}$
11. 0.04, 4% **13.** 0.2, 20% **15.** 0.65, 65%
17. $\dfrac{2}{3}$ **19.** $\dfrac{8}{25}$ **Written   1.** $\dfrac{1}{8}$ **3.** $\dfrac{1}{2}$ **5.** $\dfrac{7}{8}$
7. $\dfrac{3}{8}$ **9.** $\dfrac{1}{8}$ **11.** $\dfrac{1}{6}$ **13.** $\dfrac{1}{2}$ **15.** $\dfrac{5}{6}$ **17.** $\dfrac{1}{52}$
19. $\dfrac{1}{13}$ **21.** $\dfrac{1}{4}$

Pages 355-356   Lesson 11.2
Exploratory   1. $\dfrac{1}{6}$ **3.** $\dfrac{1}{3}$ **5.** 1 **7.** 0 **9.** no
Written   1. $\dfrac{1}{5}$ **3.** $\dfrac{1}{5}$ **5.** $\dfrac{1}{2}$ **7.** $\dfrac{1}{2}$ **9.** $\dfrac{1}{26}$ **11.** $\dfrac{1}{13}$
13. $\dfrac{21}{26}$ **15.** $\dfrac{2}{13}$ **17.** $\dfrac{1}{26}$ **19.** $\dfrac{1}{4}$ **21.** $\dfrac{2}{13}$ **23.** $\dfrac{2}{5}$
25. 0 **27.** $\dfrac{1}{10}$ **29.** $\dfrac{1}{5}$ **31.** $\dfrac{7}{10}$ **33.** 1 **35.** 0
37. $\dfrac{3}{10}$

Page 360   Lesson 11.3
Exploratory   1. no **3.** $\dfrac{5}{7}$ **5.** $\dfrac{3}{7}$ **7.** 0 **9.** $\dfrac{1}{7}$
11. $\dfrac{5}{7}$ **Written   1.** $\dfrac{1}{13}$ **3.** $\dfrac{7}{26}$ **5.** $\dfrac{3}{13}$ **7.** $\dfrac{5}{13}$
9. $\dfrac{7}{10}$ **11.** 1 **13.** $\dfrac{7}{10}$ **15.** 0.6 **17.** 0.09
19. 0.1 **21.** $\dfrac{1}{2}$ **23.** $\dfrac{2}{5}$ **25.** $\dfrac{1}{5}$ **27.** $\dfrac{1}{10}$ **29.** 0
31. $\dfrac{1}{5}$

Page 363   Lesson 11.4
Exploratory   1. All elements in set D are
contained in set E. **7.** $\{H, T\}$ **9.** {blue, red,
white} **Written   1.** $\{A, I\}$ **3.** $\dfrac{2}{5}$ **5.** $\{A, B, C,$
$D, E, I, O, U\}$ **7.** $\emptyset$ **9.** 5 **11.** 8 **13.** $\dfrac{4}{13}$
15. $\dfrac{6}{13}$ **Mixed Review   1.** $\dfrac{2}{5}$ **2.** $\dfrac{17}{20}$ **3.** $\dfrac{3}{5}$
4. $\dfrac{11}{20}$ **5.** 0.8 **6.** $\dfrac{1}{3}$ **7.** $1 - x$ **8.** $y - 2x$
9. 84% **10.** $2h^2 + 4h$

Pages 366-367   Lesson 11.5
Exploratory   1. (H, T), (T, T), (H, H), (T, H)
3. (H, H) **5.** (H, T) and (T, H) **7.** $\dfrac{1}{2}$

Written 3. $\frac{2}{9}$ 5. $\frac{5}{6}$ 7. $\frac{1}{36}$ 11. 1, $\frac{1}{16}$; 2, $\frac{2}{16}$; 3, $\frac{3}{16}$; 4, $\frac{4}{16}$; 3, $\frac{3}{16}$; 2, $\frac{2}{16}$; 1, $\frac{1}{16}$ 13. $\frac{7}{16}$ 15. $\frac{3}{8}$ 17. $\frac{3}{4}$ 19. $\frac{3}{16}$ 21. $\frac{3}{16}$

Page 370   Lesson 11.6
Exploratory 3. $\frac{1}{2}$ 5. $\frac{1}{6}$ 7. no

Written 3. 24 5. 504 11. $\frac{1}{8}$

Page 373   Lesson 11.7
Exploratory 3. $\frac{25}{49}$ 5. $\frac{20}{49}$ 7. 1 **Written**
1. $\frac{8}{125}$ 3. $\frac{4}{25}$ 7. $\frac{3}{10}$ 9. $\frac{3}{5}$ 11. $\frac{5}{18}$ 13. $\frac{5}{9}$ 15. $\frac{15}{28}$ 17. $\frac{3}{7}$ 19. $\frac{4}{9}$ 21. $\frac{8}{27}$ 23. $\frac{2}{5}$ 25. $\frac{1}{5}$

Pages 376-377   Lesson 11.8
Exploratory 1. 120 3. 5040 5. 12 7. 2016 9. 24 11. 720 13. 6 15. 22
Written 1. false 3. true 5. false 7. 120 9. 362,880 11. 12 13. 12 15. 6, 5, 4 17. 20 19. 42 21. 336 23. 60 25. 90 27. 990 29. 222 31. 362,824

Page 379   Problem Solving Application
1. 4 3. $\frac{6}{49}$ 5. 5 7. 12 9. $\frac{1}{2}$

Pages 380-381   Chapter Review
1. $\frac{1}{5}$ 3. $\frac{3}{5}$ 5. 0 7. $\frac{5}{26}$ 9. $\frac{5}{26}$ 11. $\frac{4}{13}$ 13. $\frac{1}{2}$ 15. $\frac{1}{4}$ 17. 0 19. $\frac{5}{36}$ 21. $\frac{2}{9}$ 23. 6 25. $\frac{1}{6}$ 27. $\frac{1}{21}$, $\frac{10}{21}$ 29. 24 31. 6 33. 60 35. 6 37. 720

CHAPTER 12   INTRODUCTION TO STATISTICS

Pages 385-386   Lesson 12.1
Exploratory 1. smaller, representative subset of general population 3. No. The people would favor one candidate. 5. Yes. These people use detergent. 7. Yes. People who prefer all different carpet colors are found here. 9. No. Only students would be surveyed.
Written 1. Latin students may not be representative of the school population. 3. Parents may be unaware of truth, or seek to change it. 5. People are too busy to talk. Also, people in New York City may not be representative. 7. The sample is not representative. People who are at work are not reached. 9. Responses limited to those with children. 11. Populations of localities may not be representative. Also, responses will be received only from those concerned enough to reply.

Pages 389-390   Lesson 12.2
Exploratory 1. 2.5 3. −1.2 5. 100 7. 0 9. 1.845 11. $7\frac{1}{3}$ 13. 2 15. 2 17. 2 19. 2
Written 1. 46; 42; 42 3. −25; −50; −50 5. $\frac{76}{7}$; 11; 10 and 11 7. 3 9. 9 11. 2 13. 3.5 19. 7.85 21. 77

Pages 393-395   Lesson 12.3
Exploratory 1. 625 3. about $\frac{8}{11}$ 5. 8,000 7. 24,000 **Written** 9. 35% 11. 30% 13. $1,800 15. $2,200 17. $2,750 19. $8,250 **Mixed Review** 1. 0 or $\frac{5}{2}$ 2. −3 3. 8 or −8 4. −$\frac{64}{9}$ 5. 4 or −15 6. 9 or −2 7. $\frac{3}{13}$ 8. $\frac{5}{13}$ 9. $\frac{9}{26}$ 10. $\frac{3}{13}$ 11. 15; 14; 14 12. 28; 27; 28 13. 88; 88; 72 or 96 14. 2 m 15. $\frac{43}{91}$ 16. base = 8 cm; altitude = 5 cm

Pages 398-399   Lesson 12.4
Exploratory 1. The bars of a histogram are right next to each other. 3. histogram 5. histogram 7. circle graph 9. 50; subtract shortest lifetime from longest and divide by 7. 11. 10 13. 36% 15. 320 17. 88%

Pages 402-403   Lesson 12.5
Written 1. 25, 75, 50 3. 50 5. 500

7. 150 **9.** 2 **11.** A majority of the parts lasted between 2300 and 2399 hours. **13.** Yes. 600 on the cumulative frequency scale is 60% of the cumulative relative frequency. **15.** 2380 **17.** 2480 **19.** first quartile, 1001; median, 1081; third quartile, 1196

Page 405 Lesson 12.6
Exploratory 1. $\frac{1}{4}$ **3.** $\frac{1}{16}$ **5.** $\frac{1}{13}$ **7.** $\frac{1}{169}$ **9.** $\frac{4}{11}$
11. 1 **13.** $\frac{1}{2}$ **15.** $\frac{6}{73}$ **Written**
1. theoretically **3.** empirically **5.** empirically **7.** too few trials **9.** 10 **11.** Toss coin many times and find *P(tails)*.

Pages 408-409 Lesson 12.7
Written 3. It is not a fair die. *P(rolling a 6)* $\approx$ $\frac{1}{2}$ **5.** 0.68 **7.** 1000 **9.** cannot predict
Mixed Review 1. $\frac{5}{4}$ **2.** $\frac{2n^2 - 11n - 15}{10n^2}$
3. $\frac{2(4a - 1)}{3a(a - 6)}$ **4.** $\frac{3}{10}$ **5.** $\frac{1}{2}$ **6.** $\frac{1}{5}$ **7.** $\frac{4}{5}$ **8.** $\frac{7}{10}$
9. $\frac{2}{9}$ **10.** $\frac{2}{15}$ **11.** $\frac{1}{3}$ **12.** $\frac{14}{45}$ **13.** 30,240 zip codes **14.** 288 permutations **15.** 20 ways **16.** 120 combinations

Page 411 Problem Solving Application
1. 3-18 eggs, 2-12 eggs; 1-18 eggs, 5-12 eggs **3.** $2.38 **5.** 131,072 **7.** AAADF, AABCF, AABDD, AACCD, ABBBF, ABBCD, ABCCC, BBBBD, BBBCC **9.** $-40°$

Page 413 Chapter Review
1. People are in a hurry. **3.** Apartments often forbid pets. **5.** 5, 6, 8 **11.** empirically **13.** theoretically

CHAPTER 13 INTRODUCTION TO COORDINATE GEOMETRY

Pages 417-418 Lesson 13.1
Exploratory 1. right 2, up 5 **3.** left 2, up 5 **5.** right 5, up 2 **7.** left 5, up 2 **13.** (0, 0)

15. (2, 3) **17.** (4, 0) **19.** $(-2, 0)$
21. $(-3, -3)$ **23.** $(3, -3)$ **Written 13.** yes **15.** no **17.** yes **19.** yes **21.** 6 **23.** 15 **25.** 9 **27.** 1

Pages 421-422 Lesson 13.2
Exploratory 1. yes **3.** no **5.** yes **7.** no **9.** yes **11.** yes **13.** c **15.** b **Written 19.** b **21.** c **23.** none **25.** a

Page 426 Lesson 13.3
Exploratory 1. no slope **3.** positive slope **5.** x and y lose 1; slope 1 **7.** x gains 1, y gains 2; slope 2 **Written 1.** 1 **3.** 0 **5.** 1 **7.** $\frac{1}{4}$
9. $-\frac{5}{2}$ **11.** $\frac{7}{2}$ **13.** $\frac{11}{9}$ **15.** 10 **17.** 0 **19.** 0
21. -1 **23.** -4 **25.** no slope **27.** $-\frac{1}{2}$
29. 1 **31.** 4 **33.** 5 **35.** 6 **37.** 4 **39.** 4

Page 429 Lesson 13.4
Exploratory 1. $-2, 5$ **3.** 2, 15 **5.** $\frac{1}{3}$, 5
7. $0.2, -5$ **9.** $\frac{5}{3}, -0.2$ **11.** no slope, no y-intercept **13.** $y = 5x - 3$ **15.** $y = \frac{1}{3}x + \frac{8}{3}$
17. $y = 2x - 3$ **Written 1.** $y = -x + 3$
3. $y = 2x + 1$ **5.** $y = \frac{1}{4}x - 3$
7. $y = \frac{1}{3}x - 1$ **9.** $y = -2x + 3$
11. $y = -\frac{1}{2}x + 4$ **13.** 1, 2 **15.** 2, 1
17. $\frac{1}{5}, 0$ **19.** $\frac{3}{4}, -\frac{1}{4}$ **21.** 8, 1 **23.** 6, 15
25. $y = x + 1$ **27.** $y = -2x + 2$
29. $y = 8x - 29$ **31.** $x - y = 4$
33. $y = 4 - \frac{1}{5}x$ **35.** $8x + 13y = -1$
37. $y = -2x + 7$ **39.** $y = \frac{5}{2}x - \frac{1}{2}$
41. $y = -\frac{2}{3}x + \frac{13}{3}$ **43.** $y = -\frac{3}{4}x + \frac{11}{4}$
45. $y = \sqrt{2}$

Pages 432-433 Lesson 13.5
Exploratory 1. no solution **3.** exactly one point **5.** b **7.** d **9.** a **Written 1.** (4, 2)

3. $(2, -2)$ **5.** $(-1, 2)$ **7.** no solution
9. $(1, 2)$ **11.** $(0, 2)$ **13.** $(1, 9)$ **15.** $(4, -1)$
17. $(-2, -4)$ **19.** $(1, 3)$ **21.** $(3, 0)$ **Mixed**
Review 1. $\frac{4}{3}$ **2.** $\left(0, \frac{43}{4}\right)$ **3.** $y = \frac{1}{7}x + \frac{19}{7}$
4. $y = -7x + 42$ **5.** $\left(\frac{11}{2}, \frac{7}{2}\right)$ **6.** 11 **7.** 2
8. 10 **9.** 4 **10.** 15 **11.** $\frac{11}{21}$ **12.** $\frac{10}{21}$

Pages 435-436   Lesson 13.6
Exploratory 1. $y = x - 3$ **3.** $y = 2x + \frac{3}{4}$
5. $y = -4x - 2$ **7.** yes **9.** yes **11.** yes
Written 1. $(-1, 8)$ **3.** $(3, -1)$ **5.** $(14, 4)$
7. $(2, 2)$ **9.** $\left(\frac{4}{3}, \frac{4}{3}\right)$ **11.** $(0, 0)$ **13.** $(8, -1)$
15. $(0, 3)$ **17.** $(6, -2)$ **19.** $(5, -1)$
21. $(2, 2)$ **23.** $\left(\frac{3}{4}, \frac{2}{3}\right)$ **25.** $\left(\frac{1}{2}, \frac{4}{3}\right)$
27. $(2.25, 3.6)$ **29.** $(1.25, 2.\overline{75})$

Pages 440-441   Lesson 13.7
Exploratory 1. $y = x + 3, x + y = 33$
3. $y = 2x - 1, 2x + 2y = 22$
5. $5x + 3y = 34, 3x + 6y = 33$
7. $y = 3x - 20, x + y = 180$ **9.** $t + u = 7$,
$10t + u = 5u$ **Written 1.** 18, 15 **3.** 7 ft,
4 ft **5.** \$5, \$3 **7.** 50°, 130° **9.** 25
11. 20 cm, 27 cm **13.** 5, 9 **15.** \$11.95,
\$7.95 **17.** $37\frac{1}{2}$ 1b, $12\frac{1}{2}$ 1b

Page 444   Lesson 13.8
Exploratory 1. A, B, E, G **3.** A, G **5.** B, C
7. C, D, F **9.** E

Pages 446-447   Problem Solving Application
Exploratory 1. yes, 3 **3.** no **5.** no **7.** no
9. yes, π **11.** yes, $-\frac{15}{8}$ **13.** $d = 55t$
Written 1. 28 **3.** $-9\frac{1}{3}$ **5.** 15 **7.** $y = 4x$
9. $y = \frac{3}{7}x$ **11.** $y = \frac{32}{3}x$ **13.** doubles
15. halves **17.** 9750 g **19.** \$467.50 **21.** $10\frac{2}{3}$
gal **23.** 14 hours 24 minutes **25.** 4776
bushels **27.** 132 volts

Page 449   Chapter Review
5. b **7.** $5x + 2y = 3$ **9.** -3
11. $y = -\frac{x}{5} + 2$ **13.** $y = -\frac{5x}{7} + \frac{19}{7}$
15. $(1, 2)$ **17.** $(3, -1)$ **19.** $(-1, 5)$
21. 13, 7 **23.** 72

CHAPTER 14   INTRODUCTION TO TRANSFORMATION GEOMETRY

Pages 454-455   Lesson 14.1
Exploratory 1. yes; domain = {100, 35, 20,
0}, range = {212, 95, 68, 32} **3.** no **5.** yes;
domain = {5, 9, 17}, range = {8, 6} **7.** yes
9. no **Written 1.** $\{-3, -2, -1, 0, 1, 2\}$
3. -1 **5.** -5 **7.** -3 **9.** $-1, 1$ **11.** -5
13. 1 **15.** -11 **29.** yes **31.** yes **33.** yes
35. no **37.** yes **39.** no **41.** yes **43.** yes

Pages 458-459   Lesson 14.2
Exploratory 1. A **3.** $\overline{AC}$ **5.** L **7.** $\overline{JL}$ **9.** $\overline{HG}$
11. N **Written 13.** none **23.** Yes. This is a
one-to-one mapping.

Pages 461-462   Lesson 14.3
Exploratory 1. y-axis **3.** neither **5.** both
7. y-axis **9.** both **Written 1.** $A'(-2, 3)$
3. $C'(0, 6)$ **5.** $E'(-3, -5)$ **7.** $G'(2, 4)$
9. $W'(1, -2)$ **11.** $K'(-2\frac{1}{2}, 7)$
13. $M'(1, -3)$ **15.** $P'(-3, -4)$ **17.** $R'(5, 0)$
19. $T'(-1, 4)$ **21.** $V'(-5, 7)$ **23.** $Y'(-8, -3)$
25. $A'(7, 4)$ **27.** $C'(10, -4)$
29. $E'(4, -6)$ **31.** $G'(-3, -8)$ **33.** $M'(1, 2)$
35. $K'(-1, -5)$ **37.** (y, x) **39.** $A'(2, -1)$,
$B'(5, 2), C'(9, -4), D'(6, -7)$
41. $A'(-6, 1), B'(-9, -2), C'(-13, 4)$,
$D'(-10, 7)$ **47.** y-axis **49.** $x = -2$

Pages 465-466   Lesson 14.4
Exploratory 1. K **3.** $\overline{GH}$ **5.** $\angle FJK$ **7.** C
9. $\overline{BC}$ **11.** $\angle FGC$ **Written 9.** $P'(0, 12)$
11. $P'(2, 1)$ **13.** $P'(-9, 10)$
15. $P'(-5, -1)$ **17.** $P'(2, 9)$ **19.** $P'(9, -2)$
21. $T_{1,4}$ **23.** $T_{5, -12}$ **25.** $T_{-11, -4}$ **27.** $2\sqrt{5}$

29. 5 **33.** yes **35.** yes **Mixed Review 1.** c
2. b **3.** a **4.** c **5.** c **6.** c

Pages 469-470 Lesson 14.5
Exploratory 1. M **3.** E **5.** 45°
Written 13. $(-4, 0)$ **15.** $(3, 2)$ **17.** $(-4, 0)$
19. $(-3, 1)$ **21.** $(-y, x)$ **23.** $(-x, -y)$
25. no **27.** yes, $\text{Rot}_{O,72°}$ **29.** yes, $\text{Rot}_{O,72°}$
31. yes, $\text{Rot}_{O,60°}$ **33.** yes, $\text{Rot}_{O,20°}$ **35.** no
37. no **39.** no **41.** yes **43.** yes **45.** yes
47. yes

Page 474 Lesson 14.6
Exploratory 1. S **3.** Q **5.** A **7.** T **9.** D

11. J **Written 1.** 2 **3.** $1\frac{1}{2}$ **5.** $\frac{1}{3}$
23. $A'(-9, 12)$ **25.** $A'(-42, 56)$ **27.** yes
29. yes **31.** false **33.** false **35.** true

Page 476 Problem Solving Application
1. yes **3.** yes **5.** yes, from N to S **7.** 70

Pages 478-479 Chapter Review
1. yes; domain = {4, 5, 6}, range = {a, b}
3. no **5.** {4, 3, -1} **7.** -2, 7 **17.** (1, 5)
19. (5, -7) **21.** (5, 8) **23.** (-10, 5)
25. $A'(-7, 8)$, $B'(-8, 10)$, $C'(-4, 12)$,
$D'(-2, 7)$ **27.** $(-3, -1)$ **29.** (3, 1) **31.** yes
33. no **35.** $(-8, 24)$ **37.** $D_{O,\frac{1}{2}}$

Index

A

Absolute value, 48
Acute angle, 146
Acute triangle, 159
Addition
    associative property, 61
    commutative property, 61
    identity property, 62
    integers, 48, A2-A3
    inverse property, 62
    on a number line, 46
    polynomials, 291
    probability, 357
    property for equations, 80
    property for inequalities, 116
    radicals, 227
    rational expressions, 309
    solving inequalities, 116
Adjacent angles, 149
Algebraic expressions, 38
    coefficient, 75
    evaluate, 38
    writing, 74
Alternate interior angles, 155
Alternative Assessment, 33, 69,
    102, 138, 170, 202, 238,
    277, 315, 346, 380, 412,
    450, 477
Altitude, 247, 254
Angles, 144
    acute, 146
    adjacent, 149
    alternate interior, 155
    bisector, 149
    complementary, 149
    congruent, 191
    consecutive, 165
    corresponding, 155
    degree, 145
    included, 191
    obtuse, 146
    opposite, 165
    quadrilateral, 165
    radians, 148
    reflex, 146
    right, 146
    sides, 144
    straight, 146
    supplementary, 149
    triangle, 162, A7
    vertex, 144
    vertical, 150

Antecedent, 16
Applications
    agriculture, 241, 439
    architecture, 139
    art, 451
    automobile design, 175
    aviation, 415
    biology, 281
    communications, 415
    design, 451
    drafting, 175
    ecology, 37
    economy, 383
    engineering, 205
    finance, 383
    foreign languages, 139
    history, 37
    literature, 1
    marine biology, 319
    medicine, 105
    meteorology, 73
    politics, 1
    physics, 319
    science, 105
    softball, 205
    sports, 241, 439
    wildlife management, 281
Area, 246
    circle, 261
    parallelogram, 248
    rectangle, 246
    surface area of sphere, A10
    trapezoid, 254
    triangle, 251, A9
    with coordinates, 437
Arithmetic mean, 387
ASA, 194
Assertion, *see statement*
Associative property, 61
Average, 387
    mean, 387
    median, 388
    mode, 387
Axioms, 154

B

Bar graphs, 391
    histogram, 396
Bases
    exponents, 39

percent proportion, 184
    prism, 265
    trapezoid, 254
    triangle, 251
Biconditionals, 28
    equivalences, 29
Bimodal, 388
Binomials, 289
    FOIL, 298
    multiplication, 297, A12-A13
Bisector, angle, 149

C

Calculators
    Scientific, 41, 48, 56, 67,
        176, 208, 222, 223, 226,
        229, 232, 236, 249, 258,
        262, 275, 375
    Graphing, 208, 258, 375,
        396, 419, A14, A15, A16
Capacity, liter, 272
Census, 385
Center
    circle, 257
    dilation, 472
    rotation, 467
Centimeter, 242
Chapter Review, 34, 70, 103,
    136, 172, 203, 239, 278,
    316, 347, 380, 413, 449, 478
Chapter Summary, 33, 69, 102,
    136, 170, 202, 238, 277,
    315, 346, 380, 412, 448, 477
Chapter Test, 36, 72, 104, 138,
    174, 204, 240, 280, 318,
    348, 382, 414, 450, 480
Chord, 257
Circle graph, 146, 393
Circles, 257
    area, 261
    center, 257
    chord, 257
    circumference, 257
    diameter, 257
    pi (π), 258
    radius, 257
Circumference, 257
Coefficient, 75
Collinear points, 144

Photo Credits

Cover, Philip Bekker

v, Fascimile reproduction of original illustration by John Tenniel; **vi**, John Zillioux/ Gamma-Liaison Network; **vii**, Fujiphoto/The Image Works; **viii**, Joe Devenney/The Image Bank; **ix**, G & M David De Lossy/The Image Bank; **x**, Bryan Yablonsky/ Duomo; **xi**, c1994 M.C. Escher/Cordon; Art-Baarn-Holland. All rights reserved; **xii**, file photos; **1**, Fascimile reproduction of original illustration by John Tenniel; **2**, David R. Frazier; **16**, courtesy Ralston Purina Company; **20**, courtesy Ohio Historical Society; **23, 32**, Light Source; **37**, Gale Zucker/Stock Boston; **65**, Ted Rice; **68**, Spencer Grant/Photo Researchers; **73**, Jeff Bates; **83, 87**, First Image; **93**, Ken Van Dyne; **105**, John Zillioux/Gamma-Liaison Network; **116**, Dan Erickson; **121**, Light Source; **125**, R.H./LPI/FPG International; **139**, Fujiphoto/The Image Works; **140**, **(tl)** Gerard Photography, **(tr)** Tim Bieber/The Image Bank, **(bl)** Alvis Upitis/The Image Bank, **(br)** Aaron Haupt; **154**, First Image; **158**, Joe Sohm/ The Image Works; **175**, Peter Yates/Saba Press Photos; **176**, Dennis Brack/Black Star; **182**, First Image; **187**, Library of Congress; **205**, Joe Devenney/The Image Bank; **220**, First Image; **241**, Tim McCabe/USDA; **246**, Collier/Condit; **257**, First Image; **267**, Ken Lambert/FPG International; **272**, First Image; **281**, G & M David De Lossy/The Image Bank; **306**, DOE; **319**, Tim Davis/Photo Researchers; **320**, First Image; **349**, Bryan Yablonsky/Duomo; **357**, **(t)** Larry Hamill, **(b)** Hickson-Bender Photography; **364**, First Image; **375**, Larry Hamill; **383**, Murray Alcosser/ The Image Bank; **384**, Doug Martin; **385, 396**, First Image; **415**, Steve Northup/ Black Star; **423, 430, 438**, First Image; **451**, c1994 M.C. Escher/Cordon Art-Baarn-Holland. All rights reserved; **456**, George H. Matchneer; **463**, Thomas Zimmerman/FPG International; **467**, Doug Martin; **471**, First Image; **A1**, file photo; **B1**, Gabe Palmer/The Stock Market; **B2**, White House Historical Association; **B3**, **(t)** Joseph DiChello, **(c)** Latent Image, **(b)** Doug Martin; **B5**, Mak-1; **B6**, Aaron Haupt; **B7**, **(t)** Doug Martin, **(cl)** StudiOhio, **(cr)** Bachman/ Uniphoto, **(bl)** Ed Young/The Stock Market, **(br)** George Anderson; **B8-B9**, StudiOhio; **B10**, Smithsonian Institution; **B11**, Mak-1; **B12**, **(t)** Aaron Haupt, **(bl)** Skip Comer, **(br)** Aaron Haupt; **B13**, Hal Lynes; **B15**, Johnny Johnson; **B16**, **(t)** Aaron Haupt, **(b)** Tim Courlas.

Manipulative Lab Activities

Understanding Through Hands-On Experiences

GLENCOE
Macmillan/McGraw-Hill

LAB 1 Adding and Subtracting Integers

Use with: Lessons 2.5 and 2.6, pages 48–51
Materials: counters, integer mat

You can use counters to help you understand addition and subtraction of integers. In these activities, counters with a "+" represent positive integers, and counters with a "−" represent negative integers.

Integer Models

- A *zero-pair* is formed by pairing one positive counter with one negative counter.

- You can remove or add zero-pairs to a set because removing or adding zero does not change the value of the set.

Activity 1: Use counters to find the sum –3 + (–2).

- Place 3 negative counters and 2 negative counters on the mat.

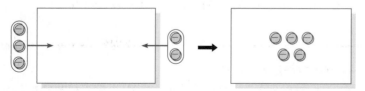

- Since there are 5 negative counters on the mat, the sum is –5. Therefore, –3 + (–2) = –5.

Activity 2: Use counters to find the sum –2 + 3.

- Place 2 negative counters and 3 positive counters on the mat. It is possible to remove 2 zero-pairs.

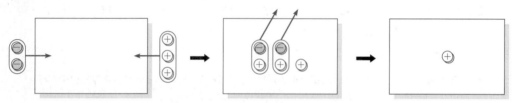

- Since 1 positive counter remains, the sum is 1. Therefore, –2 + 3 = 1.

Activity 3: Use counters to find the difference −4 − (−1).

- Place 4 negative counters on the mat. Remove 1 negative counter.

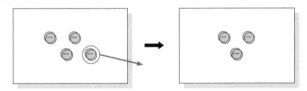

- Since 3 negative counters remain, the difference is –3. Therefore, −4 − (−1) = −3.

Activity 4: Use counters to find the difference 3 − (−2).

- Place 3 positive counters on the mat. There are no negative counters, so you can't remove 2 negatives. Add 2 zero-pairs to the mat. Remember, adding zero-pairs does not change the value of the set. Now remove 2 negative counters.

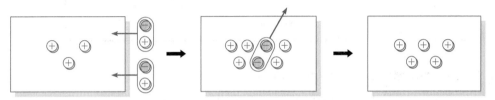

- Since 5 positive counters remain, the difference is 5. Therefore, 3 − (−2) = 5.

Model **Find each sum or difference using counters.**

1. 4 + 2	**2.** 4 + (−2)	**3.** −4 + 2	**4.** −4 + (−2)
5. 4 − 2	**6.** −4 − (−2)	**7.** 4 − (−2)	**8.** −4 − 2
9. 1 − 4	**10.** 2 − (−7)	**11.** −3 + 6	**12.** −3 + 3

Draw **Tell whether each statement is true or false. Justify your answer with a drawing.**

13. 5 − (−2) = 3   **14.** −5 + 7 = 2   **15.** 2 − 3 = −1   **16.** −1 − 1 = 0

Write **17.** Write a paragraph explaining how to find the sum of two integers without using counters. Be sure to include all possibilities.

LAB **2** Solving One-Step Equations

Use with: Lessons 3.3, 3.4, pages 80–86
Materials: cups, counters, equation mat

You can use cups and counters as a model for solving equations. In this model, a cup represents the variable, counters with a "+" represent positive integers, and counters with a "−"represent negative integers. After you model the equation, the goal is to get the cup by itself on one side of the mat, by using the rules stated below.

Equation Models

- A zero-pair is formed by pairing one positive identical counter with one negative counter.

- You can remove or add the same number of identical counters to each side of the equation mat.

- You can remove or add zero-pairs to either side of the equation mat without changing the equation.

Activity 1: Use an equation model to solve $x + (-3) = -5$.

- Place 1 cup and 3 negative counters on one side of the mat. Place 5 negative counters on the other side of the mat. The two sides of the mat represent equal quantities.

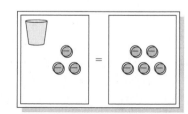

- Remove 3 negative counters from each side to get the cup by itself.

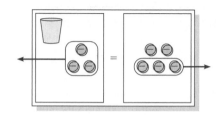

- The cup on the left side of the mat is matched with 2 negative counters. Therefore, $x = -2$.

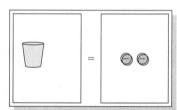

Activity 2: Use an equation model to solve $2p = -6$.

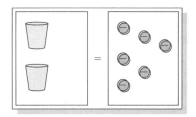

- Place 2 cups on one side of the mat. Place 6 negative counters on the other side of the mat.

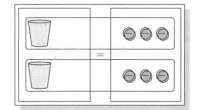

- Separate the counters into 2 equal groups to correspond to the 2 cups.

- Each cup on the left is matched with 3 negative counters. Therefore, $p = -3$.

Activity 3: Use an equation model to solve $r - 2 = 3$.

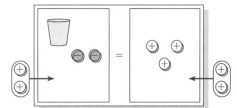

- Write the equation in the form $r + (-2) = 3$. Place 1 cup and 2 negative counters on one of the side mat. Place 3 positive counters on the other side of the mat. Notice that it is not possible to remove the same kind of counters from each side. Add 2 positive counters to each side.

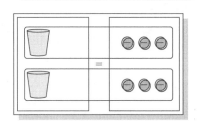

- Group the counters to form zero-pairs. Then remove all zero-pairs.

- The cup on the left is matched with 5 positive counters. Therefore, $r = 5$.

Model   Use an equation model to solve each equation.

     1. $x + 4 = 5$ **2.** $y + (-3) = -1$ **3.** $y + 7 = -4$ **4.** $3z = -9$

     5. $m - 6 = 2$ **6.** $-2 = x + 6$ **7.** $8 = 2a$ **8.** $w - (-2) = 2$

Draw   Tell whether each number is a solution of the given equation. Justify your answer with a drawing.

     9. $-3; x + 5 = -2$ **10.** $-1; 5b = -5$ **11.** $-4; y - 4 = -8$

Write   **12.** Write a paragraph explaining why you use zero-pairs to solve an equation such as $m + 5 = -8$.

LAB 3 Solving Multi-Step Equations

Use with: Lessons 3.6, pages 90–92
Materials: cups, counters, equation mat

You can use an equation model to solve equations with more than one operation or equations with a variable on each side.

Activity 1: Use an equation model to solve $2x + 2 = -4$.

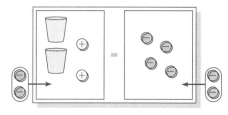

- Place 2 cups and 2 positive counters on one side of the mat. Place 4 negative counters on the other side of the mat. Notice it is not possible to remove the same kind of counters from each side. Add 2 negative counters to each side.

- Group the counters to form zero-pairs and remove all zero-pairs. Separate the remaining counters into 2 equal groups to correspond to the 2 cups.

- Each cup is matched with 3 negative counters. Therefore, $x = -3$.

Activity 2: Use an equation model to solve $w - 3 = 2w - 1$.

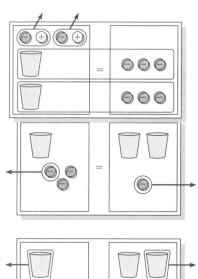

- Place 1 cup and 3 negative counters on one side of the mat. Place 2 cups and 1 negative counter on the other side of the mat. Remove 1 negative counter from each side of the mat.

- Just as you can remove the same kind of counter from each side of the mat, you can remove cups from each side of the mat. In this case, remove 1 cup from each side.

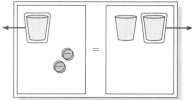

- The cup on the right is matched with 2 negative counters. Therefore, $w = -2$.

Model Use an equation model to solve each equation.

1. $2x + 3 = 13$	**2.** $2y - 2 = -4$	**3.** $-4 = 3a + 2$
4. $3m - 2 = 4$	**5.** $3x + 2 = x + 6$	**6.** $3x + 7 = x + 1$
7. $3x - 2 = x + 6$	**8.** $y + 1 = 3y - 7$	**9.** $2b + 3 = b + 1$

LAB 4 The Isosceles Triangle Theorem

Use with: Lesson 5.5, pages 158–161
Materials: two straws, straight pin, protractor

Isoceles triangles have at least two sides congruent. Compete the following activity to discover an important property of isosceles triangles.

Activity: Determine if the angles of an isosceles triangle have any special properties.

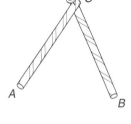

- Connect two straws of the same length with a pin so that they form an angle. Think of the angle as ∠C and the straws as sides $\overline{AC}$ and $\overline{BC}$ in triangle ABC.

- Lay the straws on your desk and position them so that the edge of the desk represents line AB.

- Use the protractor to position the straws so that ∠C has each of the measures listed below. For each measure of ∠C, find the measures of ∠A and ∠B.

  1. m∠C = 90   **2.** m∠C = 60
  3. m∠C = 50   **4.** m∠C = 120

Write   **1.** Do you see a pattern in the measures of angles ∠A and ∠B for each measure of ∠C?

2. Write a conjecture about the measure of the angles opposite congruent sides of a triangle.

3. Write a proof for your theory.

Side-Side-Angle as a Test for Triangle Congruence

LAB 5

Use with: Lesson 6.5, pages 191–193
Materials: two straws, scissors, straight pin, tape, straightedge, protractor, paper, pencil

You have studied different methods for proving that two triangles are congruent. Is proving that two sides and a non-included angle of two triangles are congruent, or SSA, sufficient for proving the triangles congruent?

Activity

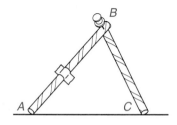

- Using a straightedge, draw a horizontal line on your paper.

- Connect two straws of unequal length with a pin, then position the straws so that the two straws form a triangle with the segment. Anchor one of the straws with tape, and label the vertices of the triangle *A*, *B*, and *C*.

- Measure ∠A and ∠C.

- Now swing the straw representing $\overline{BC}$ so that the end of the straw touches the segment again. Label this point D.

- Measure ∠A and ∠BDA.

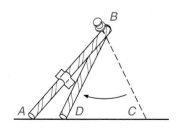

Write

1. List the corresponding parts of *ABC* and *ABD* that are congruent.

2. If SSA is a test for triangle congruence, would you be able to say that $\triangle ABC \cong \triangle ABD$?

3. Tell whether you think that SSA is a valid test for triangle congruence and justify your answer.

Model

4. What other tests for triangle congruence exist? Make a model to demonstrate each one.

LAB 6 Area of a Triangle

Use with: Lesson 8.3, pages 251–253
Materials: pencil, grid paper, scissors, straightedge

You know formulas for the areas of squares and rectangles. Use the following activity to discover the formula for the area of a triangle.

Activity

- Draw a triangle on grid paper so that one edge is along a horizontal line as shown at the right. Label the vertices of the triangle *A*, *B*, and *C*.

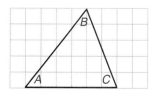

- Draw a line perpendicular to $\overline{AC}$ through *A*.

- Draw a line perpendicular to $\overline{AC}$ through *C*.

- Draw a line parallel to $\overline{AC}$ through *B*.

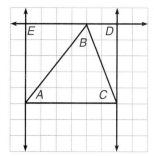

- Label the points of intersection of the lines drawn as *D* and *E* as shown.

- Find the area of rectangle *ACDE* in square units.

- Cut out rectangle *ACDE*. Then cut out △*ABC*. Place the two smaller pieces over △*ABC* to completely cover the triangle.

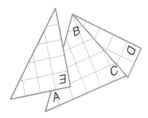

Write

1. Write a sentence comparing the area of △*ABC* and the sum of the areas of △*ABE* and △*BCD*.

2. How does the area of △*ABC* compare to the area of rectangle *ACDE*? How are the bases and the heights of △*ABC* and rectangle *ACDE* related?

3. Write a formula for the area of a triangle, *A*, in terms of the lengths of its base, *b*, and height, *h*.

LAB 7 Surface Area of Spheres

Use with: Lesson 8.8, pages 268–271
Materials: styrofoam ball, scissors, tape, straight pins, paper, pencil

A styrofoam ball is a model of a sphere. Use the following activity to find a formula for the surface area of a sphere.

Activity

- Cut the styrofoam ball in half. Trace around the edge to draw a circle. Then cut out the circle.

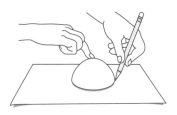

- Fold the circle into eighths. Then unfold and cut the eight pieces apart. Tape the pieces back together in the arrangement shown at the right.

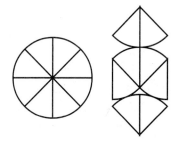

- Use tape or a straight pin to put the two pieces of the sphere back together. Then use pins to attach the pattern to the sphere.

Write

1. How much of the surface of the sphere was covered by the pattern?

2. What is the area of the pattern in terms of the radius of the sphere, r?

3. Write a formula for the surface area of the sphere.

LAB 8 Multiplying a Polynomial by a Monomial

Use with: Lesson 9.6, pages 295–296
Materials: algebra tiles, product mat

You can use rectangles to model multiplication. In this activity, you will use algebra tiles to find the product of simple polynomials. The width and length of the rectangle will each represent a polynomial. So, the area of the rectangle represents the product of the polynomials.

Activity 1: Use algebra tiles to find the product $x(x + 2)$.

- The rectangle has a width of x units and a length of $x + 2$ units. Use your algebra tiles to mark off the dimensions on a product mat.

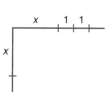

- Using the marks as a guide, make the rectangle with algebra tiles.

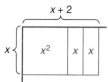

- The rectangle has 1 x^2-tile and 2 x-tiles. The area of the rectangle is $x^2 + 2x$. Thus, $x(x + 2) = x^2 + 2x$.

Activity 2: Use algebra tiles to find the product $x(x - 1)$.

- The rectangle has a width of x units and a length of $x - 1$ units. Mark off the dimensions on a product mat as shown at the right.

- Make the rectangle with a yellow x^2-tile and a red x-tile. The red tile is used because the length is –1.

- The area of the rectangle is $x^2 - x$. Thus, $x(x - 1) = x^2 - x$.

Model   Use algebra tiles to find each product.

  1. $x(x + 3)$ **2.** $x(x - 2)$ **3.** $2x(x + 1)$ **4.** $3x(2x - 1)$

Draw   Tell whether each statement is true or false. Justify your answer with a drawing.

  5. $x(2x + 3) = 2x^2 + 3x$ **6.** $2x(3x - 4) = 6x^2 - 4x$

Write   **7.** Suppose you have a square garden plot that measures x feet on a side. If you double the length of the plot and increase the width by 3 feet, how large will the new plot be? Write your solution in paragraph form, complete with drawings.

LAB **9** Multiplying Polynomials

Use with: Lesson 9.7, pages 297–299
Materials: algebra tiles, product mat

You can find the product of simple binomials using algebra tiles.

Activity 1: Use algebra tiles to find the product $(x + 1)(x + 2)$.

• The rectangle has a width of $x + 1$ units and a length of $x + 2$ units. Use your algebra tiles to mark off the dimensions on a product mat.

• Using the marks as a guide, make the rectangle with algebra tiles.

• The rectangle has 1 x^2-tile, 3 x-tiles, and 2 1-tiles. The area of the rectangle is $x^2 + 3x + 2$. Thus, $(x + 1)(x + 2) = x^2 + 3x + 2$.

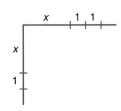

Activity 2: Use algebra tiles to find the product $(x + 2)(x - 1)$.

• The rectangle has a width of $x + 2$ units and a length of $x - 1$ units. Use your algebra tiles to mark off the dimensions on a product mat.

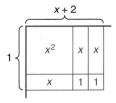

• Begin to make the rectangle as shown at the right. You need 1 positive x^2-tile, 1 negative x-tile, and 2 positive x-tiles. Now you need to determine whether to use 2 positive 1-tiles or 2 negative 1-tiles to complete the rectangle.

• Remember that the numbers at the top and on the side give the dimensions of the tile needed. The area of each tile is the product of –1 and 1, or –1. This is represented by a negative 1-tile. Fill in the space with 2 negative 1-tiles to complete a rectangle.

• You can rearrange the tiles to simplify the polynomial you have formed. Notice that a zero pair is formed by the x-tiles. In simplest form, $(x + 2)(x - 1) = x^2 + x - 2$.

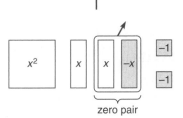

zero pair

Activity 3: Use algebra tiles to find the product $(x - 1)(x - 1)$.

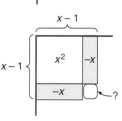

- The rectangle has a width of $x - 1$ units and a length of $x - 1$ units. Use your algebra tiles to mark off the dimensions on a product mat.

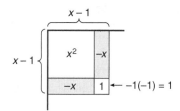

- Begin to make the rectangle as shown at the right. You need 1 positive x^2-tile and 2 negative x-tiles. Now you need to determine whether to use a positive 1-tile or a negative 1-tile.

- The area of the tile is the product of -1 and -1. Since $-1 \cdot -1 = 1$, the tile needed is a positive 1-tile.

- The area of the rectangle is $x^2 - x - x + 1$. In simplest form, $(x - 1)(x - 1) = x^2 - 2x + 1$.

Model   Use algebra tiles to find each product.

1. $(x + 2)(x + 2)$ **2.** $(x + 1)(x + 3)$ **3.** $(2x + 1)(x + 2)$

4. $(x + 1)(2x + 3)$ **5.** $(x + 2)(x - 3)$ **6.** $(x - 3)(x + 1)$

7. $(2x + 1)(x - 2)$ **8.** $(x - 1)(x - 2)$ **9.** $(x - 2)(x - 3)$

Draw   Tell whether each statement is true or false. Justify your answer with a drawing.

10. $(x + 3)(x + 2) = x^2 + 6$ **11.** $(x + 3)(x - 2) = x^2 - x - 6$

12. $(x + 2)(x - 2) = x^2 - 4$ **13.** $(x - 1)(x - 3) = x^2 + 4x + 3$

Write   **14.** You can also use the distributive property to find the product of two binomials. The figure at the right shows the model for $(x + 2)(x + 1)$ separated into four parts. Write a paragraph explaining how the model shows the use of the distributive property.

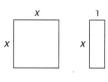

15. Write a paragraph describing the visual patterns possible in the rectangle representing the product of two binomials. Include drawings in your answer.

LAB 10 Solving Quadratic Equations

Use with: Lesson 10.5, pages 332–335
Materials: graphing calculator

You can use a graphing calculator to solve quadratic equations graphically.

There are three possible outcomes when solving a quadratic equation. The equation will have either two real roots, one real root, or no real roots. A graph of each of these outcomes is shown below.

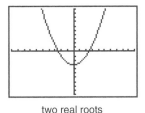

two real roots

two equal roots

no real roots

Activity: Use a graphing calculator to solve $3x^2 + 6x - 1 = 0$.

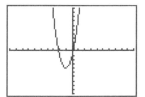

- Graph the related function, $y = 3x^2 + 6x - 1$. Make sure your calculator is set for the standard viewing screen, [–10, 10] by [–10, 10].

- Press [TRACE] and then use the arrow keys to estimate the x-intercepts. These x-intercepts are the roots of the equation.

- To determine the x-intercepts with greater accuracy, use the ZOOM feature. Set the cursor on the x-intercept and observe the value of x. Then zoom-in and place the cursor on the intersection point again. Any digits that are unchanged since the last trace are accurate. Repeat this process of zooming-in and checking digits until you have the number of accurate digits that you desire.

X=.15296053 Y=-.0024802

- The graph shown at the right indicates that one root of $3x^2 + 6x - 1 = 0$ is close to 0.153. Actually, the root is approximately 0.1547. Another root is about –2.1547.

Model Use a graphing calculator to find the roots of each quadratic equation accurate to four decimal places.

    1. $2x^2 - 11x - 15 = 0$ **2.** $0.2x^2 - 0.250x - 0.0738 = 0$
    3. $1.2x^2 - 3.6x + 5.8 = 0$ **4.** $3x^2 + 3.08x - 1.36 = 0$
    5. $x^2 + 31.54x + 229.068 = 0$ **6.** $35x^2 + 66x - 65 = 0$

Write   **7.** Write a paragraph to explain how graphing the related function can help find the roots for a quadratic equation.

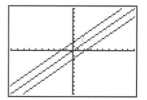

LAB 11 Intercepts

Use with: Lesson 13.4, pages 427–429
Materials: graphing calculator

Activity 1: Use a graphing calculator to graph $y = x$, $y = x + 2$, and $y = x - 2$ on the same screen. Describe the similarities and differences among the graphs.

- Graph each of the equations. Make sure your calculator is set for the standard viewing screen, [–10, 10] by [–10, 10].

- The graphs are all lines with a left-to-right upward slope, and all lines have the same slope. The lines pass through different points on the y-axis. The graph of $y = x$ passes through the origin, (0, 0), the graph of $y = x + 2$ passes through (0, 2), and the graph of $y = x - 2$ passes through (0, –2).

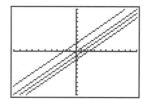

Activity 2: For each equation, predict what its graph will look like. Then graph each equation on the same screen as the equations in Activity 1.

a. $y = x + 3$

The graph of $y = x + 3$ is a line that passes through (0, 3) with the same slope as $y = x$, $y = x + 2$, and $y = x - 2$.

b. $y = x - 1$

The graph of $y = x - 1$ is a line that passes through (0, –1) with the same slope as $y = x$, $y = x + 2$, and $y = x - 2$.

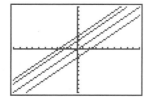

Draw

1. Graph $y = -x$, $y = -x + 2$, and $y = -x - 2$ on the same screen. Sketch the graphs. Describe the similarities and differences among the graphs.

2. For each equation, predict what its graph will look like. Then graph the equation on the same screen with those in Exercise 1. Describe the similarities and differences among the graphs.
   a. $y = -x + 3$ b. $y = -x - 1$

Write

3. The slope-intercept form of an equation of the line is $y = mx + b$, where m is the slope and b is the y-intercept. Write a paragraph explaining how the values of m and b affect the graph of the equation. Include several drawings with your paragraph.

LAB 12 Systems of Equations

Use with: Lesson 13.5, pages 430–433
Materials: graphing calculator

Since you can graph several functions on the screen at one time, you can use a graphing calculator to solve systems of equations. Use the following activity to explore how to solve a system of equations using a graphing calculator.

Activity: Find the locus of points that satisfy both $y = 3.4x + 2.1$ and $y = -5.1x + 8.3$.

- Set the viewing window of your calculator to the standard viewing window of [–10, 10] by [–10, 10].

- Graph both functions.

- Use the trace function to place the cursor at the point of intersection and read the coordinates.

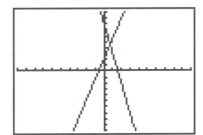

- Use the zoom feature of the calculator to find the coordinates of the intersection point more accurately. Repeatedly tracing and zooming allows you to find very accurate coordinates.

Write

1. According to your explorations on the graphing calculator, what are the coordinates of the point of intersection to the nearest hundredth?

2. Solve the system of equations algebraically. Is your solution close to the results found on the graphing calculator?

3. When do you think finding a solution to a system of equations with a graphing calculator is appropriate? Does this method of solution always give an accurate solution?

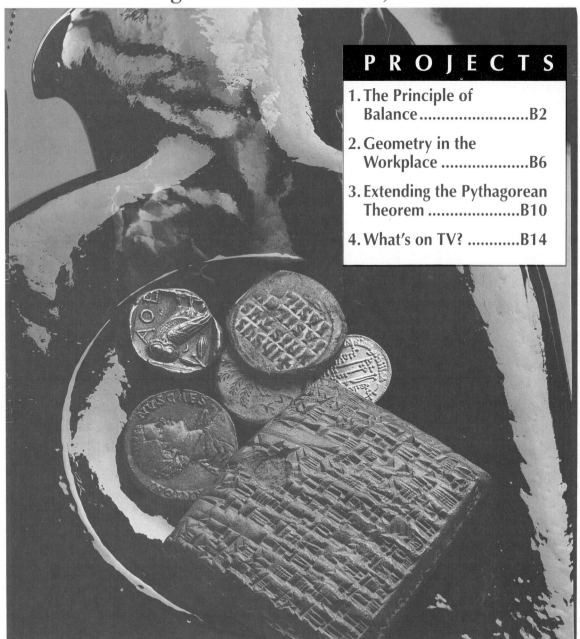

GLENCOE

EXTENDED PROJECTS

for Integrated Mathematics, Course 1

PROJECT **1** The Principle of Balance

Chapter 3 defines four properties of equality. They are the addition, subtraction, multiplication, and division properties of equality. Each definition begins with an equation in *balance*.

$$a = b$$

Each property states that if something is done to one side of the equation, the equation can be kept in balance by doing the same thing to the other side of the equation.

$$a + 3 = b + 3 \qquad a - 7 = b - 7 \qquad 5a = 5b \qquad \frac{a}{2} = \frac{b}{2}$$

This is the basic principle of balance. Two quantities that are in balance will remain that way until something is done to only one of the quantities. To bring the quantities back into balance once something has been done to one side, you must do the same thing to the other side of the equation. Likewise, you could bring the two sides back into balance by undoing what was done to the first side.

The word *balance* has numerous meanings in addition to the meaning in algebra. World War I officially began on July 28, 1914, when Austria-Hungary declared war against Serbia after the June assassination of Archduke Francis Ferdinand of Austria-Hungary. Before it ended, 28 nations would become involved in this global conflict.

When President Woodrow Wilson asked Congress for a declaration of war on April 2, 1917, he said, "It is a fearful thing to lead this great peaceful people into war, into the most terrible and disastrous of all wars, civilization itself seeming to be in the balance." Wilson meant that by avoiding war, the United States might help to secure victory for an uncivilized enemy. The way to avoid such a calamity was to reverse that possibility by defeating the enemy. At the end of the war on November 11, 1918, Wilson declared that America's participation in the effort had helped bring civilization back into balance.

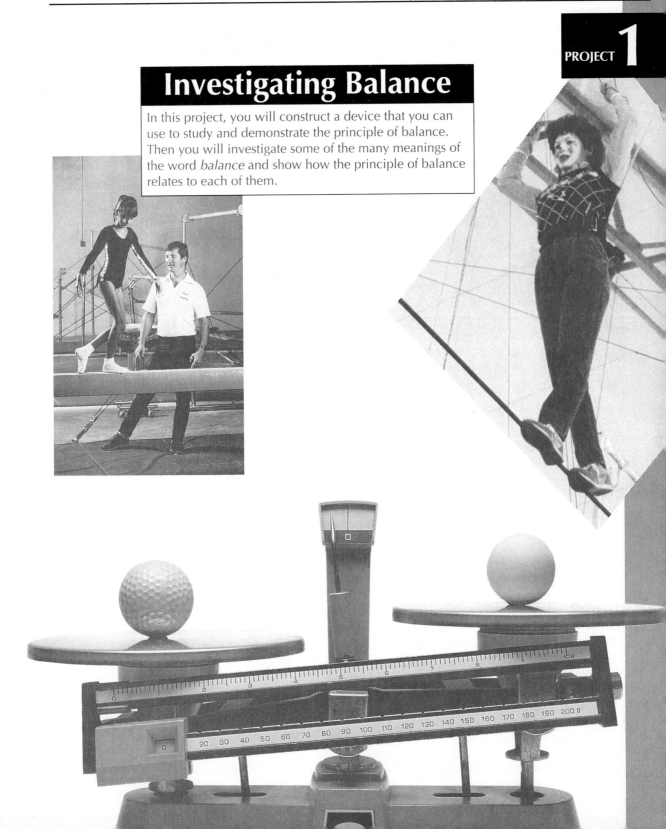

Investigating Balance

In this project, you will construct a device that you can use to study and demonstrate the principle of balance. Then you will investigate some of the many meanings of the word *balance* and show how the principle of balance relates to each of them.

PROJECT **1** Getting Started

Some balances are so precise that they can detect a difference in weights no greater than the weight of the period at the end of this sentence. But all balances, from the simplest to the most precise, have certain features in common. A horizontal bar is balanced on a vertical support. Pans are suspended from the ends of the bar. A pointer attached to the bar moves along a scale on the support, indicating when the pans are balanced.

Follow these steps to carry out your project.

■ Design a balance that you can make with materials you can easily find.

■ Build your balance. Keep in mind that the precision of the balance depends more on the care and accuracy with which you build it than it does on the materials you use.

■ Outline a plan you can follow to demonstrate the principle of balance. Your plan should include demonstrations of various ways that a balanced system can be thrown out of balance and steps that can be taken at such times to return the system to a state of balance.

■ Carry out your plan. Take accurate notes so that you can describe your work later.

■ Discuss your results with your group. Did you learn anything about balance that surprised you? How could you improve the construction of your balance to ensure more accurate results?

■ Determine how you might use your balance to represent an equation. Include ways to solve the equation using a balance. Discuss the limitations of your balance.

■ Apply what you have learned to interpreting the principle of balance as it is found in the real world.

Listed below are six everyday phrases containing the word *balance:*
a. *balance* the budget
b. *balance* of trade with Japan
c. *balance* of nature
d. the government system of checks and *balances*
e. *balance* the tires
f. *balance* your checkbook

For each use of the word balance, carry out the following directives:

■ Research the meaning of the word.
■ Research the principles governing the working of the system suggested by the phrase.
■ Determine what it means to be "in balance" in the system.
■ Determine how the system can be thrown out of balance.
■ Determine how an out-of-balance system can be returned to a state of equilibrium.

Extensions

1. Research and report on additional meanings of the word *balance* that are not included in the above list.
2. Make a list of properties of equality not mentioned in Chapter 3. State each property.
    Example: The Squaring Property of Equality
    Statement: For any numbers *a* and *b*, if $a = b$, then $a^2 = b^2$.

Culminating Activities

Show what you have learned in this project by completing one of the following activities.

1. Write a report explaining how you built your balance, outlining the method you used to demonstrate the principle of balance, and describing the results of your research into the various meanings of the word balance. Supplement your report with diagrams, graphs, illustrations, or any other means that might help a reader better understand your research.
2. Create a "balance exhibit." Use displays, artwork, props, experiments, music, or any other materials to produce an exhibit for other students to view that will illustrate some of the many ways that the idea of balance can be applied.

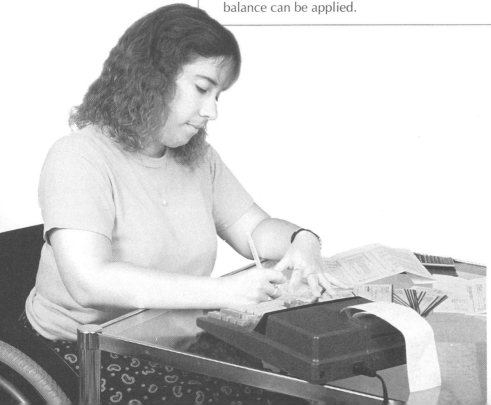

PROJECT 2 Geometry in the Workplace

- A rough diamond will be worth $50,000 if the diamond cutter measures the angles and cuts the diamond precisely. However, it will be nearly worthless if she makes an error in measurement.

- A trial lawyer uses deductive reasoning to lead the jury step by step through the lawyer's argument. From hypothesis—that the defendant did not commit the crime—to conclusion, the lawyer contends that the jury should find the defendant innocent.

- A product engineer for a sporting goods manufacturer uses his knowledge of surface area and volume to design a strong yet light motorcross helmet.

Not long ago, geometry was taught as "pure" or abstract mathematics—mathematics that had little or no relationship to the real world. Students studied the subject solely to learn about the deductive system employed in Euclidean geometry and to study the principles of logic.

There was merit in that approach, and the abstractions of geometry still are incorporated in high school mathematics today. But educators now know the value in including everyday applications of geometry in their lessons, and relating the subject to life outside the classroom. That is why this text emphasizes applications in addition to theory. Today's mathematics courses demonstrate that geometry has theoretical importance and, as you will show in this project, importance in everyday life.

Write a Career Profile

In this project, you will conduct research into a profession that interests you. Your goal is to find connections between that profession and geometry. Finally, you will write a career profile that highlights the connections and applications you have uncovered.

PROJECT **2** # Getting Started

Follow these steps to carry out your project.

1. Choose a career that interests you. You might already have an idea of the career path you would like to pursue. If not, spend some time thinking about a career. For now, consider a career solely from the standpoint of what you think might be an enjoyable and fulfilling line of work a few years from now. (There is nothing wrong with not knowing at this point what you want to do after you graduate. Some very successful people did not make their ultimate career choices until they were in their twenties or thirties, or even later.)

You might ask yourself:

- Which occupation(s) would enable me to do what I like to do most?
- Which occupation(s) might provide the opportunity to best fulfill my potential?
- Which occupation(s) would allow me time to spend with the kind of people I enjoy being with? to work with the kinds of things I like to work with?
- Which occupation(s) offers benefits I might enjoy, such as travel or the opportunity to make a significant contribution to society?

If you are still baffled, you might get some ideas by studying this list compiled by the U.S. Department of Labor. It lists occupations expected to grow more than 40% (in number of job openings) during the 1990s.

Medical assistants	Radiologic technicians
Home health aides	Data processing equipment repairers
Medical secretaries	Operations research analysts
Physical therapists	Surgical technicians
Travel agents	Computer systems analysts
Occupational therapists	Paralegals
Computer programmers	Child-care workers

2. Talk to people in the profession you have decided to research. Ask them about ways they might use geometry in their work. (Listen for even vague applications of geometry; they might lead to more fully developed ideas later on.) Before you meet with someone, prepare a list of ideas they might not think of as "geometrical," but that might help them find geometrical content in their work. Do they ever need to make logical deductions? Do they use measurements or proportions in their work? With reasonable effort, you should be able to uncover applications of geometry in nearly every profession.

3. Read about the profession. Check the library for books about the profession or related professions. Your local or state employment office also might have booklets or statistics that relate to the occupation.

Extensions

1. Consult the *U.S. Statistical Abstract,* a government publication that is produced annually. (Your public library should have a copy.) Find statistics on changes in earnings and numbers of jobs in major areas of employment in recent years. Prepare a visual that summarizes your findings.

2. Sherlock Holmes is perhaps the most famous literary character who used geometry—in the form of deductive logic—in his work. Read a Sherlock Holmes mystery (written by Arthur Conan Doyle) and report to the class on Holmes' use of deductive reasoning and logic.

Culminating Activities

Show what you have learned in this project by completing one of these activities.

1. Write a career profile entitled "Geometry and _____." Fill in the blank with the name of the profession you have investigated. In your profile, describe the kind of work that is done in the profession. Detail the educational background necessary to enter the field. Include the amount of mathematics that people in the profession typically have studied. Then describe your findings on relationships between geometry and the profession.

2. Participate in a "Career Day" with classmates. Prepare an exhibit that highlights interesting aspects of the profession you have researched, illustrating connections between the profession and geometry. During Career Day, visit the exhibits of other students and be prepared to answer questions about your exhibit. Discuss both the career itself and applications of geometry to the career.

Extending the Pythagorean Theorem

The Pythagorean Theorem, discussed in Lesson 12-3, may be one of the most elegant, amazing, and useful results in all of mathematics. Mathematicians Gary Musser and William Burger have characterized the theorem as "perhaps the most spectacular result in geometry." Some mathematicians consider it a mark of honor to discover their own proofs of the theorem. In the classic 1907 book *The Pythagorean Proposition,* author Elisha Loomis gave more than 370 different proofs of the theorem, including one devised by President James Garfield. The theorem has wide applications and appears in nearly every branch of mathematics. What continues to intrigue and delight people who enjoy mathematics, even thousands of years after its discovery, is the utter simplicity of the Pythagorean relationship, combined with its far-reaching implications. No other relationship combines these two features so elegantly.

The Pythagorean Theorem states that in a right triangle, where a and b are the measures of the legs and c is the measure of the hypotenuse, then $c^2 = a^2 + b^2$.

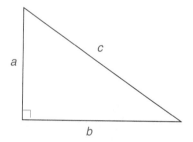

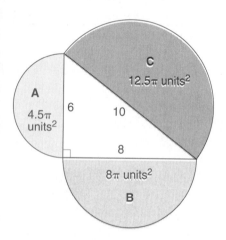

What is not so well known, and what is likely to increase your appreciation for the Pythagorean relationship considerably, is that the theorem given above is only one of many closely related Pythagorean theorems. Consider a 6–8–10 right triangle with semicircles constructed on each side.

Notice that the radius of each semicircle is half the measure of that side of the triangle. By using the formula for the area of a circle ($A = \pi r^2$), we can determine the areas for the semicircles constructed on the legs. They are $\frac{1}{2}(\pi \cdot 3^2)$, $\frac{1}{2}(\pi \cdot 4^2)$, and $\frac{1}{2}(\pi \cdot 5^2)$, or 4.5π, 8π, and 12.5π. Notice that $4.5\pi + 8\pi = 12.5\pi$.

At least in this example, we have a *Pythagorean Theorem for Semicircles:*

In a right triangle, if A and B are the areas of semicircles constructed on the legs and C is the area of a semicircle constructed on the hypotenuse, then $C = A + B$.

Is this true for all right triangles? What about the area of equilateral triangles constructed on the legs and hypotenuses of right triangles? What about pentagons? Hexagons? Is there a Pythagorean Theorem for volumes or surface areas of cubes, cylinders, or pyramids constructed on the legs and hypotenuses of right triangles? The answer to many of these questions and many others like them is: Yes!

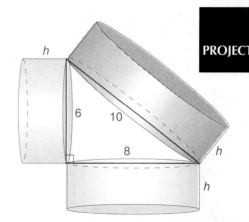

PROJECT **3**

Build a PythagoLab

In this project, you will attempt to discover some of the little known but fascinating variations of the Pythagorean Theorem. You and the members of your group will choose a set of variations of the theorem that you would like to investigate. You will use calculators, paper and pencil, and 2-dimensional and 3-dimensional models of geometrical shapes to test the proposed variations and then summarize your results.

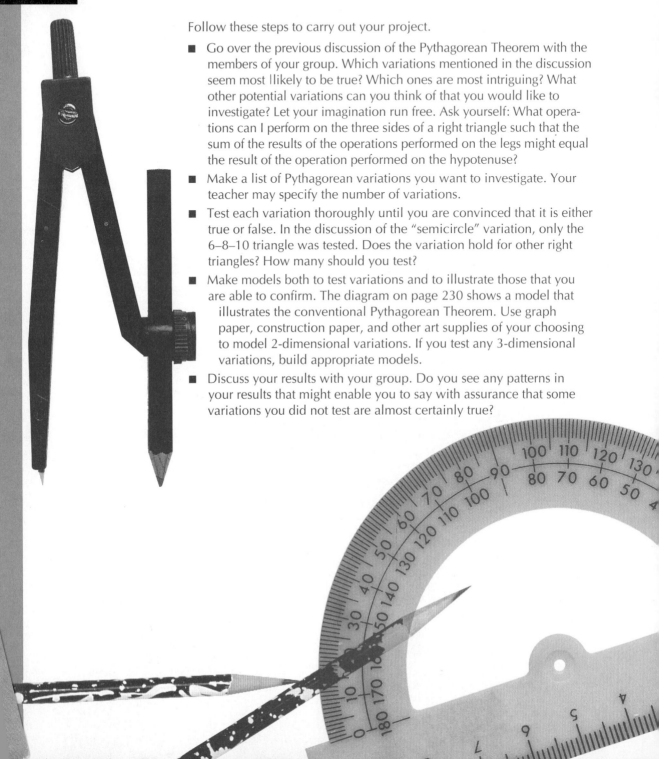

PROJECT **3** # Getting Started

Follow these steps to carry out your project.

■ Go over the previous discussion of the Pythagorean Theorem with the members of your group. Which variations mentioned in the discussion seem most llikely to be true? Which ones are most intriguing? What other potential variations can you think of that you would like to investigate? Let your imagination run free. Ask yourself: What operations can I perform on the three sides of a right triangle such that the sum of the results of the operations performed on the legs might equal the result of the operation performed on the hypotenuse?

■ Make a list of Pythagorean variations you want to investigate. Your teacher may specify the number of variations.

■ Test each variation thoroughly until you are convinced that it is either true or false. In the discussion of the "semicircle" variation, only the 6–8–10 triangle was tested. Does the variation hold for other right triangles? How many should you test?

■ Make models both to test variations and to illustrate those that you are able to confirm. The diagram on page 230 shows a model that illustrates the conventional Pythagorean Theorem. Use graph paper, construction paper, and other art supplies of your choosing to model 2-dimensional variations. If you test any 3-dimensional variations, build appropriate models.

■ Discuss your results with your group. Do you see any patterns in your results that might enable you to say with assurance that some variations you did not test are almost certainly true?

Extensions

1. Find as many examples of practical applications of the Pythagorean Theorem as you can. You may wish to talk to science teachers, surveyors, engineers, architects, and others in professions that rely heavily on mathematics.

2. Research and report on Pythagoras, the Order of the Pythagoreans, and the discovery of the Pythagorean Theorem.

3. Research and report on Fermat's Last Theorem.

4. *Pythagorean triples* are triples of whole numbers that satisfy the Pythagorean relationship. The numbers 3, 4, and 5 form a Pythagorean triple because $3^2 + 4^2 = 5^2$. Find other examples of Pythagorean triples.

5. Write a proof or detailed explanation to verify the results you found in your PythagoLab.

Culminating Activities

Show what you have learned in this project by completing one of the following activities.

1. Put together an exhibit that summarizes your investigation into variations of the Pythagorean Theorem. The design of the exhibit is up to you, but you will probably want to include your 2- and 3-dimensional models, an explanation of the procedure that you followed, and a summary of your conclusions.

2. Create a detailed checklist that investigators could follow to test proposed variations of the Pythagorean Theorem.

4 What's on TV?

There were only about 100 television sets in Great Britain when the world's first television broadcasting service began in London on November 2, 1936. Regular commercial broadcasts began in the United States three years later for an audience not much larger than the one that began in London.

From those modest beginnings, television has grown in less than 60 years into one of the most influential forces in the world. Today, 98% of American households have at least one TV set. Nearly two-thirds have two or more sets. In 1950, the average set was on about $4\frac{1}{2}$ hours per day. In 1983, the average surpassed 7 hours per day for the first time, and it has remained near that level ever since. Where once newspapers and magazines provided the main sources of news and information for Americans, today television serves that function. Americans also obtain most of their information about consumer products from television commercials. In response, advertisers spend billions of dollars annually on TV advertising. Car makers alone spend more than $3 billion each year on television commercials. The cost of a 30-second advertisement on a much-watched program like the Super Bowl can run as high as $800,000.

Given the influence of television on the lives of most people, it is no surprise that pollsters are continually surveying the tastes, opinions, and viewing habits of television viewers. Perhaps you have participated in a survey. Even if you have not, you will have an opportunity to do so now.

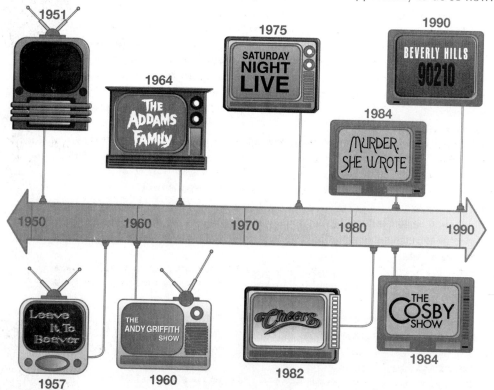

Conduct a TV Survey

PROJECT 4

In this project, you and the members of your group will decide on an issue relating to television that you would like to investigate. You will write, conduct, and analyze the results of a television survey intended to shed light on the issue you have chosen. Within certain broad limitations, the design of the survey will be left to you so that you can investigate the questions that interest you most. After you conduct your survey, you will analyze your results using some of the statistical methods discussed in Chapter 12.

Getting Started

Follow these steps to carry out your project.

■ With your group, discuss the issues relating to television that interest you most. Try to move beyond a comparison of favorite programs to deeper issues involving television. Is television generally a positive influence on young people? Do people watch too much television? Is there too much violence on television? Are commercials truthful?

■ After you have talked about some of these issues and identified areas of common interest, choose a broad question about television that intrigues you. You can choose one of the above questions or another one that better reflects your own interests. Be sure that the question you choose is one you can shed light on by surveying viewers.

■ Create a survey consisting of at least eight questions. Each question should be designed to furnish data that will help you answer the main question your group has chosen to investigate. Spend a good deal of time on this step. Your questions should form an integrated whole aimed at revealing key viewer attitudes and habits. Try to break away from tired survey questions like "What is your favorite program?" Instead, devise creative questions that will illuminate the central issue you have chosen to explore. "How many people do you usually watch television with?" "Of the following products advertised on television, which ones would you not absolutely buy?"

■ Think about the people you will be surveying. How many should you survey? Should they be chosen from the general population or should you target a small, precisely defined group? How will you record the responses when considering the sex, age, marital status, income, and so on of the interviewee?

■ It may be necessary for people you intend to survey to keep track of certain information for a period of time—programs watched, length of time, and so on. If so, design a "viewer data sheet" and distribute it to the people you will be interviewing.

■ Gather your data.

■ Analyze your data. Use plots, measures of central tendency, graphs, scatter plots, regression lines drawn with graphing calculators, or any other statistical tools that seem useful.

■ Discuss these questions with your group:
  1. What conclusions can we draw, based on the results of our survey?
  2. How has our survey shed light on the central question we posed at the beginning of the project?
  3. What questions has the survey raised that should be asked in a follow-up survey?

PROJECT 4

Extensions

1. Networks and sponsors closely watch the results of the Nielsen ratings. A low rating often means the television show will be canceled. A high rating ensures another season of shows and often gives the network validation for charging sponsors more for the commercials that appear during the show. Contact the A.C. Nielsen Company to learn about the methods they use to survey public viewing habits. Its address is Nielsen Media Research, 1290 Avenue of the Americas, New York, NY 10104.

2. Research and report on changes in viewer tastes in television programming over the past several decades.

3. Contact an advertising agency to find out how it makes decisions on where to advertise its client's products (television, radio, newspapers, and so on).

Culminating Activities

Show what you have learned in this project by completing one of the following activities.

1. Write a report summarizing the work of your group. State the question you chose to answer and describe how you decided on which survey questions to ask. Explain how you analyzed your data and detail your conclusions. Include with your report any graphs, plots, displays, or other supporting material you feel is needed to clarify your work.

2. Create a series of posters designed around the issue raised by your main question. In your display, describe your survey, detail your results, and summarize your conclusions. Include graphs, plots, photos, art work, or any other visual aids that might help viewers understand your survey.